FIRST WORLD PETRO-POLITICS

The Political Ecology and Governance of Alberta

Edited by Laurie E. Adkin

Despite fluctuations in oil prices, oil revenues remain an important part of Alberta's – and Canada's – economy. The global demand for oil shows no signs of abating and Alberta is and will continue to be one of the world's richest sources of oil and other fossil fuels. The production and delivery of Alberta's oil, however, present significant political, economic, and environmental challenges. Not only are pipelines across Canada and the United States proving to be expensive to build and politically contentious, the extraction and refinement of bitumen from Alberta's oil sands – the world's second-largest reserve of oil – is the fastest-growing source of CO_2 emissions in Canada, and the greatest obstacle to meeting the country's obligations to reduce greenhouse gas emissions.

Alberta thus plays a pivotal role in the global political-economic system sometimes referred to as 'fossil capitalism.' *First World Petro-Politics* surveys Alberta's environmental conflicts and political economy, and uses the Alberta case study to make a broader contribution to the development of political ecology as a theoretical approach and to its application in a first world context. Essays cover a wide range of related topics, including the effects of successive neoliberal governments on development and environmental regulation, 'rentierism,' responses of First Nations to extractive development, corporate licensing strategies, social movements, the role of the media, and access to information. Alberta, as a first world 'petro-state,' represents the conflicts and choices that we confront, as a species, on a global scale. This ambitious volume offers an integrated, transdisciplinary analysis of the province's economy, politics, and ecology and how they relate to deeper questions of democracy and citizenship.

LAURIE E. ADKIN is an associate professor in the Department of Political Science at the University of Alberta.

First World Petro-Politics

The Political Ecology and Governance of Alberta

EDITED BY LAURIE E. ADKIN

UNIVERSITY OF TORONTO PRESS
Toronto Buffalo London

Toronto Buffalo London
www.utppublishing.com

ISBN 978-1-4426-4419-9 (cloth)
ISBN 978-1-4426-1258-7 (paper)

Library and Archives Canada Cataloguing in Publication

First world petro-politics : the political ecology and governance of Alberta / edited by Laurie E. Adkin.

Includes bibliographical references and index.
ISBN 978-1-4426-4419-9 (cloth). ISBN 978-1-4426-1258-7 (paper)

1. Petroleum industry and trade – Political aspects – Alberta. 2. Petroleum industry and trade – Government policy – Alberta. 3. Petroleum industry and trade – Social aspects – Alberta. 4. Petroleum industry and trade – Environmental aspects – Alberta. 5. Alberta – Politics and government. 6. Alberta – Economic policy. I. Adkin, Laurie Elizabeth, 1958–, author, editor

HD9574.C23F57 2016 338.2'7282097123 C2016-902524-1

University of Toronto Press acknowledges the financial assistance to its publishing program of the Canada Council for the Arts and the Ontario Arts Council, an agency of the Government of Ontario.

Canada Council for the Arts Conseil des Arts du Canada

Funded by the Government of Canada Financé par le gouvernement du Canada | Canadä

This book is dedicated to the activists,
researchers, educators, and non-Idlers
who animate, inform, and sustain Alberta's civil society:
You are the midwives of our green future.

Contents

Figures

Tables and Appendices

Tables

Appendices

Preface

The seeds of this book were planted many years ago. I was hired at the University of Alberta in 1991 and worked here through the Klein years, during which a radical neoliberal project coincided with the rapid take-off of oil sands mining after 2000. Although my own research and teaching did not focus on Alberta, by the mid-2000s I had begun to wonder why so little was being written by university scholars about the increasingly pivotal place of this province in the global political ecology of fossil capitalism. At the same time, I noted the absence of scholarly accounts of the nature of the province's political institutions – including the ad hoc, government-initiated, consultation processes through which citizens were attempting to influence environmental policy. There were on-the-ground accounts of such experiences by Albertans who were trying to engage in these processes and with institutions like the Energy and Utilities Board, accounts that repeatedly suggested Alberta's citizens were dealing with what political economists refer to as a 'captured state' – in this case, a provincial state whose policy logic appeared to be dictated by the ruling party's perceptions of fossil fuel industry interests. During the 2002 'Kyoto debate' in Canada, the government of Alberta – supported by a big-business coalition – vociferously opposed ratification of the Kyoto Protocol by the federal Liberal government. The 'climate change' policy emerging from the Alberta government at the end of 2002 was clearly not going to make much of a dent in the province's greenhouse gas emissions, notwithstanding the government's sustainable development rhetoric. A handful of ENGOs and research institutes were trying to inform the public and the media about the environmental impacts of industrial and resource-based development in the province.

My initial probing of the question 'where is the research?' elicited some responses that are indicative of the relationship between a petro-state and its universities: with so much research (and capital) funding flowing to selected branches of the universities from corporations in the resource extraction sectors, and from governments committed to the extractive model of development, few university researchers in these branches were willing to risk the loss of research funding that might ensue from engaging such questions publicly and critically. At the same time, critical political ecology had by the mid-2000s barely established a toehold in the social sciences and humanities in Canada.

By 2007 I had undertaken a project that examined cases of environmental conflict in Canada from a democratic theory perspective. In the process of recruiting work for that book, I discovered that there existed a smattering of scholars around the country whose work related to Alberta, and so the contents of a mental folder labelled 'Alberta book' began to grow. The book took further shape with a call for chapter proposals, a successful SSHRC grant application, a book proposal, and an authors' workshop held in Edmonton in November 2008. James McCarthy, now a professor of geography at Clark University, inspired us all with his keynote presentation on the origins of political ecology and its emerging directions.

Since 2009, this volume has shape-shifted significantly, becoming organized around the idea of Alberta as a first world petro-state and all of the questions to which this label gives rise. The contributors were asked to situate their case studies within multiple scales and to address the implications of their policy analyses for questions about the nature of Alberta's state and civil society, the quality of its democracy, and the possibilities for practising and deepening ecological citizenship. The researchers – few of whom had worked, initially, within the framework of critical political ecology – come from diverse academic backgrounds. They hold varying views about necessary policy reforms, and tackle these from different methodological starting points. They agree, however, on the importance of bringing interdisciplinary knowledge and critical theoretical insights to bear upon the enormous societal and ecological challenges facing future generations.

There is, of course, much more work to be done; many important questions, policy areas, and relationships could not be addressed in the pages of this volume. I hope that *First World Petro-Politics* will encourage new scholars to contribute to the intellectual tasks necessary to building an ecologically sustainable, democratic, and socially just future to succeed fossil capitalism.

The environment for such a project changed suddenly in May 2015 with the election in Alberta of a social democratic party to government. When the contributions to this book were first written, the Progressive Conservative Party regime that had ruled for four decades was still in office, notwithstanding the many fractures and ruptures in its support base that our analyses were identifying. Demonstrating the unpredictability and power of political agency, as well as the need for further research into the changes in Alberta's civil society over the last decade, the PC Party was swept from office on 5 May. As I write these words, the New Democratic Party government has been in office for less than a year. We hope that the analyses of petro-politics and fossil capitalism provided in this book will enrich public debate as Albertans consider the possibilities for the future political-ecological development of their province, their country, and their planet.

Laurie Adkin
Edmonton, March 2016

Acknowledgments

The editor and authors of this work are grateful to the Social Sciences and Humanities Research Council of Canada (SSHRC) and to the Universities of Alberta and Calgary for funding an authors' workshop that allowed us to develop shared objectives for this project. In addition to receiving a Scholarly Workshops and Conferences grant (SSHRC 646-2008-47), we were supported by the Department of Political Science and the Associate Dean of Research in the Faculty of Arts at the University of Alberta, by the Department of Rural Economy and the Associate Dean of Research in the Faculty of Agriculture, Life, and Environmental Sciences (ALES) at the University of Alberta, and by the Urban Studies Program at the University of Calgary. The kind of mutual knowledge that we were able to build through the workshop is especially important for an interdisciplinary group of scholars. Our understanding of political ecology was enriched by the contributions of our keynote speaker, Professor James McCarthy. We laud the important role of the SSHRC in supporting independent social science research for the public good. The VP Research and Associate Dean of Arts (Research) at the University of Alberta also provided funding from their jointly adjudicated Publication Subvention Program to assist with the production of this book. We are very grateful for this support.

Randy Haluza-DeLay (The King's University, Edmonton), Naomi Krogman (University of Alberta), and Byron Miller (University of Calgary) helped to shape this project in its early stages, took leading roles in the workshop sessions, and provided comments on some of the chapters. Dr Krogman also assisted with the preparation of the SSHRC grant application. We have all benefited from their contributions. Thank you also to the anonymous readers of the manuscript, who offered valuable suggestions for improvements.

The editor would like to express her deep appreciation to all of the authors for their work on successive drafts and for their commitment to this 'mega-project' during its long gestation. It is important, too, to acknowledge that this book builds on the work of earlier scholars who have sought to explain the nature of Alberta's political-economic development. In particular, we owe a great debt to Lawrence (Larry) Pratt (1944–2012) for his path-breaking contributions to the study of the political economy of Alberta.

On a personal note, I am enormously grateful to Mustafa Kaya for all of the domestic, caring, and parenting labour that kept our household going during repeated periods of intensive work on this project, and for his moral support through many setbacks. Thank you, once again, Olivier, for being such a forgiving child. Seeing projects like this through takes irreplaceable time from our most important relationships, and one hopes that the resulting contribution to the general good makes such losses at least meaningful and comprehensible, if no easier, for those who bear them.

FIRST WORLD PETRO-POLITICS

The Political Ecology and Governance of Alberta

1 Ecology and Governance in a First World Petro-State

LAURIE E. ADKIN

Alberta plays a pivotal role in the global political-economic system that is sometimes referred to as 'fossil capitalism.' In addition to the production of coal, conventional oil and gas, and, more recently, coal bed methane and shale gas, large corporations are developing the world's second-largest reserve of oil, the Alberta bituminous sands (also known as the oil sands, or the tar sands).[1] The Alberta Energy Regulator predicts that bitumen production will grow from approximately 2.1 million barrels per day (bpd) in 2014 to 4.1 million bpd by 2023.[2] The extraction and refinement of bitumen from the tar sands is the fastest-growing source of CO_2 emissions in Canada, and the greatest obstacle to meeting our country's obligations to reduce greenhouse gas emissions. Canada is also a major (and at the time of writing, the largest) supplier of crude oil to the United States, which situates the Alberta oil and gas industry within the economic and military security sphere of the American state. Thus, Alberta's energy exports and environmental externalities make the political ecology of the province globally and regionally significant. Looked at from another perspective, global networks encompassing markets, commodity chains, and organizations, as well as flows of information, capital, and people, play important roles in shaping the *local* political ecology, with all of its consequences for ecosystems, livelihoods, citizenship, and the struggles of indigenous peoples.

Alberta, as a first world 'petro-state,' represents the conflicts and choices that we confront, as a species, on a global scale. Is it possible, in

The author thanks Eric Abrahams for his assistance with the research for table 1.1.

this context, to turn away from a heavy reliance upon fossil fuels and the consumerism they support, and to choose an ecologically sustainable path? Ecological economist Elmar Altvater (2006) sees the paths before us as, essentially, a 'solar revolution' that creates possibilities for more decentralized and egalitarian social relationships, or an increasingly authoritarian form of global apartheid, in which the consequences of climate change and other ecological and social catastrophes will disproportionately be borne (at least initially) by the world's poorest populations. Sociologists Debra Davidson and Michael Gismondi (2011, 205), like many ecologists before them, argue that our current path of human population growth, resource consumption, and production of wastes has led us to a 'precipice' that we cannot leap with technological innovation; this is a moment to pause and to change course – toward a 'post-carbon' future. Viewed in this way, the alternatives to the current model of development in Alberta are not just a slowing down of the rate of bitumen extraction, or the use of more energy-efficient technologies to reduce GHG emissions from fossil fuel extraction and combustion, but a moratorium on further expansion, a shrinking of economic growth, and a major societal investment in renewable energies and energy conservation. There are social and economic costs associated with such directions, and hence important questions about their equitable distribution, and how to ensure sustainable livelihoods. Possible ecological futures, transition strategies, and 'resources of hope' for social and political change (Williams 1989) are the core concerns of this book. We examine the political ecology of a particular place, but always with the understanding that these are questions that must be answered at multiple scales.

Despite the significance of the Alberta case for our understanding of how fossil capitalism is reproduced in the global north, and of where the fracture lines and fragilities of this system lie, this jurisdiction has been surprisingly absent from comparative studies of petro-states, and its political economy has received little scholarly attention in recent decades (with some exceptions, which are discussed below). It is time for a renewed effort to identify what has driven Alberta's political-economic development of the past twenty-five years, as well as to identify the potential for change in a more ecological and egalitarian direction. The authors of this work aim to make a substantial (if necessarily incomplete) contribution to mapping the political ecology of a Canadian province that – after decades of neoliberal governance – is confronting multiple social and ecological crises. In so doing, we attempt to think through and beyond the characterization of Alberta as a first world petro-state by drawing upon the multiple approaches and insights of political ecology to identify further dimensions of Alberta's place within fossil capitalism and by examining related questions of democracy and citizenship.

In this chapter, I introduce political ecology as an analytical framework and highlight the particular interest of the authors of this book in questions of citizenship and democracy. I then explain why Alberta may be viewed as falling into the 'petro-state' subset of rentier states. Throughout, I discuss the contributions of earlier work to these questions, and indicate how our authors seek to build on these foundations.

Political Ecology

Political ecology, as an integrative theoretical framework, had its early roots in human geography, informed by political economy, ethnography, and ecological science. Since the 1980s it has expanded theoretically to encompass feminist, post-structuralist, and post-colonial approaches.[3] Jill Belsky describes political ecology as a 'transdisciplinary perspective' – a 'hybrid' that 'struggles to "bridge" different disciplines, ontologies, and the theory-praxis nexus' (2002, 271). While perhaps most familiar to geographers, political ecology is making inroads into political science and sociology. Most of the scholarship associated with political ecology has drawn upon empirical research and fieldwork carried out in countries and regions of the global south; only recently is political ecology being employed as a framework for the analysis of cases in the global north, or 'first world.'[4] This collection represents the first attempt to bring authors from diverse disciplines together to examine the first world jurisdiction of Alberta from the multiple perspectives of political ecology.

In these pages, we adopt a broad definition of political ecology approaches as those which see 'the natural world and social relations as mutually constitutive' (Belsky 2002, 270) or as 'co-determining' (Heynen et al. 2006, 11). Just as the economy is embedded in natural processes and ecosystems, nature is continually reshaped by human activities. All of the crises confronting us today – from global poverty to climate change – are dual in this sense: they are simultaneously, and dynamically, both social and ecological. Economic or technical solutions that do not comprehend this interdependent relationship may at best alter one or another aspect of these interlinked crises. At worst, they may reproduce, prolong, or compound these crises.

Viewed as a theoretical framework for making sense of the political, economic, social, and environmental conditions in which we find ourselves, political ecology encompasses not only a structural or institutional level of analysis (political-economic) but also a sociological, actor-centred level of analysis, as well as ecological science and cultural studies. Further, as with any empirically founded, comparative analysis, the phenomena we study must be understood historically. In lieu of highly general explanations for phenomena like human overuse of resources or of the 'commons,' (e.g., 'human nature' claims), political ecologists look for the specific historical conditions that have given rise to various practices (or outcomes) in different contexts (Neumann 2005, 6).

Moreover, political ecology analyses typically take into account the ways in which different spatial scales (from the local to the global) are interconnected, or nested within one another. We can hardly think of any economic or environmental conflict, for example, that is purely 'local.' Whether we are looking at a corporate investment, a plant closure, a strike, the year's farm income, the collapse of a fishery, or an environmental approval process, we will have difficulty comprehending outcomes in the absence of an analysis of the influences of institutions on multiple levels (juridical, governmental, economic), of actors and their networks and resources (market, political, expert/scientific, civil society), as well as ecosystem dynamics.

In regard to the political ecology of Alberta, we see that the extraction of bitumen from the tar sands has given rise not only to local struggles about the survival of Aboriginal communities, the defence of boreal forest habitat and watersheds, and the infrastructural needs of the city at the heart of the boom economy (Fort McMurray), but also to struggles that criss-cross the country, the continent, and the globe. As mentioned above, Alberta crude oil fuels economic growth in the USA and Europe, enriching the multinational corporations (MNCs) that refine and export it from various locations in Canada and the USA. China has been looking increasingly to Canadian sources of oil, liquefied natural gas, and coal to fuel its economic growth. The exports of Alberta bitumen and crude oil have effects on the value of the Canadian dollar, affecting other regions of the country where manufacturers strive to compete in export markets. Demand for workers draws the underemployed from other parts of the country to Alberta and has led to changes in provincial and federal labour and immigration laws to facilitate the in-migration of temporary foreign workers.[5] Components for oil sands production are manufactured as far away as South Korea and must be shipped across the Pacific Ocean to ports in the United States, and then transported by land to Alberta, generating opposition every step of the way, as described in chapter 12 of this book. Suppliers of components are looking to Arctic routes as alternatives.[6] The hydrocarbon-based products have to be transported by pipelines – east to the Great Lakes region, west to the Pacific coast, south to the Gulf of Mexico – once again multiplying environmental and social risks and generating opposition to pipeline and refinery expansions. Citizens in North America and Europe are called upon to boycott Alberta's 'dirty' oil (so called because of its carbon intensity and social conditions of production); the European Union has allocated an unfavourable fuel rating to fuel derived from Alberta's oil sands. The calculations of global energy market suppliers and investors take into account the conditions at multiple sites of production around the globe. The web connecting Alberta's political ecology to these many nodal points of the global political economy and to transnational social movements is complex and constantly changing. This collection of essays draws attention to some of the intersections of this web, recognizing the impossibility of providing a comprehensive mapping of a continually re-forming reality.

The Role of Discourse Analysis in Political-Ecological Theory

Increasingly, political ecology studies have employed discourse analysis in addition to political-economic and scientific-ecological modes of analysis. Discourse analysis is used extensively in this volume to help get at questions like why certain ideas or knowledge claims are widely accepted among a particular segment of the population, and where the challenges to these 'hegemonic' ideas come from. For this reason, I devote some space here to setting out the key elements of discourse analysis.

It is through language as well as symbolism that meaning is constructed, and much of what we understand as 'politics' is the competition among different actors to establish a dominant, or 'hegemonic,' interpretation of a given issue – one that better serves the interests actors are defending or advancing. We commonly refer to the 'framing' of issues, meaning the practices of drawing attention to particular associations among objects (and, in the process, obscuring other possible associations). For example, are the oil sands a 'natural resource' whose 'development' provides 'economic prosperity' achieved within environmental limits (the government's 'sustainable development' discourse)? Or are they a source of mainly corporate enrichment leading to 'ecological destruction' and 'social injustice' (the environmental movement's discourse)?

Actors assert (or assume) the existence of certain entities that are foundational to how we see the world. In doing so, actors draw upon or appeal to different kinds of knowledge. Developments in scientific knowledge, for example, continually transform our ontologies (our assumptions about entities that have the status of factuality, or existence). Conservation science produced the concept of an ecosystem – an area or a set of interrelationships that provides the conditions for a reasonably predictable, cyclical reproduction of certain populations. 'Ecosystem' entered popular – and political – discourse and now influences how we understand what is at stake in diverse societal conflicts. In general, studying the discourses used by actors in a given debate or conflict helps us to understand how they comprehend what is at stake and what solutions are possible – or how they want others to understand these things (their 'discursive strategies').

The example of 'ecosystem' as an ontological concept that came into existence at a certain point in time and gained popular currency raises a question about how political ecologists should understand the difference between what 'really exists' and what exists only because we humans have 'discursively constructed' a concept. What do we understand, for example, by 'nature'? Do we think of nature as 'something out there' – akin to wilderness – that is separate from (or even in opposition to) humans, or to the cultures and environments that humans have created? Does 'nature' have any agency of its own – dynamics (or 'laws') that are beyond the control or intervention of humans – or is nature merely inert 'natural resources' or aesthetic landscapes? Other examples of ontological differences include: forest vs. boreal ecosystem; whale as source of food and oil vs. whale as endangered, sentient being; unproductive swamp vs. wetland that provides 'ecosystem services.' Given that our understandings of the same objects vary historically and culturally, how do we know which meanings are closest to 'the truth'? For the most part, political ecologists adopt what is called a 'critical realist' perspective that 'acknowledges the ontological independence of the biophysical world while at the same time recognizing that our *understanding* of the natural world is partial, situational, and contingent' (Neumann 2005, 10–11, italics added).

Why does this matter? Ontological beliefs underpin our worldviews – what we can 'see' and what we cannot 'see.' They are permanently contested by social actors. Whether or not I believe that global climate change is actually occurring (as argued by various fields of science, as well as by Aboriginal cultures relying upon careful observation of their environment over generations) is the first step in a series of unfolding possibilities for my behaviour as an individual. If I accept that climate change is occurring, the next question is 'why'? The answer to this opens up the field of possible solutions or rational responses. Here again, multiple actors intervene to try to establish the dominant interpretation of the causes of, and solutions to, climate change.

In accepting that the natural or biophysical world (the tree outside your window, your body's needs for oxygen and nutrition) has its own processes, cycles, or dynamics – independent of whatever we humans may believe about these – critical realists are inclined to accept the weight of scientific knowledge (laws of physics, earth and biological sciences) about the natural world. However, they do not do so uncritically, as they are aware of changes in scientific paradigms, the social embeddedness of science, and the many critiques of scientific knowledge from cultural, feminist, and postmodern perspectives. As Roderick Neumann observes, every kind of knowledge is 'situated' historically and culturally and therefore is only a 'partial' perspective. The challenge for us is to be critically aware of the sources, interests, and limits of knowledge claims, and of our own criteria in choosing among them. So, from a critical realist perspective it is possible to recognize that the concept 'climate change' has been 'constructed' by humans at a certain point in time, drawing upon particular cultural-scientific forms of knowledge (and excluding or marginalizing others); in this sense, discourse constructs the reality that we are capable of 'seeing,' conceptualizing, or arguing about. At the same time, there are biophysical processes occurring – some of which are described or captured by climate science – that we may partially understand, but that do not depend upon our perceptions of them in order to exist.

Political conflict is organized around ontological categories like 'the economy,' 'markets,' 'society,' 'class,' 'gender,' 'race,' 'ecosystem,' or 'capitalism.' Political discourses jostle to establish hegemonic interpretations of such concepts as 'sustainable development,' 'human nature,' or 'democracy.' Seeking justification of their interpretations, actors appeal to different types of evidence or reasoning, which may be experiential, religious, cultural, or 'scientific.' (Most environmental conflicts involve competing uses of the natural sciences or branches of knowledge with non-experimental methodologies.)

Critical Political Ecology's Normative Orientation

Importantly, political ecologists combine careful empirical research with critical analysis (including of scientific knowledge), guided by *an explicit normative framework*. Paul Robbins describes this approach as

> empirical, research-based explorations to explain linkages in the condition and change of social/environmental systems, with explicit consideration of relations of power ... As critique, political ecology seeks to expose flaws in dominant approaches to the environment favored by corporate, state, and international authorities, working to demonstrate the undesirable impacts of policies and market conditions, especially from the point of view of local people, marginal groups, and vulnerable populations. It works to 'denaturalize' certain social and environmental conditions, showing them to be the contingent outcomes of power, and not inevitable. (2004, 12)

Studies by political ecologists of environmental degradation in the global south have emphasized inequalities in the distribution of natural resources (Hartmann 2009; Peluso and Watts 2001). With regard to first world political ecology, we see a similar normative orientation, influenced by Marxism, feminism, post-colonial, and other critical perspectives. Most political ecology research is critical in these ways, and seeks to conceptualize alternatives to anti-ecological, anti-democratic systems.

Jean-Guy Vaillancourt (1995, 27) notes that, although different disciplines adopt different 'angles,' political ecology addresses common research questions. One of these questions is the relationship between environmental risks and social equity, or social justice concerns. Environmental justice studies in the US (and, recently, in Canada) have informed first world political ecology (Aygeman et al. 2009; Bryant and Goodman 2008, 711; Castree 2007; Heynen et al. 2007; Mascarenhas 2012; McCarthy 2005; Prudham 2004; Schroeder et al. 2006). In their research on environmental health, Robert Brulle and David Pellow (2006, 116) point out that the operations of the state may 'systematically create and maintain environmental inequality' through their intended or unintended policy outcomes. Julian Agyeman (2005, 3) comments: 'The purveyors of environmental bads, such as large multinationals, are favoured in pluralistic decision-making processes because of their disproportionate influence, economic muscle, and knowledge.' Much of this literature has foregrounded the relationships between neoliberal political economy and intensified ecological destruction over the past four decades (Harvey 2005; Heynen et al. 2007; Lipietz 1987; Peck 2006). The political-ecological critique of neoliberalism is particularly pertinent to analysis of the Albertan case, as argued by a number of authors in this book.

What Holds It All Together?

Notwithstanding a shared normative stance, few authors integrate all of the analytical approaches (e.g., multi-scalar political economy, gender analysis, discourse analysis, ecological science, ethnography) in their work. Roderick Neumann (2005, 6) observes that few political ecologists 'incorporate all of these methodologies, or do so in the same relative proportions.' To some extent, this may be because different problems lead us to focus on different axes of analysis or on different levels of analysis. Some studies aim to provide detailed accounts of a phenomenon at one scale (e.g., local actions), while others examine the same phenomenon as the outcome of factors at a different (regional, national, or international) scale. In other words, 'the geographic scales at which each methodology is engaged may vary greatly within and among studies' (Neumann 2005, 6). One recent work that comes very close to drawing upon all of the approaches utilized by political ecologists, in analysing the meanings of the Alberta oil sands, is *Challenging Legitimacy at the Precipice of Energy Calamity*, by Debra J. Davidson and Mike Gismondi (2011). These authors combine, in particular, a political economy analysis with an actor-centred discourse analysis to delve into questions of ideological hegemony and the multi-scalar potential for social-ecological transformation. In many ways, our book picks up where these authors leave off, continuing their analysis of Alberta's political ecology as a 'dynamic, complex system defined by unpredictability and interconnected networks' (2011, x), as well as their aspiration to inform current counter-hegemonic struggles.

Given that our authors are drawn from many different disciplinary backgrounds, the essays that make up this collection exhibit variations in focus and methodology (political economy, discourse analysis, ecological science, spatial/mapping technologies, institutional analysis) as well as variations in the scale of analysis. The authors do share, however, a critical view of Alberta's hydrocarbon-based model of development. They interrogate its consequences for local and global environments, for indigenous peoples, for social equality and well-being, and for Albertans' global citizenship. At the same time, they seek to identify ecologically sustainable and socially just alternatives, and the conditions in which these may be realized.

Political Governance and Democratic Theory

A recent turn in the field of political ecology is the development of linkages between ecology, democracy, and citizenship, drawing on radical democratic theory (Adkin 1998, 2003, 2009; Heynen et al. 2006, 2). Analysis of environmental policy making offers us a window into how political, legal, and media institutions restrict or facilitate citizen participation and informed public debate, as well as frame and delimit societal alternatives and environmental futures. Such work identifies the ways in which existing political decision-making processes are undemocratic and anti-ecological, and proposes democratizing reforms. However, this is not just a problem of procedural design; the systematic subordination of ecological and egalitarian criteria in policy processes reflects power relations at the societal level (Adkin 2009). Moreover, these relations extend far beyond the provincial jurisdiction, as market actors now operate in networked fashion on a global scale. Governments have signed on to international agreements that constrain domestic policy in multiple ways, and local economies can be gutted overnight by the decisions of bond-holders, currency speculators, and investors. Actors draw upon all kinds of resources (juridical, financial, organizational, knowledge, social media) to influence policy processes and outcomes. Questions about democracy, citizenship, and governance take us beyond the state to look at relations between the state and non-state actors who play roles in many domains of economic and societal regulation.

A starting point in the effort to understand how these dynamics constitute democracy and produce the environments in which we live is to document the workings of regulatory processes (the framing of policy options by government agencies, environmental impact assessments, public consultations, the constitution of regulatory bodies, and so on). We do this with a view to making more transparent the predominant norms, interest representation (or exclusion), and criteria and resources for citizen participation that characterize these processes. The chapters in this book contribute to recent work on Alberta (Bowness and Hudson 2013; Davidson and Gismondi 2011; Hierlmeier 2008; Hoberg and Phillips 2012) that analyses the growing contestation of political institutions and regulatory processes that have systematically produced anti-ecological and undemocratic decisions. We see that government agencies responsible, in principle, for protecting human health and the environment have been able to avoid fulfilling these mandates by declining to conduct scientific assessments of environmental health problems,[7] censuring medical authorities who have raised concerns about illnesses linked to environmental causes,[8] or weakening environmental protection legislation.

More generally, a number of authors have identified a growing 'democratic deficit' in Alberta and Canada associated with heavy dependence on oil revenue (Shrivastava and Stefanick 2012, 2015). The 'democratic deficit,' of course, is predicted by petro-state theory. In the chapters that follow, we ask how Alberta's reliance upon oil and gas revenue (particularly the expanding extraction of bitumen for export) has driven bureaucratic restructuring, government investment priorities, public policy, political rights and representation, and citizenship. However, it will also be important to examine a fuller range of explanations for these phenomena.

The governance lens provides significant insights not only into state-society relationships in the province, but also into the meaning of, and conditions for, ecological citizenship. The authors ask which societal interests predominate in the determination of economic and environmental policy in the province, and under what conditions. How do regulatory processes function to sustain the hegemony of these interests? What room to manoeuvre exists within these processes for the social actors who contest the values, priorities, and consequences of Alberta's 'governance regime'? What are the relative strengths and weaknesses of the various social actors seeking to influence the direction of the province's ecological and societal development? What kinds of institutions would be more democratic, and how might the democratization of the political system and policy-making processes affect the outcome of 'environment-versus-development' conflicts? In other words, what is the relationship between the political-institutional regime and the dominant model of development? This question does not assume, of course, that the Government of Alberta is supremely sovereign in relation to economic actors and institutions. Provincial policy is often formed in the context of federal-provincial jurisdictional conflicts or ambiguities, as well as Federal Court and Supreme Court rulings.

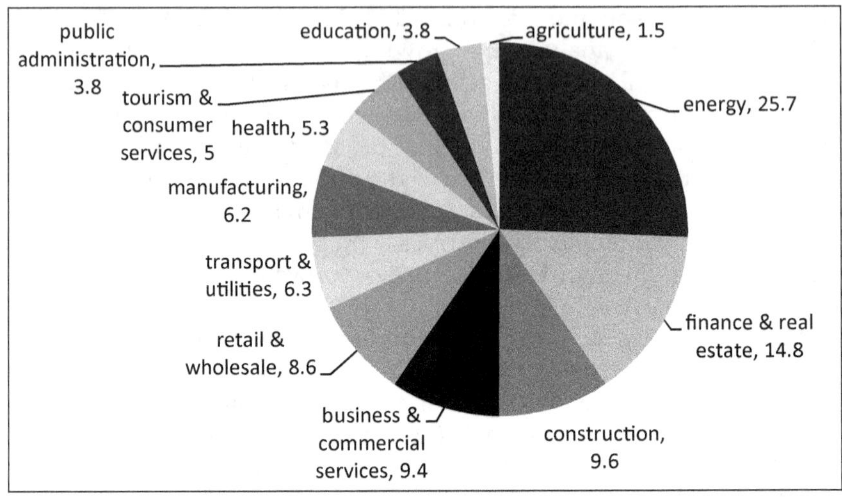

Figure 1.1. Percentage distribution of Alberta GDP by economic sector, 2010. (*Source*: Government of Alberta 2011, p. 9, with data from Statistics Canada and Alberta Treasury Board and Enterprise.)

The resources and influence of major economic interests in the province lie also in the global economy and the institutional agreements that underpin it. These realities must be part of any political-ecological analysis of state-society relations in Alberta.

Petro-Politics

Alberta is not only a first world state; it is also a first world petro-state. By the latter term we mean that the government's revenue draws substantially upon rents from sales of hydrocarbon resources, a large portion of provincial GDP comes from oil and gas exports, and investment in the sector generates a lot of direct and indirect employment and tax revenue. Over the last four budget years for which data were available when this chapter was written (2010–11, 2011–12, 2012–13, 2013–14), the share of non-renewable resource revenue in the province's total revenue was 21.7, 29.6, 20.1, and 21.2 per cent, respectively (figures from Government of Alberta budget documents). In 2010, gross revenues from all hydrocarbons were $73.2 billion, with revenues from crude oil (including the oil sands) accounting for almost two-thirds of the total (Government of Alberta 2011, 14). In figure 1.1 we see that energy pro-

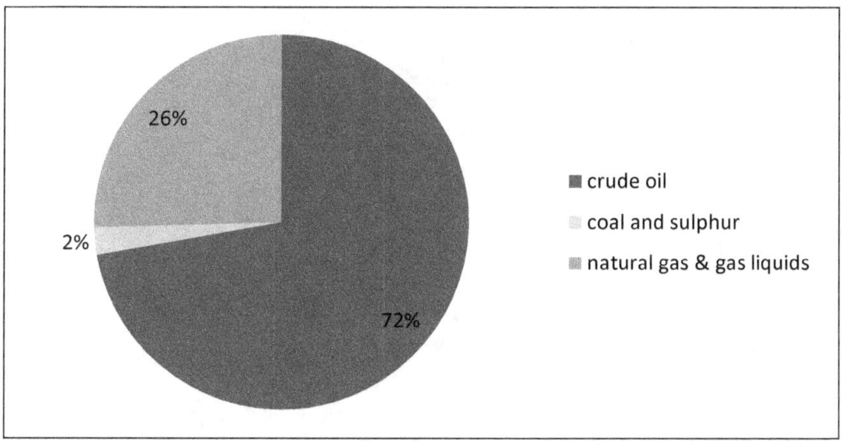

Figure 1.2. Energy products as percentages of Alberta's total energy exports 2010. Energy exports accounted for $53.8 billion in 2010. (*Source*: Government of Alberta 2011, p. 14, using data from Statistics Canada, Energy Resources Conservation Board, Alberta Treasury Board and Enterprise.)

duction accounts for the largest share of provincial GDP. Construction, which takes third place, is substantially driven by the expansion of oil sands exploitation. It may surprise many Canadians to see that agriculture accounts for less than 2 per cent of the province's GDP, despite the popular association of Alberta with the wide open spaces of agriculture and livestock pasture.

In figure 1.2 we see the shares of crude oil, coal and sulphur, and natural gas and gas liquids in the province's total energy exports, valued at $53.8 billion in 2010. Crude oil has by far the largest share (72 per cent). If we add refined petroleum products to energy exports ($1.4 billion in 2010), these total $55.2 billion, or 71 per cent of the value of all of Alberta's exports of goods ($77.8 billion) in 2010 (see figure 1.3.)

Alberta's oil sands – estimated to contain 169 billion barrels of crude bitumen (ERCB 2012, 2) – represent 95 per cent of Canada's oil reserves. As shown in figure 1.4, oil sands production more than doubled from 2000 to 2009 to 1.3 million barrels per day (bpd), reaching 1.9 million bpd in 2012. In addition, Alberta accounts for 75 per cent of Canada's natural gas production (and Canada is the world's third-largest producer of natural gas, after the USA and the Russian Federation). The province is also rich in coal, with estimates of remaining reserves at 33

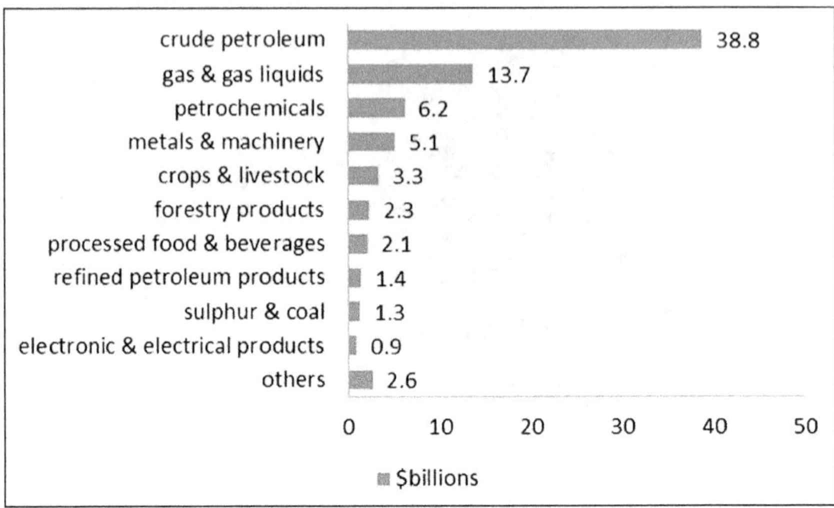

Figure 1.3. Alberta's major exports 2010 ($ billions). Total exports of goods = $77.8 billion. Note that exports of services are not included in this estimate. (*Source*: Government of Alberta 2011, p. 11.)

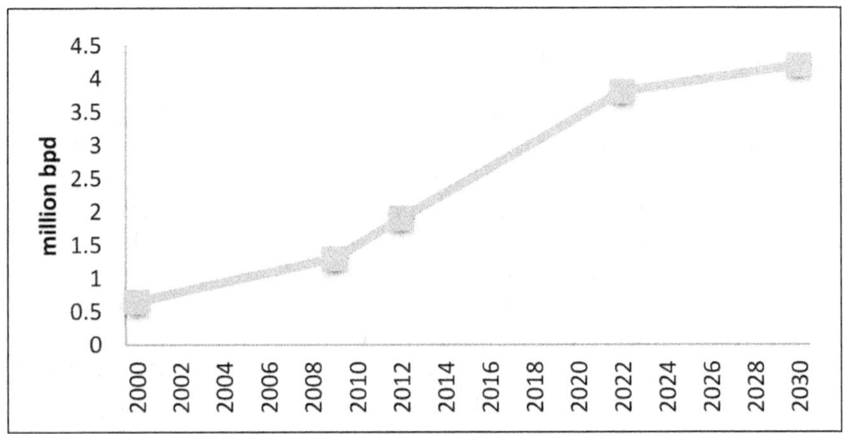

Figure 1.4. Output of crude oil from the Alberta tar sands projected to 2030. (*Source:* Energy Resources Conservation Board 2012; Alberta Energy, http://www.energy.alberta.ca/OilSands/oilsands.asp [accessed June 2013]).

billion tons (ERCB 2012). As we see in chapter 15 of this volume, Alberta has been a large consumer of the coal mined in the province; 59 per cent of the province's electricity was fuelled by coal in 2011 (Alberta Energy 2012b). There is also significant potential for gas production from coal-bed methane – up to 500 trillion cubic feet, according to the Alberta Geological Survey (2012) – and from shale (Rokosh et al. 2009). Other resource sector exports (particularly pulp, paper, and wood products, and beef) are also important to the provincial economy, but crude oil, bitumen, and natural gas exports and prices determine whether or not the economy is in boom or bust, and (especially since the expansion of oil sands mining after 2000) hydrocarbons have enmeshed the province in a global web of economic, ecological, and political relationships.[9] As we see in table 1.1, the importance of oil and gas revenue to the provincial economy puts Alberta in the same league as Russia or Norway.

The federal Conservative Party, since it formed a minority government in 2006 and was subsequently re-elected with a majority in 2008, has sought to make Canada an 'energy superpower.' The Conservatives' failure to implement substantial greenhouse gas emission reduction measures, increasingly authoritarian responses to civil society and parliamentary opposition, systematic undermining of environmental legislation as well as ministerial research and enforcement capacities,[10] and refusal to respond to increasing evidence that the manufacturing sector is suffering from 'Dutch Disease' (Beine et al. 2012; Lemphers and Woynillowicz 2012; Morison 2012; Mourougane 2008) led a growing number of commentators to ask if the Canadian state, too, was not becoming a petro-state (Homer-Dixon 2013; Monbiot 2009; Nikiforuk 2007a, 2011, 2013; Oremus 2012).

On the other hand, there are striking differences between Alberta/Canada and other petro-states – even other first world petro-states (as we see in the comparisons with Norway). Thus, this volume explores what kind of petro-state Alberta is – how its geology, history, and positioning within multiple 'scales' (the Canadian federation, the American 'backyard,' the global political economy) are shaping its state and civil society, and how its current institutions and hegemonic politics are determining ecological and democratic futures for its own citizens, as well as for people beyond its borders. However, as Angela Carter and Anna Zalik explain in their chapter in this volume, we do not confine our analysis to the relationships and concepts that have received most attention in the petro-state literature to date. Instead, we try to grasp the dynamics shaping the province's politics and environment within the broader framework of political ecology.

Table 1.1. Comparison of Four Petro-States (Data for the Most Recent Year Available)

Jurisdiction	Oil and gas production as % of GDP	Oil and gas sales as % of export earnings	Oil and gas rents/taxes as % of total state revenue	Per capita income (current $US)	Public social expenditure as % of GDP[a]	% of oil sector owned by state	Income inequality (Gini coefficient[b])	UN HDI[c] 2011 (ranking)	Voter turnout (%)
Alberta	25.7 (2010)	61.4 (2010)	35 (2011)	67,340 (2009)	16.9 (2007)[f]	0	0.32 (2010)	Alberta (3) [Canada 0.908 (6)]	57 (2012)
Norway	21 (2010)	47 (2010)	26 (2010)	98,102 (2011)	20.8 (2007)	80 (2012)	0.244[e] (2012)	0.943 (1)	76 (2009)
Russian Federation	25 (2010)	66 (2011)	50 (2010–11)	13,089 (2011)	12.0 (2007)	c. 40 (2008)	0.425 (2011)	0.755 (66)	65 (2012)
Venezuela	33 (2009)	80 (2009)	50 (2009)	10,809 (2011)	13.6 (2006)	c. 60 (2013)[d]	0.39 (2011)	0.735 (73)	78.7(2013)

[a] The OECD calculates public social spending as 'the amount of resources committed by the government in the areas of pensions, benefits (social support) and health.' For more details, see OECD, *Society at a Glance 2011: OECD Social Indicators*, sec. 5, 'Public Social Spending,' http://www.oecd-ilibrary.org/docserver/download/fulltext/8111041ec020.pdf?expires=1342425722&id=id&accname=guest&checksum=7EE4710FAAB3A2655388E84DDF6050A4.

[b] The Gini coefficient is defined as the area between the Lorenz curve (which plots cumulative shares of the population, from the poorest to the richest, against the cumulative share of income that they receive) and the 45° line, taken as a ratio of the whole triangle. The values of the Gini coefficient range between 0, in the case of 'perfect equality' (i.e., each share of the population gets the same share of income), and 1, in the case of 'perfect inequality' (i.e., all income goes to the individual with the highest income).

[c] Data from United Nations Development Program, 'Table 1. Human Development Index and Its Components' (figures for 2011), http://hdr.undp.org/en/media/HDR_2011_EN_Table1.pdf (accessed 15 July 2012), except for Alberta.

[d] The government of Venezuela owns 100 per cent of the shares of Petroleos de Venezuela (PDVSA), which is the third-largest state-owned oil company in the world. However, not all oil produced in Venezuela is produced by PDVSA. Exxon Mobil and ConocoPhillips were expropriated in 2007 when they refused to renegotiate their contracts with the Chavez government. Venezuelan law requires foreign investors to form partnerships with PDVSA in which the state-owned oil company has 60% ownership. Foreign companies' revenues are subject to a 50% tax rate and a 33% royalty (tax). Thirty-six lease blocks in the Orinoco oil field have been auctioned off to twenty-seven companies from twenty-one nations, including state-owned companies from Iran, Belorussia, and Cuba, and big publicly traded oil companies like Spain's Repsol, Brazil's Petrobras, Italy's Eni, and France's Total. Other transnational companies retained their minority interests in Venezuelan oil production after 2007, including Statoil and Chevron.

[e] Gini co-efficient after taxes and transfers.

[f] Includes spending on social services, education, and health.

Sources: See appendix 1.1.

The petro-state literature's criticisms of petro-elites, while often trenchant, stop short of identifying global capitalism, fuelled by petroleum – or 'fossil capitalism' (Elmar Altvater 1998, 2006) – as fundamentally antithetical to democracy and to ecological sustainability. Political ecology adopts the more critical view that the capitalist model of development – fuelled, predominantly, by oil since WWII – is necessarily anti-democratic and ecologically ruinous for the planet's ecosystems (Lipietz 1992; O'Connor 1991, 1998; Rogers 2006). Thus, the solutions to the grave social and ecological crises we confront today will need to be more far-reaching than restoring the rule of law or liberal democratic institutions to particular capitalist states. Moreover, the petro-state literature's focus is not the roots of ecological crisis, and so it gives little attention to questions of ecological sustainability – for example, to the question of whether this can be achieved through technological innovation or whether more fundamental social and economic reorganization is required. Finally, in its search for general laws – such as the existence of a necessary relationship between a substantial state dependence on revenue generated by oil production, on the one hand, and authoritarianism, on the other, the petro-state literature tends to bypass contextual, socio-historical analyses of state formation and state-society relations.

One of the problems we face when trying to discern the relationship between heavy reliance upon hydrocarbon extraction and export and democratic deficit is that of separating out the effects of 'petro-politics laws' from other factors, such as ruling party ideology, the characteristics of civil society, or the strategies adopted by corporate actors in different contexts. Petro-politics alone cannot explain the forms of disciplining of dissent that we have witnessed in Alberta since the early 1990s, and that have characterized neoliberal regimes (including non-oil-producing states) elsewhere (Fairbrother et al. 1997; Gamble 1988, Harvey 2005; Leys 1989). Interestingly, few observers referred to a 'democratic deficit' in Alberta back in 1985, when energy resources accounted for 36.1 per cent of the province's GDP (compared, e.g., to 25.7 per cent in 2010) (Government of Alberta 2011, 9). Indeed, as argued more fully in the concluding chapter of this book, a key lesson of the comparative study of petro-states seems to be that their politics following the discovery of oil reserves are determined as much by the historical formation of their social classes and political institutions as they are by the effects of heavy economic reliance upon oil rents.

Alberta provides a case study of a first-world petro-state governed politically by neoliberal ideologues for more than two decades. The governments of Premier Ralph Klein, in office from 1992 to 2006, ushered in a radical experiment in neoliberal governance modelled on those in the United Kingdom, under Margaret Thatcher, and in New Zealand, in the form of Rogernomics.[11] The provincial energy company Encana was privatized by the Klein government, and Alberta remains one of the few oil-producing countries that do not have a publicly owned oil corporation. Alberta is, however, joined in that club by the United States. Interestingly, only recently has the USA been characterized as a petro-state, although it has, historically, been one of the world's largest oil and gas producers.

The well-known American economist Jeffrey Sachs, quoted in a 2010 article in the *Washington Post*, agreed that the United States could certainly be characterized as a petro-state:

> 'We have lots of the characteristics of petro-states ourselves even though we use that term for others,' he says. He cites our overdependence on oil and a tax policy that keeps oil relatively cheap. Moreover, he adds, 'big oil plays an unnatural role in our politics ... Oil elects presidents, drives our foreign policy, our domestic policy, our climate change policy ... It's led us to terrible energy policies and a breakdown of regulation. We look to the Niger Delta as an example of what an oil state does to its own environment, but it's precisely what we're doing to our own environment.' (quoted in Mufson 2010)

Political economist Michael Klare asked, in a 2012 article, if pressures from the oil and gas industry to remove barriers to high-risk oil and gas extraction (through fracking, offshore drilling) would result in the USA and Canada becoming more like 'Third World petro-states,' with 'eviscerated' environmental protection and authoritarian measures to quell citizen protests. The 15 February 2014 issue of *The Economist* featured a picture of President Obama garbed as an Arab oil sheik, below the heading 'The Petrostate of America.' In this issue, *The Economist* pointed out that the United States, or 'Saudi America,' had 'surpassed Russia as the world's largest producer of oil and gas,' and that by 2020 'it should have overtaken Saudi Arabia as the largest pumper of oil.' Not surprisingly, *The Economist* characterized the United States as a 'liberal democracy,' although many critical political economists today see instead a plutocratic system of government and widening social inequalities (Harvey 2010; Krugman 2006, 2013, 2014; Piketty 2014; Reich 2008, 2010).[12]

A more insightful article, by Mark Hertsgaard in the 27 February 2014 issue of BloombergBusinessweek,[13] explains the long and close relationship of big oil companies with the US state, and the enormous influence they wield over US government policy. However, Hertsgaard makes an additional argument regarding the influence of oil interests that resonates strongly in the Albertan context: 'For nearly a century now, broad swaths of the populace and powerful individuals in government, finance, and other key sectors have seen oil as indivisible from national interest.' That identification of the interests of oil producers with the interests of the citizenry as a whole – one both actively promoted by governments and (the same) corporations and passively internalized by citizens as consumers of downstream products and as automobile owners – operates as powerfully in Alberta as it does in most parts of the USA. Alberta is a subnational jurisdiction whose population of four million is only 1.3 per cent of that of the USA; the influence of Alberta in the world is hardly comparable to that of the USA. Yet there are similarities in political culture, consumption norms underpinned by oil production and consumption, and the political influence of the large resource-extractive corporations vis-à-vis governments. Further research on the political economies of oil- and gas-producing states in the USA, along with new work on Alberta, may help us to get at some of the difficult questions about what shapes political cultures, institutions, and regulatory norms in first world hydrocarbon-extraction-dependent economies (see Carter, this volume; Davidson and Gismondi 2011; Goldberg et al. 2009; Kusnetz 2014; Watts 2012; Zalik 2012).

Diana Gibson, a former research director for the Parkland Institute and author of a 2007 report on the province's royalty regime, observed that 'with over 80% of the world's oil locked up by national oil companies, the rest of the world has realized that this resource is too valuable and strategic to be handed over to foreign multinationals. Canada, on the other hand, has allowed majority foreign ownership of both oil and gas' (2007, 5). She also noted that nationally owned oil companies from Norway, China, Korea, Japan, and Abu Dhabi have been investing in the Alberta tar sands. Moreover, a series of studies have made clear that Alberta's share of oil rents compares very poorly with the rents collected by other petro-states (Alberta Royalty Review Panel 2007; Boychuk 2010; Campanella 2012; Gibson 2007; Macnab et al. 1999; Warnock 2006). While some argue that the owners of the resource should receive 100 per cent of the rent, this is far from being the case in Alberta.

Somewhat problematic about this argument is the suggestion that it is the 'foreign-ness' of the multinational corporation having ownership of Canadian resources that is the cause of the Government of Alberta's poor record as manager of the publicly owned resource. Yet many of the MNCs operating in the Alberta oil sands (e.g., Chevron, Shell, BP, Imperial Oil/Exxon) also operate in other jurisdictions (including the USA) where government takes are higher. Here again, further comparisons of Alberta's royalty and taxation system with those in the oil-producing US states might prove informative. Notably, Alberta's Royalty Review Panel report found that Alberta is the lowest tax and royalty jurisdiction *in North America* as well as one of the lowest in the world (Gibson 2007, 6).

Alberta is repeatedly, and unflatteringly, compared to Norway with regard to management of oil wealth. For example, Bruce Campbell of the Canadian Centre for Policy Alternatives has observed that

> the Norwegian government owns 80% of petroleum production, and retains roughly 85% of the net petroleum revenues mainly through a 78 per cent company tax and through direct access mechanisms ... Norway's Petroleum Savings Fund has amassed [since 1991] over $664 billion in assets, all invested abroad, with only the return used for domestic spending. It not only ensures the future of social welfare benefits, but also helps to offset upward pressure on its currency and mitigate potential Dutch Disease effects. Alberta's Heritage Savings Fund [established in 1976] now contains $16 billion, just 2% of Norway's fund, and a miniscule share of the petroleum revenue that has flowed into Alberta over the last 36 years. (2013)[14]

Thus, in regard to its rentier[15] state behaviour, Alberta is an exceptional first world petro-state in a number of ways. Compared to counterparts in North America, it has negotiated poorer deals with multinational corporations, and compared to Norway, it has saved very little of the royalty wealth it has collected. Saskatchewan, coming late onto the scene with regard to the extraction of its oil sands, has had to compete for capital investment with Alberta and its exceptionally 'generous' royalty and taxation regime (see chapter 3's discussion of the 'Alberta Advantage').

Petrofied Climate Policy

Under the governments led by Premier Ralph Klein, Alberta's official discourse and policies on climate change resembled closely those of the similarly oriented George W. Bush administration in the USA. The federal Conservatives, in government from 2006 to 2015 under the leadership of Stephen Harper, share ideological roots with the Alberta [Progressive] Conservatives and have adopted tandem national energy and climate change policies. Their Promethean and neoliberal ideological orientations go a long way towards explaining these governments' dogged commitment to fossil capitalism. With the election of the like-minded Liberal-National coalition government in Australia in 2013, fossil capitalism (in particular, the coal industry) found new political defenders in that country as well.

While European and other first world capitalist states (including some subnational jurisdictions in the USA and Canada) have recognized the necessity of shifting to lower-carbon economies, the Conservative governments of Alberta and Canada defended and promoted (until their electoral defeats in 2015) Alberta's position in the global economy as a major exporter of fossil fuels, while resisting demands for even market-based environmental regulation. Both levels of government delayed measures like energy taxes or a national cap-and-trade system that might bring about absolute emissions in greenhouse gases (GHGs) and set the country on the path to meeting significant reduction targets. Instead, the Conservatives' position (federally and in Alberta) has been that the Kyoto target of a 6 per cent reduction below 1990 levels by 2020 would necessitate substantial damage to the economy. Structural dependence on energy sector revenue is certainly part of the explanation for this intransigence,[16] but the outcomes should not therefore be taken as a fait accompli. In other words, it is possible that different political leadership could initiate a transition to a post-carbon economy.

At the federal level – in the context of heightened public concern about climate change and their (then) minority government's vulnerability – the Conservatives committed in 2007 to a national GHG reduction target of 20 per cent below a 2006 baseline by 2020 (which translates to a level of emissions that is 3 per cent above the 1990 level) (Auditor General of Canada 2012, 2.31). However, the government reduced this target further, at the 2009 Copenhagen CoP, to a 17 per cent reduction from 2005 levels by 2020, matching the target adopted by US president Barack Obama. In December 2011, after obtaining a majority government, the Conservatives announced that Canada would be formally withdrawing from the Kyoto Protocol. Data subsequently released by Environment Canada, reviewed in the spring 2012 report of the Auditor General of Canada, revealed that in 2020 Canada's GHG

Table 1.2. Energy Efficiency and Carbon Intensity Performance for Four Petro-States

Country	Energy consumption per unit of GDP (2008)[a]	Carbon intensity[b] (2007)	Carbon intensity (2011)	GHGs per capita (tonnes) (2011 or latest available year)[c]
Canada	0.17	0.48	0.44	20.35
Norway	0.08	0.18	0.2	10.77
Russian Federation	0.21	0.79	0.85	n/a
Venezuela	n/a	0.48	0.56	n/a

[a] Measured by the International Energy Agency in tonnes of oil equivalent (toe) per thousand 2005 US dollars of GDP calculated using Purchasing Power Parities (PPPs). *Source*: OECD, *Economic Surveys*: *Russian Federation 2011*, fig. 5, 'Total Energy Consumption per Unit of GDP' (Paris: OECD, December 2011), p. 12. The figure takes data from IEA, *World Energy Statistics* database and World Bank, WDI database.

[b] US Energy Information Administration, 'International Energy Statistics,' http://www.eia.gov/cfapps/ipdbproject/IEDIndex3.cfm?tid=91&pid=46&aid=31.

[c] *OECD Factbook 2014*: *Economic, Environmental, and Social Statistics*, 'Greenhouse Gas Emissions: Tonnes per Capita,' DOI: 10.1787/factbook-2014-graph164-en, May 2014. The OECD average was 12.76 tonnes per capita.

emissions would be 7.4 percent above 2005 levels instead of 17 percent below; Canada would need to reduce emissions by 178 million tonnes between 2012 and 2020 to meet the Conservatives' Copenhagen target. The Auditor General did not see a regulatory framework in place to achieve these reductions. In particular, long-promised regulations for the oil and gas sector were nowhere in sight. In 2014, it was impossible to find on Environment Canada's website any record of the government's Copenhagen commitment, and it was no longer referred to by the Conservative government.

As we see in table 1.2, Canada is performing poorly in the carbon intensity of its GDP, as compared to another first world petro-state, Norway – although the post-2008 recession has brought about some decline in Canada's GHG emissions.

The positions taken by Alberta's governments with regard to the Kyoto targets have played a preponderant role in obstructing progress on GHG emissions reductions in Canada. In this regard, Canadian political scientist Douglas Macdonald characterized Alberta as a 'highly motivated veto state' (2009, 153). Alberta's climate change plans, adopted in 2002 and 2008 and discussed briefly in chapter 6, have relied heavily upon an ineffectual 'cap-and-trade' system (with no hard caps) for large emitters (over 100,000 tons CO_2e per annum) and promises of future emission reductions from carbon capture and sequestration technology to achieve the weak target of a 14 per cent reduction below a 2005 baseline by 2050. Since the 1990s, Alberta has not, in any meaningful sense, had a 'climate change' policy. Instead, proposed GHG reduction measures in Alberta have largely been a 'social license' adjunct to energy resource development priorities (Adkin 2014). The NDP government elected in May 2015 has initiated new policy measures, but as of spring 2016 it is not evident how effective these will be in reducing the province's greenhouse gas emissions.

The drive to extract Earth's remaining fossil fuel reserves has become a critical threat to the viability of ecosystems that sustain millions of people. In his 19 July 2012 article in *Rolling Stone* entitled 'Global Warming's Terrifying New Math,' author Bill McKibben cited climate change experts from multiple research institutes in support of the conclusion that 'we have five times as much oil and coal and gas on the books as climate scientists think is safe to burn. We'd have to keep 80 percent of those reserves locked away underground to avoid that fate.' Since Alberta's governments have been building the province's economic future on the promise of the exploitation of its non-conventional oil and gas reserves, the stage has been set for environmental conflict at multiple scales.

Research on Alberta, and a Little History

Alberta has been the subject of a number of historical studies that are helpful in understanding the formative influences upon its contemporary political culture. The best known of these is Crawford B. Macpherson's *Democracy in Alberta: Social Credit and the Party System*, first published in 1953. Macpherson argued that Alberta's early political economy – one dominated by small commodity producers and positioned as a periphery vis-à-vis central Canada (i.e., as a 'quasi-colonial' society) – had given rise to a 'quasi-party system' that 'concealed' class differences.[17] The relatively small industrial working class and the comparative absence of 'polarized class politics' have been viewed as partial explanations of the populist nature of political discourse in the province, and of Albertans' historical tendencies to stick with one party for long periods of time and to show little interest in party politics or elections (see McCormick 1980; Wiseman 2013, xiii). A number of historians suggest that one of the enduring legacies of this quasi-colonial status, which persisted well into the twentieth century, has been the 'provincial sovereigntist' discourse of the province's ruling parties that is discussed in chapter 6 of this book in terms of 'nativism.'

Macpherson's party system theory has been challenged or modified on a number of grounds. Seymour Martin Lipset (1950, 1954) observed that Saskatchewan and Alberta had shared similar class structures, yet Saskatchewan's populist political culture had taken a 'collectivist' direction, in contrast to Alberta's liberal-individualist direction. Cultural factors, therefore, had to play a more significant role in explaining the development of political cultures and party systems than Macpherson's approach had considered. Historian Nelson Wiseman (1981, 2007a, 2007b) has emphasized the importance of the values American immigrants carried to Alberta between the 1890s and the 1920s in explaining the right-wing, libertarian variant of populism and Christian conservatism that came to predominate in Alberta's political culture after the mid-1930s (under the Social Credit governments). However, Wiseman also draws attention to the existence of cultural divisions *within* Alberta, first between the south – heavily influenced by American settlers with their 'populist-liberal-evangelical' beliefs (2007b, 245) – and the northeast (adjacent to Saskatchewan) – 'settled by continental Europeans,' including Ukrainians (249). A second division is between rural areas and the largest cities, the former having been strongly influenced by the US-inspired populist and religious movements and the latter by more traditional class identities brought by British and other European working-class immigrants (248–9).[18]

Looking at political-economic developments in Alberta in the decades following the discovery of oil at Leduc in 1947, economist Ed Shaffer remarked upon the significant growth of the non-agricultural workforce in the 1960s, mainly due, in his analysis, to the Social Credit government's spending of oil revenues on public services (1984, 182–3). In Shaffer's view, this trend 'undermined the rural political base of the Socreds,' creating the pro-industrialization and pro-economic-diversification middle strata who came to support the modernizing Progressive Conservative Party by the end of the decade. This picture was confirmed by political economist Larry Pratt, who observed that at the beginning of the 1970s 'a major part of Alberta's population worked in white collar occupations for large private or public institutions' and had 'secular values' (1984, 204). Pratt further argued that the 1970s 'province-building' initiatives of the Lougheed Conservatives represented the interests of 'an ascendant class of indigenous businessmen, urban professionals, and state administrators' who shared a vision of an independent, diversified economy that would no longer be subject to the self-serving policies of the central Canadian bourgeoisie. In Shaffer's view, however, it was clear by the mid-1980s that a manufacturing sector independent of the oil and gas industry had not really been established in Alberta; rather, the main industry, petrochemicals production, remained dependent upon oil and gas production, and the smaller metal fabrication and machinery industries were 'satellites' of the energy sector. This conclusion was supported by other analysts (Norrie and Percy 1981; Pratt 1984). Shaffer predicted that the 'growing dependence of the Alberta work force on the oil industry' would make the province even more vulnerable to fluctuations in the price of crude and would reduce its bargaining power with the multinational companies (1984, 176). He further observed that the share of corporate profits in Alberta's GDP had risen sharply over the 1970s, reaching 22 per cent in 1979, as compared with the Canadian average of 10 per cent (180).

By the end of the 1980s, the PC government was no longer talking about economic diversification in the sense of secondary manufacturing; however, the Getty government allocated $2.35 billion to forestry projects in the late 1980s, including the Alberta-Pacific pulp mill (Pratt and Urquhart 1994, 5; see also Richardson et al. 1993). With the rise of world oil prices after 1990, the election of a group of free-marketeers to the legislature in 1993, and the implementation of a royalties and tax regime (in 1996) geared to attracting investment in the oil sands, resource extraction continued to drive the economy. The Canadian Association of Petroleum Producers (CAPP 2005, 1) reported corporate investment in the oil sands from 1996 to 2004 amounting to $36 billion (with investment being defined as 'purely a development expenditure, which increases the production level of a project').

While most of the major scholarly studies of Alberta's political economy were published before the 1990s (Leadbeater 1984; Pratt 1976, 1977; Richards and Pratt 1979), a number of later scholarly works sought to make sense of the ruling party's turn to neoliberal governance in the early 1990s, including its implications for the terms of natural resource exploitation (Harrison 2005; Harrison and Laxer 1995; Pratt and Urquhart 1994). In addition to the scholarly literature,[19] think tanks like the Pembina Institute, the Parkland Institute, the Canada West Foundation, and the Canadian Institute of Resources Law, as well as non-governmental organizations (NGOs) like the Canadian Parks and Wilderness Society and Sierra Club Canada, have published a number of important studies on energy policy, the royalties regime, land use regulation, water policy, greenhouse gas emissions, and other environmental questions. A few journalists have also written accounts of oil sands development and the environmental effects of the oil and gas industry more generally (Marsden 2007; Nikiforuk 2008).

Class Structure in the 2000s

Since this book presents new research on the province's political economy since the 1990s, I will not attempt a detailed review of the last twenty years here. A few paragraphs are warranted, however, to update the account of changing class structure, as this is important context for the subsequent analyses. Figure 1.5 provides a snapshot of the provincial workforce by Statistics Canada's 'industry groups' as of May 2015. From this figure, we see that services in the private sector account for nearly half of existing jobs, while jobs in the public sector (health, education, public administration) account for about 20 per cent. The two largest categories of employment are wholesale and retail trade (services) and construction. In the first category – especially in the retail sales clerks and cashiers categories – there is a predominance of low-paid, female, and often part-time or casual labour. The average hourly wage for women working in 'sales and services' in 2013 was reported by the provincial government (Government of Alberta 2014, 18) as being $10.42, which was just slightly above the minimum wage. (As of July 2014 Alberta had the lowest minimum wage of all jurisdictions in Canada.) In the second category, the workforce is almost entirely male, and commands higher wages.[20] A number of studies have documented a widening income inequality gap in the province, reflecting the polarization (and gendered and racialized nature) of the labour market (Campanella et al. 2014; Dorow 2015; Gibson 2012; Lahey 2015). Using data from Statistics Canada, the Public Interest Alberta research and advocacy organization has calculated that 71 per cent of minimum-wage workers are women (Kleiss 2014). Of the 23 per cent of Alberta workers who are low-wage earners (under $16.00/hr), 61 per cent are women (PIA 2014).

Temporary foreign workers (TFWs) have been brought to Alberta by employers in both the retail and construction sectors – particularly in the Fort McMurray region (Foster and Barnetson 2015; Foster and

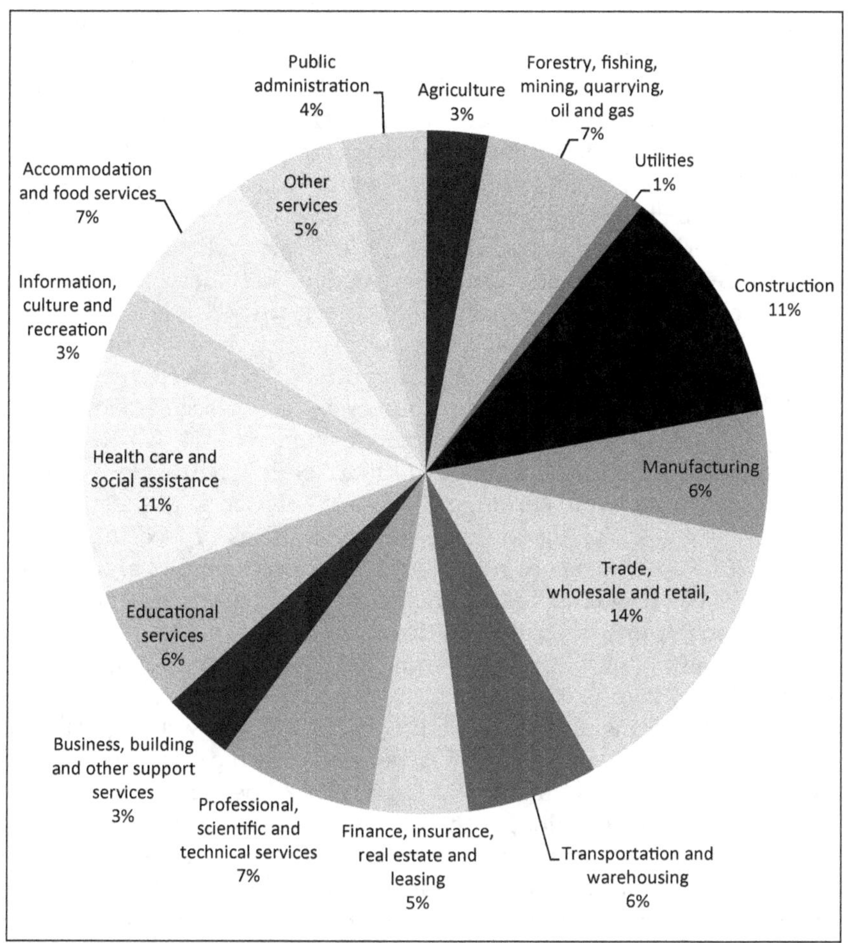

Figure 1.5. Employment in Alberta, 2015. *Source*: Statistics Canada, CANSIM Table 282-0088, 'Employment by Major Industry Group, Seasonally Adjusted, by Province [Monthly] [Alberta].'

Taylor 2013; O'Shaughnessy and Doğu, this volume). The Alberta Federation of Labour (AFL) estimated the number of TFWs in Alberta in December 2008 to be 57,843 'despite the global economic collapse that struck in the fall of that year – an increase of 55% in one year and a quadrupling of the program in five years.'[21] In December 2013 the federal government put the number of TFWs in Alberta at 85,000 – more than in any other province (Steward 2014). An audit of the federal government's TFW Program conducted by the AFL in spring 2014 found that 'employers were routinely allowed to pay TFWs minimum wage in industries and occupations that are supposed to pay higher rates.'[22]

Only 7 per cent of the workforce is employed in 'natural resource' jobs, which includes forestry, fishing, mining, quarrying, and oil and gas field workers. The average wage for 'primary industry' workers in Alberta given by Statistics Canada is about $31.00/hour, but this category includes farm labourers as well as mega-truck drivers and supervisors, so we can assume that hourly wages for workers in the oil and gas sector are much higher.[23] The Government of Alberta reported in 2014 that workers in the natural resources extraction sectors earned, on average, $41.00/hr in 2013 (Government of Alberta 2014, 17). Journalist Gillian Steward reported in June 2014 that, while food service workers in Fort McMurray were earning $400 to $500 per week 'at the most,' in the oil and gas sector the average weekly wage as of December 2013 was $2,067, 'up 73 per cent from 2001.' The decline in global oil prices since 2014 has triggered yet another 'bust' in Alberta's economy, contributing to the loss of 73,000 full-time, private-sector jobs in the province between January 2015 and January 2016 (Statistics Canada 2016). Hardest hit by layoffs were men in the fifteen-to-twenty-four-years age group. Some of this unemployment was offset by growth in part-time employment, which increased by 10 per cent over the same period, and by public sector employment, which grew by 13 per cent.

Clearly, Alberta is no longer the society of small commodity producers that C.B. Macpherson wrote about in 1953, when the recent discovery of oil at Leduc was already transforming the provincial economy. Today, only 3 per cent of Alberta's workers are employed in agriculture while as much as 45 per cent of the province's jobs are directly or indirectly related to oil and gas (Mansell and Schlenker 2006, 29). This transformation has very significant implications for questions of sustainable agricultural production for local markets – objectives of green transition in Alberta.

Post-colonial Character of Alberta's Political Economy

The literature referred to so far focuses predominantly on the formation of settler culture and class structure, as well as the province's political party system. Another important dimension of the political ecology of Alberta, however, is the concurrent history of dispossession and political regulation of the Aboriginal population. The development of successive phases of commodities production and resource extraction (fur trade, grains, cattle, coal, forestry, conventional oil and gas, bitumen and other non-conventional fossil fuels) required the forcible imposition of a British colonial system of property rights and political institutions – including the reserve system – upon the indigenous inhabitants of the Dominion. The political ecology of Alberta is rooted in this historical dispossession, and continues to be shaped by the struggles of Aboriginal communities to protect or secure the land base, resources, and sovereignty rights necessary to ensure their cultural survival and well-being. While this is not the place to unfold a review of the literature in multiple domains (anthropological, sociological, political, legal) that speaks to the history of Aboriginal peoples in Alberta, the situation described by one legal scholar in 2002 continues to be of central concern to the authors of this book:

> The centre of the Aboriginal peoples' livelihood and worldview is their special relationship with the land and its resources. However, increasingly rapid resource development threatens their future relationship with the land and the environment. Despite the Governments of Canada owing both private and public fiduciary duties to Aboriginal people, the devastation of the land continues without their rights and views being fully considered. Nowhere is this more prevalent than in the province of Alberta. While the Law regarding the duty to consult has been spoken to by the Supreme Court of Canada and various courts and tribunals outside of Alberta, establishing that the duty exists within Alberta in the natural resource context continues to be a battle ... The author infers that the profitability of Alberta's plentiful resources and the conservativeness of the territory, exemplified most strongly by the government, have thus far prevented recognition of the duty. The law and justice demand more. (Szatylo 2002, 202–3)

Multiple Perspectives on the Political Ecology of Alberta in recent scholarship

Scholarship in the conservation sciences, resource management, and environmental and indigenous legal studies has been very important for political ecologists seeking to understand the Alberta case. Scientists such as Bradshaw et al. (1997), Schindler and Adamowicz (2007), and Schneider and Dyer (2006), along with environmental management experts (Natcher 2000; Ross et al. 2006; Schneider and Walsh 2005;

Sherrington 2005; Spaling et al. 2000; Timoney and Lee 2001) and environmental law scholars (Keeping 2004; Kennett and Ross 1998; Percy 1997; Ross and Potes 2007; Sharvit et al. 1999), have begun to produce significant bodies of work dealing with environmental, economic, and legal aspects of resource exploitation and land use in the province, including in relation to Aboriginal rights.

A recent collection (Dorow and O'Shaughnessy 2013a) has used Fort McMurray, Alberta, as the nodal point for an extended examination of the concept of community. It offers rich detail concerning the sociology of this 'boomtown,' while demonstrating the complexity of the multi-scalar networks that have brought to this place 'more than twenty thousand mobile workers' and intensified 'battles over Aboriginal rights and environmental impacts' (Dorow and O'Shaughnessy 2013b, 127, 129). While the focus of these authors is not political ecology, environmental policy, or democratic theory, per se, they draw upon multiple sister approaches (political economy, ethnography, discourse analysis, cultural and social theory) to provide enormous insight into questions of hegemony and counter-hegemony. (See also Dorow and Doğu 2011).

A number of cultural theorists have recently turned their attention towards Alberta's petro-politics and, in particular, the conflicts surrounding development of the oil sands. It is particularly in this field – rather than in the more economics-based field of the petro-politics literature – that we are seeing the development of an international network of scholars concerned with understanding the many meanings of the Alberta oil sands (Pendakis and Wilson 2012; Szeman 2012).[24]

From the perspective of democratic theory, a number of recent works are beginning to draw attention to the ways in which the Albertan governments and hegemonic economic interests have designed policy-making processes so as to protect or advance particular societal priorities, while marginalizing or excluding other interests and priorities (Bowness and Hudson 2014; Fluet and Krogman 2009; Hierlmeier 2008; Hoberg and Phillips 2012; MacKendrick 2005; Parkins 2006; Parkins and Davidson 2008). Comparatively absent has been work that attempts to integrate political, economic, social, *and* ecological analysis in order to critically examine both the forces and the processes that produce policy directions and the social and ecological outcomes of these directions. (A notable exception is Davidson and Gismondi's *Challenging Legitimacy at the Precipice of Energy Calamity*, mentioned above.)

The petro-state frame has influenced a handful of analyses of Alberta (and Canadian) politics published in non-academic venues (Nikiforuk 2007a, 2012; Urquhart 2010). Shrivastava and Stefanick (2012) employed the petro-state frame to argue that first world states are not immune to the corrosive effects of oil revenue dependence on democracy, while acknowledging the ways in which Alberta and Canada fit the petro-state model only imperfectly. Shrivastava and Stefanick's *Alberta Oil and the Future of Democracy in Canada* (2015) examines a range of policy areas through the prisms of petro-state and liberal democratic theory. Apart from these publications, repeated searches of scholarly journals have yielded no works that theorize Alberta's political economy as a case of a first world petro-state. The authors in this volume take up this challenge, while offering a comparative examination of corporate strategies and of social resistance to oil exploitation in contexts ranging from other North American jurisdictions to Africa. Thus, we connect developments in Alberta to the global conflicts at the heart of fossil capitalism.

A New Approach

While the petro-state literature (reviewed further in chapters 2 and 17) identifies trends or correlations that have some resonance in the Alberta (or Canadian) context (e.g., the perceived positive correlation between state revenue from oil rents and a democratic deficit), and while Alberta's political economic profile allows the province to be placed within the petro-state category, comparative research suggests problems and limitations with petro-state theses. Our approach in this book, then, is to keep in play the multiple political, economic, cultural, and institutional factors necessary to explain why states that govern great reserves of fossil fuels, and are affected by similar global market pressures, may nevertheless have significantly different political ecologies. Thus, while the authors in this volume believe that the label 'petro-state' draws attention to aspects of Alberta's political ecology that might otherwise be overlooked, and prompts us to look for certain patterns, they do not assume that all of the phenomena they study can be understood solely as the products of 'laws of petro-politics.' Instead, we need to dig into the specific historical formation of Alberta's political ecology – in all its complexity – to fully understand the institutions and relationships that we see today.

APPENDIX 1.1. Sources for Table 1.1

Alberta

Alberta College of Social Workers. 2010. 'Closing the Disparity Gap Phase II.' Discussion paper. In *ACSW Social Policy Framework 2010: Visioning a More Equitable and Just Alberta*, p. 33. http://www.parklandinstitute.ca/acsw_social_policy_framework_2010.

Briggs, Alexa, and Celia R. Lee. 2012. 'Poverty Costs: An Economic Case for a Preventative Poverty Reduction Strategy in Alberta' (Calgary: Vibrant Communities Calgary/Action to End Poverty in Alberta), p. 10. http://vibrantcanada.ca/files/poverty-costs_feb06-2012.pdf.

Government of Alberta. 2011, November. 'Alberta International Trade Review: 2010' (Edmonton: Government of Alberta).

Government of Alberta. 2011, October. 'Facts on Alberta: Living and Doing Business in Alberta' (Edmonton: Government of Alberta).

Government of Alberta. 2011. 'Highlights of the Alberta Economy' (Edmonton: Government of Alberta, December), p. 9.

Government of Alberta. 2012. 'Industry and Economy.' http://alberta.ca/industryandeconomy.cfm (accessed 11 July 11 2012).

Hazell, Elspeth, Kar-Fai Gee, and Andrew Sharp. 2012, May. *The Human Development Index in Canada: Estimates for the Canadian Provinces and Territories, 2000–2011* (Ottawa: Centre for the Study of Living Standards).

Norway

Norwegian Petroleum Directorate. 2012. 'Facts 2012.' http://www.npd.no/en/Publications/Facts/Facts-2012/Chapter-3/.

OECD. c. 2008. Social Expenditure database, 'Public and Private Social Expenditure in Percentage of GDP in 2007.' http://www.oecd.org/document/9/0,3746,en_2649_33933_38141385_1_1_1_1,00.html (accessed 15 July 2012).

OECD. Dataset: Income Distribution and Poverty Income Distribution – Inequality Country Tables (Norway). http://www.oecd.org/els/soc/income-distribution-database.htm (accessed 11 July 2012).

World Bank. c. 2012. 'GPD per Capita (Current US$).' http://data.worldbank.org/indicator/NY.GDP.PCAP.CD?order=wbapi_data_value_2011+wbapi_data_value+wbapi_data_value-last&sort=desc (accessed 11 July 2012).

Russian Federation

Abelsky, Paul, and Anna Ulaeva. 2010. 'Russia Sees Oil, Gas Share of GDP

Falling to 14%,' *Bloomberg*, January. http://www.bloomberg.com/apps/news?pid=newsarchive&sid=aCfUrLO_ySjc.

Denisova, Irina. 2012. 'Income Distribution and Poverty in Russia.' OECD Social, Employment and Migration working papers, no. 132 (OECD), p. 8. http://dx.doi.org/10.1787/5k9csf9zcz7c-en.

Heuty, Antoine. 2012, February. *Russia's Management of Oil and Gas*. Revenue Watch Institute. http://www.resourcegovernance.org/sites/default/files/documents/rwi_russia_mgmnt_2012_01.pdf

Kurlyandskaya, Galina, Gleb Pokatovich, and Mikhail Subbotin. 2010. 'Framework Paper: Oil and Gas in the Russian Federation.' Paper presented at World Bank Conference on Oil and Gas in Federal Systems, 3–4 March. http://siteresources.worldbank.org/EXTOGMC/Resources/336929-1266445624608/Framework_Paper_Russian_Federation2.pdf.

Lazko, Elena, and Alexey Nesterenko. 2012. 'An Overview of the Oil and Gas Sector in the Russian Federation – 2012.' *Hydrocarbon World* 7 (1), pp. 12–14.

OECD. 2011. 'Panel A. Public Social Spending by Broad Policy Area and Total Net Social Spending, in 2007, in Percentage of GDP.' In *Society at a Glance 2011 – OECD Social Indicators*. www.oecd.org/els/social/indicators/SAG.

OECD. 2005. *Economic Survey of the Russian Federation 2006: Expanding State Ownership in the Russian Federation* (Paris: OECD), p. 2. http://www.oecd.org/dataoecd/27/44/37732242.pdf.

Treisman, Daniel. 2010. 'Rethinking Russia: Is Russia Cursed by Oil?' *Journal of International Affairs* 63 (2), pp. 85–102.

World Bank. 2012. *Russian Federation: Russia Overview*. http://www.worldbank.org/en/country/russia/overview (accessed 11 July 2012).

World Bank. 2011. 'GPD per Capita (Current US$).' http://data.worldbank.org/indicator/NY.GDP.PCAP.CD?order=wbapi_data_value_2011+wbapi_data_value+wbapi_data_value-last&sort=desc (accessed 11 July 2012).

Venezuela

Alvarez, Cesar, and Stephanie Hanson. 2009, February. 'Backgrounder: Venezuela's Oil-Based Economy' Council on Foreign Relations. http://www.cfr.org/economics/venezuelas-oil-based-economy/p12089.

CIA World Fact Book. 2013. https://www.cia.gov/library/publications/the-world-factbook/rankorder/2172rank.html?countryName=Venezuela&countryCode=ve®ionCode=soa&rank=69#ve (accessed 18 April 2013).

Lopez, Virginia, and Jonathan Watts. 2013. 'Nicolás Maduro Declared Venezuela Election Winner by Thin Margin,' *The Guardian*, 15 April. http://www.guardian.co.uk/world/2013/apr/15/nicolas-maduro-wins-venezuela-election.

Sanati, Cyrus. 2013. 'Chavez's Death Won't Spur New Venezuela Oil Drilling.' *CNNMoney*, 6 March. http://finance.fortune.cnn.com/2013/03/06/hugo-chavez-death-oil/.

Weisbrot, Mark, and Luis Sandoval. 2007, July. *The Venezuelan Economy in the Chavez Years* (London: Centre for Economic Policy Research). http://venezuelanalysis.com/files/pdf/cepr_economy_venezuela_07_07.pdf.

NOTES

1 In lieu of the geological term 'bituminous sands,' most media and political actors use the terms 'tar sands' or 'oil sands' to describe the area where bitumen extraction is taking place, or the resource itself. Bitumen, mixed with sand, clay, and water, underlies a large area of northeastern Alberta (about 141,000 square kilometres). While the distinctions are discursively significant (see Davidsen, this volume), some authors use the terms interchangeably. The authors in this volume make their own choices with regard to the moniker for the bituminous/oil/tar sands.

2 Alberta Energy Regulator, *Alberta's Energy Reserves 2013 and Supply/Demand Outlook 2014–2023*, Statistical Publication No. ST98-2014 (Calgary: AER, May 2014), http://www.aer.ca/documents/sts/ST98/ST98-2014.pdf (accessed 28 August 2014), pp. 3–20.

3 An excellent overview of the development of political ecology as an integrative approach is provided by Neumann (2005). The essays in Gale and M'Gonigle (2000) consider the relationship – past, present, and future – between Marxist political economy and ecological thought. For an example of a post-colonial, ethnographic approach to political ecology, see Escobar (1998).

4 We use this term to refer to the early industrializing capitalist countries, many of which were also colonizing powers or settler colonies.

5 Gary Lamphier, 'As Oilsands Activity Heats Up, Alberta Employers Brace for Soaring Costs, Worker Shortages,' edmontonjournal.com, 7 March 2012.

6 Machinery destined for the oil sands mines is already being disembarked at Thunder Bay, and there are reports that suppliers are considering the possibility of barging modules down the Mackenzie River. See Richard Gilbert, 'Oilsands Equipment for Northern Alberta Will Move via Thunder Bay, Ontario,' *Daily Commercial News*, 27 August 2008, http://184.106.34.132/article/id30239; Shipping Online.CN, 'Mackenzie River via Arctic May Ship Canada's Oil Sands,' http://www.shippingonline.cn/news/newsContent.asp?id=17756 (accessed 3 October 2012). In the wake of the

opposition to transportation of megaloads through Idaho and Montana, the port of Prince Rupert on BC's coast has been mentioned as another entry point for large components destined for the Alberta oil sands. See 'Ports Invest for Growth,' *Heavy Lift & Project Forwarding International* 27 (July/August 2012), 71, http://content.yudu.com/Library/A1xq2y/HLPFIJuly-August2012/resources/73.htm.

7 Residents of 'Upgrader Alley' near Edmonton (called the 'Alberta Industrial Heartland' by the provincial government), or communities like Fort Chipewyan that are downstream of the tar sands – as well as rural communities living near intensive livestock operations – have had to hire scientific consultants to investigate their health concerns. A study of water quality in the Athabasca delta by a consultant was commissioned by the Nunee Health Authority in Fort Chipewyan (Timoney 2008). Likewise, Northeast Sturgeon County Industrial Landowners hired Dr Donald R. Blake, from the University of California, Irvine, to sample air quality. See Nikiforuk 2008, chap. 8; Blake 2005, 2007; and Griffiths and Dyer 2008.

8 This happened in Alberta in the cases of GP John O'Connor (Fort Chipewyan), who linked cancers in his patients to contamination of the Athabasca River downstream of the tar sands, and Dr David Swann, medical officer of health for the Palliser Health Authority, who publicly supported implementation of Kyoto Protocol GHG reduction targets. See Nikiforuk (2007b); 'Kyoto Views Get Medical Officer Fired,' *CBC News*, 5 October 2002, http://www.cbc.ca/news/canada/story/2002/10/04/docfired021004.html.

9 For figures on oil sands production see table 6.1 in Adkin and Stares, this volume.

10 The federal Conservative government forced through Parliament a huge omnibus Bill C-38 in June 2012 which amended multiple environmental laws. In Chapter 12, Adkin and Courteau observe a similar government response to statutory protection for environmental rights in the case of Montana, currently undergoing a hydrocarbon-led economic boom.

11 This is the popular name given to the brand of neoliberal and libertarian political economy advanced by Roger Douglas while finance minister in the Labour government of 1984–88, and subsequently, by the Association of Consumers and Taxpayers, formed in 1993. Douglas was invited to speak to the Reform Party in 1991 (Manning 1991, 276) and to an Alberta Conservative Party Caucus meeting in Red Deer in 1993. His book *Unfinished Business* (Random House, 1993) is said to have been highly influential among Alberta Conservatives. Douglas was also a mentor for the leader of the Ontario Conservatives, Mike Harris, who implemented a neoliberal

agenda in that province in the mid-1990s. Preston Manning and Mike Harris co-authored 'Vision for a Canada Strong and Free,' published by the neoliberal think-tank Montreal Economic Institute in 2007.

12 *CBS News* commissioned a survey in 2013 that produced some deeply concerning results concerning growing poverty in the United States. See '80 Percent of U.S. Adults Face Near-Poverty, Unemployment, Survey Finds,' *CBS News*, 28 July 2013, http://www.cbsnews.com/news/80-percent-of-us-adults-face-near-poverty-unemployment-survey-finds/ (accessed 30 July 2014).

13 http://www.businessweek.com/articles/2014-02-27/oil-industrys-power-in-u-dot-s-dot-petro-state-shapes-keystone-xl-debate#p1.

14 Similar comparisons are found in Anielski 2002 and Campbell 2012.

15 The term 'rentier' refers to states that derive a large proportion of their revenue not from income and corporate taxes but from resource 'rent.' Rent is what is left over after the company's production costs and a reasonable rate of profit have been deducted from its gross revenue. The chairperson of the Royalty Review Panel, appointed in 2007 by the Ed Stelmach government, was among those expressing the view that Albertans should be receiving 100 per cent of rent from oil production (Bill Hunter, quoted in 'Premier Won't Be Bullied into Royalty Decision,' CBC News, 20 September 2007, http://www.cbc.ca/canada/edmonton/story/2007/09/20/stelmachoyalty.html [link defunct]).

16 Economic modelling done by researchers at the Canadian Energy Research Institute found that, using the fiscal and royalty regimes in place in 2005, the federal government was receiving 41 per cent of oil sands revenue from income tax, royalties, corporate tax, provincial sales tax, GST, property tax, and the like. The Alberta government's share was 36 per cent, and other provincial governments and municipalities were securing a combined 23 per cent (Timilsina et al. 2005, 97–9).

17 On Alberta's early class development, see also Bell (1993), Caragata (1984), Leadbeater (1984), and Richards and Pratt (1979).

18 A good general history of Alberta is Palmer and Palmer 1990.

19 Because space is limited I have mentioned only book-length publications in this very brief review (with the exception of Pratt 1977), leaving out the more extensive literature on Alberta's politics to be found in journals and edited collections.

20 Statistics Canada data on wages in Alberta as of June 2014 showed the average hourly wage for retail service workers as $15.72 and for skilled trades workers $31.56. The table that reports labour force data is 282-0069, 'Labour Force Survey Estimates (LFS), Wages of Employees by Type of

Work, National Occupational Classification for Statistics (NOC-S), Sex and Age Group, Unadjusted for Seasonality, Monthly (Current Dollars),' CANSIM, http://www5.statcan.gc.ca/cansim/a26?lang=eng&id=2820069. These averages include both full- and part-time employees; we can assume that wages for part-time workers are lower. Another Stats Can report (Pyper c. 2008) states that 'in keeping with the strength of Alberta's oil and gas sector, nearly one in four persons employed as plumbers, pipefitters and gas fitters worked in Alberta in 2007.' Ninety-seven per cent of skilled trades workers were employed full time. One skilled trades information site listed average hourly wages for electricians and plumbers in Alberta as around $44.00/hour in July 2014 (Da Silva-Powell 2014).

21 Alberta Federation of Labour, 'Temporary Foreign Workers,' http://www.afl.org/index.php/Temporary-Foreign-Workers/overview.html (accessed 29 July 2014).

22 Alberta Federation of Labour, 'MEDIA ADVISORY: Government Allows Thousands of Employers to Underpay TFWs,' 26 May 2014, http://www.afl.org/index.php/Press-Release/media-advisory-government-allows-thousands-of-employers-to-underpay-tfws.html (accessed 29 July 2014).

23 Statistics Canada, table 282-0069, 'Labour Force Survey Estimates.'

24 Szeman is the cofounder of the Petrocultures Research Cluster, which has held two conferences (2012 and 2014) bringing together researchers from oil-producing countries. See http://www.crcculturalstudies.ca/research/petrocultures.

REFERENCES

Adkin, Laurie E. 1998. 'Ecological Politics in Canada: Elements of a Strategy of Collective Action.' In *Political Ecology: Global and Local*, ed. Roger Keil et al., 292–322. London: Routledge.

Adkin, Laurie E. 2003. 'Ecology, Political Economy, and Social Transformation.' In *Changing Canada: Political Economy as Transformation*, ed. Wallace Clement and Leah F. Vosko, 393–421. Montreal and Kingston: McGill-Queen's University Press.

Adkin, Laurie E., ed. 2009. *Environmental Conflict and Democracy in Canada*. Vancouver: UBC Press.

Adkin, Laurie E. 2014. 'Making Climate Change Policy in Alberta.' Paper presented to the Canadian Political Science Association, Brock University, St. Catharines, ON, 29 May.

Agyeman, Julian. 2005. *Sustainable Communities and the Challenge of Environmental Justice*. New York: New York University Press.

Agyeman, Julian, et al., eds. 2009. *Speaking for Ourselves: Environmental Justice in Canada*. Vancouver: UBC Press.

Alberta Energy. 2012a. 'Quick Facts.' (Statistical Information provided by the Alberta Energy Resources Conservation Board as of June 2010.) Edmonton: Department of Energy, Government of Alberta, February. http://www.energy.gov.ab.ca/Org/pdfs/QuickFacts.pdf. Accessed 11 July 2012.

Alberta Energy. 2012b. 'Coal.' http://www.energy.alberta.ca/coal/coal.asp. Accessed 11 July 2012.

Alberta Geological Survey. 2012. 'Coal Bed Methane.' Last modified 17 May 2012. http://www.ags.gov.ab.ca/energy/cbm/index.html. Accessed 12 July 2012.

Alberta, Ministry of Environment. 2002. *Albertans and Climate Change: Taking Action*. 17 October. http://environment.gov.ab.ca/info/library/6123.pdf.

Alberta Royalty Review Panel (William M. Hunter, Evan Chrapko, Judith Dwarkin, Ken McKenzie, André Plourde, and Sam Spanglet). 2007. *Our Fair Share: Report of the Alberta Royalty Review Panel*. Edmonton. September. http://www.albertaroyaltyreview.ca/panel/fi nal_report.pdf/.

Altvater, Elmar. 1998. 'Global Order and Nature.' In *Political Ecology: Global and Local*, ed. Roger Keil et al., 19–44. London: Routledge.

Altvater, Elmar. 2006. 'The Social and Natural Environment of Fossil Capitalism.' In Panitch and Leys, *Socialist Register 2007*, 36–59.

Anielski, Mark. 2002. 'The Norway Advantage.' Op-ed. *Edmonton Journal*, 26 March.

Beine, Michel, Charles Bos, and Serge Coulombe. 2012. 'Does the Canadian Economy Suffer from Dutch Disease?' *Resource and Energy Economics* 34 (4): 468–492.

Bell, Edward. 1993. *Social Classes and Social Credit in Alberta*. Montreal/Kingston: McGill-Queen's Press.

Belsky, Jill M. 2002. 'Beyond the Natural Resource and Environmental Sociology Divide: Insights from a Transdisciplinary Perspective.' *Society and Natural Resources*, 15: 269–280.

Blake, Donald R. 2005. *Air Sampling in North-Central Alberta*. Irvine, CA: University of California-Irvine, 26 March.

Blake, Donald R. 2007. *Air Sampling in North-Central Alberta*. Submission to the Energy and Utilities Board on behalf of Northeast Sturgeon County Industrial Landowners and Citizens for Responsible Development, for public hearing on North West Upgrading Inc. application, April.

Bowness, Evan, and Mark Hudson. 2014. 'Sand in the Cogs? Power and Public Participation in the Alberta Tar Sands.' *Environmental Politics* 23 (1): 59–76.

Boychuk, Regan. 2012. 'Misplaced Generosity: Extraordinary Profits in Alberta's Oil and Gas Industry.' Edmonton: Parkland Institute, 25 November.

Brulle, RobertJ., and David N. Pellow. 2006. 'Environmental Justice: Human Health and Environmental Inequalities.' *Annual Review of Public Health* 27: 103–24.

Bryant, Raymond, and Michael K. Goodman. 2008. 'A Pioneering Reputation: Assessing Piers Blaikie's Contributions to Political Ecology.' *Geoforum* 39 (2): 708–15.

Campanella, David. 2012. *Misplaced Generosity: Update 2012: Extraordinary Profits in Alberta's Oil and Gas Industry*. Edmonton: Parkland Institute, 15 March.

Campanella, David, Bob Barnetson, and Angella MacEwen. 2014. *On the Job: Why Unions Matter in Alberta*. Edmonton: Parkland Institute, 21 May.

Campbell, Bruce. 2012. *Managing Oil Wealth: The Alberta/Canada Model vs. the Norwegian Model.* Ottawa: Canadian Centre for Policy Alternatives, 30 April. http://www.policyalternatives.ca/publications/commentary/managing-oil-wealth.

Canadian Association of Petroleum Producers (CAPP). 2005. *Backgrounder: Oil Sands Economic Impacts Across Canada – CERI Report*. September. http://cagc.ca/resources/studies/capp_edms.pdf.

Caragata,Warren. 'The Labour Movement in Alberta.' In *Essays on the Political Economy of Alberta*, ed. David Leadbeater, 99–137. Toronto: New Hogtown Press.

Castree, Noel. 2007. 'Review Essay: Making First World Political Ecology.' *Environment and Planning A* 39 (8): 2030–6.

Conca, Ken. 2013. 'Complex Landscapes and Oil Curse Research.' *Global Environmental Politics* 13 (3): 131–7.

Da Silva-Powell, Elias. 2014. *Average Unionized Skilled Trades Wages in Western Canada*. http://talentegg.ca/incubator/2013/03/12/western-canadas-salaries-skilled-trades/#sthash.q0prbChC.dpuf. Accessed 29 July 2014.

Davidson, Debra J., and Mike Gismondi. 2011. *Challenging Legitimacy at the Precipice of Energy Calamity*. New York: Springer Press.

Dorow, Sara, and Göze Doğu. 2011. 'The Spatial Distribution of Hope in and beyond Fort McMurray.' In *Ecologies of Affect: Placing Nostalgia, Desire, and Hope*, ed. Tonya K. Davidson, Ondine Park, and Rob Shields, 271–92. Waterloo: Wilfrid Laurier University Press.

Dorow, Sara, and Sara O'Shaughnessy, eds. 2013a. *Fort McMurray, Wood Buffalo, and the Oil/Tar Sands: Revisiting the Sociology of 'Community.'* Special issue, *Canadian Journal of Sociology* 38 (2). http://ejournals.library.ualberta.ca/index.php/CJS/issue/view/1380.

Dorow, Sara, and Sara O'Shaughnessy. 2013b. 'Introduction.' In Dorow and O'Shaughnessy, *Fort McMurray, Wood Buffalo, and the Oil/Tar Sands,* 121–40.
Dorow, Sara. 2015. "Gendering Energy Extraction in Fort McMurray.' In *Alberta Oil and the Future of Democracy in Canada,* ed. Meenal Shrivastava and Lorna Stefanick, 275–94. Edmonton: University of Athabasca Press.
Energy Resources Conservation Board (ERCB). 2012. *ST98-2012: Alberta's Energy Reserves 2011 and Supply/Demand Outlook 2012–2021.* Calgary: ERCB, June.
Escobar, Arturo. 1998. 'Whose Knowledge, Whose Nature? Biodiversity, Conservation, and the Political Ecology of Social Movements.' *Journal of Political Ecology* 5: 53–82.
Escobar, Arturo. 2001. 'Culture Sits in Places: Reflections on Globalism and Subaltern Strategies of Localization.' *Political Geography* 20: 139–74.
Fairbrother, Peter, Stuart Svensen, and Julian Teicher. 1997. 'The Ascendancy of Neo-Liberalism in Australia.' *Capital & Class* 21 (3): 1–12.
Fluet, Colette, and Naomi Krogman. 2009. 'The Limits of Integrated Resource Management in Alberta for Aboriginal and Environmental Groups: The Northern East Slopes Sustainable Resource and Environmental Management Strategy.' In *Environmental Conflict and Democracy in Canada,* ed. Laurie E. Adkin, 123–39. Vancouver: UBC Press.
Foster, Jason, and Bob Barnetson. 2015. 'Exporting Oil, Importing Labour, and Weakening Democracy: The Use of Foreign Migrant Workers in Alberta.' In *Alberta Oil and the Future of Democracy in Canada,* ed. Meenal Shrivastava and Lorna Stefanick, 249–74. Edmonton: University of Athabasca Press.
Foster, Jason, and Alison Taylor. 2013. 'In the Shadows: Exploring the Notion of "Community" for Temporary Foreign Workers in a Boomtown.' In Dorow and O'Shaughnessy, *Fort McMurray, Wood Buffalo, and the Oil/Tar Sands,* 167–89.
Gale, Fred P., and R. Michael M'Gonigle, eds. 2000. *Nature, Production, Power: Towards an Ecological Political Economy.* Cheltenham, UK: Edward Elgar.
Gamble, Andrew. 1994. *The Free Economy and the Strong State: The Politics of Thatcherism.* 2nd ed. London: Palgrave Macmillan.
Gibson, Diana. 2007a. *Selling Albertans Short: Alberta's Royalty Review Panel Fails the Public Interest.* Edmonton: Parkland Institute.
Gibson, Diana. 2007b. *The Spoils of the Boom: Incomes, Profits and Poverty in Alberta.* Edmonton: Parkland Institute.
Gibson, Diana. 2012. *A Social Policy Framework for Alberta: Fairness and Justice for All.* Edmonton: Parkland Institute / Alberta College of Social Workers, 1 November.
Goldberg, Ellis, Erik Wibbels, and Eric Mvukiyehe. 2009. 'Lessons from

Strange Cases: Democracy, Development, and the Resource Curse in the U.S. States.' *Comparative Political Studies* 41 (4–5): 477–514.

Government of Alberta. 2008a. *Climate Change Strategy: Responsibility/Leadership/Action Plan*. 24 January. http://environment.gov.ab.ca/info/library/7894.pdf.

Government of Alberta. 2008b. *Provincial Energy Strategy*. December. http://www.energy.gov.ab.ca/Org/pdfs/AB_ProvincialEnergyStrategy.pdf.

Government of Alberta. 2011. *Highlights of the Alberta Economy 2011*. Edmonton: Government of Alberta, December. http://www.albertacanada.com/documents/SP-EH_highlightsABEconomy.pdf.

Government of Alberta. 2014. 'Women 2013.' Alberta Labour Force Profiles. March. http://work.alberta.ca/documents/labour-profile-women.pdf. Accessed 3 September 2014.

Griffiths, Mary, and Simon Dyer. 2008. *Upgrader Alley: Oil Sands Fever Strikes Edmonton*. Drayton Valley, AB: Pembina Institute, June.

Haluza-DeLay, Randolph B., and Debra Davidson. 2008. 'The Environment and a Globalizing Sociology.' *Canadian Journal of Sociology* 33 (3): 631–56.

Harrison, Trevor, William Johnston, and Harvey Krahn. 1996. 'Special Interests and/or New Right Economics? The Ideological Bases of Reform Party Support in Alberta in the 1993 Federal Election.' *Canadian Review of Sociology & Anthropology* 33 (2): 159–79.

Harrison, Trevor, William Johnston, and Harvey Krahn. 2003. *Trouble in Paradise? Citizens' Views on Democracy in Alberta*. Edmonton: Parkland Institute.

Harrison, Trevor, and Harvey Krahn. 1995. 'Populism and the Rise of the Reform Party in Alberta.' *Canadian Review of Sociology & Anthropology* 32 (2): 127–50.

Harrison, Trevor, and Gordon Laxer, eds. 1995. *The Trojan Horse: Alberta and the Future of Canada*. Montreal: Black Rose.

Harvey, David. 2005. *A Brief History of Neoliberalism*. Oxford: Oxford University Press.

Harvey, David. 2010. *The Enigma of Capital and the Crises of Capitalism*. Oxford: Oxford University Press.

Hartmann, Betsy. 2010. 'Rethinking the Role of Population in Human Security.' In *Global Environmental Change and Human Security*, ed. Richard Matthew et al., 193–214. Cambridge, MA: MIT Press.

Heynen, Nik, Maria Kaika, and Eric Swyngedouw. 2006. 'Urban Political Ecology.' In *In the Nature of Cities*, ed. N. Heynen, Maria Kaika, and Eric Swyngedouw, 1–20. London: Routledge.

Heynen, Nik, James McCarthy, Scott Prudham, and Paul Robbins, eds. 2007.

Neoliberal Environments: False Promises and Unnatural Consequences. London: Routledge.

Hierlmeier, Jodie L. 2008. '"The Public Interest": Can It Provide Guidance for the ERCB and NRB?' *Journal of Environmental Law and Practice* 18 (3): 280–311.

Hoberg, George, and Jeffrey Phillips. 2011. 'Playing Defence: Early Responses to Conflict Expansion in the Oil Sands Policy Subsystem.' *Canadian Journal of Political Science* 44 (3): 507–27.

Homer-Dixon, Thomas. 2013. 'The Tar Sands Disaster.' *New York Times*. Op ed. 31 March. http://www.nytimes.com/2013/04/01/opinion/the-tar-sands-disaster.html?_r=0.

Keeping, Janet. 'Human Rights Law, Cultural Integrity, and Oil & Gas Development.' *Law Now* 1 October 2004, 20–2.

Kennett, Steven A. 2002. *Integrated Resource Management in Alberta: Past, Present and Benchmarks for the Future*. CIRL occasional paper no. 11. Calgary: Canadian Institute of Resources Law.

Kennett, Steven A., Shelley Alexander, et al. 2006. *Managing Alberta's Energy Futures at the Landscape Scale*. Report prepared for the Alberta's Energy Futures Project, Institute for Sustainable Energy, Environment and Economy (ISEEE), University of Calgary, 29 June.

Kennett, Steven A., and Richard Schneider. 2008. *Alberta by Design: Blueprint for an Effective Land-Use Framework*. Edmonton: Canadian Parks and Wilderness Society/Pembina Institute for Alternative Development.

Klare, Michael T. 2012. 'A New Energy Third World in North America?' *Huffington Post*, 1 April. http://www.huffingtonpost.com/michael-t-klare/oil-lobby-us-_b_1395440.html.

Kleiss, Karen. 2014. 'Most Low-Wage Alberta Workers Are Women, Reports Show.' *Edmonton Journal*, 2 September. http://www.edmontonjournal.com/Most+wage+Alberta+workers+women+reports+show/10165344/story.html.

Krugman, Paul. 2006. 'The Great Wealth Transfer.' *Rolling Stone*, no. 1015 (14 December), 44–50.

Krugman, Paul. 2013. 'Why Inequality Matters.' *New York Times*, 16 December, A25.

Krugman, Paul. 2014. 'On Inequality Denial.' *The New York Times*, 2 June, A21.

Kusnetz, Nicholas. 2014. 'How Oil and Gas Firms Gained Influence and Transformed North Dakota.' Report of the Center for Public Integrity, Washington, DC, 21 July. http://www.publicintegrity.org/2014/07/21/15107/how-oil-and-gas-firms-gained-influence-and-transformed-north-dakota.

Lahey, Kathleen. 2015. *The Alberta Disadvantage: Gender, Taxation, and Income Inequality*. Edmonton: Parkland Institute. March. http://s3-us-west-2.amazonaws.com/parkland-research-pdfs/thealbertadisadvantage.pdf.

Laxer, Gordon, and Trevor Harrison, eds. 2005. *The Return of the Trojan Horse: Alberta and the New World (Dis)Order*. Montreal: Black Rose.

Leadbeater, David, ed. 1984a. *Essays on the Political Economy of Alberta*. Toronto: New Hogtown Press.

Leadbeater, David. 1984b. 'An Outline of Capitalist Development in Alberta.' In *Essays on the Political Economy of Alberta*, ed. David Leadbeater, 1–76. Toronto: New Hogtown Press.

Lemphers, Nathan, and Dan Woynillowicz. 2012. *In the Shadow of the Boom: How Oil Sands Development is Reshaping Canada's Economy*. Drayton Valley, AB: Pembina Institute, May.

Leys, Colin. 1989. *Politics in Britain: From Labourism to Thatcherism*, rev. ed. Toronto: University of Toronto Press.

Lipietz, Alain. 1992. *Towards a New Economic Order: Postfordism, Ecology, and Democracy*. New York: Oxford University Press.

Lipset, Seymour Martin. 1950. *Agrarian Socialism: The Cooperative Commonwealth Federation in Saskatchewan: A Study in Political Sociology*. Berkeley: University of California Press.

Lipset, Seymour Martin. 1954. 'Democracy in Alberta.' *Canadian Forum*, December, 175–7.

Macdonald, Douglas. 2009. 'The Failure of Canadian Climate Change Policy: Veto Power, Absent Leadership, and Institutional Weakness.' In *Canadian Environmental Policy and Politics: Prospects for Leadership and Innovation*, 3rd ed., ed. Debora L. VanNijnatten and Robert Boardman, 152–66. Don Mills, ON: Oxford University Press.

Mackendrick, Norah. 2005. 'The Role of the State in Voluntary Environmental Reform: A Case Study of Public Land.' *Policy Sciences* 38 (1): 21–44.

Macnab, Bruce, James Daniels, and Gordon Laxer. 1999. *Giving Away the Alberta Advantage: Are Albertans Receiving Maximum Revenues from Their Oil and Gas?* Edmonton: Parkland Institute, November.

Manning, Preston. 1991. *The New Canada*. Macmillan.

Mansell, Robert, and Ron Schlenker. 2006. *Energy and the Alberta Economy: Past and Future Impacts and Implications*. Paper no. 1 of the Alberta Energy Futures Project, Institute for Sustainable Energy, Environment and Economy, 15 December. Calgary: University of Calgary.

Marsden, William. 2007. *Stupid to the Last Drop: How Alberta is Bringing Environmental Armageddon to Canada (and Doesn't Seem to Care)*. Toronto: Alfred A. Knopf.

Mascarenhas, Michael. 2012. *Where the Waters Divide: Neoliberalism, White Privilege, and Environmental Racism in Canada.* Lanham, MD: Lexington.

McCarthy, James. 2005. 'First World Political Ecology: Directions and Challenges.' *Environment and Planning A* 37 (6): 953–8.

McCullum, Hugh. 2006. *Fuelling Fortress America: A Report on the Athabasca Tar Sands and U.S. Demands for Canada's Energy.* Edmonton: Parkland Institute.

McGirk, James. 2013. 'Petro State.' *This Land,* 18 December. http://thislandpress.com/12/18/2013/petro-state/?read=complete.

Mitchell, Timothy. 2011. *Carbon Democracy: Political Power in the Age of Oil.* London: Verso.

Monbiot, George. 2009. 'Canada's Image Lies in Tatters: It Is Now to Climate What Japan Is to Whaling.' *The Guardian,* 30 November. http://www.guardian.co.uk/commentisfree/cif-green/2009/nov/30/canada-tar-sands-copenhagen-climate-deal. Accessed 11 July 11.

Mufson, Steven. 2010. 'Oil Spills. Poverty. Corruption. Why Louisiana is America's Petro-State.' *Washington Post,* 18 July. http://www.washingtonpost.com/wp-dyn/content/article/2010/07/16/AR2010071602721.html?sid=ST2010071605299.

Neumann, Roderick P. 2005. *Making Political Ecology.* London: Hodder Arnold.

McKibben, Bill. 2012. 'Global Warming's Terrifying New Math.' *Rolling Stone.* 19 July. http://www.rollingstone.com/politics/news/global-warmings-terrifying-new-math-20120719.

Morison, Ora. 'Sovereign Fund Would Take Care of Dutch Disease: OECD.' Interview with economist Peter Jarrett. *Globe and Mail,* 15 June. http://www.theglobeandmail.com/report-on-business/economy/economy-lab/sovereign-fund-would-take-care-of-dutch-disease-oecd/article4265658/. Accessed 15 July 2012.

Mourougane, A. 2008. 'Achieving Sustainability of the Energy Sector in Canada.' *OECD Economics Department Working Papers,* no. 618. Paris: OECD, 27 June. DOI: http://dx.doi.org/10.1787/241263284264.

Nikiforuk, Andrew. 2007a. 'Is Canada the Latest Emerging Petro-Tyranny?' *Globe and Mail,* 11 June.

Nikiforuk, Andrew. 2007b. 'Backlash against a Whistle-Blower.' *Globe and Mail,* 19 May, F3.

Nikiforuk, Andrew. 2008. *Tar Sands: Dirty Oil and the Future of a Continent.* Vancouver: David Suzuki Foundation/Greystone.

Nikiforuk, Andrew. 2011. 'Biggest Silent Election Issue: Oil's Erosion of Canada.' *The Tyee,* 21 April. http://thetyee.ca/Opinion/2011/04/21/SilentElectionIssue/ (accessed 11 July 2012).

Nikiforuk, Andrew. 2012. 'Petro State Per Usual: Reading Alberta's Election.'

The Tyee, 25 April. http://thetyee.ca/Opinion/2012/04/25/Reading-Albertas-Election/ (accessed 11 July 2012).

Nikiforuk, Andrew. 2013. 'Oh, Canada: How America's Friendly Northern Neighbor Became a Rogue, Reckless Petrostate.' *Foreign Policy*, 24 June. http://www.foreignpolicy.com/articles/2013/06/24/oh_canada.

Norrie, Ken, and Mike Percy. 1981. 'Westward Shift and Interregional Adjustment: A Preliminary Assessment.' Economic Council of Canada, discussion paper no. 201, May.

O'Connor, James. 1991. 'On the Two Contradictions of Capitalism.' *Capitalism, Nature, Socialism* 2 (3): 107–9.

O'Connor, James. 1998. *Natural Causes*. New York: Guilford.

Oremus, Will. 2012. 'Saudi Arabia. Nigeria. Venezuela. Canada? Is Our Neighbor to the North Becoming a Jingoistic Petro-State?' *Slate Magazine*, January 20. http://www.slate.com/articles/news_and_politics/politics/2012/01/canadian_tar_sands_is_our_neighbor_to_the_north_becoming_a_jingoistic_petro_state_.2.html. Accessed 11 July 2012.

Palmer, Howard, and Tamara Palmer. 1990. *Alberta: A New History*. Edmonton: Hurtig.

Panitch, Leo, and Colin Leys, eds. *Socialist Register 2007: Coming to Terms with Nature*. London: Merlin.

Parkins, John. 2009. 'Managing Conflict in Alberta: The Case of Forest Certification and Citizen Committees.' In *Environmental Conflict and Democracy in Canada*, ed. Laurie E. Adkin. 174–90. Vancouver: UBC Press.

Parkins, John R. 2006. 'De-centering Environmental Governance: A Short History and Analysis of Democratic Processes in the Forest Sector of Alberta, Canada.' *Policy Sciences* 39 (2): 183–202.

Parkins, John R., and Davidson, Debra J. 2008. 'Constructing the Public Sphere in Compromised Settings: A Case Study of Environmental Decision-Making in the Alberta Forest Sector.' *Canadian Review of Sociology* 45 (2): 176–96.

Parkland Institute, Committee on Alberta's Finances. 2006. *Fiscal Surplus, Democratic Deficit: Budgeting and Government Finance in Alberta*. Edmonton: Parkland Institute.

Peck, Jamie. 2007. 'Neoliberal Hurricane: Who Framed New Orleans?' In Panitch and Leys, *Socialist Register 2007*, 102–29.

Peet, Richard, and Michael Watts, eds. 2004. *Liberation Ecologies: Environment, Development, Social Movements*. 2nd ed. New York: Routledge.

Pellow, D.R., and R. Brulle. 2005. *Power, Justice, and the Environment: A Critical Appraisal of the Environmental Justice Movement*. Cambridge, MA: MIT Press.

Peluso, Nancy Lee, and Michael Watts. 2001. *Violent Environments*. Ithaca: Cornell University Press.

Pendakis, Andrew, and Sheena Wilson, eds. 2012. 'Sighting Oil.' Special issue, *Imaginations: Journal of Cross Cultural Image Studies* 3 (2).

Piketty, Thomas. 2014. *Capital in the Twenty-First Century.* Trans. Arthur Goldhammer. Cambridge, MA: Belknap.

Pratt, Larry. c. 1976. *The Tar Sands: Syncrude and the Politics of Oil.* Edmonton: Hurtig.

Pratt, Larry. 1977. 'The State and Province-Building: Alberta's Development Strategy.' In *The Canadian State: Political Economy and Political Power*, ed. Leo Panitch, 133–62. Toronto: University of Toronto Press.

Pratt, Larry, and Ian Urquhart. 1994. *The Last Great Forest: Japanese Multinationals and Alberta's Northern Forests.* Edmonton: NeWest.

Prudham, Scott. 2004. 'Poisoning the Well: Neoliberalism and the Contamination of Municipal Water in Walkerton, Ontario.' *Geoforum* 35: 343–59.

Public Interest Alberta (PIA). 2014. '2014 Statistics of Low-Wage Workers in Alberta.' Media release. 3 September. http://pialberta.org/sites/default/files/Documents/2014%20Alberta%20Low%20Wage%20Fact%20Sheet%20-%20Calgary.pdf.

Pyper, Wendy. c. 2008. 'Skilled Trades Employment.' Statistics Canada Perspectives. http://www.statcan.gc.ca/pub/75-001-x/2008110/article/10710-eng.htm#a9. Accessed 29 July 2014.

Reich, Robert B. 2008. *Supercapitalism: The Transformation of Business, Democracy, and Everyday Life.* New York: Vintage.

Reich, Robert B. 2010. 'Inequality in America and What to Do about It.' *Nation* 291 (3/4): 13–15.

Richards, John, and Larry Pratt. 1979. *Prairie Capitalism: Power and Influence in the New West.* Toronto: McClelland and Stewart.

Richardson, Mary, Joan Sherman, and Michael Gismondi. 1993. *Winning Back the Words: Confronting Experts in an Environmental Public Hearing.* Toronto: Garamond.

Robbins, Paul. 2004. *Political Ecology: A Critical Introduction.* Oxford: Blackwell.

Rogers, Heather. 2007. 'Garbage Capitalism's Green Commerce.' In Panitch and Leys, *Socialist Register 2007*, 231–53.

Rokosh, C.D., J.G. Pawlowicz, H. Berhane, S.D.A. Anderson, and A.P. Beaton. 2009. 'What Is Shale Gas? An Introduction to Shale-Gas Geology in Alberta.' ERCB/AGS Open File Report 2008-08. Edmonton: Energy Resources Conservation Board/Alberta Geological Survey, July.

Ross, Monique, and Veronica Potes. 2007. 'Crown Consultation with Aboriginal Peoples in Oil Sands Development: Is It Adequate, Is It Legal?' Occasional paper no. 19. Calgary: Canadian Institute of Resources Law.

Schindler, David, and Victor Adamowicz. 2007. *Running out of Steam? Oil Sands Development and Water Use in the Athabasca River-Watershed: Science and Market Based Solutions*. Environmental Studies and Research Centre, University of Alberta/Munk Centre for International Studies, University of Toronto, May.

Schroeder, Richard A., Kevin St. Martin, and Katherine E. Albert. 2006. 'Political Ecology in North America: Discovering the Third World Within?' *Geoforum* 37: 163–8.

Sharvit, Cheryl, Michael Robinson, and Monique M. Ross. 1999. 'Resource Developments on Traditional Lands: The Duty to Consult.' Canadian Institute of Resource Law occasional paper no. 6.Calgary: CIRL.

Shrivastava, Meenal, and Lorna Stefanick. 2012. 'Do Oil and Democracy Only Clash in the Global South?: Petro Politics in Alberta, Canada.' *New Global Studies* 6 (1): 1–27.

Shrivastava, Meenal, and Lorna Stefanick, eds. 2015. *Alberta Oil and the Future of Democracy in Canada*. Edmonton: University of Athabasca Press.

Steward, Gillian. 2014. 'Oilsands Wages Driving Push for Temporary Foreign Workers.' *Toronto Star*, June 23.http://www.thestar.com/opinion/commentary/2014/06/23/oilsands_wages_driving_push_for_temporary_foreign_workers.html.

Szatylo, Deborah M.I. 2002. 'Recognition and Reconciliation: An Alberta Fact or Fiction?' *Indigenous Law Journal* 1 (Spring): 201–36.

Szeman, Imre. 2012. 'Crude Aesthetics: The Politics of Oil Documentaries.' *Journal of American Studies* 46 (2): 423–39.

Timilsina, Govinda R., Nicole LeBlanc, and Thorn Walden. 2005. *Economic Impacts of Alberta's Oil Sands*. Study no. 110, vol. 1 (October). Calgary: Canadian Energy Research Institute.

Timoney, Kevin P. 2007. *A Study of Water and Sediment Quality as Related to Public Health Issues, Fort Chipewyan, Alberta*. Conducted for Nunee Health Board Society, Fort Chipewyan, Alberta. Sherwood Park, AB: Treeline Ecological Research, 11 November.

Timoney, Kevin, and Peter Lee. 2001. 'Environmental Management in Resource-Rich Alberta, Canada: First World Jurisdiction, Third World Analogue?' *Journal of Environmental Management* 63: 387–405.

Turner, Terisa E., and Diana Gibson. 2005. *Back to Hewers of Wood and Drawers of Water: Energy, Trade and the Demise of Petrochemicals in Alberta*. Edmonton: Parkland Institute.

Urquhart, Ian. 2010. 'Petrostate?' *Alberta Views*, October, 27–33. http://www.albertaviews.ab.ca/2011/10/04/petrostate/.

Vaillancourt, Jean-Guy. 1995. 'Sociology of the Environment: From Human Ecology to Ecosociology.' in *Environmental Sociology: Theory and Practice*, ed. M.D. Mehta and E. Ouellet, 3–31. North York, ON: Captus.

Walker, Peter A. 2003. 'Reconsidering 'Regional' Political Ecologies: Toward a

Political Ecology of the Rural American West.' *Progress in Human Geography* 27 (1): 7–24.
Warnock, John W. 2006. *Selling the Family Silver: Oil and Gas Royalties, Corporate Profits, and the Disregarded Public.* Edmonton: Parkland Institute.
Watts, Michael. 2012. 'A Tale of Two Gulfs: Life, Death, and Dispossession along Two Oil Frontiers.' *American Quarterly* 64 (3): 437–67.
Williams, Raymond. 1989. *Resources of Hope: Culture, Democracy, and Socialism.* Ed. Robin Gable. London: Verso.
Wiseman, Nelson. 1981. 'An Historical Note on Religion and Parties on the Prairies.' *Journal of Canadian Studies* 16 (2): 109–112.
Wiseman, Nelson. 2007a. 'Five Immigrant Waves: Their Ideological Orientations and Partisan Reverberations.' *Canadian Ethnic Studies* 39 (1/2): 5–30.
Wiseman, Nelson. 2007b. 'The Far West: Parvenu Political Culture.' In *In Search of Canadian Political Culture,* 237–309. Vancouver: UBC Press.
Wiseman, Nelson. 2014. Introduction to *Democracy in Alberta: Social Credit and the Party System* by C.B. Macpherson, 3rd ed., vii–xxv. Toronto: University of Toronto Press.
Zalik, Anna. 2012. 'The Race to the Bottom and the Demise of the Landlord: The Struggle over Petroleum Revenues Historically and Comparatively.' In *Flammable Societies: Studies on the Socio-economics of Oil and Gas,* ed. John-Andrew McNeish and Owen Logan, 267–86. London: Pluto.

2 Fossil Capitalism and the Rentier State: Towards a Political Ecology of Alberta's Oil Economy

ANGELA V. CARTER AND ANNA ZALIK

During Barack Obama's first presidential trip abroad to Canada in February 2009, he was greeted by the Mikisew Cree and Athabasca Chipewyan First Nations of Alberta and by ForestEthics, a prominent US and Canadian environmental organization, with a simple and stark message: 'Canada's Tar Sands: the dirtiest oil on earth.' The statement appeared in a full-page *USA Today* advertisement that read 'You'll never guess who's standing between us and our new energy economy' and depicted an oil-splattered map of Canada oozing south across the US border. A year later, in response to ForestEthics' lobbying campaign of Fortune 500 companies, two popular US retailers, Bed Bath and Beyond and Whole Foods, announced they would no longer use fuel from the Canadian tar sands in their operations. The *Financial Times* described the retailers' decision as 'underlin[ing] how industry is moving to fill the void left by inaction at Copenhagen and the failure of the US Congress to limit carbon emissions' (McNulty 2010). Yet within days, the retailers weakened their position, confronted by the painful reality that their suppliers at all levels were likely purchasing fuel from the tar sands; the North American capitalist economy is far too interwoven, and oil industry supply far too complex, for an uncomplicated boycott of tar sands fuel. As these American retailers applied pressure on tar sands operations in February 2010, the American National Petrochemical Refiners Association (NPRA) was launching a legal suit against the California state government's 2007 Low Carbon Fuel Standard (LCFS),

The authors thank Laurie Adkin, Byron Miller, André Plourde, and Diana Gibson for reviewing material in this chapter.

which effectively banned tar sands fuels from the California market. Including oil giants BP and Shell among its members, the NPRA suit alleged that the LCFS puts 'undue and unconstitutional burdens on interstate commerce' (Macalister 2010).[1] Shortly thereafter, in 2008, the US Conference of Mayors resolved to reduce the use of tar sands fuel due to its high carbon emissions. Yet in the aftermath of the 2010 BP spill, business press commentators on both sides of the border were advocating for an expansion of tar sands mining to reduce the need for ultra-deep drilling offshore. By 2011, there were massive protests against the development of the Keystone XL pipeline (estimated to allow a tripling of tar sands extraction) that resulted in the arrest of leading Canadian social and environmental justice activists in Washington. At the same time, the Canadian federal government had ramped up its 'Pan-European Oil Sands Advocacy Strategy' to protect the tar sands from European fuel-quality legislation.[2] By 2013, as the Canadian Minister of Natural Resources lobbied officials across the European Union, European environmentalists engaged in acts of civil disobedience to protest tar sands projects and block imports.

This sample of events, often occurring far beyond the boundaries of Alberta and Canada, suggest the need for a 'multi-scalar' approach to the study of Alberta. They highlight the importance of broadening accounts of the state as the locus for control and resolution of contradictions associated with oil extraction. Attuned to these wider levels of analysis, this chapter suggests two approaches to understanding the political economy of oil in Alberta. The first, using theories of 'rentierism,' emphasizes the internal politics of Alberta as a petro-polity within the Canadian federal context. An analysis of socio-environmental regulation in Alberta demonstrates that the province conforms to key aspects of 'rentier' theories.

However, we argue that this perspective is limited by its narrow state-based focus; hence our introduction of a second approach that situates Alberta's rentierist attributes within a broader, historical (Arrighi 1994; Wallerstein 1974) and multi-scalar context (Amin 1990; Keil et al. 1998; Peet and Watts 2004). This approach – which we will refer to here as critical political ecology – acknowledges that all local or regional objects of study are in fact constituted by socio-political relations encompassing diverse actors and sites.[3] Or, to quote Mahon and Keil, a given 'scale' is 'socially produced and reproduced through myriad, sometimes purposeful, sometimes erratic, social economic, political and cultural actions' (2009, 8).

Figure 2.1. Alberta premier Ed Stelmach meeting with Dr Sultan Ahmed Al Jaber of Masdar, Abu Dhabi's Future Energy Company, during the World Future Energy Summit held 16–21 January 2010 in Abu Dhabi. They signed a Memorandum of Understanding committing Masdar and the Alberta ministry of energy to explore potential joint ventures related to carbon capture and storage. (*Source*: Government of Alberta news release, 14 January 2010, http://www.gov.ab.ca/acn/201001/276392E1AF7E6-0AEA-06BF-ED79E1F-B0049A9E7.html. *Photo*: Government of Alberta, Ministry of International and Intergovernmental Relations, http://international.alberta.ca/images/Premier_Stelmach_and_Dr_Sultan_Ahmed_Al_Jaber_of_Masdar.jpg.)

Critical political ecology, informed by the Marxist theorization of natural resources (Altvater 2006; Amin 1991; Harvey 2006; Foster 1999; Martinez Alier 2002; O'Connor 1991), as well as by post-colonial accounts that view the oil economy as centrally linked to imperialism and accumulation (Bromley 2006; Coronil 1997; Labban 2008; Mitchell 2002; Nore and Turner 1980; Turner 2004; Vitalis 2009), offers important insights into the social and ecological consequences of capitalist accumulation. Thus, the political space 'Alberta' cannot be understood apart from the socio-economic, colonial (historical), and political relationships that have shaped it and that connect it to a range of actors outside its borders. Indeed, the political economy of oil has been transnational for over a century.

We proceed by first situating Alberta as a rentier state. Then, after elaborating the key limitations to rentier theory, we indicate how Alberta's rentier attributes shape, and are the products of, its contemporary role in international environmental politics and the global 'fossil capitalist' oil economy. Provincial economic relationships and social conditions are linked to a 'global' network of oil industry strategy. We hope to demonstrate that political ecology – informed by critical political economy, cultural geography, colonial history, and environmental sociology – offers crucial theoretical insights and methodological tools for understanding contemporary Alberta.

Rentierism and Alberta

Theories of rentierism, focusing on the level of the nation state, hold that mineral- or natural-resource-dependent states derive their economic power mainly from 'rent,' in contrast with states dependent upon industrial or agricultural production, which require, and are shaped by, a more diverse range of state-labour-capital relations in which governments rely on taxes paid by citizens. Rent includes the financial payments to states as landlords over territory and over the natural resources embedded within it (Mommer 2002). In some cases these states have failed to diversify their economies, and their reliance on exports (hydrocarbons in the case of Alberta) contributes to what has been labelled the *resource curse* (Auty 1993). This literature stresses that oil exports typically generate a surplus on the national current account and inflate the national currency, thereby disguising structural weaknesses in the domestic economy. An overvalued currency makes other exports of the oil-producing country relatively expensive in international markets, which may lead to a decreased market share for domestic manufacturers. The circulation of oil revenue in the national economy, the so-called Dutch Disease, prompts further consumption of imported goods at prices with which domestic manufacturers cannot compete. These effects undermine efforts to diversify the economy's productive base. Indeed, they may intensify reliance on oil exports to generate state revenues and employment. Such conditions are said to typify a number of important natural resource exporters in the global south.

In relation to state theory, rentierism was first applied to Iran and other Middle East oil-producing states (Beblawi and Luciani 1987; Chaudhry 1989, 1994; Mahdavy 1970), in which it was observed that major nationalizations of oil production had provided central states with an inflow of foreign exchange that made them less reliant upon, and thus less accountable to, their citizens (hence the authoritarian tendencies of rentier states). Instead, oil states tend to support and promote the rent-providing industry. Theories of rentierism have been employed to explain various 'democratic deficit' countries in sub-Saharan Africa, Latin America, and Asia, and – given the major focus of this literature on oil producers – has led to the related concept of 'petro-state' (Karl 1997; Omeje 2008; Ross 2001). The latter term, now commonly applied, has been employed in the analysis of economic and political developments in the Russian Federation and other USSR-successor states in central Asia (see Labban 2008 for a critical view on some of this literature).

In some crucial respects, as discussed below, Alberta's political economy shares the 'typical' attributes of rentierism, including the province's dependence on oil rents, its government's support for the oil industry (exemplified by subsidies to the industry and the lenient royalty regime), and its 'democratic deficits.' These three elements of the Albertan rentier state are elaborated below.

Alberta's Dependence on Oil Rents

Alberta's reliance on the tar sands industry is the most recent example of the province's historical dependence on natural resource extraction for exportation that began with the eighteenth-century westward expansion of the fur trade, shifting to the "wheat boom" in the twentieth century (Richards and Pratt 1979). But the discovery of the Leduc oil field in 1947 and other major fields set off an unprecedented boom that was a turning point in Alberta's transformation from an agricultural to an oil economy (Mansell and Schlenker 2006, 10). As Laird notes, petroleum resources 'fuelled Alberta's long run of prosperity that spanned much of the 20th century ... Energy built modern Alberta' (2005, 156). The (predominantly foreign) capital investment, technology, and skilled labour required for developing Alberta's oil extraction, upgrading, and refining economy stimulated regional development and urbanization, and provided revenues to government independent of taxing citizens (Mansell and Schlenker 2006).

The Alberta government is now strongly dependent on revenue from fossil fuel energy exploitation. Revenues from non-renewable resources accounted for approximately 30 per cent of all provincial revenue over the 1996–2008 period, and as much as 51 per cent (in 2000).[4] Longitudinal studies have reported significant economic impacts. From 1971 to 2004 the oil and gas industry produced approximately $280 billion in government revenues, over $1.5 trillion in GDP or value added, and $600 billion in employment income (almost 12 million person-years of employment).[5] All told over this period, the oil and gas industry is estimated to have contributed – directly and indirectly – 42 per cent of the provincial GDP (Mansell and Schlenker 2006, ii–iii).

Alberta's dependence on energy rents is increasingly based on tar sands production, which in 2012 represented 61 per cent of the province's total liquid hydrocarbon production (CAPP 2013, table 3.7e). Bitumen production has been projected to surpass, by 2014, the combined production of all energy forms in the province (renewable, natural gas liquids, coalbed methane, natural gas, conventional oils and coal) (Energy Resources Conservation Board 2008). Rising oil prices and a highly favourable royalty regime have attracted significant investment, with indirect 'spinoff' effects felt across the province and country. According to the Canadian Association of Petroleum Producers (CAPP 2013, table 4.16b), in 2011 the oil sands industry spent $41 billion on capital and operating costs and paid $4.5 billion dollars in royalties. The Alberta government forecasts significant expansions in the tar sands, predicting as much as $2 trillion in 'investments, reinvestments, and revenues' to come from new projects from 2011 to 2035.[6]

Subsidizing the Tar Sands

Although now extraordinarily lucrative, tar sands extraction would not have been possible without early and significant provincial and federal investments (Chastko 2004; Clarke 2008; Nikiforuk 2008) that began in the 1930s (Pratt 1976). In 1974, Alberta founded the Alberta Oil Sands Technology and Research Authority (AOSTRA), replaced in 2000 by the Alberta Energy Research Institute (AERI). AOSTRA was the provincial government's 'proactive response' to 'develop oil sands technologies that would allow Alberta's vast resources to be exploited at relatively low costs.' Provincial funding to AOSTRA spiked in the mid-1980s to approximately $75 million annually (AERI 2009a, 2009b). One recent estimate indicates that over one billion dollars of public money has been invested in research that led to the technology to mine for bitumen (Boychuk 2010).

Substantial direct subsidies to the tar sands continue today. The Global Subsidies Initiative estimates that the federal government and the provincial governments of Alberta, Newfoundland and Labrador, and Saskatchewan spent $2.8 billion in 2010 to support the oil industry. Subsidies included government spending (such as research programs benefiting the oil industry and low-cost loans and insurance), as well as government revenue not collected due to tax breaks for the industry. The study noted that "most of these subsidies seek to increase exploration and development activity, with a focus on reducing the costs of exploration, drilling and development through a mix of tax breaks and royalty reductions" – and Alberta gets the "vast share" (73 per cent) of these subsidies (EnviroEconomics Inc. et al. 2010, 15). However, note that these discussions of direct and implicit subsidies omit a major subsidy given to the tar sands in the form of the very low price assigned to the industry's greenhouse gas emissions and the province's permissive land remediation standards. These uncounted 'externalities' are additional subsidies provided to the tar sands industry by current and future citizens in Alberta and far beyond its borders.

Examples of special injections of research funding to resolve new industrial challenges include $19 million in 2006 from the Canada Foundation for Innovation for the University of Calgary's Institute for Sustainable Energy, Environment and Economy (ISEEE) projects for in situ extraction research ('University of Calgary Receives' 2006); $5 million in 2008 from the federal budget for carbon capture and storage (CCS) research ('U of C Institute Expanding' 2008); $2 billion promised in 2008 by the provincial government for CCS projects (directed by the CCS Development Council led by former Syncrude president Jim Carter); $1.5 million from the Alberta government in 2009 for research into tar sands reclamation at the University of Alberta's School of Energy and the Environment; and $9.3 million from the Canada Foundation for Innovation in 2010 for the University of Calgary's In Situ Energy Centre (which has also received funding from the Alberta government, the ISEEE, and oil companies) ('U of C Opens' 2010). Also in 2010, the provincial government invested $25 million in Carbon Management Canada, a new research network supported by ISEEE (Harrison 2010).[7]

Both provincial and federal levels of government have also provided a welcoming investment and taxation environment for tar sands developers. Comparisons of 'frontier oil' developments internationally (including oil sands, heavy oil, and offshore oil production) conducted by Alberta's Royalty Review Panel in 2006 showed that Alberta's 2007 take, ranging from approximately 39 per cent to 47 per cent, was well below that of Norway (at around 76 per cent), Venezuela (at approximately 71 per cent), and California, Angola, and Alaska (64–67 per cent) (Alberta Royalty Review Panel 2007, 23–34). Among the legislation criticized for low payout is the Oil Sands Generic Royalty Regime established in 1997. This regime, which collected 1 per cent until 'project payout,' then 25 per cent of net revenues, acted as an incentive to keep reinvesting in further oil sands development projects rather than reaching payout. For this reason, between 1997 and 2008, government royalties were slow or stagnating even as oil sands production increased (Woynillowicz, Severson-Baker, and Raynolds 2005, 62). The government's goal – to stimulate investment in the oil sands – was met, but with profits accruing to the operators rather than the province's residents (Taylor and Raynolds 2006; see also chapters 6 and 16, this volume).

Oil Change's 2006 campaign against US government subsidies to the energy sector – calling for a separation of 'oil and state' – could apply to Alberta as well as to the Bush White House. Indeed, since the mid-2000s, coalitions of environmental organizations, religious organizations, research institutes, and former senior-ranking Canadian officials have been formally petitioning the auditor general of Canada to investigate and phase out federal subsidies to the oil and gas industry. In 2010, Ecojustice filed yet another petition (with Kairos, Friends of the Earth, Pembina Institute, and other organizations) to end federal subsidies to oil and gas producers.

Of course, as in other rentier spaces, the Alberta government's tendency to support and respond to industry needs is reinforced by oil industry lobbying associations and by the revolving door for Conservative politicians and senior civil servants between energy sector companies and government posts. Corporate attempts to influence provincial and federal levels of government – and, more recently, American and international organizations – are discussed further in chapter 5 of this volume.

Alberta's 'Democratic Deficits'

As suggested by rentier theory, high dependence on oil rents results in a deterioration of the link between the state and citizens as oil rents release states from reliance upon tax revenue. In Alberta, rents from energy (increasingly from tar sands production) have allowed the government to reduce taxes to the lowest rate in Canada; the province boasts of its 'Alberta Advantage' of low personal income and corporate tax rates and no general sales tax (Harrison 2005; chapters 3 and 16, this volume). Oil rents have freed the provincial government from relying on a more diversified tax base. Conventional state theory would indicate this is problematic because it orients the state towards pandering to oil industry interests rather than towards meeting the broader needs of citizens; this is political scientist Terry Karl's (1997) 'oil-based social contract,' in which government institutions become more accountable to private developers than to citizens. Then, when revenues from oil are disrupted – as occurred in Alberta during the recent (post-2008) recession – the tax base is undermined to the point that the state cannot maintain previous levels of spending. In Alberta, such dependence also makes the state especially wary of any ecological criticisms of the tar sands that might threaten future production.

Have democratic deficits followed in Alberta? Conservative governments have ruled the province for decades, facilitated by the first-past-the post system, although during the 1980s and 1990s the majority of voters in various ridings voted for parties to the left of the PCs. Since the Social Credit party of William Aberhart defeated the agrarian populist United Farmers of Alberta in the middle of the 1930s depression, provincial politics have been dominated by the right. Studies have identified a growing depoliticization of citizens across the country due in part to the absence of proportional representation in the voting system.

Arguably, oil-rent dependence and anti-democratic trends intensified after the end of Peter Lougheed's relatively Keynesian tenure as premier, as the global economic crises of the 1980s manifested themselves in Alberta. Following upon Lougheed, the subsequent premiership of Don Getty was viewed as inept; in the 1989 provincial elections Getty lost his seat in Edmonton, forcing him to run in a by-election in Stettler. Ultimately, he was replaced as leader by Ralph Klein, previously mayor of Calgary, spelling the full-on implementation of neoliberalism in Alberta. This neoliberal deepening did not go uncontested. In the 1993 provincial election Klein retained a Conservative majority, but his party received less than 50 per cent of the popular vote.[8]

In Alberta (and since 2006 at the federal level), anti-democratic tendencies have been increasingly manifest. As described in greater detail in chapter 17, this is visible in the government's punitive responses to critics, its clientelist practices, its restriction of access to decision-making processes, its withholding of information from the public,[9] and the constraints placed on opposition parties, as well as the previously referenced distortions of the electoral system (see, for example, Adkin 1995; Brownsey 2005; Dabbs 2006; Nikiforuk 2008; Soron 2005; Urquhart 2010). The muzzling of government-employed scientists and, indeed, extreme cutbacks to such science have occurred both provincially and federally.

However, as Adkin argues in chapter 3, to more fully understand Alberta's recent politics – in particular, its anti-democratic tendencies – we need to take into account the variant of neo-liberalism that has been practised by the ruling party since the 1980s, and particularly since Ralph Klein became premier in 1993. As described by Byron Miller (2007), neoliberal governance in Alberta has involved creeping privatization and the triumph of market ideology alongside a decline in democratic ideology. This decline, and the active promotion of self-interested instrumentalism, was enforced in part through the deliberate defunding of organizations critical of the Conservative government. While the presence of oil rents may have facilitated voter apathy, government cutbacks to social services were justified through the promotion of market values. This was exacerbated from 1993 onward through Premier Ralph Klein's campaign against 'out of control' spending and the government's ideological decision to eliminate the provincial debt through cuts to public services rather than through increasing revenue from taxation and oil and gas royalties (Flanagan 2005).[10]

Government policy on oil development runs counter to public opinion in Alberta in key respects. At the peak of the tar sands investment boom in the mid-2000s, public polling showed that over 70 per cent of Albertans surveyed wanted the provincial government to 'suspend new oil sands project approvals until environmental and infrastructure issues have been resolved' (Pembina Institute 2007, 3). More than 80 per cent wanted 'increased government investment in environmental protection in the oil sands' (5), and over 90 per cent wanted requirements for greenhouse gas reductions from oil sands plants (with 70 per cent of those surveyed wanting *absolute* reductions in GHG emissions as opposed to the per-barrel intensity-based target approach of the government) (8). Concern has been widespread about the pace of development. Yet rather than responding to these concerns, the Alberta government launched public relations campaigns to 're-brand' the province and 'greenwash' the tar sands (these efforts are described in detail in chapter 6). Rather than duly considering citizens' concerns about the tar sands, provincial leaders have been adamant that tar sands expansion will continue, a point made obvious in Premier Stelmach's often-repeated statement that the government will not be 'touching the brake' on tar sands development (Stelmach 2007; see also McLean 2006).

The views of Albertans are only partially reflected by the party in power due to the longstanding problems in the province's electoral and party funding system. While the Progressive Conservative Party and the Social Credit Party before it have held 'landslide' majority governments in Alberta for seventy-five years (in terms of seats in the legislature), this is not a clear indication of popular support for the winning party. For example, in three out of the six provincial elections since 1989, the Progressive Conservatives received less than 50 per cent of the vote. Thus, the appearance of broad public support for the Progressive Conservatives in Alberta (as elsewhere in Canada) is produced both by the first-past-the-post electoral system[11] and by low voter turnout due to the disenfranchisement felt by many citizens (Soron 2005).

Rentier state theory is a useful way of understanding some aspects of Alberta's political economy through comparisons with other petro-states. It allows us to identify how Alberta exhibits central features of a rentier state in terms of dependence on oil rents (hence the government's explicit support of the industry) and democratic deterioration. With regard to regulatory capture, Alberta (and Canada more broadly) exhibits characteristics common to oil exporters, which – as revealed by analyses following the Deepwater Horizon disaster – have typified the US Minerals Management Service, as well (Freudenburg and Gramling 2011). At the same time, it is evident that the Alberta institutional setting exhibits characteristics of a stronger state than many of the southern exporters typically identified as rentier states. In comparison to petro-states in the global south, Alberta's economy, infrastructure, and social services are relatively 'developed.'

Rentier concepts hone our analysis of the Albertan petro-state. However, focusing on Alberta's conformity with the rentier model can reduce our understanding of the province's political ecology to an overly state-centric framework, as analysts of other oil complexes have also argued (Lowi 2004; Omeje 2008; Watts 2005). As emphasized at the outset of this chapter, our frame of analysis must be much broader.

Limitations of Rentier Theory

Rentier state literature has a tendency to derive state or regime type from resource dependency, or to view resource dependency as a root cause of poor governance (Auty 1993; Karl 1997; Ross 2001). It tends to reify state forms, taking the formal, structural aspects of the state *a priori* – that is, the state is understood as a pre-given structure rather than a set of institutions that is constituted through social relations. Yet Alberta's contemporary politics cannot be understood as simplistically 'structurally determined' by the rentier form of the state. To grasp the particular political dynamics and institutions of any rentier state, we need to identify the ways in which such a state is shaped by global, regional, and, in the case of Alberta, national relationships (or 'scales') over time. With regard to understanding the political economy of oil, state-centric accounts generally omit the broad historical conditions influencing oil development, key non-state actors, and the international or transnational context.

First, rentier-state explanations of regime types are relatively ahistorical. A focus on Alberta's contemporary politics as 'structurally determined' by the rentier form of the state underemphasizes the historical factors that have allowed Alberta – and Canada as a whole – to employ such revenues in a more diversified economy than the 'classic' rentier form. Since the 1960s, work inspired by Harold Innis (1962) has characterized Canada as a 'staples economy'; this theorization was co-constituted, in part, with explicitly anti-colonial work associated with dependency theory (Girvan 1987; Polanyi-Levitt). In Canadian political economy, debates arising from the so-called 'staple theory of Canadian economic growth' (Watkins 1963) have at times criticized this school of thought for its tendency to fetishize the extracted commodity and to obscure the role of labour exploitation in capital accumulation (McNally 1987). Arguably, the role of organized labour in a range of sectors in the Canadian economy – and the industrialization with which labour organization is associated – distinguishes Canada from the 'typical' oil state model. In other words, significant diversification of the economy has taken place in Canada, albeit in a regionalized fashion. Nevertheless, staples theory continues to be an influential interpretation of Canadian political economy, pertinent to contemporary discussions of the role of oil and gas exports and so-called Dutch Disease (Stanford 2013).

In addition to Canada's long history of economic dependence on raw commodities exportation, its specific colonial history is important to understanding its contemporary political ecology. Generally speaking, to understand the formation of specific petro-states we need to reconstruct the ways in which colonialism and imperialism have facilitated access to particular territories by major oil companies (Mitchell 2002, 2011; Okonta 2008; Rodney 1972; Santiago 2006; Vitalis 2009). As Labban (2008) argues, the global oil market reflects the spatial and economic relations of exploitation that were entrenched through colonialism. In Alberta, as elsewhere, the oil industry's protagonists did not explore under built environments in urban centres, but rather sought out 'frontier zones.' The Canadian judicial institutions that have facilitated and legitimated territorial dispossession (i.e., land tenure arrangements and treaty systems) remain a central locus of contestation for access to hydrocarbon revenues, against the expansion of oil mining, and for greater transparency in revenue allocation and ecological monitoring. Major related legal cases include Beaver Lake vs. the Crown, Prairie Chipewyan vs. the Province of Alberta, and various cases mounted by Aboriginal nations and environmental NGOs in British Columbia and Alberta.[12]

Second, rentier theory does not attend to actors who are co-constituted with state agents, a point that is better captured in historical-sociological or Gramscian explanations of state regulatory approaches (Abrams 1988; Adkin 1998; Gramsci 1971; Tilly 1985). In particular, we emphasize the importance of studies of the firm and of social movements pressing for industrial and regulatory reform. With regard to the firm, the multinational operating companies – major protagonists in the tar sands – remain a 'black box' in much academic analysis (Zalik 2012). In the case of Alberta, their interests have often coincided with those of both the former crown corporation Petro-Canada[13] and the national/regional elites closely tied to the extracted resource. In addition, social movements and community responses to regulatory processes have a significant role to play – at multiple scales – in delimiting the options available to rentier states and to elite interests. These actors do receive attention in the critical literature on the petro-states, and are sometimes singled out by 'security'-oriented approaches as targets for surveillance. So, the state must be understood as socio-historically constituted through various forms of popular mobilization and other factors shaping the domestic and the global economy (McMichael 1990).

A third problem with an exclusively state-centred approach to understanding the impact of reliance upon oil extraction is that it neglects the global or transnational context of Alberta's tar sands development.

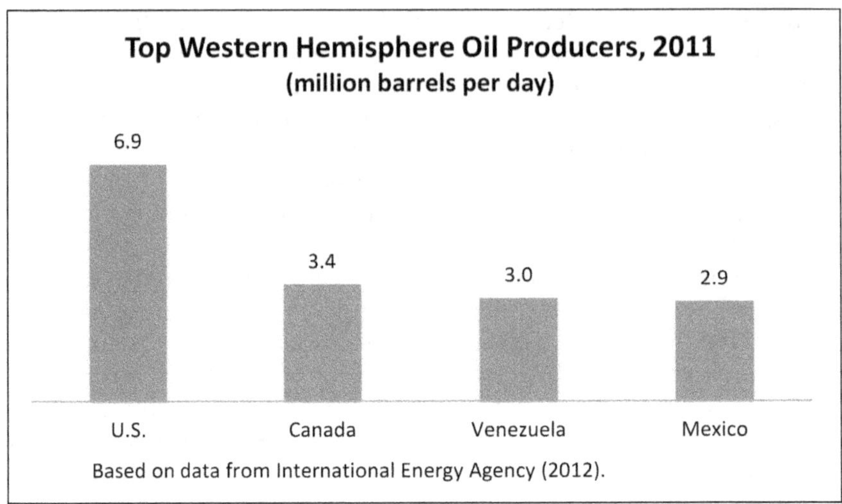

Figure 2.2. Top Western hemisphere oil producers, 2011. Based on data from International Energy Agency 2012.

Yet the political economy of oil has been essentially transnational since the industry's beginnings (Yergin 1991). Moreover, many of the 'externalities' of hydrocarbon mining are also global: economic recession triggered by rising oil prices, wars, and climate change. A transnational context draws attention to how Alberta's tar sands reserves, hailed as among the greatest oil reserves in the world, have been positioned as critical to national and continental energy security. Alberta has been the largest oil producer by far in Canada, with tar sands production accounting for nearly 50 per cent of total Canadian crude oil and equivalent (Alberta Energy 2009). Rough estimates indicate at least 70 per cent of this bitumen flows south to the American market. As figures 2.2 and 2.3 depict, the Canadian tar sands rank as the key strategic supply of oil to the United States, ranking above Mexico and Saudi Arabia. Canada ranks second to the United States among top oil producers in the Western hemisphere. Pressure to develop Alberta's tar sands is explained by its increasingly important position in the global political economy of energy.

There is also a decidedly 'micro' element to the rentier literature that examines the penetration of an 'anti-productive' culture associated with a cash-flooded economy. As a result of large oil rents, groups that

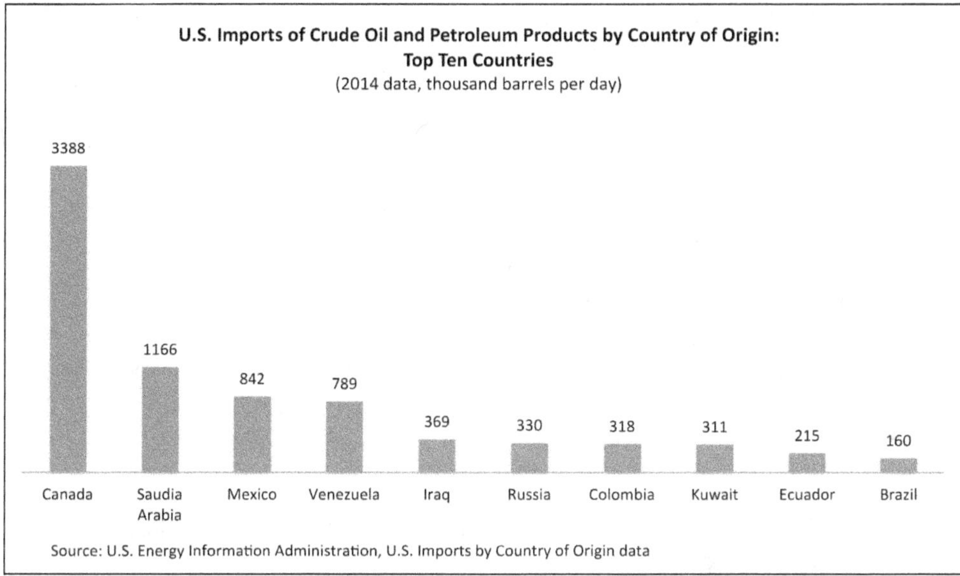

Figure 2.3. US imports of crude oil and petroleum products by country of origin: top ten countries (2014 data, thousand barrels per day). (*Source*: US Energy Information Administration, 'U.S. Imports by Country of Origin,' http://www.eia.gov/dnav/pet/pet_move_impcus_a2_nus_ep00_im0_mbblpd_a.htm.)

are largely marginalized from the oil industry, including rural communities and youth unable to secure paid employment, seek out or are offered means of accessing a small share of rents to ensure social consent for oil development. Such actions are often described through the lens of 'corruption' and theorized as 'rent-seeking' behavior. From the alternative perspective on the oil industry that we discuss below, however, these behaviours are not anti- or counter-productive because they do not promote good labouring subjects of capitalism; rather, such payments, contracts, or philanthropic contributions foster relations with the industrial sector that fracture unified demands for a greater share of revenue by historically unified communities. Benefits from energy sector companies may undermine social resistance to industrial proposals – originating from civil society sector organizations seeking environmental and climate justice, First Nations, and other affected communities (see Gibson 2007 and chapters 10, 11, and 14 in this volume). The divisive effects of industrial strategy on those making social claims are clearly apparent in the Alberta oil patch.

Thus, while rentier theory offers some insights into the political economy of Alberta, it misses significant elements offered by a multi-scalar, historical analysis that demands attention to the relationships between the global political economy of oil (Bridge 2008; Labban 2008; Mitchell 2002, 2011; Vitalis 2009) and local political ecologies (Peluso 1992; Watts 2000). While these approaches inform one another, they draw our theoretical and methodological attention in slightly different directions. The former centres upon the role of the state and the corporation in the constitution of transnational industry and the local political economy; the latter attends to localized and transnationalized resistance by multiple non-state actors, the historical context shaping current resource development, and the global political-ecological implications of fossil capitalism (to which we now turn).

Resituating Alberta in a Multi-scalar Context: Fossil Capitalism and Political Ecology

Theories of socio-ecological change arising from critical geography, Marxist and neo-Marxist political economy, and the sociology of environment and science consider historically, culturally, and locally contingent events and outcomes, as well as the material factors shaping the characteristics of the petro-state. Rather than naturalizing the so-called rentierist or resource curse structure, we draw upon a critical political ecology approach to broaden our understanding of state-society relations in Alberta and their ecological conditions and consequences.

For Altvater (2006) the term 'fossil capitalism' is a way of capturing the linkages between oil and the ecological impacts of the intensive regime of capitalist accumulation.[14] The fossil capitalism concept situates hydrocarbon energy as the very foundation of capitalist expansion since World War II, and the key commodity in capitalist markets; from coal to oil, industry and imperial history have been wholly dependent upon fossil fuels (Mitchell 2011). Oil fulfils, in Altvater's assessment, the 'requirements of the capitalist process of accumulation' because it provides flexible energy for the geographic expansion of capitalism globally through transportation and communications, and for the temporal expansion of capitalism (fossil energy permits constant energy for continuous work, as opposed to waiting on irregular biotic energy) (2006, 41; see also Alvater 1998, 41n17). The significant role fossil energy has played in the development of capitalism may be seen in the 'perfect correlation between the given measures of growth of industrial capital stocks and the consumption of energy from fossil fuels both in the USA and in the UK' (Martinez-Alier, quoted in Altvater 1998, 40). Fossil fuel provided the energy for a 'quantum leap in the speed and reach of human activities' as well as for an 'enormous growth in labour productivity and social surplus production' (23). Therefore, research on the relationship between capitalism and nature must examine the energy fuelling the interaction: 'At the centre of the analysis of capitalism's relation to nature is its inherent and unavoidable dependence on fossil fuels, and particularly on oil' (Altvater 2006, 39).

Given oil's importance to capitalism, the circumstances surrounding this fuel also pose special challenges to capitalism (and opportunities to those who would advocate a different system altogether), for as capitalism has expanded (an expansion permitted by oil), the pace of oil extraction and consumption has increased, leading to the dual problems of ultimately limited supplies and environmental degradation. Earlier, eco-Marxist theorist James O'Connor (1991) argued that capitalism generates not only the social and economic crises that Marx identified in the nineteenth century, but also profound ecological contradictions. These arguments have been borne out in powerful fashion by the contemporary crisis of climate change.

Conclusion: Political Ecology's Alternatives to Fossil Capitalism

Critical political ecology's primary theoretical contribution is its understanding of capitalism as systematically creating environmental crisis, the impacts of which are unequally borne, due to its drive for continuous growth and extraction and its externalization of environmental costs. These tendencies contribute to crises in capitalism itself (increasing costs or restrictions on natural resources, pollution, deteriorating ecosystems, as well as oppositional social movements) to which capitalism readjusts. The state form that, in Alberta, has permitted the deepening of these crises is not only one that may be characterized as rentierist, but also one that has been 'hollowed out'[15] by neoliberalism. The political ecology of the province today manifests clearly its embeddedness within global fossil capitalism and its colonial history, as well as the profound conflict between capitalist accumulation and the ecological conditions that sustain biodiversity and cultural diversity on our planet.

What solutions are evident to the problems in the current system? Given their understanding of the interaction between social relations and ecological impacts,[16] political ecologists argue for the necessity of political democratization. This is Laurie Adkin's crucial point: environmental campaigns must avoid a global apartheid of inequity between rich and poor countries and they must not be construed as racist, Eurocentric, and authoritarian. Instead, Adkin advocates building *the most inclusive solidarities possible,* crossing class, gender, race, and generational divides (2000, 68–73) and working towards 'deepened and broadened participation by individuals and groups in decisions about the direction of society' (78).

In terms of economic alternatives, contrary to the current capitalist paradigm, some ecological economists and political ecologists promote a systematic transition towards 'steady state,' 'circular,' or 'post-carbon' economies (Daly and Farley 2010; M'Gonigle 2000, 10; Victor 2008). Such models would require reining in consumerism in the global north, rapidly replacing fossil fuels with ecologically sustainable alternatives (see Weis et al., this volume) and 're-embedding' economic activities in local communities (Alperovitz et al. 2000, 167–71). Also, since poverty and economic marginalization are both result and *cause* of environmental degradation, income security must be a central feature of this alternative; otherwise there will always be pressure to accept development at any environmental cost (Alperovitz et al. 2000, 167).

Alternatives to fossil capitalism are available at local, regional, and global scales. The lack of change in the direction of more ecologically sustainable and equitable societies appears to be less attributable to the absence of technological means or organizational models than to the political-economic and cultural relations of power with which the authors of this volume grapple. This critical examination of various aspects of the relationships that have given rise to, and sustained, Alberta's political ecology and – more narrowly – its petro-state aims to uncover the obstacles to, and the potentials for, a just ecological transition.

NOTES

1 As stated by the NPRA, 'The California LCFS is unlawful for a number of reasons, including the fact that it violates the Commerce Clause of the United States Constitution by imposing undue and unconstitutional bur-

dens on interstate commerce. California's LCFS also would have little or no impact on GHG emissions nationwide and would harm our nation's energy security by discouraging the use of Canadian crude oil – our nation's largest source of crude – and ethanol produced in the American Midwest.'

2 In January 2014 it was revealed that the Canadian government had committed itself to spending up to $22 million to promote Canadian energy exports in the United States, Europe, and Asia. See Alex Boutilier, 'Ottawa Hires Ad Firm for $22 Million Oilsands Campaign,' *Toronto Star*, 9 January 2014, http://www.thestar.com/news/canada/2014/01/09/ottawa_hires_ad_firm_for_22_million_oilsands_campaign.html.

3 As summarized by Mahon and Keil, 'scale is ... not understood cartographically, as the relation between distance on a map and distance on the ground, but [is] socially constructed [and] needs to be understood in terms of its relations to other scales' (2009, 8).

4 1996–2007 data on Alberta's total provincial revenues are summarized from Statistics Canada's table 385-0002 (in real dollars); data on non-renewable resource revenue were provided by Joe Miller, executive director of policy, planning and external relations with Alberta Energy (Joe Miller, e-mail to authors, 5 November 2007). Following the standards used by Alberta Energy, 'non-renewable revenues' include revenues from natural gas and byproducts, conventional crude oil, synthetic crude oil and bitumen, coal, bonuses from sale of Crown leases, rentals and fees, minus Alberta Royalty Tax Credit (ARTC) and 'Special Royalty Features.' Data for 2007–8 are from Plourde (2010, 8).

5 According to provincial government statistics, employment gains, while more modest than GDP impacts, are still significant, with estimates for direct and indirect jobs from the energy industry at approximately 14 per cent of total employment (data from the Government of Alberta 2008b and Statistics Canada 2008).

6 This figure is provided on the Government of Alberta's oil sands website, http://www.oilsands.alberta.ca/ (accessed 10 August 2011).

7 For a critical assessment of CCS projects in Alberta, see Le Billon and Carter (2012). See also table 6.2 in chapter 6, this volume.

8 This in part reflected the tensions between Calgary as oil city and bureaucratic '(R)Edmonton' of that period. The then provincial Liberal leader, Laurence Decore, had been mayor of Edmonton.

9 The report of Alberta's Royalty Review Panel (Alberta Royalty Review Panel 2007, 5) states the problem explicitly in regard to information about oil revenue: 'There is an absence of accountability from the government to

the owners of the resource. Even with substantial effort, Albertans cannot determine whether their interests are being well served.' Sheila Pratt recounts the government's response to the auditor general, who made similar criticisms in 2006, in *Alberta Views* ('An Audit Too Far,' October 2010, 34–9).

10 A number of authors have stressed that Alberta's fiscal deficit in the late 1980s and 1990s was in large part the result of *corporate subsidies.* According to Kevin Taft, from 1986/7 to 1992/3 'the corporate sector in Alberta was a net drain on the provincial taxpayer of $5.3 billion – the very period Alberta's debt increased so rapidly' (2010). Thus, 'corporate welfare, rather than social spending, was the primary cause of Alberta's fiscal crisis' (Miller 2007, 239).

11 See Fair Vote Alberta at www.fairvote.ca/Alberta. For popular vote results, see the Election Almanac data at www.electionalmanac.com/canada/alberta/popularvote.php.

12 These cases resulted in a range of successes and some defeats. See, for instance, the Ecojustice Kearl tar sands cases focusing upon climate regulation, their Shell Jackpine mine case, various cases by West Coast Environmental Law, and a number of cases regarding hydrocarbon expansion in BC, including West Moberly's recent success concerning a proposed coal mine.

13 PetroCanada was created by the federal government in 1975, largely privatized after 1991, and merged with the private corporation Suncor in 2009.

14 The French economists who became known as the School of Regulation – in particular, Michel Aglietta and Alain Lipietz – theorized the transition to mass consumer capitalism in America, following World War I and the Great Depression, as a transition from an 'extensive regime of accumulation' to an 'intensive regime of accumulation.' For an introduction to this political economy approach, see Lipietz (1987, 1992).

15 This 'hollowing' process is typified by government authority being transferred to other levels of governance (global, regional, local) or to non-state actors – which is often a transfer of political authority to business and markets (Jessop 1994, 2002).

16 As Watts notes, the '*pressure of production on resources* is transmitted through social relations which impose excessive demands on the environment' (2000, 262, italics in original).

REFERENCES

Abrams, Philip. 1988. 'Notes on the Difficulty of Studying the State (1977).' *Journal of Historical Sociology* 1 (1): 58–89.

Adkin, Laurie E. 1998. *The Politics of Sustainable Development: Citizens, Unions and the Corporations*. Montreal: Black Rose.

Adkin, Laurie E. 2000. 'Democracy, Ecology, Political Economy: Reflections on Starting Points.' In *Nature, Production, Power: Towards an Ecological Political Economy*, ed. Fred P. Gale and R. Michael M'Gonigle, 59–81. Cheltenham: Edward Elgar.

Aglietta, Michel. 2000. *A Theory of Capitalist Regulation: The US Experience*. London: Verso.

Alberta Energy. 2009. *Oil Sands*. http://www.energy.gov.ab.ca/OurBusiness/oilsands.asp. Accessed 24 July 2009.

Alberta Energy Research Institute. 2009a. *Alberta Energy Research Institute: History*. http://www.aeri.ab.ca/sec/abo/his_001_1.cfm. Accessed 1 June 2009.

Alberta Energy Research Institute. 2009b. *Alberta Oil Sands Technology and Research Authority (AOSTRA): History of AOSTRA and Accomplishments*. http://www.aeri.ab.ca/sec/abo/content/docs/history_on_the_web_020130.pdf. Accessed 1 June 2009.

Alberta Royalty Review Panel (William M. Hunter, Evan Chrapko, Judith Dwarkin, Ken McKenzie, André Plourde, and Sam Spanglet). 2007. *Our Fair Share: Report of the Alberta Royalty Review Panel*. Edmonton. September. http://www.albertaroyaltyreview.ca/panel/fi nal_report.pdf/.

Alperovitz, Gar, Thad Williamson, and Alex Campbell. 2000. 'Ecological Sustainability: Some Elements of Longer-Term System Change.' In *Nature, Production, Power: Towards an Ecological Political Economy*, ed. Fred P. Gale and R. Michael M'Gonigle, 159–171. Cheltenham: Edward Elgar.

Altvater, Elmar. 1998. 'Global Order and Nature.' In *Political Ecology: Global and Local*, ed. Roger Keil, David Bell, Peter Penz and Leesa Fawcett, 19–44. London: Routledge.

Altvater, Elmar. 2006. 'The Social and Natural Environment of Fossil Capitalism.' In *Socialist Register 2007: Coming to Terms with Nature*, ed. Leo Panitch and Colin Leys, 36–59. London: Merlin.

Amin, Samir. 1990. *Maldevelopment: Anatomy of a Global Failure*. London: Zed Books.

Arrighi, Giovanni. 1994. *The Long Twentieth Century*. New York: Verso.

Auty, Richard M. 1993. *Sustaining Development in Mineral Economies: The Resource Curse Thesis*. London: Routledge.

Beblawi, Hazem, and Giacomo Luciani. 1987. *The Rentier State*. London: Croom Helm.

Boychuk, Regan. 2010. 'Misplaced Generosity: Extraordinary Profits in Alberta's Oil and Gas Industry.' Edmonton: Parkland Institute.

Bridge, Gavin. 2008. 'Global Production Networks and the Extractive Sector: Governing Resource-Based Development.' *Journal of Economic Geography* 8: 389–419.

Bromley, Simon. 2005. 'The United States and the Control of World Oil.' *Government and Opposition* 40 (2): 225–55.

Brownsey, Keith. 2005. 'Ralph Klein and the Hollowing of Alberta.' In Trevor W. Harrison, *The Return of the Trojan Horse*, 23–36.

Bryant, Raymond, and Michael K. Goodman. 2008. 'A Pioneering Reputation: Assessing Piers Blaikie's Contributions to Political Ecology.' *Geoforum* 39 (2): 708–15.

CAPP (Canadian Association of Petroleum Producers). 2013. *Statistical Handbook for Canada's Upstream Petroleum Industry*. Calgary.http://www.capp.ca/library/statistics/handbook/Pages/default.aspx. Accessed 1 August 2013.

Castree, Noel. 2007. 'Review Essay: Making First World Political Ecology.' *Environment and Planning A* 39 (8): 2030–36.

Chastko, Paul. 2004. *Developing Alberta's Oil Sands: From Karl Clark to Kyoto*. Calgary: University of Calgary Press.

Chaudhry, Kirsten. 1989. 'The Price of Wealth: Business and State in Labor Remittance and Oil Economies.' *International Organization* 43 (1): 100–45.

Chaudhry, Kirsten. 1994. 'Economic Liberalization and the Lineages of the Rentier State.' *Comparative Politics* 27 (1): 1–25.

Clarke, Tony. 2008. *Tar Sands Showdown: Canada and the New Politics of Oil in an Age Of Climate Change*. Ottawa: CCPA.

Coronil, Fernando. 1997. *The Magical State: Nature, Money, and Modernity in Venezuela*. Chicago: University of Chicago Press.

Dabbs, Frank. 2006. 'Ralph Klein's Real Legacy.' *The Tyee*, September 8. http://thetyee.ca/Views/2006/09/08/RalphKlein/.

Daly, Herman, and Joshua Farley. 2010. *Ecological Economics*, 2nd ed. Washington, DC: Island Press.

Energy Resources Conservation Board. 2008. *Alberta's Energy Reserves 2007 and Supply/Demand Outlook 2008–2017*. Calgary: ERCB. http://www.aer.ca/documents/sts/ST98/st98-2008.pdf.

EnviroEconomics Inc., Dave Sawyer, and Seton Stiebert. 2010. *Fossil Fuels – At What Cost?: Government Support for Upstream Oil Activities in Three Canadian Provinces: Alberta, Saskatchewan, and Newfoundland and Labrador*. Geneva: Global Subsidies Initiative/International Institute for Sustainable Development.

Fairhead, James, and Melissa Leach. 1996. *Misreading the African Landscape*. Cambridge: Cambridge University Press.

Flanagan, Greg. 2005. 'Not Just about Money: Provincial Budgets and Political Ideology.' In Trevor W. Harrison, *The Return of the Trojan Horse*, 115–35.

Foster, John Bellamy 1999. 'Marx's Theory of Metabolic Rift: Classical Foundations for Environmental Sociology.' *American Journal of Sociology* 105 (2): 366 – 405.

Freudenburg, William R., and Robert Gramling. 2011. *Blowout in the Gulf: The BP Oil Spill Disaster and the Future of Energy in America*. Boston: MIT Press.

Gibson, Diana. 2007. *The Spoils of the Boom: Incomes, Profits and Poverty in Alberta*. Edmonton: Parkland Institute.

Girvan, Norman. 1978. *Corporate Imperialism, Conflict, and Expropriation*. New York: Monthly Review Press.

Gramsci, Antonio. 1971. *Selections from the Prison Notebooks*. New York: International.

Harrison, Lynda. 2010. 'Alberta Invests $25 Million Into Clean Energy Research.' *Daily Oil Bulletin*, 16 April.

Harrison, Trevor W., ed. 2005. *The Return of the Trojan Horse: Alberta and the New World (Dis)Order*. Montreal: Black Rose.

Hartmann, Betsy. 2010. 'Rethinking the Role of Population in Human Security.' In *Global Environmental Change and Human Security*, ed. Richard Matthew, Jon Barnett, Bryan McDonald, and Karen O'Brien, 193–214. Cambridge, MA: MIT Press.

Harvey, David. 2006. *Spaces of Global Capitalism*. London: Verso.

Humphreys, Macartan, Jeffrey Sachs, and Joseph Stiglitz, eds. 2007. *Escaping the Resource Curse*. New York: Columbia University Press.

Innis, Harold A. 1962. *Essays in Canadian Economic History*. Ed. Mary Innis. Toronto: University of Toronto Press.

International Energy Agency. 2012. 'Key World Energy Statistics.' Paris. http://www.iea.org/publications/freepublications/publication/kwes.pdf.

Jessop, Robert. 1994. 'Post-Fordism and the State.' In *Post-Fordism: A Reader*, ed. Ash Amin, 251–279. Oxford: Blackwell.

Jessop, Robert. 2002. *The Future of the Capitalist State*. Cambridge: Polity.

Karl, Terry Lynn. 1997. *The Paradox of Plenty: Oil Booms and Petro-States*. Berkeley: University of California Press.

Keil, Roger, David V.J. Bell, Peter Penz, and Leesa Fawcett, eds. 1998. *Political Ecology: Global and Local*. London: Routledge.

Klare, Michael T. 2001. *Resource Wars: The New Landscape of Global Conflict*. New York: Metropolitan.

Labban, Mazen. 2008. *Space, Oil and Capital*. London: Routledge.

Laird, Gordon. 2005. 'Spent Energy: Re-Fueling the Alberta Advantage.' In Trevor Harrison, *The Return of the Trojan Horse*, 156–72.

Le Billon, Philippe, and Angela Carter. 2012. 'Securing Alberta's Tar Sands: Resistance and Criminalization on a New Energy Frontier.' In *Natural Resources and Social Conflict: Towards Critical Environmental Security*, ed. Matthew A. Schnurr and Larry A. Swatuk, 170–192. London: Palgrave Macmillan.

Lipietz, A. 1987. *Mirages and Miracles: The Crises of Global Fordism*. London: Verso.

Lipietz, Alain. 1992. *Towards a New Economic Order*. New York: Oxford University Press.

Macalister, Terry. 2010. 'Oil Groups Mount Legal Challenge to Schwarzenegger's Tar Sands Ban.' *The Guardian*, 14 February.

Mahdavy, Hussein. 1970. 'The Patterns and Problems of Economic Development in Rentier States: The Case of Iran.' *Studies in the Economic History of the Middle East*, ed. M.A. Cook, 428–67. London: Oxford University Press.

Mahon, Rianne, and Roger Keil. 2009. *Leviathan Undone? Towards a Political Economy of Scale*. Vancouver: UBC Press.

Mansell, Robert L., and Ron Schlenker. 2006. *Energy and the Alberta Economy: Past and Future Impacts and Implications*. Paper no. 1 of the Alberta Energy Futures Project, Institute for Sustainable Energy, Environment and Economy, University of Calgary. http://www.iseee.ca/media/uploads/documents/AB%20Energy%20Futures/policypapers/1-Energy%20and%20the%20Alberta%20Economy_%20Past%20and%20Future%20Impacts%20and%20Implications.pdf.

Martinez-Alier, Joan 2002. *The Environmentalism of the Poor*. London: Edward Elgar.

McCarthy, James. 2002. 'First Word Political Ecology: Lessons from the Wise Use Movement.' *Environment and Planning A* 34: 1281–1302.

McCarthy, James. 2005. 'First World Political Ecology: Directions and Challenges.' *Environment and Planning A* 37 (6): 953–8.

McLean, Archie. 2006. 'Stelmach Won't "Brake" Oil Sands Growth.' *Edmonton Journal*, December 5.

McMichael, Philip. 1990. 'Incorporated Comparison within a World Historical Perspective: An Alternative Comparative Method.' *American Sociological Review* 55: 385–97.

McNally, David. 1981. 'Staples Theory as Commodity Fetishism: Marx, Innis and Canadian Political Economy.' *Studies in Political Economy* 6: 35–63.

McNulty, Sheila. 2010. 'Suppliers of Oil Sands Fuel Shunned.' *Financial Times*, 10 February. http://www.ft.com/intl/cms/s/0/a9713b16-15e3-11df-b65b-00144feab49a.html#axzz43YrWPyoi. Accessed 21 March 2016.

M'Gonigle, R. Michael. 2000. 'A Dialectic of Centre and Territory: The Political

Economy of Ecological Flows and Spatial Relations.' In *Nature, Production, Power: Towards an Ecological Political Economy*, ed. Fred P. Gale and R. Michael M'Gonigle, 3–16. Cheltenham: Edward Elgar.

Miller, Byron. 2007. 'Green Cities Are Great Cities: Making Alberta's Cities Global Leaders in the Fight against Climate Change.' In *Alberta's Energy Legacy: Ideas for the Future*, ed. Robert Roach, 133–54. Calgary: Canada West Foundation.

Mitchell, Timothy. 2011. *Carbon Democracy: Political Power in the Age of Oil.* London: Verso.

Mitchell, Timothy. 2002. 'McJihad: Islam in the US Global Order.' *Social Text* 20 (4): 1–18.

Mommer, Bernard. 2002. *Global Oil and the Nation State*. Oxford: Oxford University Press.

National Task Force on Oil Sands. 1995. *The Oil Sands: A New Energy Vision for Canada*. Edmonton: Alberta Chamber of Resources.

Nikiforuk, Andrew. 2008. *Tar Sands: Dirty Oil and the Future of a Continent*. Vancouver: Douglas & McIntyre.

Nore, Petter, and Terisa Turner. 1980. *Oil and Class Struggle*. London: Zed Press.

O'Connor, James. 1991. 'On the Two Contradictions of Capitalism.' *Capitalism, Nature, Socialism* 2(3): 107–9.

Okonta, Ike. 2008. *When Citizens Revolt: Nigerian Elites, Big Oil and the Ogoni Struggle for Self-Determination*. Trenton: African World Press.

Omeje, Kenneth. 2008. *The Rentier Space: Extractive Economies and Conflicts in the Global South*. London: Ashgate.

Peet, Richard, and Michael Watts, eds. 2004. *Liberation Ecologies: Environment, Development, Social Movements*. London: Routledge.

Peluso, Nancy Lee. 1992. *Rich Forests, Poor People: Resource Control and Resistance in Java*. Berkeley: University of California Press.

Peluso, Nancy Lee, and Michael Watts. 2001. *Violent Environments*. Ithaca: Cornell University Press.

Pembina Institute. 2007. 'Backgrounder: Albertans' Perceptions of Oil Sands Development: Poll Part 1: Pace and Scale of Oil Sands Development.' Drayton Valley, AB: Pembina Institute. http://www.pembina.org/pub/1446.

Plourde, André. 2010. 'Oil and Gas in the Canadian Federation.' Buffet Centre for International and Comparative Studies. Evanston: Northwestern University. http://buffett.northwestern.edu/publications-projects/working-papers/energy/oil-and-gas-in-the-canadian-federation.html.

Pratt, Larry. 1976. *The Tar Sands: Syncrude and the Politics of Oil*. Edmonton: Hurtig.

Prudham, Scott. 2004. 'Poisoning the Well: Neoliberalism and the Contamination of Municipal Water in Walkerton, Ontario.' *Geoforum* 35: 343–59.

Richards, John, and Larry Pratt. 1979. *Prairie Capitalism: Power and Influence in the New West*. Toronto: McClelland and Stewart.

Robbins, Paul. 2004. *Political Ecology: A Critical Introduction*. Oxford: Blackwell.

Rodney, Walter. 1972. *How Europe Underdeveloped Africa*. London: Bogle-L'Ouverture.

Ross, Michael. 2001. 'Does Oil Hinder Demoracy?' *World Politics* 53 (3): 325–61.

Santiago, Myrna. 2006. *The Ecology of Oil: Environment, Labor and the Mexican Revolution 1900–1938*. Cambridge: Cambridge University Press.

Schroeder, Richard A., Kevin St. Martin, and Katherine E. Albert. 2006. 'Political Ecology in North America: Discovering the Third World Within?' *Geoforum* 37: 163–8.

Soron, Dennis. 2005. 'The Politics of De-Politicization: Neo-Liberalism and Popular Consent in Alberta.' In Trevor W. Harrison, *The Return of the Trojan Horse*, 65–81.

Stanford, Jim. 2013. 'The Staple Theory at 50: Introduction to a Special Blog Series.' *Rabble.ca*. 8 October. http://rabble.ca/blogs/bloggers/progressive-economics-forum/2013/10/staple-theory-50-introduction-to-special-blog-se. Accessed 13 March 2014.

Statistics Canada. 2008. *Labour Force Survey*. http://www.statcan.ca/english/Subjects/Labour/LFS/lfs-en.htm. Accessed 7 June 2013.

Stelmach, Ed, Premier. 2007. *Speeches: Canadian Association of Petroleum Producers (CAPP) Oil and Gas Investment Symposium*. http://www.premier.alberta.ca/speeches/speeches-2007-june-19-CAPP.cfm. Accessed 21 July 2009.

Taft, Kevin. 1997. *Shredding the Public Interest: Ralph Klein and 25 Years of One-Party Government*. Edmonton: University of Alberta Press/Parkland Institute.

Taylor, Amy, and Marlo Raynolds. 2006. *Thinking Like an Owner: Full Report*. Oil sands issue paper no. 13, Pembina Institute, November.

Tilly, Charles. 1985. 'War Making and State Making as Organized Crime.' In *Bringing the State Back In*, ed. Peter B. Evans, P. Dietrich Rueschemeyer, and Theda Skocpol, 167–91. Cambridge: Cambridge University Press.

Turner, Terisa, and Leigh Brownhill. 2004. 'Why Women Are at War with Chevron: Nigerian Subsistence Struggles against the International Oil Industry.' *Journal of Asian and African Studies* 39: 63–93.

'U of C Institute Expanding CCS Research.' 2008. *Daily Oil Bulletin*, 6 March.

'U of C Opens New Oilsands Facility.' 2010. *Energy Processing Canada* 42 (3): 4.

'University of Calgary Receives Funds for In Situ Research.' 2006. *Daily Oil Bulletin*, 27 November.

Urquhart, Ian. 2010. 'Petrostate?' *Alberta Views* (October): 27–33.

Victor, Peter. 2008. *Managing without Growth: Slower by Design, Not Disaster*. Cheltenham: Edward Elgar.

Vitalis, Robert 2009. *America's Kingdom: Mythmaking on the Saudi Oil Frontier*. New York: Verso.

Walker, Peter A. 2003. 'Reconsidering "Regional" Political Ecologies: Toward a Political Ecology of the Rural American West.' *Progress in Human Geography* 27 (1): 7–24.

Wallerstein, Immanuel. 1974. *The Modern World System*. New York: Academic.

Watkins, Mel. 1963. 'A Staples Theory of Economic Growth.' *Canadian Journal of Economics and Political Science* 29 (2): 141–58.

Watts, Michael. 2005. 'Righteous Oil?: Human Rights, the Oil Complex, and Corporate Social Responsibility.'*Annual Review of Environment and Resources* 30 (1): 373–40.

Woynillowicz, Dan, Chris Severson-Baker, and Marlo Raynolds. 2005. *Oil Sands Fever: The Environmental Implications of Canada's Oil Sands Rush*. Drayton Valley: Pembina Institute.

Yergin, Daniel. 1991. *The Prize: The Epic Quest for Oil, Money, and Power*. New York: Free Press.

Zalik, Anna. 2012. 'The Race to the Bottom and the Demise of the Landlord: The Struggle over Petroleum Revenues Historically and Comparatively.' In *Flammable Societies: Studies on the Socio-economics of Oil and Gas*, ed. John-Andrew McNeish and Owen Logan, 267–86. London: Pluto.

3 Alberta's Neoliberal Environment

LAURIE E. ADKIN

Alberta's oil sands are the life blood of our economy.

Premier Alison Redford, 30 May 2012[1]

In the history of Canadian politics, there has rarely been a branding exercise as powerful as the phrase 'the Alberta Advantage.' Any Canadian with a passing interest in Alberta's affairs has heard the phrase, and it immediately conjures up a picture of a resource-rich, business-friendly province with low taxes, no debt, and low unemployment. The phrase is part of a boastful narrative in which Albertans' good economic fortune is attributable not to global commodities markets but to the discipline, virtue, and common sense of free-market economics. In her 24 January 2012 Address to Albertans on the State of the Province, Premier Alison Redford repeated that Alberta has the lowest taxes in Canada, associating this fact with Albertans' high quality of life. The term 'the Alberta Advantage,' coined by the Conservative government led by Premier Ralph Klein in the 1990s, can be understood as a short-hand for neoliberal economic governance.[2] The government's 1997 budget document, entitled 'Government Business Plan Update,'[3] associated the Alberta Advantage with the government's willingness to 'work with the private sector to strengthen Alberta's economic advantages and build an economic environment conducive to investment and growth of quality jobs' (12).

This chapter focuses on the consequences of this regime – as well as the provincial government's commitment to fossil-fuel extraction and export – for environmental regulation, arguing that a 'neoliberal environment' (McCarthy, Prudham, and Robbins 2007) has indeed been the outcome.[4] While there are many dimensions to the province's neoliberal environment, here the prism is institutional; I trace the transformation of the governmental agency charged with environmental protection that has accompanied the push for investment in oil sands exploitation. Thus, this chapter contributes to studies of institutional change in petro-states, and complements the chapters in this volume by Adkin and Stares and by Carter, which also examine institutional and regulatory changes in the governance regime.

Neoliberalism

Neoliberalism is an ideology that draws upon eighteenth-century economic liberalism and philosophical libertarianism. Among the thinkers associated with neoliberalism are Friedrich Hayek, Milton Friedman,[5] Keith Joseph, and Marsden Pirie, along with the numerous think tanks in various countries that have disseminated neoliberal arguments since the 1970s (e.g., the Adam Smith Institute and the Institute of Economic Affairs in the UK, the Competitive Enterprise Institute and the Heritage Foundation in the USA, the Fraser Institute in Canada, or the Institute of Public Affairs in Australia). Neoliberalism was a core element of the New Right parties that took office in Britain in 1979 (the Thatcher-led Conservatives) and the United States in 1981 (the Reagan-led Republicans). In Canada, there was a shift to neoliberal governance at the federal level in 1983, with the election of the Mulroney Conservatives, and in New Zealand neoliberal policies were implemented with the 1984 appointment of Roger Douglas to the finance ministry (under a Labour Party government). Political parties with neoliberal economic agendas took office later in the Canadian provinces: 1993 in Alberta, 1995 in Ontario, 2001 in British Columbia. As these examples indicate, the articulations of neoliberal economics and libertarian, anti-statist beliefs to other philosophical orientations (e.g., social conservatism, the 'Moral Majority' or Christian Right, nationalism) are multiple, complex, and often contradictory. However, by the early 1990s neoliberal economic prescriptions had become dominant within international financial institutions (the 'structural adjustment policies' imposed on countries in the global south) and institutionalized in trade agreements.

In simple terms, the mantra of neoliberals is: 'markets good, states bad.' However, as many political economists have noted over the last thirty years, the imposition of neoliberal regimes has typically required a 'strong' state, deploying various forms of coercion. David Harvey (2005), one of the leading theorists of neoliberalism, views it as a political project on the part of ruling classes to push back the gains of labour and social movements that had by the 1970s come to threaten their economic and political positions of power. Colin Leys (2002) makes a similar argument with regard to Thatcherism. This project entailed breaking down the barriers to capital accumulation that had been erected by Keynesian welfare states and social democratic values. Ideologically, neoliberals characterized states as being bureaucratic, inefficient, and a threat to individual freedoms. Citizens, they argued, had become too dependent upon state welfare. Unions had too much power and were responsible for rising levels of unemployment due to excessively high wages and benefits. Citizens' entrepreneurial drives and independence needed to be revived by policies like workfare. State services and spending were 'out of control' and were `crowding out' private investment.

In policy terms, neoliberal solutions have included the privatisation (incrementally or outright) of public services, corporations, and utilities; elimination of regulations that impede the freedom of capital investment or disinvestment ('deregulation'); shifting of taxation away from corporate earnings and income taxes on the wealthy (in the name of creating 'favourable conditions for investment and job creation'); elimination or reduction of funding for social welfare programs (in the name of disciplining workers to have a stronger work ethic) and public services (e.g., health and education), entailing the familialization of welfare; reduction of public administration; weakening or elimination of workers' rights to organize and bargain collectively; delegation or 'devolution' of national or provincial government responsibilities to lower levels of government or to quasi-non-governmental organizations; and the further 'hollowing out' of states at the national level through the negotiation of international rules or agreements regulating trade and financial flows. Regarding the last point, David Harvey (2005, 33) characterizes neoliberalism as 'the financialization of everything.' The effect of all of these policies – once resistance to them was effectively defeated – was, as Harvey argues, to reassert the political hegemony of the ruling class. Neoliberal restructuring has – as three decades of evidence now make clear – restored to the ruling class of these 'Anglo-American democracies' a hugely disproportionate share of wealth.[6] This is also the case at the global level, where the income gap between the richest 20 per cent of the world's population and the poorest 20 per cent has increased (Harvey 2005, 19; Ortiz and Cummins 2011).[7]

In sum, neoliberal states have privileged the conditions for maximal, short-term returns on capital investment (and the needs of finance capital, in particular) over the interests of labour or the environment. They have intensified the rate of exploitation of labour, and allowed real incomes to fall for low- and middle-income groups. The structural adjustment policies of the 1980s, succeeded by further liberalization of trade and financial flows in the 1990s, have contributed to the rapid exploitation (for export) of the natural patrimonies of the global south, along with the transfer of highly polluting production to these regions.[8] And the global north has accumulated an enormous debt of its own – an ecological debt rooted in excessive consumption of the world's resources and disproportionate use of global commons as dumping grounds for its CO_2 emissions and industrial wastes.

With regard to local environments, neoliberal regimes have promoted commodification (as in the creation of carbon markets through cap-and-trade systems), enclosure of commons (as in the fisheries, through quota systems), privatization (of public goods such as water or parks), deregulation (the shrinkage of ministries of environment, the move from statutory environmental regulation to voluntary agreements, decreased spending on collection of data and monitoring), weaker occupational health and safety protections, and delegation of governance to private bodies (industrial associations, private laboratories) to download their responsibilities. Further, as the ecological economist Herman Daly, has long argued, globalized trade and investment have obscured ecosystem limits to growth (see Daly 1997), and corporations with global operations (like multinational mining or oil companies) operate with fewer constraints from either their 'home' or local governments insofar as their social and environmental impacts are concerned.[9] Underpinning the last three decades' massive growth in global trade has been the availability of cheap fossil fuel (Altvater 2006). The determination of powerful market actors to extend the life of this rapacious model of development – to continue accumulating short-term profits at high rates of return – now drives the reckless exploitation of unconventional reserves of fossil fuels, regardless of their environmental risks.

These global trends have provided the opportunity for the governments of Alberta to construct a local model of development that is heavily dependent upon global demand for crude oil and natural gas.

Until recently, by far the most important market has been the USA, but the province is seeking to increase its exports to China in particular. The importation of neoliberal discourse by Alberta's ruling Conservative party in the 1990s has served to legitimate the government's 'leave it to the market' approach to the pace of oil and gas exploitation and its environmental terms, as we see in the following account of the creation of Alberta's neoliberal environment.

Neoliberalism Comes to Alberta

The directions of Alberta's current model of development were set much earlier than the 1990s, under previous Conservative governments that promoted fossil fuel development, mining, and forestry as part of a modernization strategy driven by provincial 'sovereigntist' ambitions (asserted against those of the federal government) (see Leadbeater 1984; Pratt 1976, 1977; Pratt and Richards 1979; Pratt and Stevenson 1981). This strategy included pushing industrial activities further into the province's north and dispossessing First Nations of the conditions to sustain traditional cultures whenever 'development' conflicted with the claims of indigenous groups – as in the now infamous case of the Lubicon Cree (Anaya 2010, 57–64; Ferreira 1992; Goddard 1991; Huff 1999; Martin-Hill 2004; Ominayak and Thomas 2009; Venne 1992). However, these trends were accelerated after Ralph Klein became premier in 1992. At the same time that private-sector investment in resource development was promoted through an attractive tax environment, the Klein governments pursued deregulation and program spending cuts. Following their re-election in 1993, the Conservatives launched a propaganda campaign to convince the public of the necessity of immediate, drastic program spending cuts to eliminate what was presented as a menacing provincial debt and deficit. This agenda was lauded by neoliberal think tanks and lobby groups such as the Fraser Institute, the National Citizens' Coalition, and the Canadian Taxpayers Federation, which has credited Ralph Klein and his first finance minister, Jim Dinning, with slaying 'a $3 billion deficit' and 'a $20 billion debt.'[10]

As I describe below, these policies undermined state capacity for environmental regulation and intensified the ecologically destructive model of growth. The Conservatives went so far as to propose the removal of any environmental conditions on industrial activity in the northeastern part of the province, where oil sands mining is concen-

Figure 3.1. Alberta premier Ralph Klein announces that Alberta has paid its $3.7-billion debt, 12 July 2004, Calgary. The Klein government inherited over $22 billion in debt in 1993 and became the only debt-free province in the country. (*Source*: http://m.theglobeandmail.com/commentary/tom-flanagan-deficit-and-debt-fighters-of-today-should-learn-from-ralph-klein/article10586347/?service=mobile; http://beta.images.theglobeandmail.com/09a/news/national/article9987204.ece/ALTERNATES/w620/0_2004_Klein_debt.JPG. *Photo*: Jeff McIntosh, Canadian Press.)

trated, on the grounds that these areas may be 'reclaimed' or 'restored' after they have been fully mined of resources (Government of Alberta 2005). This policy led author and environmental critic Andrew Nikiforuk to describe the region as a national 'sacrifice zone.'[11]

Building the Brand: The Alberta Advantage

Alberta was the first of the Canadian provinces to embrace wholesale restructuring of government according to a model that journalist Naomi Klein calls the 'shock doctrine.' In Klein's analysis the economic and political crises, natural disasters, and military conflicts of the past four decades have been used to usher in economic policies that would otherwise never have received public approval: cuts to public services, privatization of public assets, user fees, corporate tax cuts, and deregulation. Indeed, the guru of monetarist economic theory, University of Chicago economist Milton Friedman, believed that 'only a crisis – actual or perceived – produces real change' (2002, xiv).

According to the monetarist and supply-side economists whose doctrines became widely adopted in the 1980s, government retrenchment creates room for the expansion of private markets in two ways. Existing markets are 'liberated' from the regulation that contributes to profit-squeeze. State retrenchment also opens up new opportunities for private-sector investment and capital accumulation. For example, environmental monitoring or impact assessments that were previously performed by civil servants may be 'outsourced' to private companies. The underfunding of the provision of public goods such as health care, education, or utilities generates conditions for their incremental privatization, beginning with those services that are most susceptible to profitable commodification (Leys 2002).

While Margaret Thatcher began the overhaul of the welfare state in the United Kingdom in the 1980s, New Zealand was actually the first western, developed country outside the United States to completely restructure its economy along strict neoliberal lines. In just a handful of years in the late 1980s, New Zealand's Labour government privatized most state assets, fired scores of public employees, introduced user fees for basic education, and brought in an array of private health care schemes. The rationale for such drastic action was a debt and currency crisis that – according to the authors of New Zealand's restructuring – threatened the country's solvency and the contents of every citizen's bank account. The author of New Zealand's reforms, Sir Roger Douglas, became a mentor to a new breed of Conservative politicians in Canada. In Alberta, every Conservative MLA elected in 1993 was issued a copy of Douglas' book *Unfinished Business* (1993); Douglas himself spoke to gatherings of Conservatives across Canada a number of times in the early 1990s (Alberta Hansard 1993; Schwartz 1997). Douglas' strategic advice (mirroring Margaret Thatcher's approach of the early 1980s) was to implement neoliberal policies rapidly (so as to catch the opposition unprepared), to place great emphasis upon threats to the country's fiscal solvency and 'out-of-control' government spending, and to exploit citizens' distrust of government.

Despite the absence of any imminent fiscal crisis, Conservative politicians in Alberta began a wholesale, shock doctrine–style, neoliberal restructuring of the province's governance norms. A province that had never once posted below a AA credit rating was said to have a debt crisis that threatened to topple the economy (Denis 1995; Cooper and Neu 1995; McMillan and Warrack 1995; Taft, McMillan, and Jahangir 2012). Private-sector job losses and a short recession in the early 1990s were quickly associated with runaway government spending, although spending on key portfolios had been trimmed in the 1980s, and health, education, and social services budgets were merely keeping pace with population growth, GDP, and inflation. Despite recovering rates of economic growth and resource revenues by 1994, Alberta's Conservative government dramatically slashed health, education, and social services budgets by 20 per cent in the 1993–94 budget.

An appendix to the 1997 budget, entitled 'A Plan for Change, 1993–94 to 1996–97,' describes the government's actions during this period to create a 'smaller, innovative government,'[12] among them eliminating MLA pensions; establishing the Premier's Forum on Change; cutting MLA salaries by 5 per cent; introducing a Productivity Plus program; cutting public sector salaries by 5 per cent; privatizing non-essential services;[13] reducing the public service by nearly 30 per cent; reducing school boards from 181 to 63; eliminating, merging, or privatizing 37 agencies, funds, and corporations; reducing health boards from over 200 to 17; and 'cutting red tape' and eliminating business subsidies.

Getting Away With It: Economic Growth in Alberta

By 1996, Alberta was a *cause célèbre* among neo-conservatives. Ralph Klein's swift cuts to public programs, privatization, and deregulation had resulted – according to these pundits – in a neoliberal success story. Economic growth was robust, unemployment was low, and the Progressive Conservative party cruised to easy re-election in 1997. The Alberta Advantage was lauded in the world's most respected business presses. The *New York Times*, for example, enthused in 1995 that Alberta showed that 'you can do this ... and get away with it,' adding that 'Mr. Klein has defused opposition by sparing no one and enacting his program swiftly.'[14]

Alberta is, in fact, a shining example of how to 'get away with' cuts to public services, privatization of state assets, and deregulation; there are few instances of unfettered neoliberalism being imposed with such political impunity as in Alberta. Many of New Zealand's reforms were reversed, including private health insurance. In Brazil, Argentina, Bolivia, and Chile, centre-left or Labour parties have been elected to re-stitch at least some small parts of a shredded social fabric. Yet the re-elections of Conservatives in Alberta in 1997, 2001, 2004, 2008, and 2012 have proven that politicians here can 'get away with it' given particular conditions. Among these factors is, importantly, substantial economic growth driven by fossil-fuel extraction and exports. This, in turn, has generated private-sector employment and revenue for the state that is not dependent on the typical fiscal structure of income, corporate, and other revenue sources. In addition, a range of institutional factors have helped to keep Alberta's ruling party in power, extending from the first-past-the-post electoral system (Soron 2005) to the petro-state practices described in other chapters in this volume. This is not to say that there was no opposition to the authoritarian-populist and neoliberal Klein governments; dissent was voiced by university-based intellectuals, public sector unions, other civil society actors, and the small social-democratic New Democratic Party, but no broadly-based collective action could be mobilized and sustained during this period (see Adkin 1995).

The Roaring 90s and the Environment: The Oil Boom in Context

A firm and factual representation of Alberta's spending on the environment is not a straightforward exercise. Alberta Environment became the Ministry of Environmental Protection in 1992, after Ralph Klein became premier, amalgamating the Department of Environment with the Department of Forestry, Lands and Wildlife and taking over the Parks Division from the Department of Tourism, Parks, and Recreation (Alberta Culture c. 2012). Thus, increases in the Ministry's budgets after 1992 reflect these new responsibilities. From 1993 to 1999 the Ministry of Environmental Protection was amalgamated with Forestry and Lands. In 1999 the Ministry of Environmental Protection again became the Ministry of Environment, and from 2000 to 2001 its responsibilities were narrowed due to the reorganization of ministries. Forestry was transferred to a Ministry of Resource Development. However, the latter ministry became Alberta Energy and, in 2001, some of its functions were transferred to a new ministry, Sustainable Resource Development (SRD). SRD was to administer public lands, the Surface Rights Board, and Land Compensation Board (transferred from the Department of

Agriculture, Food, and Rural Development); forest industry (transferred from the Department of Resource Development [renamed Energy]); and land and forest service and fisheries and wildlife management (transferred from the Department of Environment) (Alberta Culture c. 2012, 191). In May 2012, the ministries of Environment and Water and Sustainable Resource Development were once again amalgamated to create a new ministry, Environment and Sustainable Resource Development.) These changes affect interpretation of Alberta Environment's budgets. This examination of Environment's funding and regulatory purview focuses on the post-1994 period.

The 'Klein Revolution' of 1993–1994 did not affect Alberta Environment (Alberta Environmental Protection) budgetarily as drastically as it affected other departments. On the contrary, the Ministry enjoyed relatively stable funding between 1993 and 1995, when it amalgamated responsibilities that had previously been under the Department of Forestry, Lands, and Wildlife and the Parks Division of the Department of Tourism, Parks, and Recreation. While the cuts of the 1993 and 1994 budgets reduced government program spending by 20 per cent for most departments, Alberta Environmental Protection (AEP) suffered less than larger departments like Education, Health, and Social Services. AEP's operating budget was $374 million in 1993 and trimmed to $362 by 1995 – a cut of $8 million or 2 per cent. This is not to argue that key items were not cut from the Environment portfolio in the 1993 and 1994 budgets. Full-time equivalent employment dropped, research capacity was effectively eliminated, and multi-stakeholder decision-making bodies were stripped of their funding, reconstituted, or eliminated. In December 1994 the government announced the elimination of the Environment Council of Alberta, the Provincial Round Table on the Environment and the Economy, the Water Resources Commission, and the Alberta Environmental Trust Fund.

The initially, comparatively small cuts to the MoE's budget need to be interpreted in the context of the Ministry's relative budgetary insignificance to start with. Alberta's Environment Department has always accounted for only a tiny fraction of government programs, holding steady at just over 2 per cent of overall government program expenses throughout the 1990s, before dipping lower than 1 per cent when many of the MoE's responsibilities were transferred to the new department of Sustainable Resource Development in 1999–2000. Cuts to Alberta Environment were never really about deficit elimination – not in 1993, and certainly not in 1997. Alberta had already eliminated its deficit by

1997. Government budget estimates projected a $23 million surplus in Budget '97, and the province had already enjoyed a $1.1 billion surplus in 1996.

1996–1997: Pivotal Years in Oil and Gas Activity

In 1995 the National Oil Sands Task Force, headed by Syncrude CEO Eric Newell, released recommendations that it hoped would guide the next twenty-five years of oil sands development. Its report, entitled *The Oil Sands: A New Energy Vision for Canada,* called the oil sands 'the largest potential private sector investment opportunity for the public good remaining in Western Canada' (Alberta Chamber of Resources 1995, 4). However, in order to overcome the massive up-front capital costs for non-conventional fuel extraction, the National Oil Sands Task Force urged the province to reduce royalties to a simple, generic 1 per cent royalty until companies had paid off their capital costs. The province of Alberta adopted the Task Force recommendations with relish, and the 1 per cent royalty regime took effect in early 1996.

It is in this context that the pages of the Province of Alberta's *Budget '97* boasted of massive new private-sector investments in oil and gas activity. Looking back at 1996, the 1997 budget document indicated that '9,476 wells were completed, a 13% increase over 1995'; industry investment in oil and gas was up by 16 per cent, and 1997 drilling figures were 'expected to surpass 1996' (74). *Budget '97* noted, moreover, that much of the new oil and gas activity was in the non-conventional sector, which was responsible for all but 4 per cent of the increase in oil and gas investments. Of the twenty-four new oil and gas projects valued at over $100 million enumerated in the pages of *Budget '97*, twenty were oil sands projects (upgraders, in situ extraction, and open pit mines). The other four projects were pipelines. Of the remaining fifteen non-oil-and-gas projects valued at over $100 million, seven were petrochemical projects (three of which were in the 'industrial heartland' east of Edmonton).

At the same time, *Budget '97* contained large cuts to Alberta Environmental Protection, as the Ministry was then known. In 1996, the department's budget had been cut by $40 million (in inflation-adjusted dollars) from the previous year, with a further $22 million cut in 1997, for a total 17 per cent cut between 1996 and 1998. The Environment Department's projections for full-time-equivalent positions ('manpower' authorizations) also show their largest drop between 1996 and 1997, from 3,800 to 3,300 FTEs (Alberta Environmental Protection 1997). The number of environmental conservation officers, regulators, researchers, and monitors dropped again in 1998, to 3,100 (Alberta Environmental Protection 1998).

Thus, the Alberta Advantage – low taxes and few regulations for oil companies – began in 1996 with the introduction of the 1 per cent royalty on raw bitumen, but was accompanied by the lesser known story of hobbling Alberta Environment with deep budget cuts and layoffs. Significant cuts to the Environment portfolio – larger, even, than those that had formed part of the original 'shocks' of the early 90s – accompanied a massive surge in oil and gas activity, especially in the oil sands.

Alberta Environment Budgets from 2000–2010

Budgets at Alberta Environment stabilized after Sustainable Resource Development (primarily responsible for forestry, conservation officers, and public land management) was split from Alberta Environment's 'core businesses' of environmental monitoring, enforcement, data collection, and education. In 1999, the Environment budget was trimmed to $103 million (adjusted for inflation), at a time when the department was charged with environmental assessments for five new oil sands projects and $31 billion in investment was pouring into the oil and gas sector.

There was a $15 million increase in the 2001 operating budget for Alberta Environment. Of the increase, estimates documents indicate that $3 million was earmarked for 'fast-tracking environmental assessments for oil sands projects.'[15] The year 2001 also marked the first year that an envelope of funding was set aside to address greenhouse gas emissions: $1.5 million for 'early action initiatives' on climate change. Between 2002 and 2006, as the government tried to respond to federal climate change initiatives, the Environment budget rose 25 per cent, while Alberta's GDP grew by 43 per cent, almost entirely due to growth in the oil and gas sector.[16]

It is important to recall, however, that prior to the amalgamation of Environment with Forestry and Lands in 1992–93, the Environment budget had been $170 million (in constant 2002 dollars) with 1,174 full-time equivalent employees. At the height of the oil sands boom in 2006, the Environment budget was $130 million (in constant 2002 dollars) and 842 full-time equivalent employees worked for the department.

Capital Spending

Cuts to Alberta Environment were not confined to operating funds. Capital expenditures have taken a dramatic hit since the 1990s. In fact, from 2003 to 2008, the Government of Alberta made not a penny of investment in infrastructure managed by Alberta Environment. Alberta Environment is responsible for an array of government infrastructure, much of which is directly related to public safety. The MoE maintains dams, canals, and air and water monitoring stations. Between 1997 and 2001, total government infrastructure spending more than doubled (from $252 million to $590 million, in raw dollars) but spending on environmental infrastructure was reduced from $13 million to $8 million per year. The lack of funding was not for want of warning from the department that public safety could be at risk. The Ministry's 1999–2000 annual report indicated that water quality monitoring equipment, in particular, needed to be upgraded so that 'there are no health and safety issue concerns for Albertans' (Alberta, Ministry of Environment 2000, 24). Despite the warnings, capital investments in environmental infrastructure continued to plummet. By 2003, the MoE no longer had a line for 'capital investment' in the budget estimates that went before the Legislature. The government's investment in air and water quality equipment, data collection, and monitoring stations was zero from 2003–04 through to the end of the boom in 2008.[17]

Public Relations, Climate Change, and Environmental Underspending

Spending at Alberta Environment has been turned up in tandem with the glare of the international spotlight on the oil sands (see chapter 6, this volume). As mentioned above, in 2006 the government budgeted $130 million for the Environment Ministry; by 2009 this amount had been increased to $214 million. On paper, this was a 60 per cent increase, or $84 million. However, much of the budgeted amount was not actually spent. The 2008 actuals and the Ministry's annual report for that year show that $70 million of the allocation for 2007–2008 had not been spent. Further, several million dollars of the Ministry's budget increase had gone towards support for the minister, which includes communications and public relations.

As noted above, the province began spending modest amounts on climate change programs in 2001. By 2005, budget estimates show they planned to spend $3.4 million on these programs, but ended up spending $4.9 million. Climate change budgets have since fluctuated wildly. In each of the years 2006 and 2007, $3.6 million was budgeted for climate change initiatives. In 2008, figures show a $62 million line item for climate change initiatives, but the 2009 budget documents reveal they were forecast to spend just $10 million of this amount. In 2008, the Ministry underspent its authorized budget by $70 million, and MoE's annual report indicated that the variance reflected a deferral of Canada ecoTrust funding for Clean Air and Climate Change projects and lower-than-expected industry participation in the Climate Change and Emission Management Fund (see chapter 6). Underspending was therefore almost entirely concentrated in areas designed to curb greenhouse gas emissions. It is notable that the 2009 allocation for climate change initiatives was $77 million – roughly the same as the previous year's *unspent* funds. Thus it would seem that advertised investments in measures to reduce greenhouse gas emissions were not realized.

Large sums of money have been committed to Alberta's carbon market schemes. While a full discussion of carbon markets is beyond the scope of this chapter, Alberta's carbon market is based on large industrial operators in the oil, gas, and electricity sectors purchasing offsets from farmers, small industrial operations, and other sectors. No-till and reduced-till agricultural practices appear to be the most-used source of carbon offsets, but there is some debate as to whether these practices are actually reducing CO_2 emissions given that they are techniques that many farmers were employing anyway (Fekete 2009). Overall, while the provincial government was allocating substantial sums of money ($70 million in 2008, for example) to the operation of its 'cap-and-trade' system, external studies of the system concluded that it was doing little to reduce absolute emissions of GHGs (Pembina 2011).

Public Relations

Alberta Environment's 2008–09 annual report began by laying out two goals for the Ministry. The first goal was to 'inform Albertans on our environmental stewardship to ensure a clear provincial, national and international understanding of Alberta's leadership, commitment and action on the environment' (6). In other words, Alberta Environment's first and overarching ministerial goal was to *talk* about environmental stewardship, not to actually do it. The emphasis on public relations was reflected in departmental budgets. In 2003, the 'Communications' line under 'Departmental Support' in Alberta Environment's budget estimates was allocated $482,000. This line increased to $1.6 million in 2009 – a three-fold increase in just six years. As mentioned above, communications is part of the overall support to the minister, which has also increased significantly since 2001. Spending on supports to the minister doubled between 2001 ($5.8 million) and 2008 ($11.5 million).

In 2009, spending on 'Communications' doubled yet again, with estimated expenditures of $21 million on ministerial support, including a new $10 million line item for something called 'strategic support and integration.' Environment's 2008 annual report indicated that 'strategic support and integration' was responsible for implementing a new cumulative effects approach to land use, oil sands, and environmental assessments. As Quinn et al.'s chapter in this volume explains, this kind of integration would be an innovation that many in the environmental community would welcome. However, cumulative effects management will impose real limits on development, and the Department of the Environment has not yet been given statutory authority to limit industrial activity accordingly.[18]

No More 'Getting Away With It'? Alberta under a Global Magnifying Glass

Until the 2008–09 fiscal year Alberta's economic growth had not been significantly braked for almost fifteen years. With only a slight dip in 2002, resource revenues did not sustain any long-term decline until fall 2008, when tightening credit markets also squeezed many oil sands projects, slowing or halting construction projects throughout the sector and causing rapidly rising unemployment. A fiscal double whammy also hit Alberta in the form of much-lower-than-expected natural gas prices, which were primarily responsible for plummeting government revenues. However, global credit markets and lower oil and gas prices are only part of the story.

Alberta has come under the global magnifying glass for the lax environmental regulations that have underpinned its historic economic growth rates, as the province allowed oil sands mines, steam-assisted oil sands operations, and other oil and gas leases in northeastern Alberta to balloon into the largest industrial disturbance on earth. So far, the strategy to mitigate the impact on both the environment and on Alberta's international reputation has been either rhetorical or has amounted to government subsidies (for R & D, or in the form of emissions intensity targets that make it unnecessary for corporations to fully internalize the environmental costs of production). Carbon capture and storage schemes and carbon markets have been the favourite 'environmental sustainability' talking points for both the Alberta and Canadian governments, and both strategies involve transferring public wealth to private-sector projects – schemes that have so far evinced little likelihood of reducing greenhouse gas emissions (Campanella 2011; Pembina 2011; Thompson 2009).

There is evidence that the oil sands are already facing some degree of long-term capital divestment linked to concerns about their environmental harms. In 2008, UK-based Cooperative Asset Management, a $6 billion investment fund and division of Cooperative Financial Services, divested itself of all investments in the Alberta oil sands, warning that investing in the resource carried too many long-term risks. The large-scale construction projects, mines, and steam-assisted in situ developments, warned Cooperative, have a very real risk of becoming 'stranded' and 'uneconomic' assets.[19] The British investment firm was also not moved by the province's funding of carbon capture and storage as a way to mitigate increased CO_2 emissions from oil sands extraction. According to Niall O'Shea, engagement manager at Co-operative Asset Management, 'our doubt is that it will come soon enough, on a scale big enough to capture the emissions that will result in expansion of oil sands in the next 10 years. You would be hard pressed to find a credible authority who is confident that CCS will be effective for most oil sands operations by 2018, as Government regulation implies, if not mandates' (quoted in Phillips 2008). Cooperative was also dubious about Alberta's claims that environmental regulations are stringent enough to ensure adequate environmental protection: 'The point about legislative compliance has to have regard to the strictness of the legislation. Fines for non-compliance are small, and we believe using emissions intensity only as the key requirement sends the wrong message, or at least misunderstands the gravity of large, absolute increases in emissions.'[20] As Carter and Zalik also note, retail chains in the USA, the US Conference of Mayors, and the European Union have since adopted policies or standards that identify crude oil originating from the oil sands as incorporating a higher CO_2 emissions footprint than crude from conventional sources.

In addition to the issue of the greenhouse gas emissions produced by the extraction and refining of bitumen, and from the burning of the fuel, there are many negative 'local' consequences of oil sands mining (see chapters 5 and 10, this volume). First Nations have been reporting for many years effects on wildlife and fish as well as increased rates of human illnesses that they suspect are related to toxic byproducts of oil sands production (Buehler 2007; Meili 2007; Norwegian and Willson, 2006; Pembina 2007; Weber 2006). Research by independent scientists has been accumulating, showing that toxic contamination of water bodies caused by oil sands mining and bitumen upgrading has indeed been mounting, and may be causing deformities and tumours in fish as well as human diseases; moreover, the projected expansion of tar sands mining will most likely increase contamination to levels that exceed regulatory standards (Kelly et al. 2009, 2010; Kurek et al. 2013; Oil Sands Advisory Panel 2010; Rooney, Bayley, and Schindler 2012; Schindler 2010; Timoney 2007). The Royal Society of Canada established its own inquiry in October 2009 on the environmental and health impacts of Canada's oil sands industry, releasing its report in December 2010.

The publicity that these reports received made it impossible for Alberta and federal environmental ministries to continue using data from the Regional Aquatics Monitoring Program (RAMP) (a monitoring body made up almost entirely of representatives from energy and forest industry companies and governmental agencies) to minimize pollution concerns.[21] The ministers proceeded to appoint a number of advisory groups to further investigate the adequacy of environmental monitoring of the oil sands. The studies included:

- a federal Oil Sands Advisory Panel composed of five scientists, formed in October 2010 and reporting to the federal minister of environment in December 2010 (Dowdeswell et al. 2010);
- a panel established by the Alberta government to review the work of RAMP and to compare it with results of Kelly et al., reporting to the provincial environment minister in March 2011;
- the Alberta Environmental Monitoring Panel (AEMP), appointed by Alberta Environment Minister Rob Renner in January 2011, whose report was released by the minister in July 2011 (Kvisle et al. 2011);
- work by Environment Canada scientists to develop a monitoring plan, which was provided to the federal minister of environment in July 2011.

Notably, the AEMP (comprising seven scientists, two oil and gas industry executives, an energy economist, and a former university president) concluded, following an investigation that included public hearings, that:

> only an *independent, science-based* monitoring authority, *at arms length from government and industry*, could establish the necessary *credibility* in gathering data, carrying out analyses and making recommendations regarding local and long-term, cumulative environmental effects. A formal process termed MER: 'monitoring, evaluation and reporting,' needs to be initiated in the oil sands area, and the panel recommended a new Environmental Monitoring Commission to carry out this work. (Miall 2012; italics added)

This recommendation, combined with the scientists' judgment that the Alberta Ministry of Environment 'has no capacity to carry out credible scientific work on its own behalf' (Miall 2012), are indicative of the transformation of the provincial Ministry of Environment, over time, into an agency committed more to the promotion of oil sands development than to the protection of the environment.

With regard to the government's capacity to do effective monitoring and regulatory work, the AEMP observed:

> Of particular concern is a lack of scientific oversight of monitoring, evaluation and reporting activities, resulting in an inability to:
> - Identify critical knowledge gaps that prevent meaningful long-term monitoring and effective adaptive management;
> - Provide sufficient feedback to develop standard environmental monitoring methods, which are presently lacking, particularly in the Lower Athabasca region; and
> - Establish meaningful environmental baselines and reference conditions essential for cumulative effects monitoring; most reference stations in the oil sands area have been lost as development expanded. (Kvisle et al. 2011, 25)

Interestingly, there may be support within the oil and gas corporate leadership for a more rigorous and credible environmental monitoring system. Andrew Miall, a geologist and one of the scientists participating in the AEMP, has commented:

> My sense is that the professional and technical community comprising the petroleum industry in Calgary would welcome the establishment of

> a credible environmental management organization, as would the executives and senior managers. My many conversations with colleagues in the industry over the years have convinced me that most of the members of this group are environmentalists at heart, and many are concerned about the damage that is being done to Canada's reputation by the mismanagement that has characterized oil sands development up to now. The idea that this community could be grouped under the heading of 'Big Oil' as opposing sensible regulation of their industry is ludicrous. (Miall 2012)

Indeed, the co-chair of the AEMP, Hal Kvisle, was a former president and CEO of TransCanada Corporation and a board member of Talisman Energy, ARC Resources, and Northern Blizzard Resources. The vice president of operations for the Canadian Association of Petroleum Producers, David Pryce, was also appointed to the AEMP.

Yet Miall's view of industry support at the executive level is difficult to square with statements coming out of the Canadian Association of Petroleum Producers and other industry associations, urging the federal government to streamline environmental assessment processes, albeit, in 'environmentally responsible' ways (see chapters 5 and 13, this volume).[22] However, if there is indeed support within the corporate elite for a better environmental monitoring system, it must rest upon the belief that the data produced will not bolster demands for a moratorium on further oil sands expansion. What assumptions might underpin such a belief?

Both government, provincial and federal, and industry leaders have consistently asserted that bitumen mining, processing, and transportation can be done in an environmentally sustainable manner (see Adkin and Stares, this volume). They have argued, moreover, that future demand for oil must be met, because consumers demand it and because Canada's economic prosperity depends on energy exports. If bitumen mining and export can indeed be carried out in an 'environmentally sustainable' manner, then science is our friend, and it is only a matter of getting the technology and data systems right. Such reasoning is, indeed, frequently heard from 'the professional and technical community' referred to by Professor Miall, and is found in his own statement that 'if we don't need the resource now, we will in the near future, and so there is everything to be gained by proper environmental monitoring and management of the many extraction projects that are now underway' (2012). With 'proper' monitoring, in other words, we can continue to extract, upgrade, and export bitumen to the USA and other markets.

When this 'sustainable development' of the oil sands discourse runs up against the environmentalist argument that the fuel produced by the oil sands can be burned only at the cost of accelerated climate change and is therefore not environmentally sustainable, a detour takes us to the technological promise of carbon capture and sequestration. The Harper government, moreover, argued that the emissions intensity of oil sands production has been declining, that oil should be compared to coal as a generator of GHGs, and so on.[23]

Other developments may also reassure the oil and gas industry that the new attention to environmental monitoring will not impede the expansion of oil sands operations. The policy director of the Pembina Institute, Simon Dyer, observed that the government of Alberta approved a ninth open-pit oil sands mine in January 2011, just as it was appointing the Alberta Environmental Monitoring Panel. This approval was based on data provided by the RAMP, whose work had already been called into question by six of the studies mentioned above. Dyer commented: 'I think it [the approval] shows that despite governments in Alberta and Canada acknowledging there are real failures in environmental monitoring in the oil sands, clearly the business of managing oil-sands growth as quickly as possible is proceeding' (quoted in Wingrove 2011).

In January 2012, under intense pressure from both domestic and international critics of the existing and projected expansion of oil sands mining,[24] the federal and Alberta governments agreed to the Joint Canada-Alberta Implementation Plan for Oil Sands Monitoring.[25] This plan (not expected to be fully implemented until 2015) was favourably received by the scientists who had participated in the advisory panels. However, when ministers Kent (Environment Canada) and McQueen (Alberta Environment and Water) announced the plan to the public on 3 February, they said that it would be jointly administered by the two assistant deputy ministers from Environment Canada and Alberta Environment. This move was rejected by the scientists, led by David Schindler, who continue to argue that there must be a 'stand alone' authority (Miall 2012), which the AEMP described in its recommendations as 'a permanent, sustainably-funded, arm's length Environmental Monitoring Commission' (Kvisle et al. 2011, ii). Thus, in March 2012, the provincial minister of (the renamed) Environment and Sustainable Resource Development (ESRD) appointed another 'working group' to 'provide expert advice, viable options and recommendations ... in deciding the future governance and funding of a new provincial monitoring, evaluation and reporting system' and to report by 30 June 2012 with 'recommendations and other relevant advice that outlines and assesses governance options ranging from an arm's length public agency to a structure internal to the Government of Alberta' (AEMWG 2012, i).

Strikingly, the authors of the Alberta Environmental Monitoring Working Group report asserted that 'should Alberta choose to implement an AEM system internal to government, it would need to *create a new department of environmental monitoring, evaluation and reporting, separate from the regulatory or policy operations of government*. Such a department would still require substantive new funding and management resources' (AEMWG 2012, iv, italics added). Their preference, however, was that 'the AEM system be established as *an arm's length agency*, with a clear mandate, strong governance, strong leadership and a clear commitment to scientific monitoring. An *arm's length structure* with strong governance will most likely satisfy the key principles of legitimacy and credibility' (vi, italics added). These statements constitute further evidence that the province's Ministry of Environment has not been performing the functions that citizens might normally expect of a government agency charged with 'protecting and enhancing Alberta's natural environment, to ensure the continued enjoyment of a clean and healthy environment by all.'[26] It is telling indeed that a body charged with environmental data collection, monitoring, scientific research, and informing public 'discussion around economic development and the environment, by providing factual evidence and increasing our knowledge of environmental effects' (vi) needs to be protected from the influence of the government – including the Ministry of Environment.

In October 2012, the minister of ESRD appointed a six-member management board to propose 'how the new science-based agency will operate, long-term funding options and establishing [*sic*] a Science Advisory Board to provide input and advice on monitoring efforts' (Alberta ESRD 2012). The oil and gas industry did not (at least publicly) oppose the establishment of such an agency; on the contrary, industry representatives expressed support for a co-ordinated, 'world class' environmental monitoring system for the province (the same language as that used by Prime Minister Harper and federal Environment Minister Peter Kent in promoting their government's commitment to 'environmentally sustainable' oil sands production and the safety of oil tanker traffic in the coastal waters of BC). This apparent consensus among policy makers, corporations, and scientists about the value of scientific monitoring suggests the multiple uses of appeals to 'science-based' policy that are further examined in Stendie and Adkin, this volume.

The Alberta Environmental Monitoring, Evaluation and Reporting Agency (AEMERA) was established in 2014 by the Protecting Alberta's Environment Act, and is described by the government as 'the science-based, comprehensive, arm's-length organization responsible for co-ordinating province-wide environmental monitoring and evaluation throughout the province – beginning in the oil sands region.'[27] The first appointed chair of its seven-member board of directors was the former minister of environment in the Klein government of 2001–2004, who led the campaign against the federal government's ratification of the Kyoto Protocol.[28]

In addition, the government established the Alberta Energy Regulator (AER) under the Responsible Energy Development Act in November 2012. The AER took on the regulatory responsibilities formerly held by the Ministry of Environment (or ESRD) and the Energy Resources Conservation Board (dissolved by the act) with regard to approving energy-related projects, and will oversee public hearings. A Policy Management Office, also housed in the Ministry of Energy, will be involved in developing environmental policy and public engagement.[29] The AER got off to a rocky start, however, because the government appointed as its board chair the founding president of the Canadian Association of Petroleum Producers (CAPP) and executive VP of Encana Oil and Gas Company, Gerry Protti, who is also a former assistant deputy minister in the Alberta Energy ministry, and a former executive of several other energy companies (Henton 2013). Within days of the appointment, a coalition of more than thirty landowner, labour, environmental, and First Nation groups had formed to demand not only that 'the fox not be charged with watching the hen house,' but that the move to a single regulator for the energy sector be subjected to a public consultation.[30]

A long struggle has been waged by Aboriginal communities, environmentalists, and a growing contingent of scientists to compel public agencies to act in the public interest. The experience briefly recounted here reveals that both the public interventions of scientists drawing on their publicly funded research on the environmental impacts of the tar sands, and sustained international campaigns spearheaded by indigenous and environmental organizations (making use of the scientific research), have been necessary to wrest regulatory concessions from the governments of Alberta and Canada. It will be several years before we can more fully assess the 'legitimacy,' 'transparency,' and 'relevance' of the regulatory bodies put in place in 2013–2014, although questions have already been raised about their autonomy, priorities, and opportunities for public involvement.[31]

Conclusion

The Alberta Advantage was built on an ecological subsidy, a gift to the oil and gas industry that takes the forms of an emaciated department of environmental protection and lax environmental statutes and regulations (see Carter, this volume). When we examine government budget documents since the 1990s, it becomes clear that at least part of what we understand as the Alberta Advantage has depended upon low spending on environmental regulations, monitoring, and enforcement. In particular, Alberta's oil and gas boom of the late 1990s through to 2008 was very closely associated with fewer resources for the Department of the Environment, as the government sought to attract investment in the oil sands. The limited capacity and mandate of the Ministry of Environment also reflects the government's promotion of investment in the forestry sector and its support for coal-fired power generation (areas that we have not had space to discuss in this chapter). Overall, the Alberta Advantage has relied upon the government's refusal to apply ecosystem integrity criteria to applications for oil sands project approvals and other 'development' projects. As described in Quinn et al. (this volume), the provincial government has until recently failed to give weight to the cumulative effects of multiple industrial activities on local environments in its processes for approving new developments.

Investment in the oil sands, in turn, insulated the provincial government from political backlash against shock doctrine–style cuts to public programs and infrastructure. The oil sands boom provided near-full employment[32] and allowed some degree of reinvestment in health and education programs (but not social services) based on non-renewable resource revenues gushing into provincial coffers.

The neoliberal environment that has resulted, and is described in detail in other chapters of this volume, is one of intensive resource exploitation with the attendant environmental degradation, greenhouse gas emissions, and dispossession of Aboriginal populations. If Alberta is a petro-state, it is one in which the ruling political elite has been motivated by ideological conviction and the belief that 'there is no alternative' to dependence on private capital investment in fossil fuel exploitation as a motor of economic growth. Economic growth, in turn, is entirely unquestioned as the core 'public interest.'

Petro-states are subject to similar economic pressures emanating from the nature of global energy markets and the domestic effects of the 'resource curse.' However, differences in their domestic governance regimes yield substantially different social, political, and environmental outcomes of oil wealth. Unlike Norway, Alberta has not pursued a social democratic variant of sustainable development. Nor do its political rulers profit from their positions to enrich themselves on the scale that we see in countries like Russia, Nigeria, or the Middle East. Instead, Alberta falls into a subset of first world neoliberal petro-states that are mostly subnational and based in North America. Indeed, neoliberalism itself has been largely an Anglo-American phenomenon. It is not coincidental that in Germanwatch's 2013 climate change performance index, Australia, Canada, New Zealand, and the United States were in the bottom rungs of OECD member state rankings for 'climate change performance' (Burck et al. 2012, table 4).

Despite growing opposition to a model of development driven by intensive resource extraction and export, Alberta's Conservatives have unapologetically defended this model – as Premier Redford made clear in numerous public statements, including the one that begins this chapter. Evidence that climate effects call for low-carbon models of economic development was met by the Conservatives with a 'more of the same' approach: a conveyor belt of public money going to subsidize the fossil fuels sector (see chapters 2 and 6). In the meantime, the province's resources for undertaking a green transition remain underdeveloped due to government inaction. The resistance of the provincial and federal Conservative governments to even moderate proposals for ecological modernization (such as higher 'carbon levy') and their refusal to impose credible emission caps or targets on large emitters are explained not only by the historically close identification of interests between the large oil and gas companies and the Conservative party, but by this party's rigid commitment to a market-driven and fossil-fuelled development model since the early 1990s. The rents from oil and gas, in turn, helped to entrench one-party rule in Alberta and to insulate its governments from internal opposition.

However, the global politics of climate change, the pipeline struggles, the mobilization of environmental scientists, and the spreading Idle No More movement, combined with comparatively low oil prices since 2008, have presented new challenges to Alberta's economic growth model and to the stability of its long-standing Conservative political regime. The growing reach and influence of indigenous- and environmentalist-led campaigns has shone a bright light on the Ministry of Environment's incapacity even to collect scientifically intelligible data on the effects of oil sands exploitation, let alone to rein in the pace of development or to require higher standards of performance regarding such problems as GHG emissions, the storage of tailings, or land reclamation.

These failures have seriously undermined the credibility of the government's assurances to citizens, to investors and insurers, and to political authorities elsewhere (e.g., in the USA, European Union) that development of the oil sands is proceeding 'responsibly' and 'sustainably.'[33] The creation of a new, 'arm's length' environmental monitoring body reveals the depth of disintegration of its environmental regulatory capacity, on one hand, and recognition of the need to be seen to be doing 'credible' scientific monitoring, on the other hand. Whether the future environmental monitoring and regulatory regime will slow the rate of tar sands exploitation and increase the environmental costs of production for the industry will depend – among other factors – upon the capacity of civil society actors to mobilize public support for green transition.

NOTES

1 'Premier Alison Redford on Alberta's Oil Sands,' video produced by the Alberta government, youtube channel 'YourAlberta,' published 30 May 2012, http://www.youtube.com/watch?v=VtyfmrUVFMo&list=PL57354166730EA923.

2 The best statement of the government's use of the term 'the Alberta Advantage' may be that found in the 1997 budget (http://www.finance.alberta.ca/publications/budget/budget1997-2000/1997/ALTAADV.PDF).

3 http://www.finance.alberta.ca/publications/budget/budget1997-2000/1997/GOVBP.PDF.

4 While all areas of the province's environment have been affected by the neoliberal regulatory regime, most of our examples are drawn from oil sands regulation and climate change policy. The chapter by Quinn et al. in this volume describes the effects on landscape and wildlife habitat, and the chapters by Carter and Garvin provide further details of health and envi-

ronmental problems associated with the oil sands and gas production. For excellent accounts of developments in the forestry sector in the 1990s, see Richardson, Gismondi, and Sherman (1993) and Pratt and Urquhart (1994). See also McInnis and Urquhart (1995), and many of the chapters in Harrison (2005) and Epp and Whitson (2001).

5 In the preface to the first (1962) edition of *Capitalism and Freedom*, Milton Friedman acknowledged: 'I owe the philosophy expressed in this book and much of its detail to many teachers, colleagues, and friends, above all to a distinguished group I have been privileged to be associated with at the University of Chicago: Frank H. Knight, Henry C. Simons, Lloyd W. Mints, Aaron Director, Friedrich A. Hayek, George J. Stigler' (Friedman 2002, xvi).

6 G. William Domhoff, sociology professor at the University of California at Santa Cruz, provides an overview of recent research on wealth and income inequality trends in the USA in 'Wealth, Income and Power,' http://www2.ucsc.edu/whorulesamerica/power/wealth.html (accessed 21 January 2013).

7 Using the global accounting method, Isabel Ortiz and Matthew Cummins (2011, 11), economists at UNICEF, have recently calculated that 'as of 2007, the wealthiest 20 percent of mankind enjoyed nearly 83 percent of total global income compared to the poorest 20 percent, which had exactly a single percentage point ... Perhaps more shocking, the poorest 40 percent of the global population increased its share of total income by less than one percent between 1990 and 2007.' Using the inter-country accounting model (for 182 countries), they found similar results.

8 Friedman's 'Chicago School' experiments were first implemented in Chile, Argentina, Brazil, and Indonesia in the 1970s, supported by brutal dictatorships, torture, and the disappearances of those who opposed economic shock therapy – especially trade unionists. By the 1980s, 'shock therapy' was administered by the IMF and the World Bank, whose conditions for assisting countries reeling from currency instability and crushing foreign debts included Friedman's favoured policies of cuts to government spending, privatization of state assets, and deregulation of investment and trade. From sub-Saharan Africa to post–Soviet Russia, 'shock therapy' cut a swath through public goods and services, replacing these with school fees, high food and fuel prices, teeming cities of poor and dislocated rural migrants, health care crises (due to underfunding), and electricity shortages (due to privatization of utilities).

9 'Home' governments are reluctant to hold companies registered in their countries to local environmental and labour norms, while the 'guest' coun-

tries often have less capacity or leverage to do so, as we see with the activities of some Canadian mining companies operating outside of Canada.

10 Canadian Taxpayers Federation, 'Taxpayers Federation Honour Ralph Klein and Jim Dinning with TaxFighter Award,' article posted on taxpayer.com, April 18, 2011 (http://taxpayer.com/federal/taxpayers-federation-honour-ralph-klein-and-jim-dinning-taxfighter-award).

11 Andrew Nikiforuk, 'If Ralph's a Friend, Who Needs Enemies?' *Globe and Mail*, 28 September 2005.

12 This document conveys the impression that the government had a step-by-step, three-year plan to achieve its goals, whereas many analysts have suggested that very little planning preceded the cuts. However, it is useful as a summary of the actions that were taken at the height of the 'Klein Revolution.' The legislative framework for privatizations and delegation of government administration to non-government bodies included the Deficit Elimination Act (to permanently eliminate the annual deficit by 1996–97); Balanced Budget and Debt Retirement Act (outlawed deficits and mandated the elimination of the accumulated net debt by 2009–10); Alberta Taxpayer Protection Act (no sales tax without a referendum); Government Accountability Act (required ministries to produce three-year business plans); Business Financial Assistance Limitation Act (declared there would be no more special loans and guarantees to business); and Government Organization Act (http://www.finance.alberta.ca/publications/budget/budget1997-2000/1997/PLAN.PDF, p. 329).

13 The appendix to Budget '97 entitled 'A Plan for Change' stated: 'Government has gotten out of the business of business. Commercial entities and assets of the government have been sold, including Alberta Resources Railway, Gainers, North West Trust Company, Magnesium Company of Canada, Alberta Intermodal Services, the Lloydminster Bi-provincial Upgrader, and shares of Syncrude, Vencap Equities and the Alberta Energy Company. Some services have been privatized, including registry services, liquor sales and property assessment services. Other services are being contracted out, including highway maintenance. The unfunded liability of the Workers' Compensation Board has been eliminated and the Board has been put at arm's length from the government' (329).

14 Clyde H. Fransworth, 'Budget Cuts Paying Off for Premier of Alberta,' *New York Times*, 1 January 1995.

15 References to in-year expenditures are expressed in raw dollars.

16 'Alberta's Growth "Strongest Ever": StatsCan.' *CBC News Online*, 14 September 2006, http://www.cbc.ca/canada/calgary/story/2006/09/14/albertastats.html.

17 The department does list 'equipment and inventory purchases' as capital assets in government estimates records. Alberta Environment's equipment and inventory purchases have ranged from $685,000 (in raw dollars) in 2003 to $3.2 million in 2006. However, 2009–10 figures for amortization of Environment's capital assets show an $18 million infrastructure deficit.

18 These and other recommendations are contained in Pembina Institute 2008.

19 Cooperative Financial Services, 'Unconventional Oil Report,' 16 September 2008. http://www.goodwithmoney.co.uk/unconventional-oil-report/.

20 Niall O'Shea (engagement manager, The Cooperative Asset Management), personal communication with research assistant, 18 September 2008. Emissions intensity is a measurement of the quantity of CO_2, or other substance, that is created for each unit of production of a particular good (e.g., kgs of CO_2 per barrel of crude oil). This measurement allows us to compare the greenhouse gas 'footprints' of oil sands products over time, as well as relative to other producers (e.g., in Venezuela) or relative to other types of oil production (e.g., from conventional oil reserves). The Alberta government has mandated greenhouse gas emissions intensity targets for large emitters, but has not set absolute caps on greenhouse gas emissions.

21 See RAMP's membership at http://www.ramp-alberta.org/ramp/terms+of+reference/membership/members.aspx. According to Andrew Miall, a geology professor and one of the members of the Federal Oil Sands Advisory Panel (appointed in October 2010) as well as of the Alberta Environmental Monitoring Panel (appointed by Alberta Environment Minister Rob Renner in January 2011), the Kelly and Schindler studies had shown that 'significant quantities of polyaromatic compounds and trace metals were accumulating in the snowpack near the processing and upgrading installations, and were being released rapidly into the Athabasca River during the spring melt, at the time fish embryos were developing. These substances could only have been derived by aerial discharge from the nearby upgraders. This mode of pollution had been entirely missed by the Regional Aquatic Monitoring Program (RAMP), the independent monitoring organization established by the Alberta government in 1997. RAMP had been subject to two major independent peer reviews in 2004, and again in 2010, and both times had been severely criticized for its poor scientific methods' (http://www.es.utoronto.ca/Members/miall/oil-sands-advisory-panel).

22 See, e.g., the commentary by Dave Collyer, president of CAPP, 'One Project, One Review,' published in the *Financial Post*, 29 May 2012, as well as on CAPP's website, http://www.capp.ca/aboutUs/mediaCentre/

CAPPCommentary/Pages/one-project-one-review.aspx. Much of industry lobbying of government, of course, takes place out of public view. A freedom of information request by Greenpeace Canada provided a glimpse of the industry-government relationship in the form of a letter, dated 12 December 2011, from the Energy Framework Initiative to the federal ministers for environment and industry. The EFI comprises the Canadian Association of Petroleum Producers, the Canadian Energy Pipeline Association, the Canadian Petroleum Products Institute (now the Canadian Fuels Association), and the Canadian Gas Association. The letter was made available to the media by Greenpeace, and can be read at http://www.documentcloud.org/documents/552378-atip-industry-letter-on-enviro-regs-to-oliver.html. For the CBC's report on the letter, see http://www.cbc.ca/news/politics/story/2013/01/09/pol-oil-gas-industry-letter-to-government-on-environmental-laws.html.

23 See the letter dated 15 March 2012 from Natural Resources Canada assistant deputy minister Mark Corey, on behalf of the minister, Joe Oliver, to Dr Keith Stewart, climate and energy campaigner, Greenpeace Canada, available at http://www.greenpeace.org/canada/en/Blog/joe-oliver-distracted-by-media-reporting-on-s/blog/41823/.

24 The website of the Alberta Ministry of Energy stated in January 2013, with regard to the oil sands, 'New projects are being added every year and production is expected to increase from 1.31 million barrels per day in 2008 to 3 million barrels per day in 2018' (http://www.energy.gov.ab.ca/ourbusiness/oilsands.asp). According to the Pembina Institute, 'government regulators have already approved more than 5 million barrels per day of oilsands production, and we could reach that milestone just over two decades from now. That's triple the current rate of oilsands production, a level of expansion that has been approved despite widespread recognition that the environmental costs of our current approach to oilsands extraction are unacceptably high, and major improvements to both government oversight and industry operations are required' (Grant 2012).

25 Environment Canada, 'The Canada-Alberta Joint Implementation Plan,' 3 February 2012,http://www.ec.gc.ca/pollution/EACB8951-1ED0-4CBB-A6C9-84EE3467B211/Final%20OS%20Plan.pdf.

26 Ministry of Environment and Sustainable Resource Development, 'About Us,' http://environment.alberta.ca/01549.html (accessed 24 January 2013).

27 Ministry of Environment and Sustainable Resource Development, Government of Alberta, http://esrd.alberta.ca/focus/environmental-monitoring-in-alberta/default.aspx (accessed 12 May 2015).

28 In March 2016 the NDP government announced that it would disband AEMERA and bring its functions back into the Environment Ministry.

29 Sheila Pratt, 'New Energy Regulator Will Weaken Environmental Protection, Say Critics,' *Calgary Herald*, 18 March 2013, http://www.calgaryherald.com/business/energy-resources/Alberta+energy+regulator+will+weaken/8113914/story.html?__lsa=aff9-6d66.

30 See the coalition's letter to Premier Redford, dated 3 May 2013, http://www.greenpeace.org/canada/en/Blog/groups-demand-oil-industry-fox-get-out-of-alb/blog/45028/.

31 Space does not permit an analysis of the new provisions for public hearings. A preliminary analysis has been offered by Nickie Vlavianos, law professor at the University of Calgary, http://ablawg.ca/2012/11/15/an-overview-of-bill-2-responsible-energy-development-act-what-are-the-changes-and-what-are-the-issues/.

32 This is not to say, of course, that all population groups benefited equally from the construction boom that was generated by investments in the oil sands during this period. Strong gender inequalities continue to characterize the provincial economy (Dorow 2015; Lahey 2015). O'Shaughnessy and Doğu (this volume) observe the racialized nature of low-paid service work in Fort McMurray, and this extends to caring labour (Adkin and Abu-Laban 2008; Brickner and Straehle 2010; Taylor et al. 2012); and the influx of 'temporary foreign workers' to fill low-wage jobs in the service sector and low-skilled jobs in the agricultural and mining sectors (Alberta Federation of Labour 2007, 2009; Broadway 2012; Depatie-Pelletier 2010; Foster and Barnetson 2015; Fudge and MacPhail 2009; Hanley and Shragge 2010).

33 A German research foundation announced in March 2013 its withdrawal from a research consortium based at the University of Alberta (and receiving $25 million in funding from the Alberta government) that was examining methods for upgrading bitumen. The scientists associated with the project explained that public pressure in Germany against government support for bitumen mining, in light of climate change, and the failure of the Canadian government to act on greenhouse gas reductions, had led the Helmholtz Association to withdraw from the Alberta research. See Bob Weber, 'German Scientists Abandon Oilsands Research with Alberta, Citing "Intensive" Public Backlash,' *Calgary Herald*, 19 March 2013, http://www.calgaryherald.com/business/energy-resources/German+scientists+abandon+oilsands+research/8121477/story.html.

REFERENCES

Adkin, Laurie. 1995. 'Life in Kleinland: Democratic Resistance to "Folksy Fascism."' *Canadian Dimension* (April–May).

Adkin, Laurie, ed. 2009a. *Environmental Conflict and Democracy in Canada*. Vancouver: UBC Press.

Adkin, Laurie. 2009b. 'Democracy from the Trenches: Environmental Conflicts and Ecological Citizenship.' In *Environmental Conflict and Democracy in Canada*, ed. Laurie Adkin, 298–319. Vancouver: UBC Press.

Adkin, Laurie E., and Yasmeen Abu-Laban. 2008. 'The Challenge of Care: Early Childhood Education and Care in Canada and Québec.' *Studies in Political Economy* 81 (Spring): 49–76.

Alberta Chamber of Resources. National Task Force on Oil Sands Strategies. 1995. *The Oil Sands: A New Energy Vision for Canada*. Edmonton: Alberta Chamber of Resources. http://www.acr-alberta.com/LinkClick.aspx?fileticket=wXLRYoJxW1w%3d&tabid=205.

Alberta Culture. Provincial Archives of Alberta. c. 2012. 'An Administrative History of the Government of Alberta, 1905–2005.' Chapter 21: 'Environment: 1971–present.' Edmonton: Government of Alberta. http://www.culture.alberta.ca/paa/archives/research/adminhistory.aspx.

Alberta Environmental Monitoring Working Group (AEMWG). 2012. *Implementing a World Class Environmental Monitoring, Evaluation and Reporting System for Alberta*. Report of the Working Group on Environmental Monitoring, Evaluation and Reporting. June. http://environment.gov.ab.ca/info/library/8699.pdf.

Alberta Federation of Labour. 2007. *Temporary Foreign Workers: Alberta's Disposable Workforce. The Six-Month Report of the AFL's Temporary Foreign Worker Advocate*. Edmonton: Alberta Federation of Labour.

Alberta Federation of Labour. 2009. *Entrenching Exploitation: The Second Report of the Alberta Federation of Labour, Temporary Foreign Worker Advocate*. Edmonton: Alberta Federation of Labour, 1 January.

Alberta Hansard. 1993. Treasury Subcommittee. 17 September. http://www.assembly.ab.ca/ISYS/LADDAR_files/docs/hansards/han/legislature_23/session_1/19930917_0800_01_han.pdf.

Alberta. Ministry of Environmental Protection. 1998. *Annual Report 1997–98*. Edmonton: Government of Alberta. http://environment.gov.ab.ca/info/library/6125.pdf.

Alberta. Ministry of Environmental Protection. 1999. *Annual Report 1998–99*. Edmonton: Government of Alberta. http://environment.gov.ab.ca/info/library/6127.pdf.

Alberta. Ministry of Environment. 2000. *Annual Report 1999–2000*. Edmonton: Government of Alberta. http://environment.gov.ab.ca/info/library/6128.pdf.

Alberta. Ministry of Environment and Sustainable Resource Development

(ESRD). 2013. 'Alberta to Establish Arm's-Length Environmental Monitoring Agency.' News release, 17 October. http://alberta.ca/acn/201210/33115701C6E68-EBBC-DF50-F7CCA9C17B88AC98.html.

Altvater, Elmer. 2006. 'The Social and Natural Environment of Fossil Capitalism.' In *Socialist Register 2007: Coming to Terms with Nature*, ed. Leo Panitch and Colin Leys, 37–59. London: Merlin.

Anaya, James. 2010. *Report by the Special Rapporteur on the Situation of Human Rights and Fundamental Freedoms of Indigenous People. Addendum: Cases Examined by the Special Rapporteur (June 2009–July 2010)*. United Nations, Human Rights Council, 15th Session, 15 September, A/HRC/15/37/Add.1. http://unsr.jamesanaya.org/PDFs/Communications%20report-FINAL.pdf (accessed 8 November 2010).

Brickner, Rachel K., and Christine Straehle. 2010. 'The Missing Link: Gender, Immigration Policy and the Live-in Caregiver Program in Canada.' In 'Guest Worker Programs: Friend or Foe of an Integrated Immigration Policy,' ed. Patti Tamara Lenard and Christine Straehle, special issue, *Policy and Society* 29 (4): 309–20.

Broadway, Michael. 2012. 'Cut to the Bone.' *Alberta Views* 15 (4): 36.

Buehler, Clint. 2007. 'Fort Chip Natives Oppose More Oil Business.' *First Nations Drum* (Winter 2006–2007). http://www.firstnationsdrum.com/Fall%20 2006/HealthChip.htm.

Burck, Jan, Lukas Hermwille, and Laura Krings. 2012. 'The Climate Change Performance Index 2013: A Comparison of the 58 top CO2 Emitting Nations.' Germanwatch, December. http://germanwatch.org/en/5698.

Campanella, David. 2011. 'Fossil Intentions: Carbon Capture and Storage in Alberta.' Master's thesis, Faculty of Environmental Studies, York University.

Cooper, David, and Dean Neu. 1995. 'The Politics of Debt and Deficit in Alberta.' In Harrison and Laxer, *The Trojan Horse*, 163–81.

Daly, Herman. 1997. *Beyond Growth: The Economics of Sustainable Development*. Boston: Beacon.

Denis, Claude. 1995. 'The New Normal: Capitalist Discipline in Alberta in the 1990s.' In Harrison and Laxer, *The Trojan Horse*, 86–100.

Depatie-Pelletier, Eugénie. 2010. 'Restrictions on Rights and Freedoms of Low-Skilled Temporary Foreign Workers.' *Canadian Issues/Thèmes Canadiens* (Spring): 64–7.

Dorow, Sara. 2015. 'Gendering Energy Extraction in Fort McMurray.' In *Alberta Oil and the Future of Democracy in Canada*, ed. Meenal Shrivastava and Lorna Stefanick, 249–274. Edmonton: Athabasca University Press.

Dowdeswell, E. (chair), P. Dillon, S. Ghoshal, A.D. Miall, J. Rasmussen, and J.P. Smol. 2010. *A Foundation for the Future: Building an Environmental Moni-*

toring System for the Oil Sands. Report submitted to the Minister of Environment, Canada, December. http://www.ec.gc.ca/pollution/default.asp?lang=En&n=E9ABC93B-1.

Epp, Roger, and Dave Whitson, eds. 2001. *Writing Off the Rural West: Globalization, Governments, and the Transformation of Rural Communities*. Edmonton: University of Alberta Press/Parkland Institute.

Fekete, Jason. 2009. 'Carbon Trading Market Pays Off for Alberta Farmers.' *Calgary Herald*, 17 October. http://www2.canada.com/calgaryherald/news/story.html?id=917a3780-2a3f-4327-b54b-a32f415c6553 (accessed 20 January 2013).

Ferreira, Darlene Abreu. 1992. 'Oil and Lubicons Don't Mix: A Land Claim in Northern Alberta in Historical Perspective.' *Canadian Journal of Native Studies* 12 (1): 1–35.

Foster, Jason, and Bob Barnetson, 2015. 'Exporting Oil, Importing Labour, and Weakening Democracy: The Use of Foreign Migrant Workers in Alberta.' In *Alberta Oil and the Future of Democracy in Canada*, ed. Meenal Shrivastava and Lorna Stefanick, 249–74. Edmonton: Athabasca University Press.

Friedman, Milton. 2002. 1982 preface to *Capitalism and Freedom*. 40th anniversary ed. Chicago: University of Chicago Press.

Fudge, Judy, and Fiona MacPhail. 2009. 'The Temporary Foreign Worker Program in Canada: Low-Skilled Workers as an Extreme Form of Flexible Labor.' *Comparative Labor Law & Policy Journal* 31 (1): 5–45.

Goddard, John. 1991. *The Last Stand of the Lubicon Cree*. Vancouver: Douglas & McIntyre.

Government of Alberta. 2005. 'Mineable Oil Sands Strategy' (draft for discussion), October. http://www.energy.alberta.ca/OilSands/pdfs/MOSS_Policy2005.pdf (accessed 19 September 2011).

Grant, Jennifer. 2012. '"Responsible Resource Development" Must Be More Than a Slogan.' Commentary published in *Hill Times* (27 August), *Stratford Beacon Herald* (4 September), and available on http://www.pembina.org/oil-sands/overview.

Hall, Tony. 1995. 'Alberta's Revolution.' *Canadian Forum* 73 (836): 29.

Hanley, Jill, and Eric Shragge. 2010. 'Rights & Resistance as Canada Shifts towards the Use of Guestworkers.' *Social Policy* 40 (3): 6–13.

Harrison, Trevor W., ed. 2005. *The Return of the Trojan Horse: Alberta and the New World (Dis)Order*. Montreal: Black Rose.

Harrison, Trevor W., and Gordon Laxer, eds. 1995. *The Trojan Horse: Alberta and the Future of Canada*. Montreal: Black Rose.

Harvey, David. 2005. *A Brief History of Neoliberalism*. Oxford: Oxford University Press.

Harvey, David. 2006. 'Neo-liberalism as Creative Destruction.' *Geographic Annals* 88 B (2): 145–58.

Huff, A.I. 1999. 'Resource Development and Human Rights: A Look at the Case of the Lubicon Cree Indian Nation of Canada.' *Colorado Journal of International Environmental Law and Policy* 10 (1): 161–94.

Kelly, E.N., D.W. Schindler, P.V. Hodson, J.W. Short, R. Radmanovich, and C.C. Nielsen. 2010. 'Oil Sands Development Contributes Toxic Elements at Low Concentrations to the Athabasca River and Its Tributaries.' *Proceedings of the National Academy of Sciences USA* 107: 16178–83.

Kelly, E.N., J.W. Short, D.W. Schindler, P.V. Hodson, M. Ma, A.K. Kwan, and B.L. Fortin. 2009. 'Oil Sands Development Contributes Polycyclic Aromatic Compounds to the Athabasca River and Its Tributaries.' *Proceedings of the National Academy of Sciences USA* 106 (52): 22346–51.

Kurek, Joshua, Jane L. Kirk, Derek C.G. Muir, Xiaowa Wang, Marlene S. Evans, and John P. Smol. 2013. 'Legacy of a Half Century of Athabasca Oil Sands Development Recorded by Lake Ecosystems.' *Proceedings of the National Academy of Sciences USA* 110 (5): 1761–6. http://www.pnas.org/content/early/2013/01/02/1217675110.full.pdf+html?sid=e2cfc926-fd07-4659-a01c-c97100fccca2.

Kvisle, H. (co-chair), H. Tennant (co-chair), J. Doucet, W. Kindierski, A.D. Miall, D. Pryce, J. Rasmussen, G.R. Taylor, R. Wallace, H. Wheater, and D. Williams. 2011. *A World-Class Environmental Monitoring, Evaluation and Reporting System for Alberta.* The report of the Alberta Environmental Monitoring Panel. Government of Alberta, 30 June. http://environment.alberta.ca/03289.html.

Lahey, Kathleen. 2015. *The Alberta Disadvantage: Gender, Taxation, and Income Inequality*. Edmonton: Parkland Institute. March. http://s3-us-west-2.amazonaws.com/parkland-research-pdfs/thealbertadisadvantage.pdf.

Lipietz, Alain. 1989. *Choisr l'audace : Une alternative pour le vingt et unième siècle.* Paris: Éditions la Découverte.

Lisac, Mark. 1995. *The Klein Revolution.* Edmonton: NeWest.

Martin-Hill, Dawn. 2004. 'Resistance, Determination and Perseverance of the Lubicon Cree Women.' In *In the Way of Development: Indigenous Peoples, Life Projects and Globalization*, ed. Mario Blaser, Harvey A. Feit, and Glenn Mcrae, 313–31. London: Zed Books.

McCarthy, James, Nik Heynen, Paul Robbins, and Scott Prudham, eds. 2007. *Neoliberal Environments: False Promises and Unnatural Consequences*. New York: Routledge.

McInnis, John, and Ian Urquhart. 1995. 'Protecting Mother Earth or Business?

Environmental Politics in Alberta.' In Harrison and Laxer, *The Trojan Horse*, 239–53.

McMillan, Melville L., and Allan A. Warrack. 1995. 'One-Track (Thinking) towards Deficit Reduction.' In Harrison and Laxer, *The Trojan Horse*, 134–62.

Meili, Dianne. 2007. Oilsands Boom Creates Uneasy Wealth in North. *Alberta Sweetgrass*, May. http://www.ammsa.com/sweetgrass/Sweet-May 1-2007.html.

Miall, Andrew. 2012. 'The Oil Sands: A New Regime of Environmental Management?' http://www.es.utoronto.ca/Members/miall/oil-sands-advisory-panel. N.d. (accessed 24 January 2013).

Neu, Dean. 1996. 'Dissent and Discipline in Alberta.' *Canadian Dimension* 30 (2): 35.

Norwegian, Herb, Grand Chief, and Chief Roland Willson. 2006. 'Northern Water Rights Issue Heats Up.' *Edmonton Journal*, 16 October 16.

Oil Sands Advisory Panel. 2010. *A Foundation for the Future: Building an Environmental Monitoring System for the Oil Sands*. Report submitted to the Minister of the Environment, December. http://www.ec.gc.ca/Publications/EF00F933-88A1-4931-A5E0-1B384BD1119C/dationForTheFutureBuildingAnEMSForTheOilSands.pdf.

Ominayak, Chief Bernard, and Kevin Thomas. 2009. 'These Are Lubicon Lands: A First Nation Forced to Step into the Regulatory Gap.' In *Speaking for Ourselves: Environmental Justice in Canada*, ed. Julian Agyeman et al., 111–22. Vancouver: UBC Press.

Pembina Institute. 2007. 'Government Protects Oil Sands Industry, Fails to Protect Athabasca River.' Media release, 2 March. http://www.pembina.org/media-release/1384.

Pembina Institute. 2008. *Taking the Wheel: Correcting the Course of Cumulative Environmental Management in the Athabasca Oil Sands*. Drayton Valley, AB: Pembina Institute, August. http://www.oilsandswatch.org/pub/1677.

Pembina Institute. 2011. *Responsible Action? An Assessment of Alberta's Greenhouse Gas Policies*. Written by Matthew Bramley, Simon Dyer, Marc Huot, and Matt Horne. Drayton Valley, AB: Pembina Institute, 16 December. http://www.pembina.org/pub/2295.

Phillips, Shannon. 2008. 'Head in the Sands.' *Alberta Views Magazine*, December/January.

Pratt, Larry, and Ian Urquhart. 1994. *The Last Great Forest: Japanese Multinationals and Alberta's Northern Forests*. Edmonton: NeWest.

Richardson, Mary, Joan Sherman, and Michael Gismondi. 1993. *Winning Back the Words: Confronting Experts in an Environmental Public Hearing*. Toronto: Garamond.

Rooney, R.C., S.E. Bayley, and D.W. Schindler. 2012. 'Oil Sands Mining and Reclamation Cause Massive Loss of Peatland and Stored Carbon.' *Proceedings of the National Academy of Sciences USA* 109: 4933–7.

Schindler, David W. 2010. 'Tar Sands Need Solid Science.' Invited comment. *Nature* 468: 499–501.

Schrecker, Ted. 2001. 'Using Science in Environmental Policy: Can Canada Do Better?' In *Governing the Environment: Persistent Challenges, Uncertain Innovations*, ed. E.A. Parson, 31–72. Toronto: University of Toronto Press.

Schwartz, Herman. 1997. 'Reinvention and Retrenchment: Lessons from the Application of the New Zealand Model to Alberta, Canada.' *Journal of Policy Analysis and Management* 16 (3): 205–32.

Smith, Peter J. 2001. 'Alberta: Experiments in Governance – From Social Credit to the Klein Revolution.' In *The Provincial State in Canada: Politics in the Provinces and Territories*, ed. Keith Brownsey and Michael Howlett, 277–309. Peterborough: Broadview.

Soron, Dennis. 2005. 'The Politics of De-politicization: Neo-liberalism and Popular Consent in Alberta.' In *The Return of the Trojan Horse: Alberta and the New World (Dis)Order*, ed. Trevor W. Harrison, 65–81. New York: Black Rose.

Taft, Kevin, Melville L. McMillan, and Junaid Jahangir. 2012. *Follow the Money: Where Is Alberta's Wealth Going?* Calgary: Detselig.

Thomson, Graham. 2009. *Burying Carbon Dioxide in Underground Saline Aquifers: Political Folly or Climate Change Fix?* Report for the Munk Centre for International Studies, University of Toronto, September.

Timoney, Kevin P. 2007. *A Study of Water and Sediment Quality as Related to Public Health Issues, Fort Chipewyan, Alberta*, on behalf of the Nunee Health Board Society, Fort Chipewyan, Alberta, 11 November–5 December. http://politicalecologyofalberta.pbwiki.com/f/Timoney%20report%20Nov2007.pdf.

Venne, Sharon. c. 1992. *Lubicon Cree Nation submission to the Royal Commission on Aboriginal Peoples/Commission royale sur les peuples autochtones.* Ottawa: Royal Commission on Aboriginal Peoples, 1992–93.

Weber, Bob. 2006. 'Band Skeptical of Study Showing Normal Cancer Rate in Village.' *Native Journal*, August. http://www.nativejournal.ca/pages/2006.08.Health.html.

Wingrove, Josh. 2011. 'Professor Quits Oil-Sands Panel over Strict Confidentiality Requirements.' *Globe and Mail*, 2 February.

4 The Ecological and Political Landscapes of Alberta's Hydrocarbon Economy

MICHAEL S. QUINN, SHELLEY M. ALEXANDER, STEVEN A. KENNETT, BRAD STELFOX, MARY-ELLEN TYLER, NICKIE VLAVIANOS, MONIQUE PASSELAC-ROSS, DANAH DUKE, NOAH PURVES-SMITH

This chapter provides an interdisciplinary ecological perspective on the costs and risks created by Alberta's extractive economy, with particular attention to the effects of petroleum energy sector development on the provincial landscape.[1] The work we summarize here focuses on landscape change and habitat loss, as well as health risks for humans associated with air and water pollution. However, we do more than survey environmental and health issues in this chapter; we also want to link these to the political-institutional factors that perpetuate an unsustainable trajectory. As Carter and Zalik point out in this volume, the specific institutions and processes shaping Alberta's 'rentier' or 'petro' state characteristics need to be identified in order to inform movements for political change. In this chapter we make a contribution to this endeavour by examining the political and institutional framework that regulates energy sector land use and its environmental impacts. In so doing, we are able to draw on a substantial body of work produced by environmental scientists and other experts who have first-hand experience with environmental approval and other regulatory processes.

Alberta's energy industry has recently undergone a major expansion, and industry players expect extraction of unconventional sources of oil and gas (e.g., oil sands and low permeability reservoirs such as tight gas, shale gas, coal bed methane, and gas hydrates) to continue to grow as conventional reserves decline (IHS Global Insight 2009). Energy sector companies compete for space on a land base that is subject to increasing human demands from a multitude of industrial, agricultural, residential and recreational land uses. The ability of that land base to support these competing land uses, while sustaining the province's diverse natural ecosystems, is critically important when considering energy futures and alternatives for Alberta.

While political and social norms, values, or regulations drive ecological change on a landscape, the failures of citizens, regulators, and corporate actors to change directions may also lie in a lack of ability to 'imagine' or 'see' the long-term consequences of current growth patterns. For this reason we conducted our analysis in a spatial landscape change simulation model, based on a program called ALCES® (A Landscape Cumulative Effects Simulator; see www.alces.ca). The approach allowed us to examine how integrated and interdisciplinary research can contribute to understanding landscape change as a result of energy development in Alberta. ALCES® produces maps and statistical tabulations that can illustrate how resources extraction has changed the natural landscape. The software allows for modelling future scenarios, based on current rates of change, and cumulative effects analysis by predicting the effects of multiple land uses in time and space. The observed changes can be directly related to how certain indicators of ecological integrity will respond, which provides evidence of direct ecological consequences (such as changes in populations of sensitive wildlife species) or changes in ecological and landscape processes. While our indicators relate the effects of change to specific focal wildlife species and environmental processes, these measures are indicative of the consequences and risks that are posed to humans. For example, an increase in fragmentation as a result of new pipelines has ecological consequences for species trying to navigate a landscape while simultaneously posing a greater risk to humans of contamination or exposure to hazardous byproducts of hydrocarbon development. Lastly, we examined the political and social context that contributes to current resource development approaches.

Understanding Landscape Change

Understanding the relationship between energy futures and landscape change begins with Alberta's endowment of hydrocarbon reserves, including conventional oil and gas, bitumen in oil sands, and unconventional or 'tight' gas. We consolidated available historic-to-current maps of hydrocarbon resources that extend under much of Alberta to show the growth and spatial distribution of energy developments (e.g., oil wells and pipeline). These maps were then transferred to ALCES® in order to 'backcast' and then estimate probable landscape change. The analyses summarized here report on data up to the year 2010 and estimate growth trajectories for fifty years (i.e., to 2060). The projections into the future are based on historical rates of change and current information about planned activities. In some cases, we present future projection based on 'business as usual' scenarios compared to the implementation of potential 'best management practices.'

Geospatial Analysis: The Energy Sector's Footprint on a 'Multiple-Use' Landscape

The spatial distribution and intensity of the energy sector's current and future footprint on the surface of Alberta reflect the province's subsurface geology. Alberta's proven reserve volumes for conventional oil and gas are now declining, reflecting a pattern where annual extractions consistently exceed new discoveries. This reduction in reserves does not, however, translate into slower growth for the energy sector's footprint (i.e., total area affected by seismic lines, well sites, pipelines and associated infrastructure). In fact, this footprint is likely to increase in both extent and intensity as energy companies find it progressively more difficult to locate and produce viable reserves. More intensive exploration (e.g., three-dimensional seismic) and drilling will be necessary to extract the remaining hydrocarbons as reserves become more scarce and fragmented.

Fluctuation in the annual drilling rate is largely driven by market conditions. The oil and gas industry drilled a record 19,365 wells in Alberta in 2004. The ten-year average annual rate (2012) for new wells is 13,788/year. In addition, there is a lag in the reclamation of abandoned drilling sites. Between 1963 and 2012, there were just under 400,000 wells drilled in Alberta; 154,111 of these have been abandoned and 101,280 have been certified as reclaimed (Alberta Environment and Sustainable Resource Development 2013). These raw numbers provide an indication of the magnitude of expansion, but they do not convey much in terms of the profound landscape-scale changes – and the social-ecological implications of these changes – that result from energy development. Until decision makers have the capacity to 'see' the cumulative effects of energy development and other land uses, and are guided by ecological criteria and democratic processes, we can expect the environmental and social outcomes of future development to be unplanned, unmanaged, and undesirable.

The industry has also responded by shifting attention to non-conventional stocks, which has important implications for the energy sector's footprint. Low permeability reservoirs (e.g., tight gas, shale gas, coal bed methane, and gas hydrates) require a higher well density than most conventional gas production, and the relatively low pressures associated with these reserves means that wells and other infrastructure may be in place for longer periods of time. Hydraulic fracturing, a rapidly expanding technique to enhance hydrocarbon recovery, is also responsible for a growing footprint of petroleum extraction on the landscape. As other chapters in this book describe, there has also been significant growth of both open-pit mining and in situ recovery in the oil sands region. Landscape transformation resulting from oil sands mining is extensive, and reclamation techniques will likely be unable to recreate pre-existing boreal landscapes and ecosystems (Woynillowicz, Severson-Baker, and Raynolds 2005). In situ recovery also has significant surface impacts, given the network of injection and recovery wells, pipelines, roads and other infrastructure that is needed to extract and transport the resource. While the precise spatial configuration of the future energy footprint is difficult to predict, data are available to show how the footprint and number of facilities have increased. The maps presented in figures 4.1, 4.2, 4.3, and 4.4 provide an illustration of spatial location and distribution for energy sector activities. The associated time series graphs in figure 4.5 demonstrate that Alberta has witnessed exponential growth in energy activity on the landscape. Figure 4.6 illustrates the relative contributions of energy sector components to the total energy-related footprint.

The spatial configuration of Alberta's energy footprint is superimposed on a landscape that is being used for many other (often competing) purposes, most of which also create a unique physical footprint. Agriculture, forestry, transportation, urban development, rural residential development, and recreation are among the principal land uses that will contribute to shaping Alberta's future landscapes. A detailed review of the entire 'multiple-use' footprint is beyond the scope of this chapter, but the expansion over time can be visualized through the province's road network, which ties together human land uses at the landscape scale. Roads provide a reasonable indication of the overall human footprint because all human land uses require access and most uses rely on roads for access (figure 4.7).

In summary, the geospatial information summarized above indicates the extent and intensity of the energy sector's footprint across much of Alberta. Mapping the proliferation of well sites, pipelines, and roads is only the first stage in understanding how energy development has driven landscape-scale change and how future energy development will have a profound influence on Alberta's landscapes. The next stage

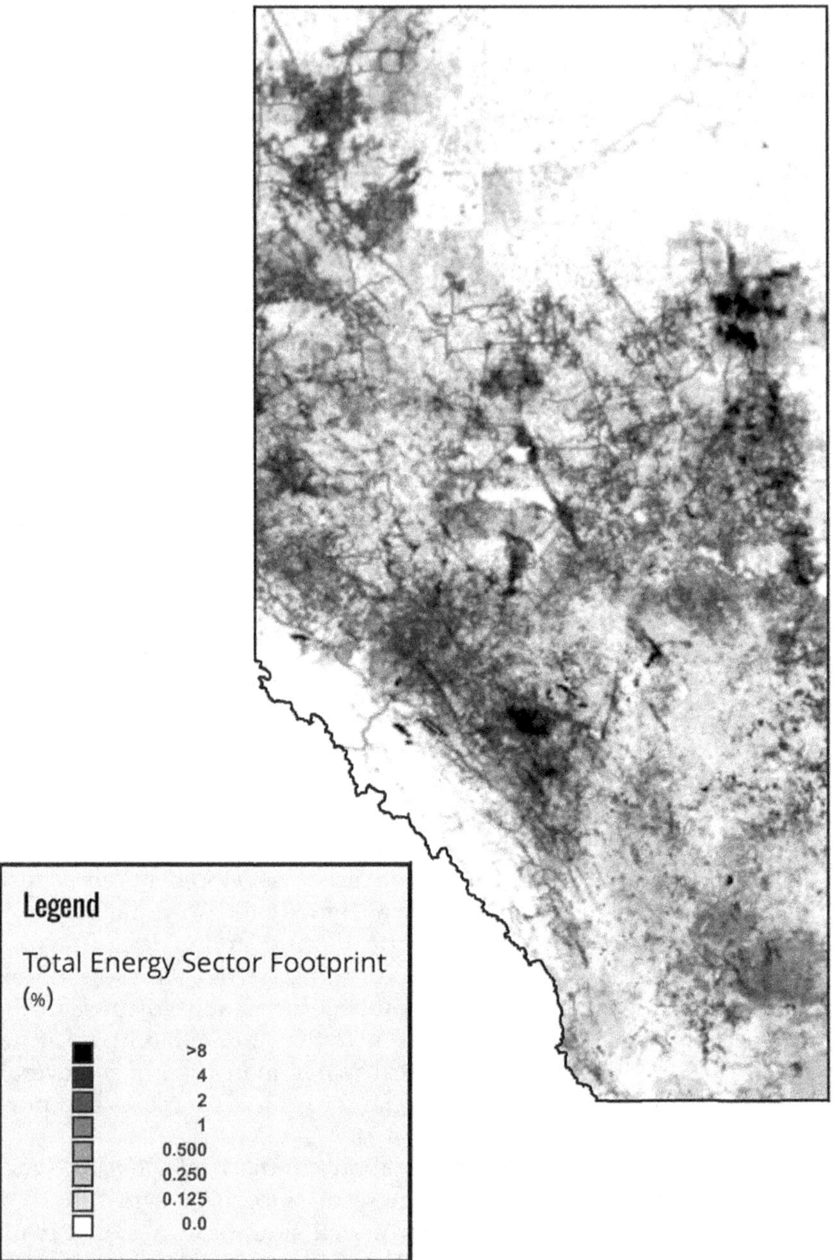

Figure 4.1. Map of Alberta showing the spatial distribution and intensity of the total energy sector footprint.

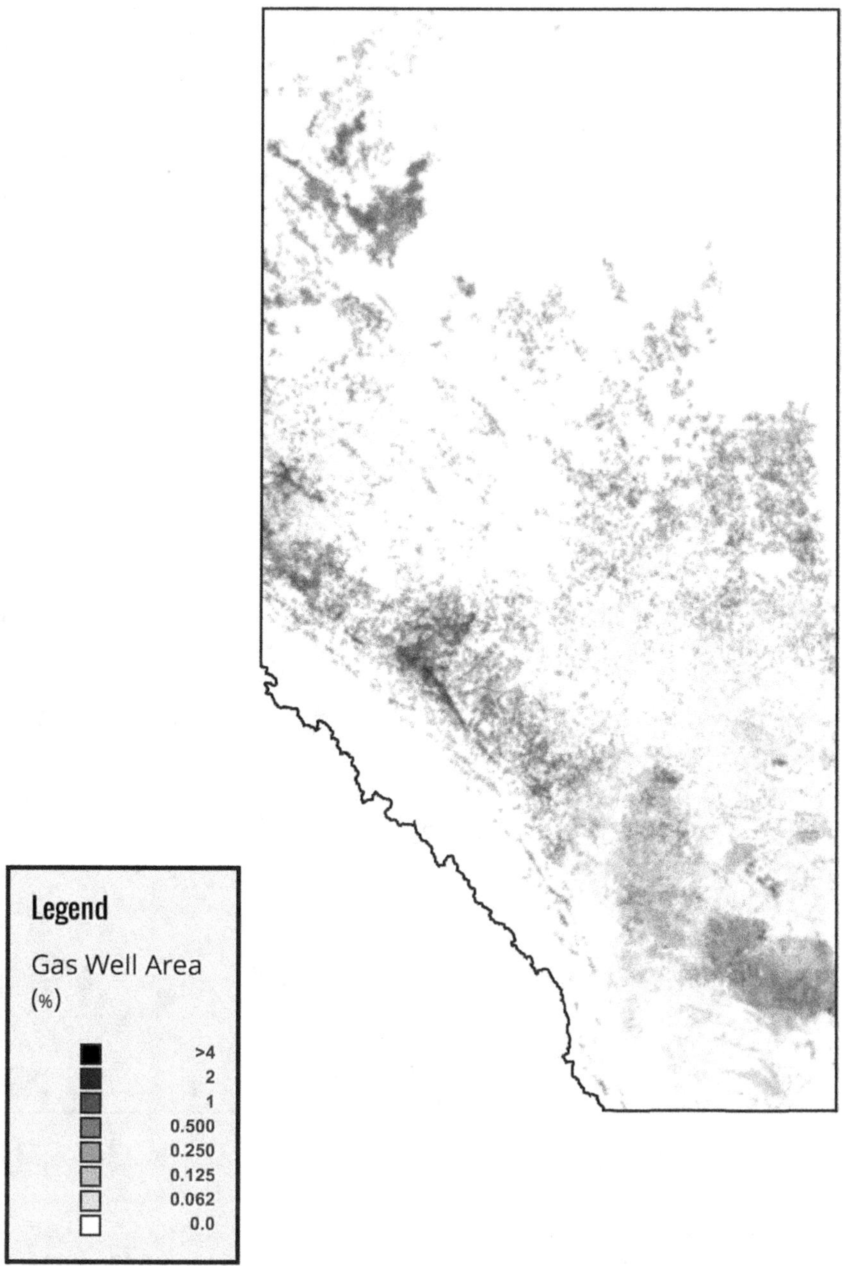

Figure 4.2. Map of Alberta showing the spatial distribution and intensity of natural gas wells.

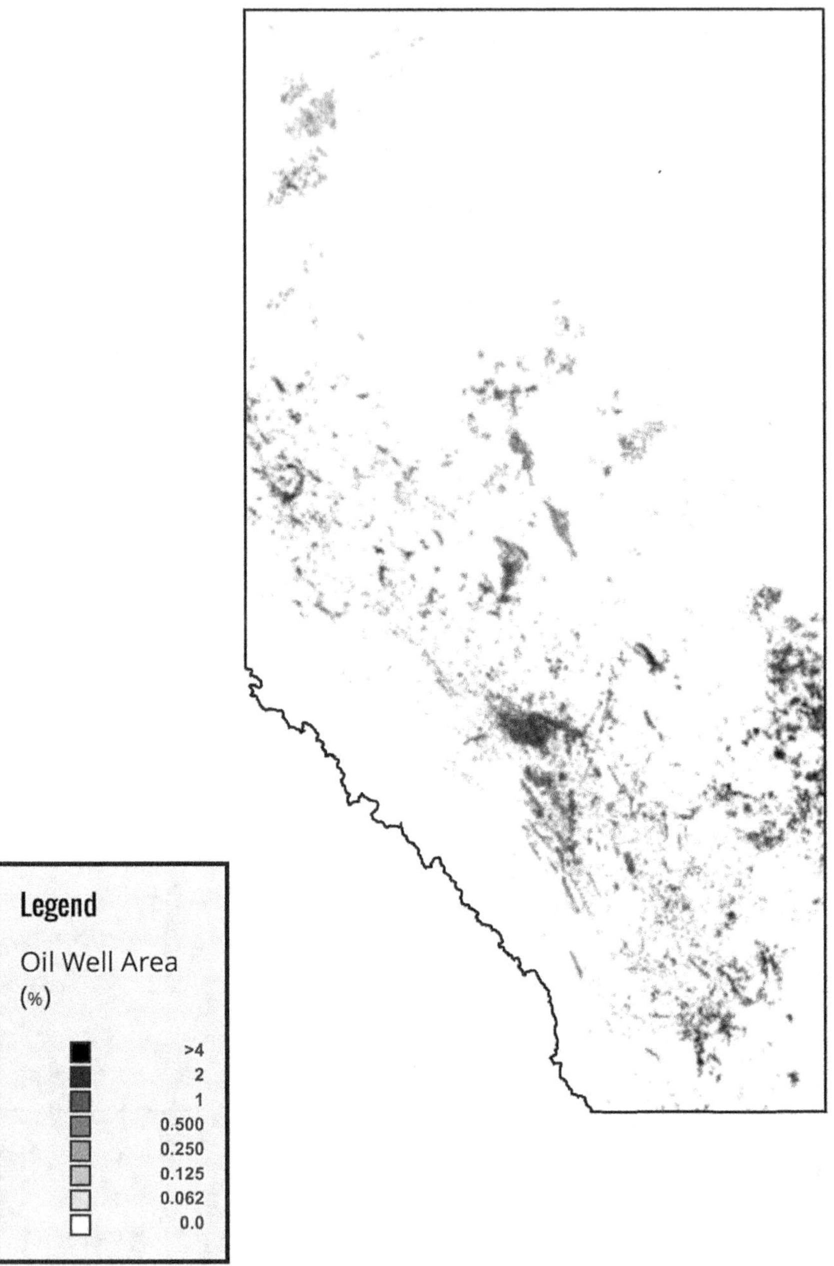

Figure 4.3. Map of Alberta showing the spatial distribution and intensity of oil wells.

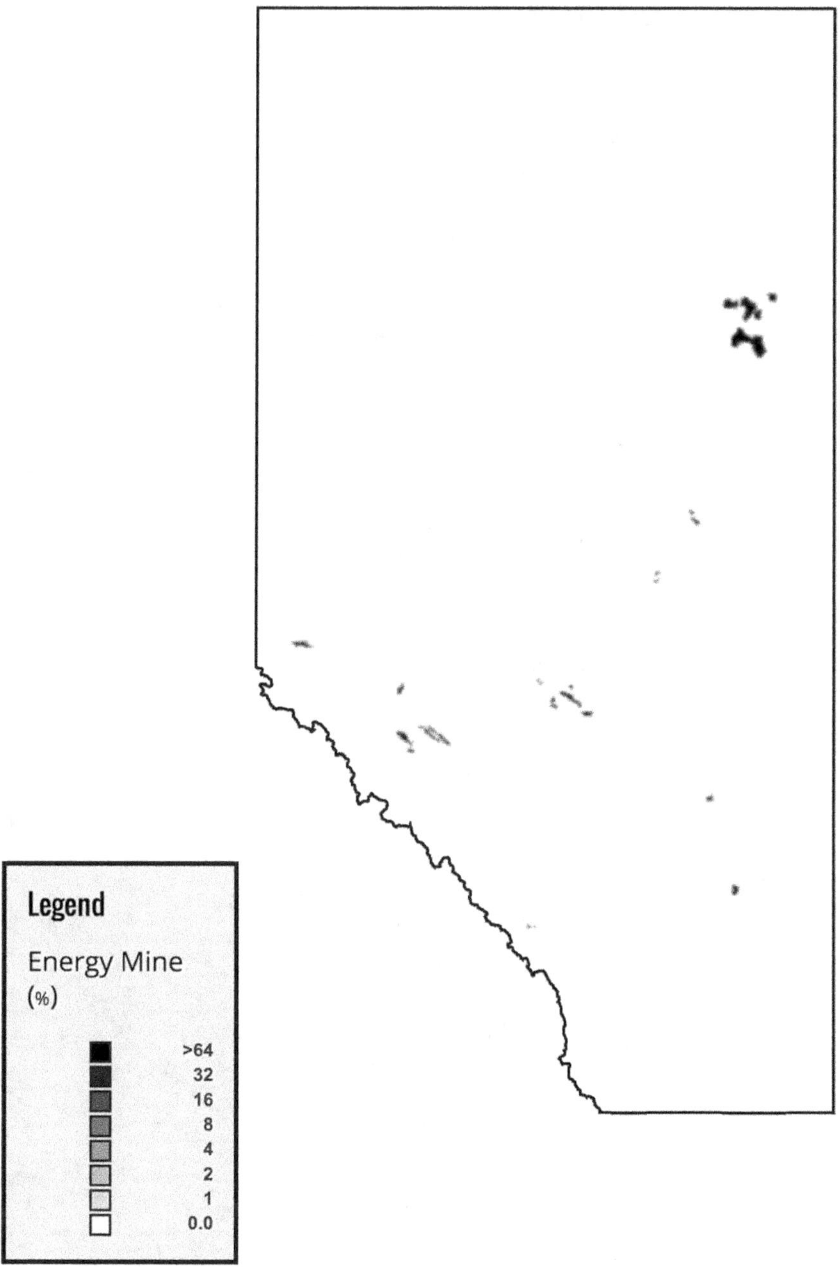

Figure 4.4. Map of Alberta showing the spatial distribution and intensity of energy mines (i.e., bitumen and coal).

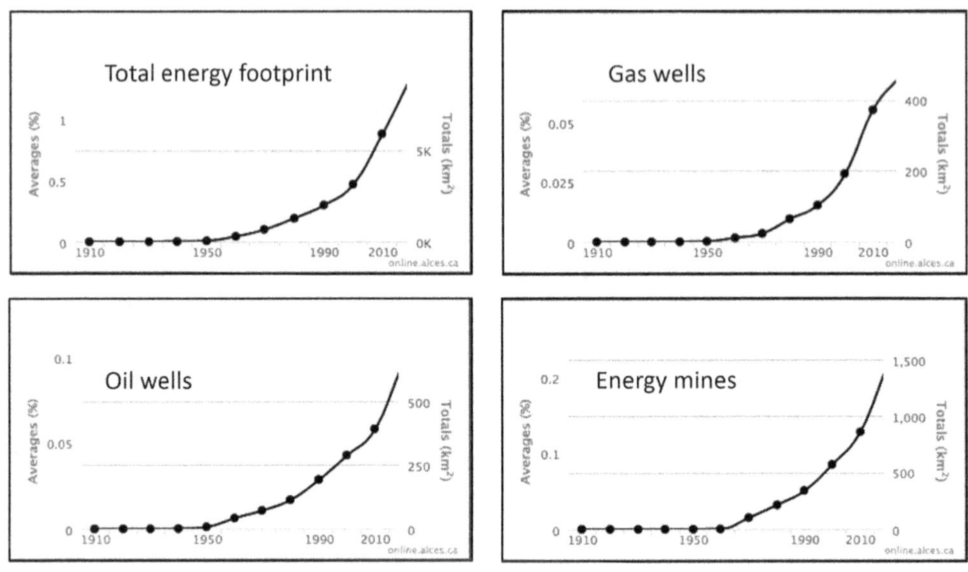

Figure 4.5. Growth rates of energy footprint area for Alberta, 1910–2010.

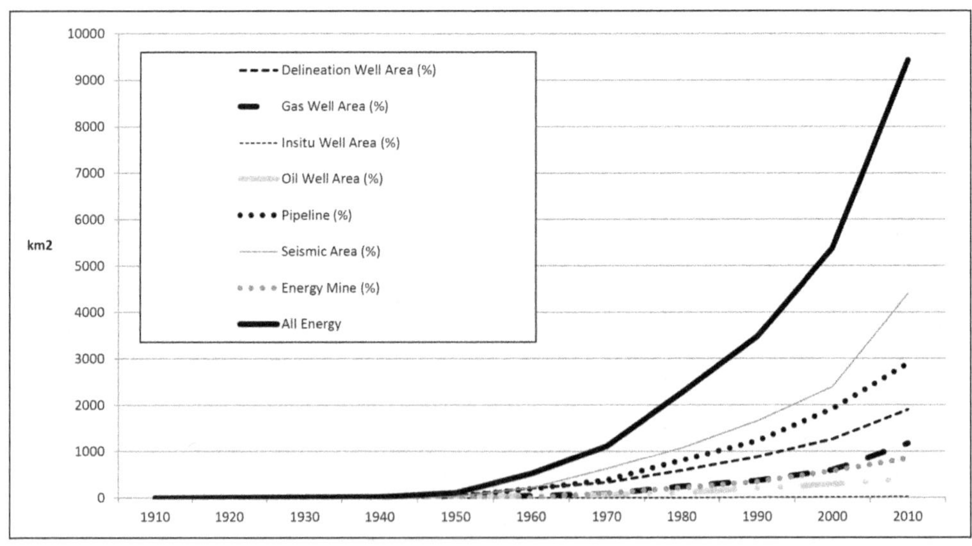

Figure 4.6. Relative contribution of different energy sector components to the total energy sector footprint in Alberta, 1910–2010.

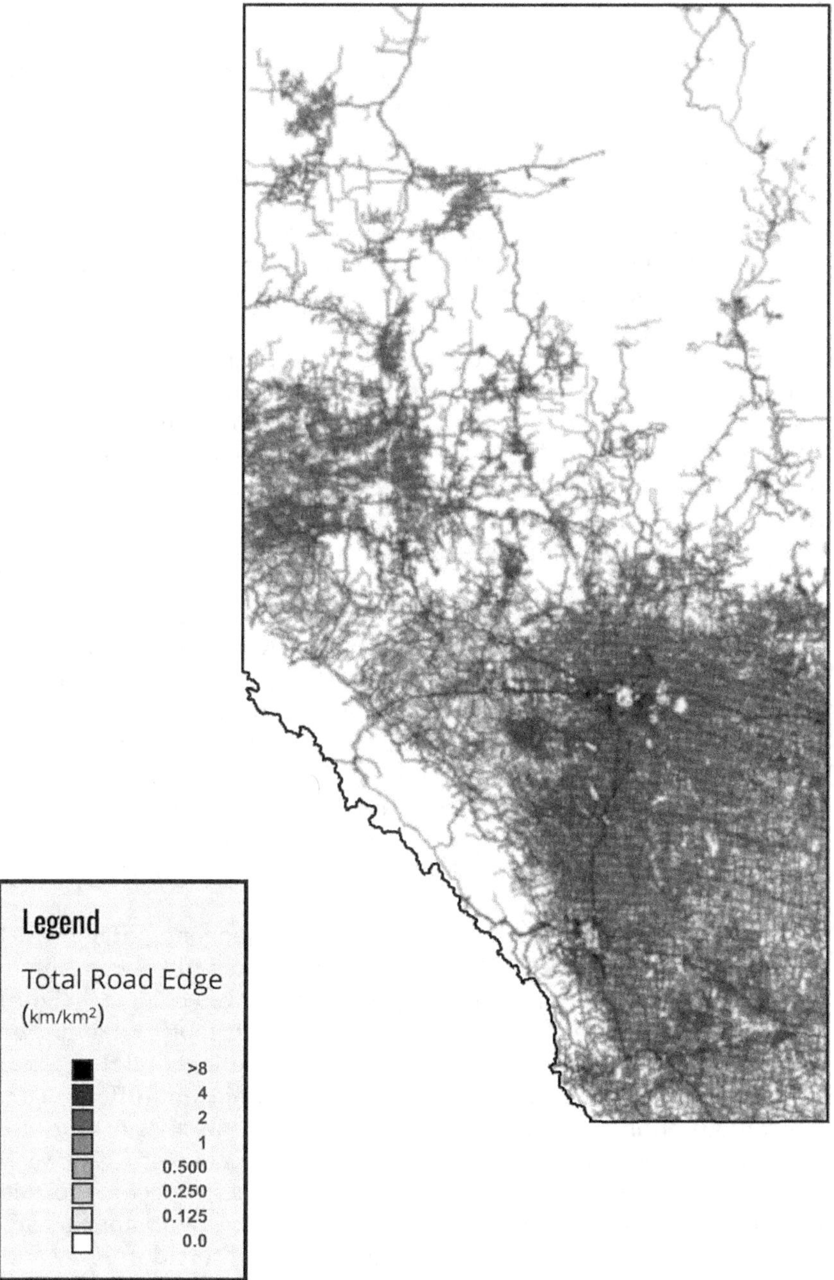

Figure 4.7. Map of Alberta showing the spatial distribution of roads (road edge).

is to move from the spatial visualization of the footprint to a measurement of changes in fundamental landscape characteristics.

Metrics of Landscape Change

The ecological consequences of the energy sector's expanding footprint can be quantified using landscape characteristics that are significant for ecological, social, cultural, and economic reasons. The landscape characteristics presented here were selected to capture those that are recognized as important to ecological science and conservation, are understandable to a broad audience, and demonstrate the absolute and relative contribution of the energy sector's footprint to landscape change in Alberta. We modelled effects at a provincial scale to 2060, which is an ecologically meaningful time frame. The disparity between meaningful economic versus ecological time frames is one of the key problems in developing sustainable economic and development practices.

The most readily observable landscape metric is the change in total footprint (area disturbed by human activity) over time. The total area used does not, however, capture all relevant information regarding the extent and impact of footprint. A key point when considering total area of disturbance is that the ecological effects of human activities may extend beyond the immediate physical footprint. For example, some wildlife species may be displaced from habitat adjacent to disturbed areas. For certain types of impacts, therefore, it is appropriate to apply a buffer area, the size of which depends on the type of disturbance, habitat type, and species of concern.

Second, any human disturbance creates anthropogenic 'edge,' which is the term used to describe the interface between natural and human land cover. Edge is a more meaningful measure of ecological significance than total area of disturbance (anthropogenic footprint) alone because it communicates information about the spatial configuration of the disturbance. The concept of edge reflects significant differences in the configuration of disturbance caused by different types of land use. While forestry and agriculture tend to create blocks of disturbance connected by a transportation infrastructure, the energy sector typically produces a multitude of much smaller disturbances with an extensive network of linear transportation and transmission corridors. Oil sands mining is an exception to this pattern within the petroleum sector. Figures 4.8, 4.9, and 4.10 illustrate the edge created by linear disturbances characteristic of the petroleum sector (seismic lines, pipelines, and roads). Figure 4.11 shows the growth of these features from 1910 to 2010.

Connectivity, the ability of organisms or ecological processes to flow unimpeded across the terrestrial and aquatic landscape, is the final metric we have chosen to include in our analysis. Terrestrial connectivity is fragmented when physical or psychological barriers are created that alter or prohibit movement (Stamps et al. 1987). Physical barriers can be a direct impediment to the movement of animals if attempts to cross result in complete avoidance or mortality. Linear disturbances, such as roads, constitute this kind of barrier for many species. Even decommissioned roads (such as those used in exploration or extraction of oil and gas) may result in the death of animals when used opportunistically by humans as access for legal or illegal hunting. There are many ecological consequences of barriers to connectivity, among them that adult animals may not be able to leave a patch to find necessary resources for survival or reproduction, and juvenile animals may not be able to disperse to new habitat, which will increase the pressure on existing patch resources, which in turn can result in changes to inter-specific and intra-specific interactions, altered predation rates, and higher extinction rates (Jackson 1999; Wilcox and Murphy 1985). Roads are now recognized as the single greatest mortality threat to wildlife; the vehicle-related mortality rate has been estimated to be several millions of animals per year (Malo et al. 2004). Impaired connectivity can also affect other ecological processes such as nutrient flows and hydrologic systems.

Aquatic connectivity allows stream-dependent species to migrate and reproduce throughout the hydrological system. Networks of roads are superimposed on an existing hydrological infrastructure, and these two networks interact in many complex ways. While larger stream channels are spanned by bridges, the majority of stream crossings involve the installation of culverts. Historically, the design and installation criteria for culverts have been dominated by hydraulic and economic efficiency; the objective was to maximize water conveyance while minimizing pipe size and cost. Until recently, little consideration was given to maintaining critical in-stream ecological conditions (Baker and Votapka 1990; Bates 2003). The installation of culverts has deleterious effects that may include direct habitat loss, changes to water quality, upstream and downstream channel effects, and impairment of ecological connectivity (Bates 2003). Maintaining aquatic connectivity

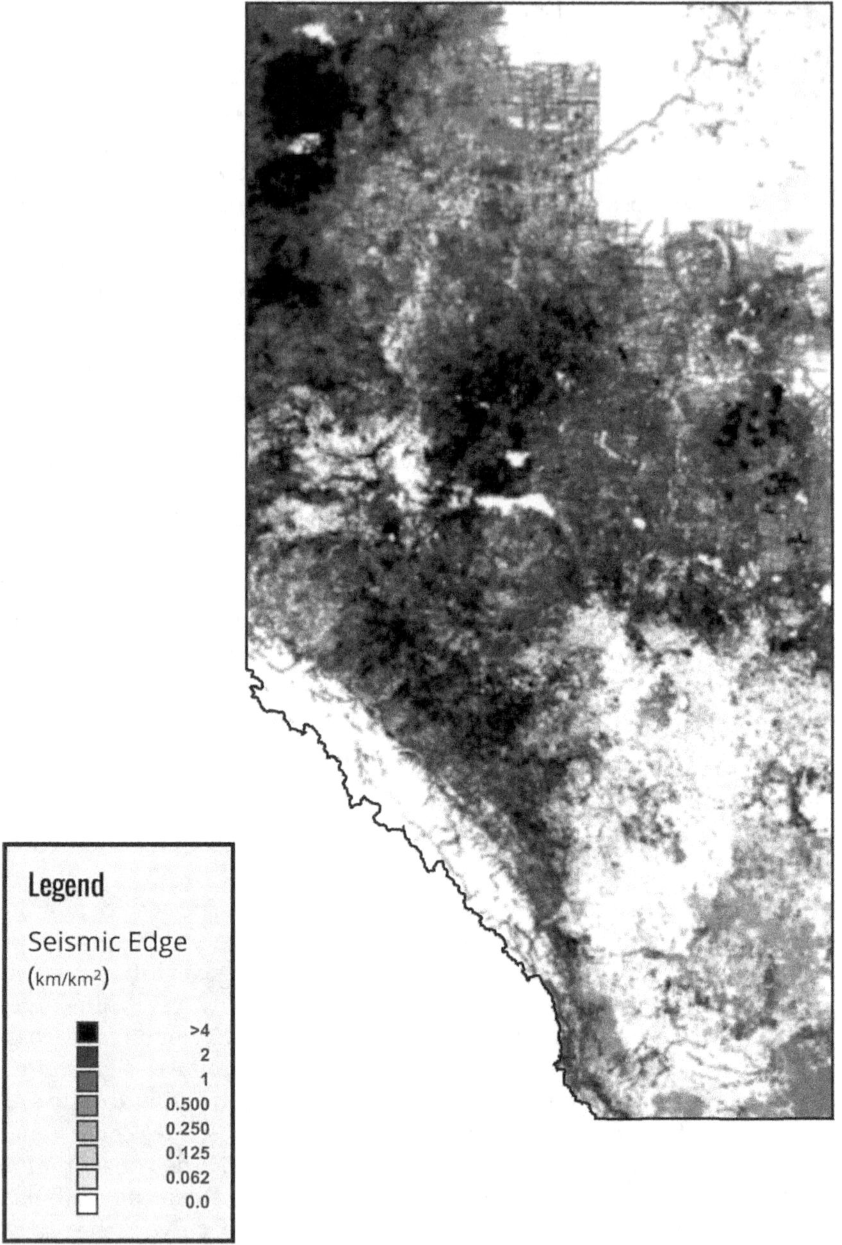

Figure 4.8. Map of Alberta showing the spatial distribution and intensity of edge created by seismic lines.

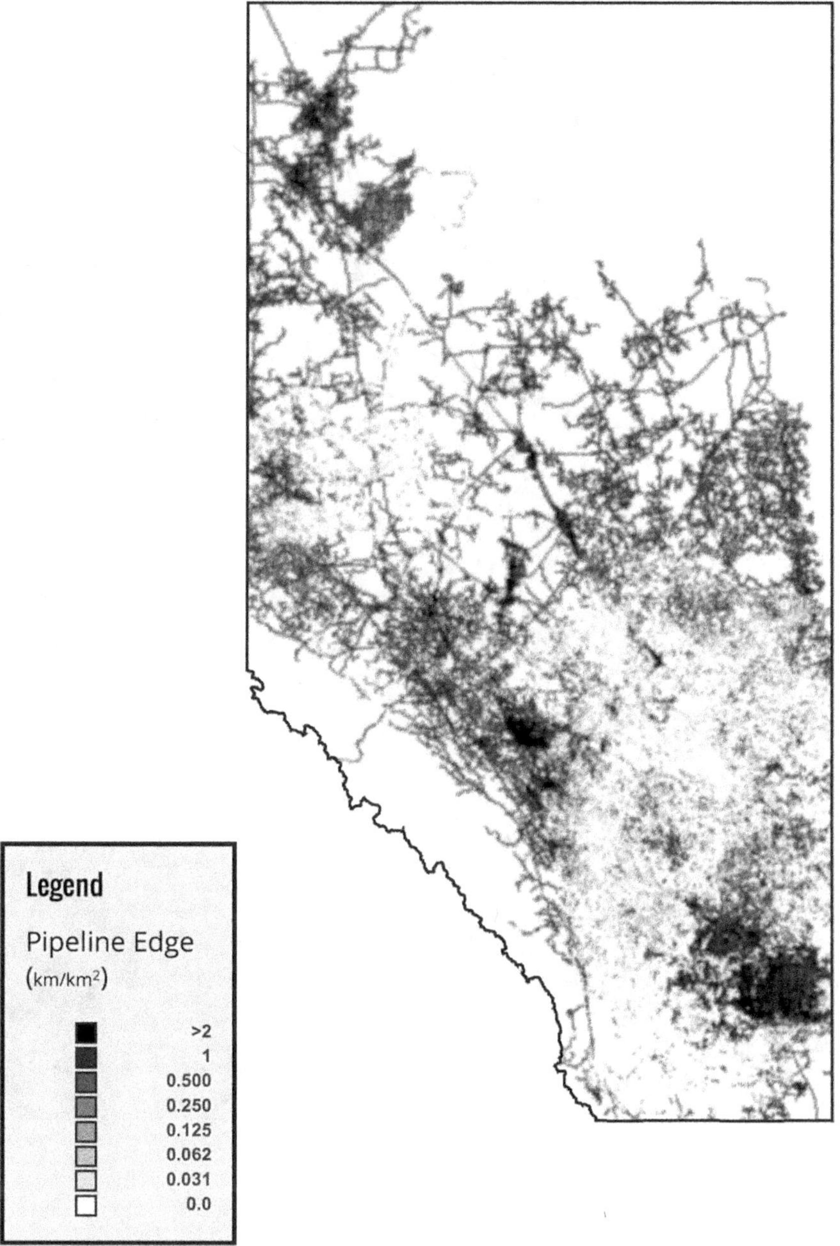

Figure 4.9. Map of Alberta showing the spatial distribution and intensity of edge created by pipelines.

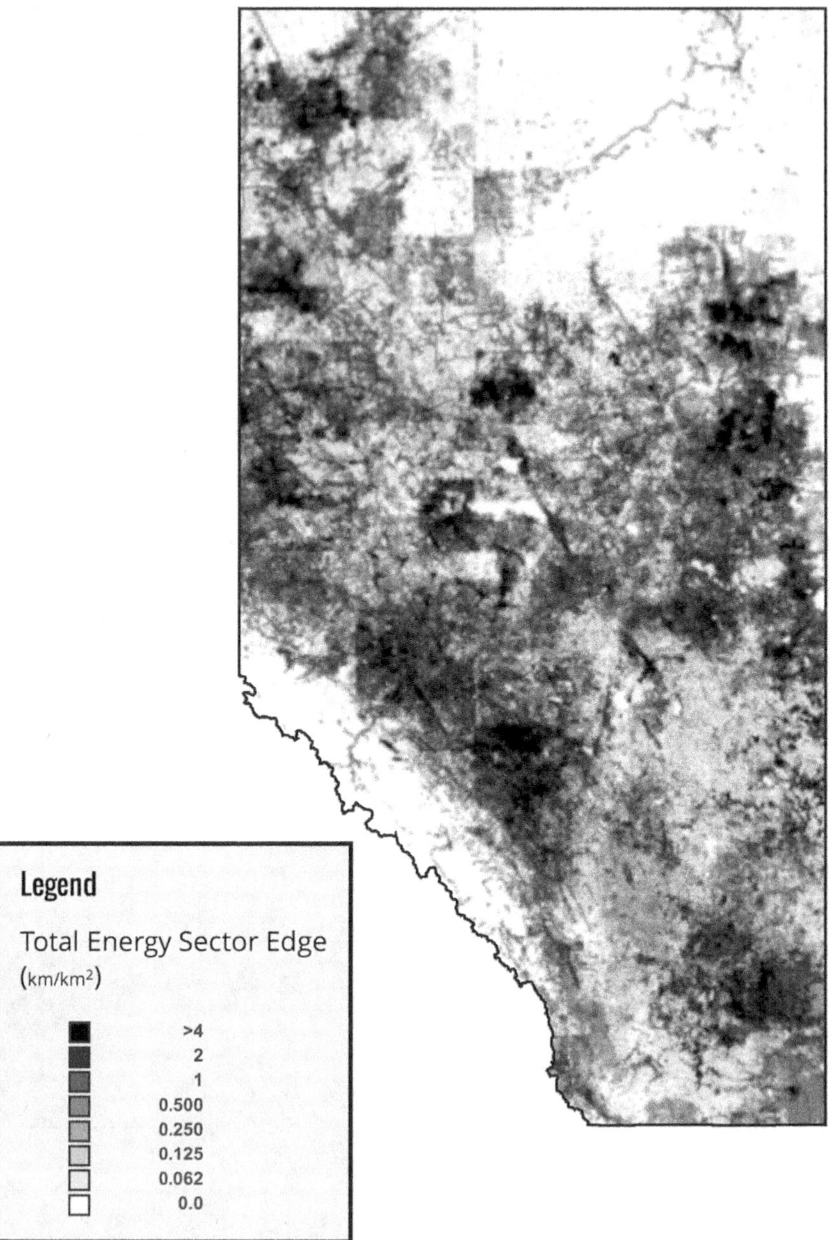

Figure 4.10. Map of Alberta showing the spatial distribution and intensity of edge created by all energy land use.

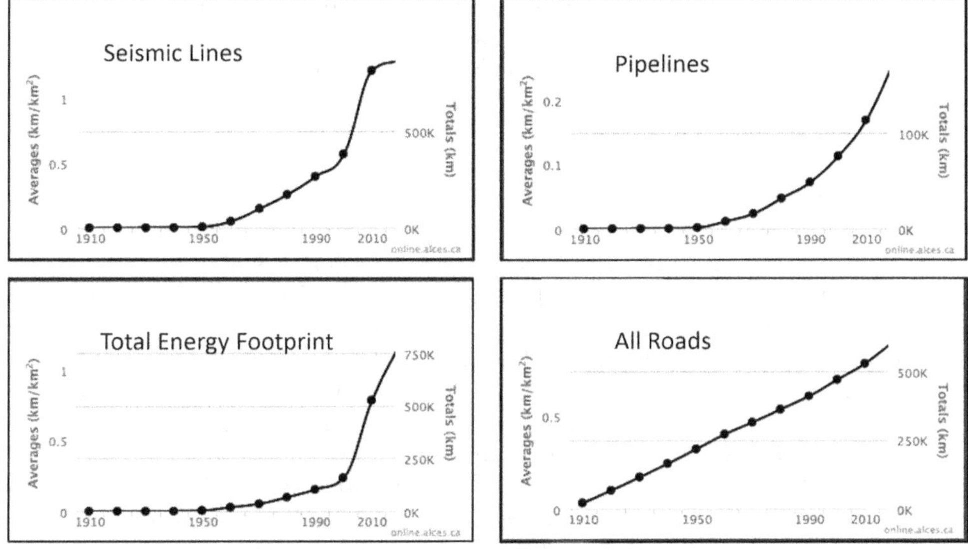

Figure 4.11. Growth rates of edge associated with seismic lines, pipelines, all energy infrastructure, and roads in Alberta, 1910–2010.

is critical to conserving the distribution and abundance of stream fish assemblages (Rieman and McIntyre 1993). For example, Eaglin and Hubert (1993) found that the biomass of brook trout (*Salvelinus fontinalis*) and brown trout (*Salmo trutta*) was negatively correlated with the number of culverts/km^2 within a drainage area, leading the researchers to conclude that culverts significantly reduce stream productivity through harmful alteration or fragmentation of habitat. Reductions in connectivity impede fish movements, alter fish community structure, and likely threaten population viability (Morita and Yokota 2002). Culverts generally have a diameter significantly less than the maximum natural stream channel and consequently alter the stream flow characteristics both through and downstream of the culvert. For instance, the increased water flow velocity associated with a culvert results in an increased resistance or barrier to upstream travel by aquatic species (Warren and Pardew 1998). Increased velocity raises the erosive potential downstream of the culvert. Such changes can have detrimental effects on the connectivity of lotic ecosystems through the creation of barriers to fish passage. These barriers may be complete, seasonal, or species/lifestage specific. Thus, improper design and maintenance of culverts can result in serious habitat discontinuities for aquatic species (Belford and Gould 1989; Thormann et al. 2004). The barrier effect here is functionally analogous to the situation described above for terrestrial species.

Indicators and Analysis of Ecological Impacts

Metrics for each of the three landscape characteristics above (anthropogenic footprint, anthropogenic edge, and landscape/aquatic connectivity) provide measures for assessing impacts associated with the exploration and development of petroleum resources. While Kennett et al. (2006) described the effects of development on three indicators (fragmentation of aquatic habitat by hanging culverts, caribou population response, and grizzly bear exposure index), the scope of this chapter is limited to a single 'focal species,' woodland caribou. The focal species approach is a 'reductionist' strategy predicated on the assumption that the species one chooses to study is a valid surrogate to understand how human disturbance affects multiple wildlife populations (Caro and O'Doherty 1999; Lambeck 1997; Landres, Verner, and Thomas 1988). For instance, a species whose habitat requirements are large and encompass those of other species is often selected as focal species. The criteria that define species' sensitivity are based on life history traits, including age of sexual maturity, number of offspring, niche requirements, and home range size, among many others (Weaver, Paquet, and Ruggiero 1996). Woodland caribou provide a good example of the complexity and interaction of landscape changes on valued ecosystem components.

Fragmentation and Focal Species Impacts

Our ALCES® modelling indicates that the total direct physical footprint of the energy sector could increase from 9,424 km^2 (~1.4 per cent of provincial area) to approximately12,000 km^2 (~1.8 per cent of provincial area) by 2060 (figure 4.12). Although the area of this footprint may seem insignificant, and it is less than that of other major land uses (e.g., forestry, agriculture), it is the spatial distribution and ecological implications of the change that make this value significant (figures 4.13 and 4.14). The energy sector's absolute change in footprint should be viewed in conjunction with other land uses in order to better under-

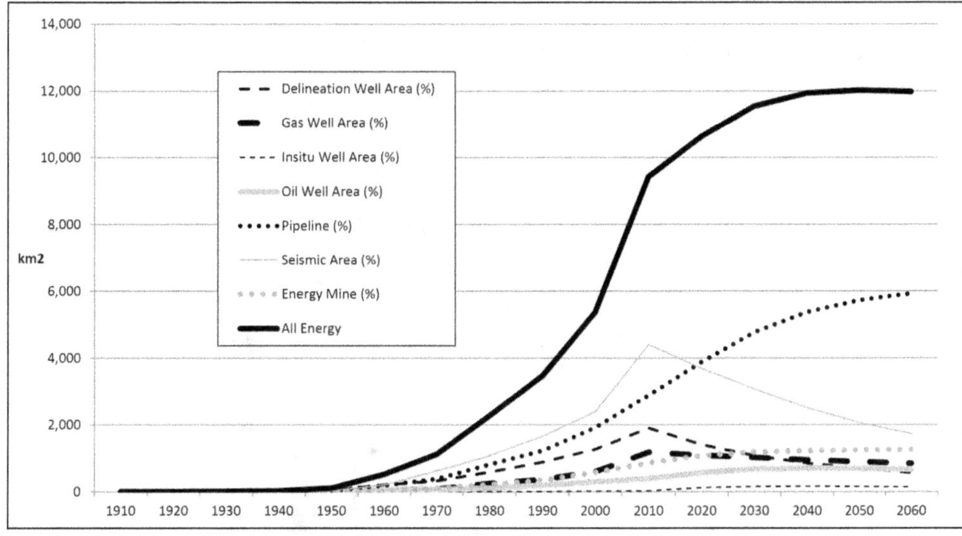

Figure 4.12. Relative contribution of different energy sector components to the total energy sector footprint in Alberta from 1910 to 2010 and projected to 2060.

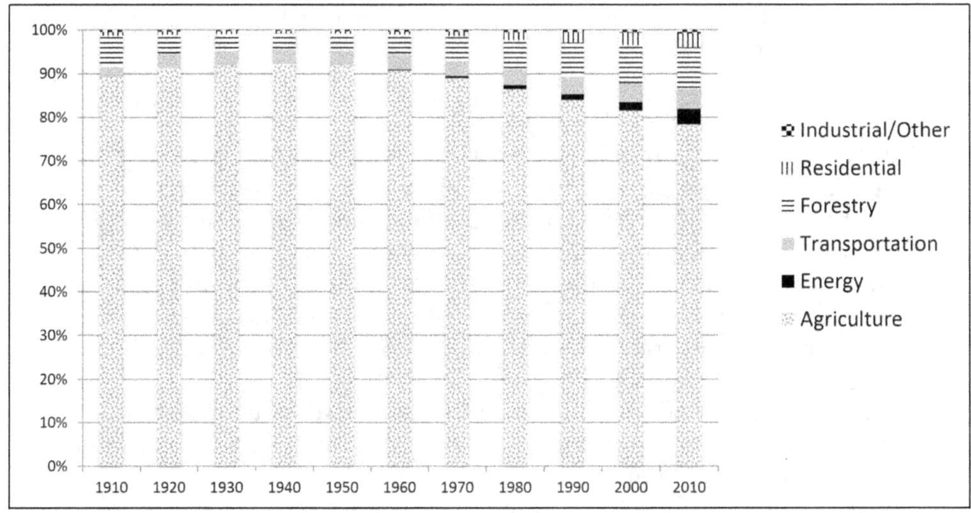

Figure 4.13. Total footprint area of energy sector compared to other major land uses in Alberta, 1910–2010.

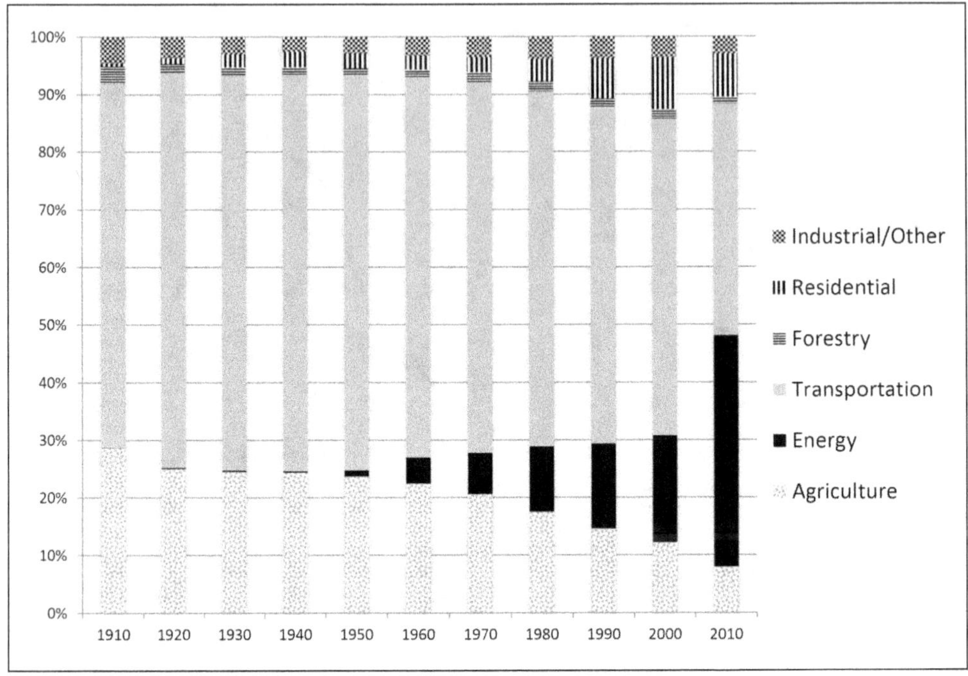

Figure 4.14. Total edge created by energy sector compared to other major land uses in Alberta, 1910–2010.

stand its contribution to landscape-scale change. Put another way, the significance of the growth in physical footprint of the energy sector must be evaluated by considering the cumulative effects of developments in other areas of land use. There are a variety of 'best management practices' that could be implemented in the future to lessen the growth of the energy sector footprint. We modelled the effect of these best management practices to assess the effect of future land disturbance (figures 4.15 and 4.16).

The dispersion of linear disturbances and small patch size disturbances associated with the energy sector results in the creation of more edge than all other major land uses combined – 1.78 million km (more than 44.5 times around the circumference of the earth). This magnitude of disturbance equates to an average of approximately 2.7 km of edge/ km^2 of the provincial area. As indicated above, the fragmentation ef-

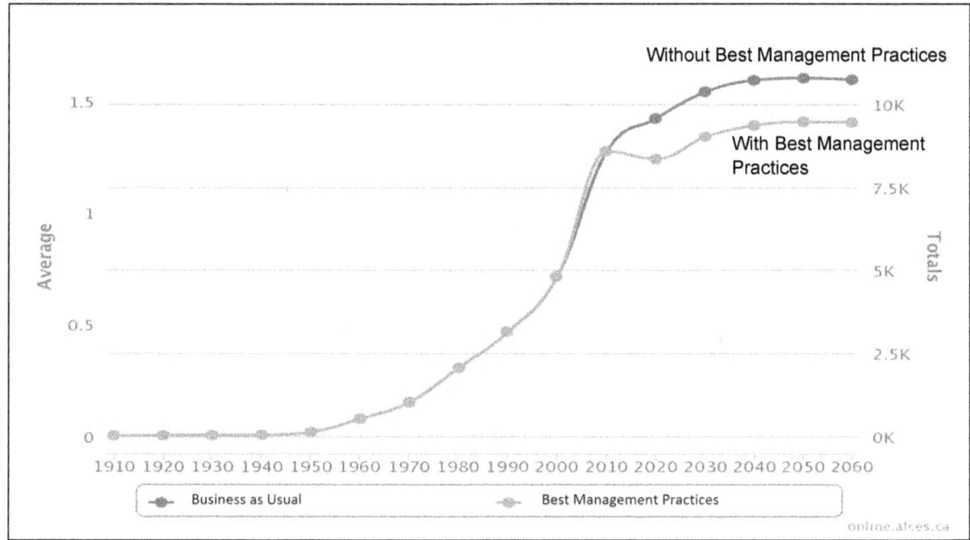

Figure 4.15. Total area of energy sector footprint from 1910 to 2010 and projected to 2060 comparing 'business as usual' to implementation of 'best management practices.'

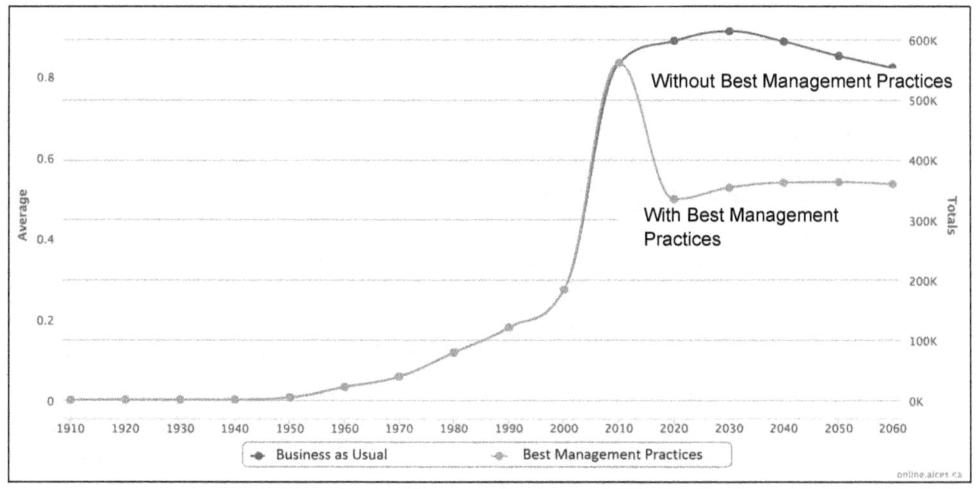

Figure 4.16. Total edge of energy sector footprint from 1910 to 2010 and projected to 2060 comparing 'business as usual' to implementation of 'best management practices.'

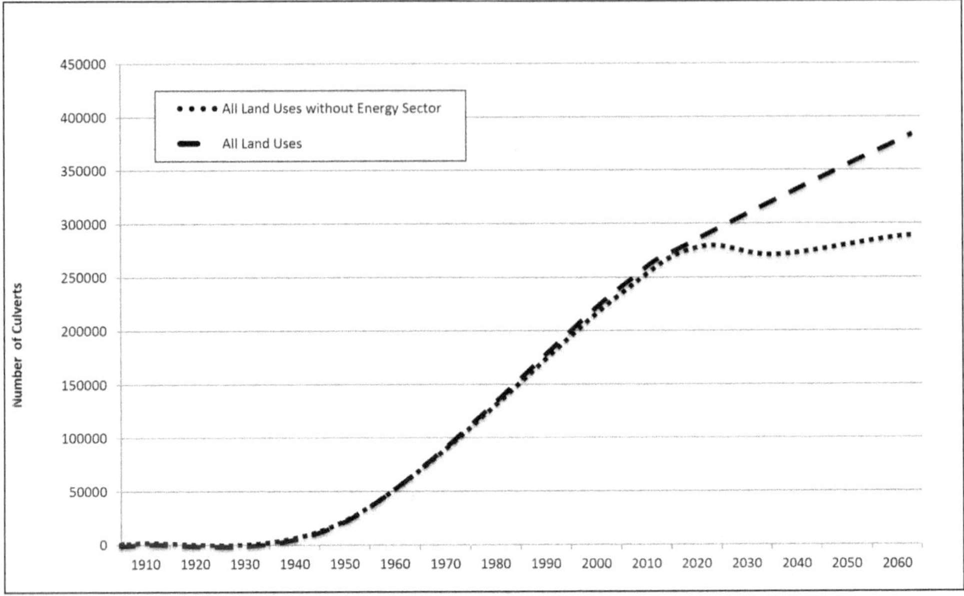

Figure 4.17. Growth in the number of culverts in Alberta from 1910 to 2010 and projected to 2060 with and without the energy sector.

fects of edge creation are highly detrimental to many species, especially those requiring interior forest habitat conditions.

Our modelling demonstrates the connectivity effects arising from both terrestrial and aquatic barriers (due to roads and culverts respectively). In 2010, there were approximately 532,400 km of roads in Alberta, and our modelling suggests that this number could increase to over 700,000 km by 2060. The linear disturbance of roads in 2010 corresponds to a direct footprint of 7,683 km^2. The total number of culverts associated with existing roads in 2010 was 319,435 and is projected to increase to 380,000 by 2060 (figure 4.17). The culverts are not all on fish-bearing streams, but the value provides an indication of the scale of the issue. We estimate that there are currently over 90,000 hanging culverts and, if current practice continues, this number will increase to over 100,000 by 2060. On average, the current situation results in a lotic discontinuity every 6 km. The significance of this value in fragmenting the habitat of fish and other aquatic biota cannot be overstated.

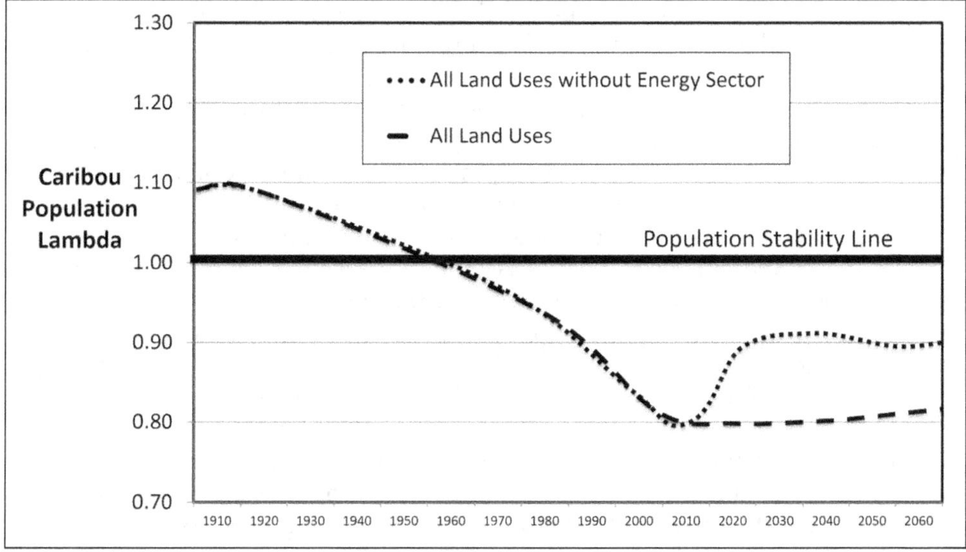

Figure 4.18. Decline in boreal caribou population performance (lambda) in Alberta from 1910 to 2010 and projected to 2060 with and without the energy sector.

Focal Species Consequences: Woodland (or Boreal) Caribou

The woodland caribou (*Rangifer tarandus caribou*) is found in the forested boreal and mountain regions of Alberta. The southern mountain and boreal populations are considered 'threatened' by the Committee on the Status of Endangered Wildlife in Canada (COSEWIC 2002) and 'threatened' under the Alberta Wildlife Act (Dzus 2001). The species has specific habitat requirements, low reproductive potential, and displays sensitivity to human disturbance. Our model results (figure 4.18) showed that woodland caribou populations are negatively affected by linear disturbances and associated human use (Dyer et al. 2001; Dzus 2001). Our simulations of caribou population dynamics, based on equations developed by the Alberta Caribou Research Committee, suggest that the viability of populations has declined significantly during the past century. The primary landscape variables correlated to this decline include the network of linear features (e.g., roads, transmission lines, seismic line) and adverse changes in forest age class structure; in particular, the loss of continuous old growth forest patches has been detrimental.

Most, but not all, of the linear features created historically in the boreal forest were created by the energy sector (followed by the forestry sector). As seen in the forecasting portion of the simulation, the further activities of the hydrocarbon sector will ensure that viability of caribou populations will remain low, and may very well lead to the extinction of most remaining boreal herds. Although the exact number of caribou in Alberta is unknown, experts concur that populations have been in decline since 1900, and future land use projections show little hope for a change in this trend (Dzus 2001). Woodland caribou naturally exist at low densities and have relatively low reproductive rates compared with other ungulates. These characteristics result in a weak ability to respond to population declines. The vulnerability of this species is particularly significant when one considers the magnitude and extent of landscape change within Alberta's forests. Applying the reported value of a 250 metre disturbance buffer to the average edge density (2 km/km^2) projected for the energy sector results in 100 per cent of the landscape being negatively affected for caribou use. This value does not take into account the cumulative effects of other land uses and synergistic effects.

The mechanisms for population effects include direct mortality, increased susceptibility to poaching, altered predator-prey dynamics, and habitat avoidance. Direct mortality as a result of vehicle collisions has been documented as a significant effect on Highway 40 in west-central Alberta. For example, thirty-two mortalities were recorded in the winters of 1991/92 and 1992/93 (Brown and Ross 1994) from an estimated total population of 150 to 200 caribou (Brown and Hobson 1998). Roads and seismic lines also provide increased potential for human access and illegal harvest. Linear disturbance and well pads have been shown to alter the movement and habitat-use patterns of both predators and prey. Caribou (and other ungulates) may be attracted to the vegetation along disturbed corridors and patches when human activity is not present. Wolves are known to preferentially select linear corridors as travel paths as they are more efficient travel routes (James and Stuart-Smith 2000; Thurber, Peterson, and Drummer 1994). This combination increases the chance of encounter between caribou and wolves.

Social, Health, Cultural, and Economic Impacts[2]

Alberta's Energy and Utilities Board (EUB) commented more than ten years ago that 'disputes between residents and petroleum companies seem to be increasing in number and intensity' (EUB 1999, p. 10).[3] In many instances the underlying concerns relate to cumulative environmental effects and to broader patterns of land use and landscape change (Kennett and Wenig 2005). Our objective here is to highlight some of the principal areas where the energy sector's development footprint is associated with impacts on other land users and values, including concerns with health risks and effects, impacts on culture or way of life, impacts on Aboriginal peoples, and economic effects and risks for Alberta's forestry sector.

Concerns with Health Risks

As Garvin's chapter in this volume documents, Albertans are increasingly concerned about the possible links between environmental effects and their health and safety. Air pollution from accidental blow-outs, gas flaring and venting, gas-processing plants, and oil sands refineries are sources of concern (Petroleum Communication Foundation 1997). The contamination of soil and water from pipeline failures, well sites, holding sites, and processing sites is another source of concern. With the proliferation of oil and gas facilities across Alberta's landscapes, these issues are receiving increased attention from landowners and other stakeholders. Work by Marr-Laing and Severson-Baker (1999) examined the air, surface, and groundwater impacts of oil and gas activities in Alberta and summarized familiar pollutants that continue to pose significant harm to human health, including:

- acute exposure to high concentrations of sulphur dioxide can irritate the upper respiratory tract and increase susceptibility to respiratory infections; long term exposure may increase the risk of developing chronic respiratory disease;
- volatile organic compounds (including compounds such as benzene) are known carcinogens and are toxic to humans;
- ground-level ozone causes adverse effects on humans, including irritation of the eyes, nose, and throat, reduced lung function, and the development of chronic respiratory disease;

- fine particulate matter has been linked to respiratory and cardiac disease because it can penetrate into the lungs and have serious effects on respiratory function; and
- air toxics such as benzene, styrene, and tolene are known carcinogens.

Effects on Culture and Way of Life

A significant number of Aboriginal peoples in Alberta live on reserves, and many still strive to maintain traditional land-based activities on what they consider to be their traditional lands (Kennett et al. 2006). The treaties that the British Crown entered into with various Indian bands in the nineteenth century guaranteed that, in exchange for the surrender of their lands, the Indians' rights to hunt, trap, and fish (in essence, their traditional livelihood) would be protected, subject only to certain limitations (Parlee, this volume). However, Aboriginal land-based rights and uses can only be realized if healthy ecosystems supporting healthy populations of wildlife are maintained.

In the Fort McMurray area, the impacts of oil sands developments on the traditional territories of Aboriginal peoples are substantial. For some communities, the opportunities to hunt, trap, and fish have vanished altogether. The health impacts are only now beginning to be understood and addressed. In the northwest corner of Alberta, the Dene Tha' First Nation have long experienced the impacts of oil and gas development and forestry on their traditional lands. The Dene Tha' took the federal government to court in 2005, alleging that they had not been adequately consulted on the environmental and regulatory review of the proposed Mackenzie Valley Pipeline, a portion of which goes across their traditional territory. They pointed out that their traditional lands have already been negatively impacted by thousands of oil and gas wells; thousands of kilometres of seismic lines, pipelines, and roads; the activities associated with this oil and gas development; and extensive timber harvesting. This resource development has made it difficult, and in some areas impossible, to exercise their rights and to maintain their way of life and culture (Kennett et al. 2006). In November 2006 the Federal Court of Canada ruled that the federal government had failed to consult Alberta's Dene Tha' about the proposed Mackenzie Gas Project, and ordered the Joint Review Panel to stop its hearings.[4]

Moreover, many other rural Albertans are concerned about the effects of oil and gas development upon their way of life, including their ability to make a living from land-based activities such as farming and ranching. There are stories emerging of fifth-generation farmers and ranchers worrying that oil and gas development is significantly reducing their ability to farm (Gregory 2006). There are three main reasons for concern: the disturbance to the landscape from oil and gas activities; the actual (or potential) contamination of air, soil, and water and the effects on livestock; and the effects of oil and gas development on property values. Ranchers in the foothills of the Rockies south of Calgary, for example, have argued that increased oil and gas activity in the area could ruin the ecological basis for their ranching way of life. There is evidence that the fescue grasses, which are essential to successful ranching in the region, have not regenerated after the drilling of earlier wells. In their view, the destruction of the native ground cover that comes with oil and gas development is incompatible with their way of life (Keeping 2004; Nikiforuk 2003).

Another concern of farmers and ranchers relates to the potential for significant decreases in property values. One study found that the mere presence of oil and gas operations on land has the effect of lowering property values by approximately 10 per cent (Molik et al. 2003). Another study concluded that neighbouring properties may be affected as well (Boxall, Chan, and McMillan 2005). In addition, stories are emerging from farmers who say they are unable to obtain loans by way of mortgage on their properties, because lenders are leery of decreasing property values and of actual or potential contamination of the land (Gregory 2006).

Economic Effects and Risks Exemplified: The Forestry Sector

The adverse impacts of petroleum exploration and extraction on Alberta's forest resources have been recognized for years (Kennett et al. 2006). As early as 1979, the Environment Council of Alberta noted that petroleum exploration had more negative impact than any other operations, and that the amount of land disturbed by seismic lines was almost equivalent to the total forest area harvested in twenty years (Environment Council of Alberta 1979).

A large proportion of oil and gas activities (62 per cent occur in the northwest area of the province) occur on productive forest lands. Some of this timber is salvaged and is made available to and utilized by surrounding sawmills or pulp and paper mills. When merchantable timber cannot be salvaged, the net effect is a reduction of the allowable annual cut (AAC) allocated to forest companies, and lost volumes must be found elsewhere to supply sawmills. Alternatively, harvest levels may have to decline or forest management may have to become more intensive.

A second issue is the contentious one of compensation paid by energy companies for timber loss or timber damage (Alberta Energy 1992; Cohen 1993). Oil and gas companies using Forest Management Area (FMA) lands for exploration and development are legally required to pay timber damages to FMA lease holders. The amount of timber damage is determined by the FMA holders and the two sectors tend to disagree about the value of the trees and the quantum of compensation to be paid (Kennett et al. 2006).

The third issue for the forest industry is the loss of productive timber lands resulting from the lack of regeneration of disturbed sites. It has been calculated that in certain provincial FMA areas, the amount of land removed for oil and gas development is almost as large as the area harvested for timber production (MacKendrick et al. 2001). Preliminary research on revegetation rates of seismic lines indicated that these linear features may persist for decades (Stelfox and Wynes 1999). Stelfox and Wynes suggested that 'the rate at which current and future land-base deletions revegetate to commercial tree species will significantly affect the long-term sustainability of current harvest levels of forest companies operating in northwest Alberta' (1999, 91).

Managing Landscape Change within the Political and Institutional Framework

Assuming that the Alberta energy sector's footprint will continue to grow, this footprint will mostly include an expanding and increasingly dense network of linear disturbances and relatively small clearings, contributing to significant changes in important landscape metrics such as forest area, edge, and connectivity (terrestrial and aquatic). From an ecological perspective, we have demonstrated that this landscape change will result in a variety of negative effects on natural ecosystems. From a resource and environmental management perspective, the landscape change is the product of a multitude of individual activities that may or may not be integrated to account for their cumulative contribution to landscape change.

Cumulative effects are therefore central to the challenge of managing landscape and energy futures in Alberta. While each individual seismic line, road, well site, and pipeline right-of-way may have a relatively small impact on regional ecosystems and other land uses, the cumulative effect of a multitude of these disturbances can change landscapes in very significant ways. Unless decision-makers have the capacity to *set and achieve landscape-scale objectives in a context where multiple human activities affect land-use values*, the cumulative effects of development are likely to be unplanned, unmanaged, and undesirable.

There are two fundamental requirements for managing landscape change. The first requirement is the institutional capacity to manage cumulative effects. Second, the decision-making processes for land and resource use must accommodate and be responsive to a broad range of interests and values that are affected by landscape change. The next section examines some important issues and options relating to both of these requirements.

Institutional Architecture for Integrated Landscape Management

The capacity of Alberta's policy, planning, and regulatory regime to manage cumulative environmental effects has been examined in several detailed studies (Barss 2003; Creasey 1998; Farr et al. 2004; Kennett and Ross 1998; Schneider 2002; Schneider et al. 2003; Timoney and Lee 2001). All of these studies highlight serious deficiencies in the ability to set and achieve landscape-scale or ecological-scale objectives. These deficiencies relate to fundamental institutional and policy problems that are well-recognized obstacles to cumulative effects assessment and management. Duinker and Greig (2006) and Kennett (1999) unanimously conclude that Alberta has lacked the substantive policy and planning direction and the integrated decision-making processes required to manage landscape change from energy development and other land and resource uses. The provincial government has recently recognized this shortcoming and has initiated research and management discussions to address this large and complex problem area (see, for example, Alberta Environment 2007).

THE POLICY CONTEXT

The Government of Alberta has established growth mandates for major land uses such as energy, forestry, and agriculture, with little specific direction on how these activities should be reconciled with each other on an increasingly crowded land base (Timoney and Lee 2001). Increasing pressures on the land are also fueled by policies that promote or accommodate the expansion of urban areas, low-density rural residential development, transportation infrastructure, and the construction of facilities and infrastructure for recreation and tourism (Kennett and Wenig 2003, 2005). While the Government of Alberta has also adopted various environmental and resource management policies, including a policy entitled 'Alberta's Commitment to Sustainable Resource and Environmental Management' (Government of Alberta 1999), these general statements have yet to translate into a comprehensive set of initiatives to manage the increasing cumulative pressures on ecosystems and other land-use values.

LAND-USE PLANNING

The ability of land-use planning in Alberta to address cumulative effects is limited by structural and operational weaknesses in the planning regime and by the types of land-use parameters that have been used for planning. The Integrated Resource Planning (IRP) process that applies to some public lands in Alberta was eviscerated by budget cuts in the 1990s (see Adkin and Phillips, this volume), such that Creasey concluded that it 'essentially ceased to exist within government policy initiatives' (1998, 79). Existing plans remain in effect, and are occasionally updated. However, the IRP process was never extended to all areas of public land, many of the plans are out of date, and there does not appear to be a systematic planning process in use for public lands. Furthermore, plans have no legal force; they are intended only to provide policy guidance. The IRP process has also proven incapable of managing cumulative effects because it embodies a 'multiple-use' zoning approach that appears to assume most activities and values can be accommodated on a given land base (Creasey 1998; Kennett and Ross 1998). Permitted activities are listed for each zone, but there is generally no guidance regarding the acceptable intensity of development, the total amount of disturbance to be allowed, or the mechanisms for coordinating activities so as to minimize cumulative footprint and impacts.

Following the IRP process, the Government of Alberta designed and launched a new land-use planning initiative: the Alberta Land-Use Framework (LUF). The LUF is an attempt to provide provincial policy direction and guidelines for growth management (Government of Alberta 2008). The LUF is guided by seven core strategies to improve land-use decision making in the province. One of the strategies encompasses a cumulative effects approach to be applied at a regional scale. The LUF delineates seven regions of Alberta, defined primarily by major watersheds and aligned with both natural regions and municipal boundaries. To address cumulative effects the LUF identifies a need for the province to 'develop a process to identify appropriate thresholds, measurable management objectives, indicators and targets for the environment (air, land, water and biodiversity), at the regional levels and, where appropriate, at local levels' (Government of Alberta 2008, 31). The framework is supported by a new piece of legislation entitled the Alberta Land Stewardship Act and accompanying regulations. The management of cumulative effects is central to the purpose of the act: 'to create legislation and policy that enable sustainable development by taking account of and responding to the cumulative effect of human endeavour and other events' (Government of Alberta 2009, sec. 1, [2] [d]). The creation of this legislation and framework is a clear indication that the province recognizes the need to address cumulative effects in land-use management; time will tell whether or not the approach succeeds.

MINERAL RIGHTS DISPOSITION

The disposition of mineral rights occurs through a sealed bidding process administered by Alberta Energy. Detailed analysis of this process from the perspective of cumulative effects management (Creasey 1998; Farr et al. 2004; Wenig and Quinn 2004) has revealed serious deficiencies, outlined as follows:

- incremental decision making of this type is severely limited in its capacity to address cumulative effects in the absence of an adequate policy and planning framework;
- the internal government mechanism for reviewing proposed mineral rights dispositions to identify potential environmental impacts, the Crown Mineral Disposition Review Committee (CMDRC), is generally viewed as unable to consider cumulative effects;
- the rights disposition process provides no opportunities for the involvement of land owners, other land users and the public at large in the decision to issue mineral rights (Kennett and Ross 1998, Wenig and Quinn 2004);
- the rights disposition in Alberta often results in a patchwork quilt of small mineral rights holdings owned by different companies (Farr et al. 2004) – this makes collaborative, integrated land-use planning highly problematic due to the competitive nature of the industry that results in lack of communication between lease holders;

- Alberta's mineral rights regime contains strong incentives to accelerate the pace of development. (Farr et al. 2004)

As this list indicates, the existing process is wholly inadequate not only on ecological grounds, but also because it unjustifiably limits citizens' rights to participate in decision-making about land use, with all of the consequences that such decisions entail.

THE EUB'S PROJECT REVIEW PROCESS

The capacity to manage cumulative effects and landscape change has also been weak. According to the EUB's own interpretation of its general statutory mandate, it fell to the EUB to deal with concerns about cumulative effects (for a recent review of energy regulation in Alberta, see Jaremko 2013). In addition, the EUB's project review process has been the most open, accessible, and transparent stage in the decision-making continuum that governed energy development until now. Yet a considerable body of research documents the deficiencies of these project review processes as instruments of cumulative effects management. As Schneider et al. (2003, 8) concluded in a paper that applied ALCES® modelling and policy analysis to a case study of cumulative effects in northeastern Alberta: 'Nothing within the current regulatory framework will prevent further increases in the cumulative industrial footprint.' The 2012 Responsible Energy Development Act created a new energy regulator, the Alberta Energy Regulator, with an expanded mandate for environmental impact management. It remains to be seen whether or not this change will help to overcome the problems of incrementalism and institutional fragmentation in Alberta that have created numerous obstacles to the management of cumulative effects and landscape change. Until these challenges are addressed, it will be difficult, if not impossible, to manage the future development of the energy sector and other land uses in a way that sets and achieves landscape-scale objectives (Farr et al. 2004; Kennett 1998; Kennett 2002).

Conclusion

Alberta's landscapes have undergone profound changes over the past one hundred years, and more change is inevitable. The energy sector is posed to be the dominant agent of landscape change over the coming decades and beyond. Although the oil sands present enormous environmental problems (examined in other chapters in this volume), the environmental and social impacts we have focused on here are linked directly to the network of linear disturbances and relatively small clearings (e.g., for well sites and other facilities) that characterize land use by the oil and gas industry.

The ALCES® modelling approach provided a starting point for quantifying and visualizing changes to landscape patterns over time (based on effects on terrestrial and aquatic indicators). To understand what is at stake it is necessary to go beyond both the spatial representation of development and the non-spatial analysis of landscape metrics – to examine their ecological and social implications. From an ecological perspective, it is clear that the goal setting required for cumulative effects management must go far beyond the multiple-use zoning for specified landscapes that has characterized land-use planning in Alberta to date (Dias and Chinery 1994; Kennett 1998; Kennett and Ross 1998). The emerging Alberta Land-Use Framework may provide opportunities for the integration of cumulative effects management into provincial regulation. In general, planning should identify values, objectives, and principles for land and resource use and then confront directly the ecological limitations on activities and the trade-offs among them that are required to reach desired end states (Court et al. 1994; Spaling and Smit 1993; Wight 1994).

Options proposed by ecologists and conservation biologists for limiting the energy sector's footprint and reducing the resulting impacts on ecological and other values include constraining the proliferation of linear disturbances and managing public access to the seismic lines, roads, and pipeline rights-of-way that are created. Several management tools are available to achieve these objectives. These include reclaiming disturbances once they are no longer needed for energy development; establishing optimal transportation grids for areas that may be subject to cumulative impacts from energy development and other land uses; requiring companies operating on the same land base to coordinate operational planning and share infrastructure; developing stronger fiscal incentives or regulatory requirements to adopt 'best practices' when creating linear disturbances; creating a policy of 'no net increase' in linear disturbance density within specified areas; adopting a 'roadless areas policy' that would identify areas with few or no roads or other access corridors and explicitly recognize the ecological value of these areas when making land-use decisions; and managing the human use of industrial access corridors once they have been created.

However, the achievement of even these reforms – let alone the environmental movement's demands for a 'green transition' – faces stiff obstacles in the current political economy of the province. The insistence upon 'ecological limits' to resource exploitation and industrial development is a recipe for conflict with the powerful interests at the core of fossil capitalism. As described in other chapters in this volume, Alberta's settler/frontier/extractive model of capitalism, articulated since the early 1990s to the ruling party's neoliberal ideology, is highly resistant to regulation that might slow the growth of hydrocarbon extraction and exports. Premier Klein's successors continued his governments' efforts to attract more energy sector investment in the extraction of the province's resources. In this context, the achievement of ecological and social justice goals is contingent upon the democratization of political decision-making processes[5] – an argument made more fully in the concluding chapter of this volume. Moreover, the political-economic and institutional analyses offered in this volume lead us to ask what kinds of conditions may nurture the informed, active citizens needed to democratize the petro-state and advance ecological alternatives.

NOTES

1 Our chapter follows from and summarizes a more extensive body of work by Kennett et al. (2006) with updates to make the information more current.
2 The following sections draw on Kennett et al. (2006); a more detailed treatment of each category of impact is provided in that report.
3 As Theresa Garvin explains in chapter 9, the EUB was split into two new regulatory bodies in January 2008: the Energy Resources Conservation Board (ERCB) and the Alberta Utilities Commission (AUC). In 2012, the ERCB was renamed the Alberta Energy Regulator with an expanded mandate for cradle-to-grave approval and supervision of all environmental aspects for oil, natural gas, coal, and oil sands activities.
4 In July 2007 the federal government and the Dene Tha' First Nation announced the signing of an agreement setting out a consultation process. See Aboriginal Affairs and Northern Development Canada, http://www.aadnc-aandc.gc.ca/aiarch/mr/nr/m-a2007/2-2894-eng.asp.
5 While this chapter has focused on provincial decision-making processes, it is clear that federal laws, regulations, and approval processes, as well as supra-national institutions (like NAFTA) and global market actors, also play important roles in determining the nature and consequences of hydrocarbon extraction.

REFERENCES

Alberta Energy, Forestry, Lands and Wildlife. 1992. 'Information Letter 92-13.' Edmonton: Government of Alberta.

Alberta Environment. 2007. 'Towards Environmental Sustainability: Proposed Regulatory Framework for Managing Environmental Cumulative Effects. http://environment.alberta.ca/documents/CEMS_Policy_Paper_Final.pdf (accessed 13 February 2014).

Alberta Environment and Sustainable Resource Development. 2013. 'Oil and Gas Wells Reclamation.' http://environment.alberta.ca/02862.html (accessed 13 February 2014).

Baker, Calvin O., and Frank E. Votapka. 1990. 'Fish Passage through Culverts.' San Dimas, CA: U.S. Department of Agriculture, Forest Service, Technology and Development Center.

Barss, Richard. 2003. 'Regional Planning and the Environment – The Municipal Contribution in Alberta: A Planning Failure.' Master's thesis, University of Calgary.

Bates, Ken. 2003. 'Design of Road Culverts for Fish Passage.' Olympia, WA: Washington Department of Fish and Wildlife.

Belford, David A., and William R. Gould. 1989. 'An Evaluation of Trout Passage through Six Highway Culverts in Montana.' *North American Journal of Fisheries Management* 9: 437–445.

Boxall, Peter C., Wing H. Chan, and Melville L. McMillan. 2005. 'The Impact of Oil and Natural Gas Facilities on Rural Residential Property Values: A Spatial Hedonic Analysis.' Paper no. 2005-03. Department of Economics, Institute for Public Economics, University of Alberta, Edmonton.

Brown, W. Kent, and David P. Hobson. 1998. 'Caribou in West-Central Alberta – Information Review and Synthesis.' Unpublished report prepared by Terrestrial and Aquatic Environmental Managers Ltd. for West-Central Alberta Caribou Standing Committee, Calgary.

Brown, W. Kent, and Ian Ross. 1994. 'Caribou Vehicle Collisions: A Review of Methods to Reduce Caribou Mortality on Highway 40, West-Central Alberta.' Prepared by Terrestrial and Aquatic Environmental Managers Ltd. for Alberta Environmental Protection, Fish and Wildlife Services and Alberta Transportation and Utilities, Research and Development Branch. Calgary.

Caro, Tim M., and Gillian O'Doherty. 1999. 'On the Use of Surrogate Species in Conservation Biology.' *Conservation Biology* 13(4): 805–14.

Cohen, Stacey. 1993. 'The Timber Battle.' *Forestry Oil and Gas Review* 5(3): 18–21.

COSEWIC (Committee on the Status of Endangered Wildlife in Canada). 2002.

'COSEWIC Assessment and Update Status Report on the Woodland Caribou *Rangifer tarandus caribou* in Canada.' COSEWIC Secretariat, Canadian Wildlife Service, Environment Canada. Ottawa.

Court, John D., Colin J. Wright, and Alasdair C. Guthrie. 1994. *Assessment of Cumulative Impacts and Strategic Assessment in Environmental Impact Assessment.* Prepared for the Commonwealth Environment Protection Agency. Canberra: Commonwealth of Australia.

Creasey, J. Roger. 1998. 'Cumulative Effects and the Wellsite Approval Process.' Master's thesis, University of Calgary.

Dearden, Philip, and Stephen Doyle. 1990. *Threats to National Parks: A Review and Synthesis of the Literature.* Gatineau: Western Region, Canadian Parks Service.

Dias, Oswald, and Brian Chinery. 1994. 'Addressing Cumulative Effects in Alberta: The Role of Integrated Resource Planning.' In *Cumulative Effects Assessment in Canada: From Concept to Practice*, ed. Alan J. Kennedy, 311–12. Calgary: Alberta Association of Professional Biologists.

Duinker, Peter N., and Lorne A. Greig. 2006. 'The Impotence of Cumulative Effects Assessment in Canada: Ailments and Ideas for Redeployment.' *Environmental Management* 37: 153–61.

Dyer, Simon J., Jack P. O'Neill, Shawn M. Wasel, and Stan Boutin. 2001. 'Avoidance of Industrial Development by Woodland Caribou.' *Journal of Wildlife Management* 65: 531–42.

Dzus, Elston. 2001. *Status of the Woodland Caribou (Rangifer tarandus caribou) in Alberta.* Wildlife status report no. 30. Edmonton: Alberta Environment, Fisheries and Wildlife Management Division/Alberta Conservation Association.

Eaglin, Gregory S., and Wayne A. Hubert. 1993. 'Effects of Logging and Roads on Substrate and Trout in Streams of the Medicine Bow National Forest, Wyoming.' *North American Journal of Fisheries Management* 13: 844–6.

Environment Council of Alberta. 1979. *The Environmental Effects of Forestry Operations in Alberta: Report and Recommendations.* Edmonton: Environment Council of Alberta.

EUB (Alberta Energy and Utilities Board). 1999. 'Regulatory Highlights for 1999.' Calgary: EUB.

Farr, Dan., Steven A. Kennett, Monique M. Ross, J. Brad Stelfox, and Marian Weber. 2004. *Al-Pac Case Study Report.* National Roundtable on the Environment and the Economy. Ottawa. http://www.nrtee-trnee.ca.

Forman, Richard T.T. 1995. 'Some General Principles of Landscape and Regional Ecology.' *Landscape Ecology* 10 (3): 133–42.

Government of Alberta. 1999. 'Alberta's Commitment to Sustainable Resource and Environmental Management: Implementation Plan.' Integrated Resource Management Implementation Committee, Alberta Environment, Edmonton.

Government of Alberta. 2008. *Land-Use Framework*. https://landuse.alberta.ca/LandUse%20Documents/Land-use%20Framework%20-%202008-12.pdf (accessed 15 April 2013).

Government of Alberta. 2009. *Alberta Land Stewardship Act*. Edmonton: Queen's Printer.

Government of Canada. 2000. *Unimpaired for Future Generations? Conserving Ecological Integrity with Canada's National Parks*. 2 vols. Ottawa: Panel on the Ecological Integrity of Canada's National Parks, Government of Canada.

Gregory, Joy. 2006. 'Farmers Worry Oilpatch Legacy Jeopardizes Agricultural Future.' *EnviroLine* 16: 14.

IHS Global Insight. 2009. *Measuring the Economic and Energy Impacts of Proposals to Regulate Hydraulic Fracturing: Task 1 Report*. Prepared for the American Petroleum Institute. http://www.api.org/Newsroom/upload/IHS_GI_Hydraulic_Fracturing_Task1.pdf (accessed 20 January 2014).

Jackson, Scott. 1999. 'Overview of Transportation Related Wildlife Problems.' Paper presented at the Third International Conference on Wildlife Ecology and Transportation (ICOWET), Missoula, MT, 13–16 September.

James, Adam R.C., and A. Kari Stuart-Smith. 2000. 'Distribution of Caribou and Wolves in Relation to Linear Corridors.' *Journal of Wildlife Management* 64: 154–9.

Jaremko, G., and Energy Resources Conservation Board. 2013. *Steward: 75 Years of Alberta Energy Regulation*. Calgary: Energy Resources Conservation Board.

Keeping, Janet. 2004. 'Human Rights Law, Cultural Integrity, and Oil and Gas Development.' *LawNow* 20,1 October.

Kennett, Steven A. 2002. 'Integrated Resource Management in Alberta: Past, Present and Benchmarks for the Future.' Occasional paper no. 11, Canadian Institute of Resources Law, Calgary.

Kennett, Steven A., Shelley Alexander, Danah Duke, Monique M. Passelac-Ross, Michael Quinn, Brad Stelfox, Mary-Ellen Tyler, and Nickie Vlavianos. 2006. *Managing Alberta's Energy Futures at the Landscape Scale*. Paper no. 18 of the Alberta Energy Futures Project. Report prepared for ISEEE, University of Calgary. http://www.alces.ca/publications/download/64/Managing-Alberta-s-Energy-Futures.pdf (accessed 13 February 2014).

Kennett, Steven A., and Monique M. Ross. 1998. 'In Search of Public Land Law in Alberta.' *Journal of Environmental Law and Practice* 8: 131–79.

Kennett, Steven A., and Michael Wenig. 2005. 'Alberta's Oil and Gas Boom Fuels Land-Use Conflicts – But Should the EUB be Taking the Heat?' *Resources* 91: 1–8.

Lambeck, Robert J. 1997. 'Focal Species: A Multi-Species Umbrella for Nature Conservation.' *Conservation Biology* 11 (4): 849–56.

Landres, Peter B., Jared Verner, and Jack W. Thomas. 1988. 'Ecological Uses of Vertebrate Indicator Species: A Critique.' *Conservation Biology* 2(4): 316–28.

MacKendrick, Norah, Colette Fluet, Debra J. Davidson, Naomi Krogman, and Monique Ross. 2001. *Integrated Resource Management in Alberta's Boreal Forest: Opportunities and Constraints.* Edmonton: Sustainable Forest Management Network, University of Alberta.

Malo, Juan E., Francisco Saurez, and Alberto Diez. 2004. 'Can We Mitigate Animal-Vehicle Accidents Using Predictive Models?' *Journal of Applied Ecology* 41: 701–10.

Marr-Laing, Tom, and Chris Severson-Baker. 1999. *Beyond Eco-Terrorism: The Deeper Issues Affecting Alberta's Oilpatch.* Drayton Valley, AB: Pembina Institute for Appropriate Development.

McLean, Archie. 2006. 'Stelmach Won't "Brake" Oilsands Growth.' *Edmonton Journal*, 5 December.

Molik, Terence E., Kevin M. Johnston, Melville McMillan, Peter Boxall, Wing Chan, John Thompson, Elaine Smith, and Michael Fujda. 2003. *Impact of Oil and Gas Activity on Rural Residential Property Values.* Calgary: Energy Utilities Board. http://www.landadvocate.org/issues/Property_Value_Study.pdf (accessed 15 April 2013)

Morita, Kentaro, and Akira Yokota. 2002. 'Population Viability of Stream-Resident Salmonids after Habitat Fragmentation: A Case Study with White-Spotted Charr (*Salvelinus leucomaenis*) by an Individual Based Model.' *Ecological Modelling* 155: 85–94.

Nikiforuk, Andrew. 2003. 'High on Grass.' *Avenue Magazine*, November, 32–41.

NRCB (Natural Resources Conservation Board). 1993. 'Application to Construct Recreational and Tourism Facilities in the West Castle Valley, Near Pincher Creek, Alberta.' NRBC decision report, application no. 9201.

Petroleum Communication Foundation. 1997. *Canada's Oil and Gas Industry and Our Global Environment.* Calgary: Petroleum Communication Foundation.

Rieman, Bruce E., and John D. McIntyre. 1993. *Demographic and Habitat Requirements for the Conservation of Bull Trout.* General technical report int-302. Ogden, UT: US Department of Agriculture – Forest Service, Intermountain Research Station.

Ross, P. Ian. 2002. 'Update COSEWIC Status Report on Grizzly Bear (*Ursus arctos*).' Committee on the Status of Endangered Wildlife in Canada. Ottawa.

Schneider, Richard R. 2002. *Alternative Futures: Alberta's Boreal Forest at the Crossroads.* Edmonton: Alberta Centre for Boreal Research.

Schneider, Richard R., J. Brad Stelfox, Stan Boutin, and Shawn Wasel. 2003. 'Managing the Cumulative Impacts of Land-Uses in the Western Canadian Sedimentary Basin: A Modeling Approach.' *Conservation Ecology* 7 (1): 8.

Spaling, Harry, and Barry Smit. 1993. 'Cumulative Environmental Change: Conceptual Frameworks, Evaluation Approaches, and Institutional Perspectives.' *Environmental Management* 17: 593–4.

Stamps, Judy A., Marybeth Buechner, and ViswanathanV. Krishnan. 1987. 'The Effects of Edge Permeability and Habitat Geometry on Emigration from Patches of Habitat.' *The American Naturalist* 129: 533–52.

Stelfox, J. Brad, and Robert Wynes. 1999. 'A Physical, Biological, and Land-Use Synopsis of the Boreal Forest's Natural Regions of Northwest Alberta.' Prepared by J.Brad Stelfox of Forem Consulting and Robert. Wynes of Daishowa-Marubeni International Ltd., Peace River, AB.

Thormann, Markus N., Pierre Y. Bernier, Neil W. Foster, David W. Schindler, and Fred D. Beall. 2004. 'Land-Use Practices and Changes – Forestry.' In *Threats to Water Availability in Canada,* 30–3. Burlington, ON: Environment Canada.

Thurber, Joanne M., Rolf O. Peterson, and Thomas D. Drummer. 1994. 'Gray Wolf Response to Refuge Boundaries and Roads in Alaska.' *Wildlife Society Bulletin* 22: 61–8.

Timoney, Kevin, and Peter Lee. 2001. 'Environmental Management in Resource-Rich Alberta, Canada: First World Jurisdiction, Third World Analogue?' *Journal of Environmental Management* 63: 387–405.

Warren, Melvin L., and Mitzi G. Pardew. 1998. 'Road Crossings as Barriers to Small-Stream Fish Movement.' *Transactions of the American Fisheries Society* 127: 637–44.

Weaver, John L., Paul C. Paquet, and Len F. Ruggiero. 1996. 'Resilience and Conservation of Large Carnivores in the Rocky Mountains.' *Conservation Biology* 10 (4): 964–76.

Wenig, Michael M., and Michael S. Quinn. 2004. 'Integrating the Alberta Oil and Gas Tenure Regime with Landscape Objectives: One Step toward Managing Cumulative Effects.' In *Access Management: Policy to Practice,* ed. H. Epp, 27–39. Edmonton: Alberta Society of Professional Biologists.

Wilcox, Bruce A., and Denis D. Murphy. 1985. 'Conservation Strategy: the Effects of Fragmentation on Extinction.' *American Naturalist* 125: 879–87.

Woynillowicz, Dan., Chris Severson-Baker, and Marlo Raynolds. 2005. *Oil Sands Fever: The Environmental Implications of Canada's Oil Sands Rush.* Drayton Valley, AB: Pembina Institute.

5 The Petro-Politics of Environmental Regulation in the Tar Sands

ANGELA V. CARTER

> The tar sands represent the bleak future that awaits the world if we refuse to listen to science and fail to make significant commitments to cut greenhouse gas emissions ... It's time to stop the tar sands. (Greenpeace Canada 2009)

> The present hegemonic regime of accumulation ... is incapable of ushering in an era of genuine sustainable development. This is because its two central institutions – the state and the corporation – benefit too greatly from unsustainable growth. (Gale 2000, 195)

Alberta's tar sands developments embody a central contemporary tension between the need for oil, a fundamental global commodity, and the health of the environment, which is put at risk along the entire oil production chain, from exploration to consumption. The surge in 'unconventional' oil projects over the last decade signals a significant shift in global oil production from relatively accessible conventional reserves

This chapter is based on research conducted over the 2006–12 period. I am grateful to the following colleagues who generously provided comments on this work: Laurie Adkin, Simon Dyer, Diana Gibson, John Peters, André Plourde, and Nickie Vlavianos. I also thank Aura Villanueva and Candice Pike for research assistance and acknowledge the organizations that funded the research leading to this chapter: the Social Sciences and Humanities Research Council, Cornell University's Centre for the Environment, Cornell University's Graduate School, Memorial University's Institute of Social and Economic Research, and the Institute for Biodiversity, Ecosystem Science, and Sustainability.

to 'frontier' oil that is farther north, farther offshore, and in ever more fragile landscapes, with ecological impacts increasing in scale, intensity, and duration.

This chapter examines the environmental regulatory system surrounding the tar sands, using the political ecology framework of analysis set out in the introduction to this volume. I survey the environmental impacts of tar sands development and briefly note who bears these impacts. I then outline the main trends in environmental regulation that permit, or do not prevent, these outcomes. These include a delayed and ineffective positioning of environmental consideration in the approval processes; important regulatory gaps or inadequacies relating to carbon emissions, freshwater extraction, reclamation, and public consultation; and analytical weaknesses regarding cumulative impacts. The analysis corroborates that of Quinn et al. (this volume) with regard to the failures of land-use management processes, their ecological and social consequences, and their democratic implications. This chapter extends that analysis, however, by explaining the province's regulatory approach in relation to what I call 'petro-politics,' or the specific form of Alberta's 'petro-polity.' The Alberta case provides an opportunity to examine the effects of oil dependence on the environmental regulatory regime. It is important to note that responsibility for environmental management of the tar sands lies predominantly with the provincial government. The Alberta Crown owns 97 per cent of Alberta's tar sands mineral rights (Alberta Energy 2006, 1-1) and, according to multiple sections of the Canadian Constitution, Alberta has 'exclusive' regulatory powers over the tar sands (Vlavianos 2007c, 4–5).

This chapter contributes to the literature on petro-states, which has traditionally centred on 'developing' states, by extending the analysis to focus on first world cases using a comparative approach. Also, this research is unique in that it emphasizes a 'subnational' case, the province of Alberta, demonstrating the applicability of petro-state theories below the level of the nation state. Finally, it enlarges the typical view of the petro-state literature beyond the domestic economic consequences of dependence on oil exports to include the ecological dimensions of this model of development.

Alberta's 'petro-political' system is marked by a symbiotic relationship between governments – specifically the long-reigning Progressive Conservative party, in power from 1971 to 2015 – and oil companies, with governments highly dependent on revenues from private oil developments and oil companies earning impressive profits from extraction on public lands. The provincial (and also federal) government ensures the continuation of the industry via funding or subsidies, by actively defending and promoting the industry at home and abroad (see Adkin and Stares, this volume), by being reluctant to dig deeper into the environmental questions raised, and by not intervening to protect the environment where regulatory authority exists to do so. The tar sands industry reinforces these governmental approaches via coordinated lobbying efforts, political financing, and media and community public relations campaigns. I argue that, driven by its single-minded prioritization of hydrocarbon extraction as an economic strategy for the province, and pressured and influenced by a powerful, globally integrated industry, the Albertan petro-state has developed environmental regulation processes and institutions that forward rapid, extensive oil development and do not meaningfully restrain the resulting environmental impacts. The shared interests of government and industry have translated into an environmental policy regime that is biased towards tar sands development. The regulatory system has been 'molded,' to use Terry Karl's (1997, 16) term, to support these developments and to restrain or impede effective environmental regulation.[1] This is the status quo regime with which the New Democratic Party (NDP) government, elected in May 2015, has to wrestle, and which it may reform.

Alberta's Petro-State

A petro-state is defined by high dependence[2] on oil resulting in a particular set of political-economic challenges captured in the 'resource curse' concept. Literature in this vein demonstrates correlations between high dependence on natural resources – impacts that are particularly exaggerated in oil-based states – and particular economic and political trends. Multiple studies confirm the negative long-term economic impacts of natural resource dependence generally and oil dependence in particular (Auty 1993; Bridge 2008; Gelb 1988; Humphreys, Sachs, and Stiglitz 2007; Nankani 1979; Neumayer 2004; Sachs and Warner 1999; Wheeler 1984), including declining per capita GDP over time, the export of development benefits (as the industry is often dominated by foreign investment by multinational corporations that do not reinvest profits into the region of extraction), and risks to other key economic sectors (the 'Dutch Disease,' where instead of invigorating other industries, the oil sector tends to inhibit them by increasing general production costs and drawing labour away from manufacturing and agricultural industries). General economic volatility is also a key problem: oil-dependent states are at the mercy of booms and busts associated with unpredictable oil prices.

These economic challenges are mirrored by political challenges that, I argue, have policy consequences in Alberta. In *The Paradox of Plenty*, Terry Karl developed the idea of the 'petro-state,' which tends to become a rentier state that replaces 'statecraft' with the spending of oil dollars (1997, 16; see also Chaudry 1997). The shift from dependence on taxes for state revenue to dependence on resource rents, and the ensuing erosion of a strong, broad-based tax system, alter governments' accountability to citizens as well as citizens' engagement with the state. When oil revenues are disrupted, or are lower than anticipated – as during the post-2008 decline in oil prices – social services are negatively affected. Karl also showed that economic growth based on resource dependence has long-term institutional 'inertia' effects that keep the state focused on oil rather than working for diverse (and more resilient) development. Resource rents empower and maintain the power of groups that have a stake in maintaining the status quo, and therefore impede diversification. Wasting of the resource rent is also common: instead of saving the windfalls from natural resources, states frequently engage in corrupt spending or overspending on projects of only short-term value (Weinthal and Jones Luong 2006, 39).

Key data substantiating many of these arguments were provided by Ross (2001) in his longitudinal regression analysis of the impact of oil on democracy on 113 states from 1971 to 1997. He found a correlation between oil dependence and authoritarianism and concluded that oil has a tendency to 'hurt democracy' (2001, 356). Similarly, Jensen and Wantchekon analysed the political impact of resource dependence in African states, finding that countries more dependent on natural resources were also 'more likely to be authoritarian' and more likely to have 'worse governance' (2004, 817).

Although resource curse research has long been focused on developing nation-state cases, the findings have been confirmed in developed state cases and at subnational levels of governance in federal states. For example, Goldberg et al. (2008) show that dependence on natural resource wealth has a 'conservative' effect on American states' politics: actors in power at the time that resource wealth reaches the state gain the resources to stay in power and reproduce the economic, political, and social status quo, thereby decreasing party competitiveness.

Many of the phenomena described above are recognizable in Alberta. As discussed in my chapter with Zalik (this volume), the impact of oil on the Albertan economy is comparable to its impact on petro-states elsewhere. However, the role of the state, or of political elites benefiting from state power, is only one dimension of the political economy of petro-states. Corporate actors also invest heavily in the shaping of the regulatory regime.

The following section discusses the environmental impacts of tar sands developments and analyses the main challenges in the environmental regulatory regime framing these projects. I then examine the role of corporations in creating and sustaining the problematic environmental regulatory norms in the province and reproducing the political and economic character of Alberta's development model.

Environmental Impacts and Regulation of the Tar Sands

Canada is an energy state, and more specifically an oil state, so much so that over the 2003–07 period oil prices and the Canadian dollar were closely correlated (Cross 2008, 3.4–3.5) – hence the 'petro-loonie' neologism. Energy is Canada's most valuable export (Cross 2008, 3.1), with oil and natural gas sales representing about 21 per cent of total 2008 merchandise exports, 'the highest value recorded in the previous two decades' (Plourde 2010, 9). Similarly, business investment is dominated by natural resource industries, in particular the energy sector, which in turn is 'driven by the oilsands' (Cross 2008, 3.1–3.5).

Alberta is the primary province driving these trends. Alberta dominates Canada's oil and gas industry by its basic share of reserves and production. Recent data show the province as accounting for 82 per cent of the country's total crude oil reserves, 69 per cent of natural gas reserves, 68 per cent of crude oil production, and 76 per cent of natural gas production (Plourde 2010, table 1). Oil and gas production dominates Alberta's economy; between 1994 and 2009, revenues from non-renewable energy resources have, on average, contributed 29 per cent of total provincial revenue, rising to over 40 per cent in 2000 and 2005 (Carter 2011, 167–9). The tar sands have become the fulcrum of this industry and economy: declines in conventional oil production in Alberta since the 1970s have been offset by tar sands production, which has outpaced conventional oil production since 2002 (Mansell and Schlenker 2006, 14). Alberta makes Canada an oil state, and the tar sands are becoming the backbone of this province's economy.

Tar sands underlie 140,200 square kilometres of boreal forest concentrated predominantly in northern Alberta in three major deposits: the Athabasca (with the greatest concentration of extraction activity), Cold Lake, and Peace River.[3] Massive investments in the tar sands supported the production of 1.9 million barrels per day in 2010 (nearly a doubling of production since 2002) (CAPP 2011, table 3.2a). Alberta Energy predicts 2020 output will be at three million barrels per day and 'possibly even 5 million barrels per day by 2030' (Alberta Energy 2008, table 3.2a).

The scale, intensity, and rapid expansion of tar sands projects have created serious environmental impacts.[4] From 1967 to 2006 tar sands developments created a 'cumulative disturbance' of 650 square kilometres (Timoney and Lee 2009), with major expansions expected as 84,000 more square kilometres have been leased for tar sands development and leasing regularly continues (Pembina Institute 2010). As for freshwater, tar sands operations had licences to divert 349 million cubic metres per year from the Athabasca River in 2008 (double Calgary's yearly volume), with new projects potentially raising this to 500 million cubic metres (Dyer et al. 2008, 3, 8). There are now serious concerns about maintaining basic in-stream flow (Griffiths and Woynillowicz 2009). Beyond water withdrawals, enormous tailings 'ponds' containing toxic substances from tar sands operations – now comprising over 170 square kilometres in total area (Energy Resources Conservation Board 2010) – pose a risk to local ecosystems due to leeching at rates of millions of litres per day (Kelly et al. 2009; Price 2008; Timoney and Lee 2009).

Alberta's tar sands operations also emit enormous volumes of air pollution (Timoney and Lee 2009, 73–4), including greenhouse gases (GHGs), which contribute to making the province the largest GHG emitter in the country (Miller 2007). Tar sands plants and upgraders, according to a report by Environmental Defence, Équiterre, and the Pembina Institute (2010, 7), are the 'fastest growing source of GHG emissions in the country,' contributing 37 million tonnes or 5 per cent of total Canadian emissions in 2008. Based on projections made prior to the recession, these emissions were expected to rise to 108 million tonnes by 2020, representing over 40 per cent of the total increase in emissions in the country over this period (Environmental Defence, Équiterre, and Pembina Institute 2010, 7; see also Bramley, Neabel, and Woynillowicz 2005, 5). Tar sands projects are, therefore, significantly exacerbating Alberta and Canada's growing contribution to the global problem of climate change. The extraction and upgrading projects also entail staggering emissions of nitrogen oxides (NO_x), sulphur dioxide (SO_2), and volatile organic compounds (VOCs) (Dyer et al. 2008, 25–33; Timoney and Lee 2009, 73–4). The result of these combined impacts has been a decline of numerous species, many endangered or threatened, for example caribou, lynx, marten, fisher, wolverine, and multiple bird species (Schneider and Dyer 2006; Timoney and Lee 2009).

Perhaps the most obvious burdens of tar sands developments are on local communities in the surrounding area and on the predominantly Aboriginal communities downstream who are at risk of compromised water, air, and subsistence food supplies. More recently, connections have been made between the environmental degradation associated with the tar sands and illness in communities downstream (Timoney 2007; Timoney and Lee 2009, 78). These local communities also see landscapes transformed and solely devoted to tar sands projects, limiting traditional uses. Broader still, tar sands developments stress a major river system in an increasingly drought-prone province and result in water pollutants being transported into the fragile inland Peace-Athabasca Delta and through the Mackenzie Basin to the Arctic Ocean (Kelly et al. 2009, 22346–51). Meanwhile, airborne pollutants increase soil and lake acidification in neighbouring Saskatchewan and Manitoba (Bytnerowicz et al. 2010; Jeffries, Semkin, and Gibson 2010). Further, the consequences of climate-change-causing GHGs will extend to future generations of Albertans, Canadians, and the global community. Emissions from the tar sands are a significant barrier to Canada's meeting its national greenhouse gas reduction commitments (Bramley, Neabel, and Woynillowicz 2005). Contrary to the claims made by Alberta's government, there is impressive evidence that these multiple, far-ranging, and long-term environmental impacts are not being effectively prevented or managed by the provincial regulatory system. Three problems with the system stand out: the timing and impact of environmental consideration, important regulatory gaps, and the management of cumulative impacts.[5]

Ineffective and Delayed Environmental Consideration

Policy makers within key departments of the Alberta government and representatives of research institutes and non-government organizations (NGOs) frequently noted during my interviews with them in 2007–09 that Alberta Environment, the department responsible for regulating environmental impacts in the tar sands, is in a structurally weak position in the decision-making and regulatory process – particularly in comparison to the department that promotes hydrocarbon extraction, Alberta Energy, and the Energy and Utilities Board (EUB), subsequently the Energy Resources Conservation Board (ERCB).[6] As discussed below, the regulatory scope of Alberta Environment is too narrow and its input occurs too late, after the leasing of land has occurred and property rights have been assumed.

At the leasing stage, environmental impacts are considered at the Crown Mineral Disposition Review Committee's (CMDRC) initial review of companies' requests for land auction, but only in a cursory manner. As Holroyd et al. note, this is 'the one and only opportunity during the tenure process to consider the environmental and social impacts of granting oil sands rights,' but this process is too narrowly focused (there is no room for a consideration of cumulative impacts), too rapid, poorly informed, and has no 'formalized' environmental assessment process (2007, 21–2). Even if the environmental analysis were to be improved here, the CMDRC's role is merely advisory. Alberta Energy ultimately decides if land requested for auction will be posted. Significantly, Alberta Energy – the primary department promoting tar sands development – has been closely aligned for decades with the oil industry, which it considers its 'principal stakeholder.'[7] Indeed, corporate requests determine the pattern and pace of land leases in the absence of any provincial land use plan.

More thorough consideration of environmental effects occurs through Alberta Environment's environmental impact assessments (EIAs). But these have limited impact on the decision-making process because the EIA results have been transferred to the (former) EUB and ERCB, where environmental considerations have been routinely overridden by other interests, particularly 'economic benefit.' In spring 2007, interviewees familiar with the workings of the former EUB noted that the board was under continual political pressure to approve the projects for which Alberta Energy had already sold access rights.[8] According to environmental lawyer Nickie Vlavianos, 'it is clear that the EIA process under EPEA [Alberta's Environmental Protection and Enhancement Act] is not a central feature of the oil and gas development process in the province' (2006, 46). The EIA 'simply provides the EUB with environmental information,' then it is the EUB 'who will make the final determination about whether a project is in the public interest or not, and environmental impacts are only one consideration in the EUB's decision' (46). The pattern of decision making observed in the EUB continued to characterize the rulings of the new ERCB. For example, in 2009 the ERCB refused the request of environmental non-government organizations (ENGOs) and policy institutes to reconvene joint panel public hearings to review approvals granted to Shell for its Muskeg River Mine Expansion and Jackpine Mine projects after the company stated it would not honour emissions reductions commitments. During project consultations, the company had agreed to lower emissions; but when Shell flatly reneged on this commitment less than two years after the consultations, the ERCB declined to intervene (Cooper 2009; Ecojustice, Toxics Watch Society, and Pembina Institute 2009). The ERCB has also been criticized for not rigorously applying standards to protect the environment. In April 2010, ERCB approved Syncrude's plans for two tailings ponds that will not meet the province's new and long-awaited Directive 074: Tailings Performance Criteria and Requirements for Oil Sands Mining Schemes (a regulation developed to phase out discharges to tailings ponds) until 2014 (Energy Resources Conservation Board 2010; see also Dyer 2010a; Obad 2010).

Overall, the ERCB's and Alberta Energy's priorities have consistently overridden any concerns raised by Alberta Environment in decisions on the tar sands. Based on interviews with those involved with the tar sands regulatory process, this trend is common knowledge both inside and outside the public service. As Alberta Environment policy makers note, even in interdepartmental initiatives that are supposed to offer a 'level playing field' for all ministries, energy interests 'typically carry the day.'[9] For its part, Alberta Environment rarely refuses approvals required by tar sands operations. When asked about Alberta Environment's apparent reluctance to slow or reject projects due to environmental impacts, policy makers noted that 'when things are good, you want to reap all the benefits you can. You don't want to stand in the way of that.'[10] Another policy maker argued that not permitting a tar sands development to occur is 'stranding' resource potential from Albertans.[11] These comments echo the position of the government as expressed by the premier and minister of energy. (See Adkin and Stares, this volume.)

At the same time, policy makers committed to environmental protection within Alberta Environment have inadequate resources and staff to monitor and enforce regulations (as Adkin found in chapter 3). Woynillowicz, using statistics from the Alberta government's fiscal plans from 2001 to 2008, also noted declines in Alberta Environment's

Figure 5.1. Syncrude tailings pond and upgrader, near Fort McMurray, April 2010. (*Photo*: Angela V. Carter.)

staff since 2000, while tar sands production was significantly expanding. The numbers suggest that 'the department's budget has not grown in parallel with its workload' (2006, 2). Similarly, Timoney and Lee note the 'paucity of relevant data available to the public due in large part to a decline in government monitoring in recent decades that has coincided with rapid and major expansion of the tar sands industry' (2009, 65). This indicates a telling discrepancy in the allocation of financial resources when it comes to tar sands development.

Given the relatively weak role of environmentally oriented government departments in Alberta, as well as the tokenistic attention given to environmental impacts in the regulatory process, it seems the overall structure of environmental regulation in this province has long been compromised. The most recent changes to regulatory leadership on energy projects further entrenched this pattern. In 2013, the Responsible Energy Development Act created the Alberta Energy Regulator (AER) to combine functions of the Alberta Environment and Sustainable Resource Development Ministry and the Energy Resources Conservation Board. AER is now the single regulator for all energy resources in the province, from initial applications for exploration to final reclamation – including hearings on development applications and appeals. Importantly, it is also tasked with administering Alberta's Environmental Protection and Enhancement Act – formerly the responsibility of Alberta Environment and Sustainable Resource Development. While the new office has been touted by the provincial government as a way to improve regulation of energy in Alberta and public involvement in the process, it has come under criticism as further restricting public participation and because it provides no additional transparency with regard to the decision-making process on energy projects in the province or on processes for appeals or decision reviews (Ecojustice 2013; see also chapter 17, this volume). Environmental lawyers have also raised concerns about the increased potential for ministerial interference in the development decisions (Harrison 2013).

Within this broad and problematic regulatory structure, there are specific issues of note in Alberta's management of tar sands development. Two outstanding problems relate to regulatory gaps and the longstanding difficulty in assessing and dealing with cumulative impacts.

Key Regulatory Gaps

The most obvious and pressing example of regulatory gaps relates to GHG emissions. Tar sands projects are a major – and the most rapidly growing – contributor to Canadian GHG emissions (Richardson 2007, 37–8). For example, in 2008 Syncrude's Mildred Lake and Aurora North Plant sites were, combined, the third-largest GHG emitter in the country (12.2 million tonnes emitted), and Suncor was the fourth-largest (8.8 million tonnes) (Environment Canada 2010). While per barrel emissions are declining, improvements are outpaced by the continuous expansion of tar sands operations.

The policy response to this situation has been notoriously weak. To meet the goal of keeping global climate within two degrees of warming using a carbon emission 'budgeting' model, the tar sands industry's proportional share of GHG emissions must be reduced from 37 million tonnes in 2008 to 24 million tonnes in 2020 (Environmental Defence, Équiterre, and Pembina Institute 2010, 7). However, current provincial targets will see GHG emissions *rising* until 2020, and only then will they begin a gradual decrease to arrive back at 2008 emission levels by 2035. Alberta's climate change policy as of spring 2015, therefore, will delay real reductions in emissions for approximately three decades. Moreover, the promised reductions are to come from carbon capture and storage (CCS) projects (Alberta Government 2008, 24), which are currently still in development and questionable in terms of their efficacy in reducing emissions within the critical time frame (see Le Billon and Carter 2012; Thomson 2009).

Policies to protect fresh water are equally problematic, especially with regard to the in-stream flow needs of the Athabasca River, the primary source for the water-intensive tar sands projects. Guided by its Water Act, Alberta Environment (or ERSD) issues licences for water withdrawals from the Athabasca River to oil sands companies. In 2007 the tar sands industry was withdrawing 349 million cubic metres annually (double Calgary's domestic withdrawals from the Bow River) (Pembina Institute 2007), and approved projects not yet in operation would double this amount (Richardson 2007, 43). Despite the problem of drought in a region that has experienced significant decreases in river flow over the last century (Griffiths, Taylor, and Woynillowicz 2006; Griffiths and Woynillowicz 2009), tar sands withdrawals from the Athabasca River have only recently become a concern for the government.

The Cumulative Environmental Management Association (CEMA), a multi-stakeholder committee created in 2000 to manage the cumulative impacts of tar sands developments, struggled for years to define in-stream flow needs to guide withdrawal policy, but it could not reach consensus by its deadline of December 2005. Therefore, the federal Department of Fisheries and Oceans and Alberta Environment developed the Water Management Framework (Alberta Environment and Fisheries and Oceans Canada 2007), which limits withdrawals during winter low-flow periods to 5.2 per cent of weekly historical median flows with maximum withdrawal caps of 15 cubic metres per second. Yet this policy runs counter to the recommendation of environmental organizations and Aboriginal communities to permit *no* withdrawal during these periods (Pembina Institute 2007). The Oil Sands Developers Group (OSDG), a regional oil industry association, predicted in 2008 that within a few years the tar sands projects alone would exceed the

5.2 per cent winter weekly withdrawal limit, growing to 6 per cent of median flows, at which level withdrawals would stabilize until 2035. In the 'growth case' scenario, 15-cubic-metre-per-second water withdrawals (the current *maximum* withdrawal cap in low-flow periods) may be *standard* by 2015 to 2030 (Irving 2008). This problem was admitted by the Alberta government's 2006 Oil Sands Ministerial Strategy Committee report, which noted that 'Alberta Environment has not had the opportunity or the resources to undertake a review to determine whether there is sufficient water available' in key rivers to permit new developments (Government of Alberta 2006, 113).

Land reclamation is a third salient regulatory gap. As Vlavianos explains, project permits are issued 'without a clear sense that reclamation is currently feasible but in the hopes that new technology will be developed that will someday allow for proper reclamation' (Vlavianos 2007c, 52). Reclamation of tailing ponds is a particularly pressing issue given the lack of proven technology and methods. Reclamation results to date are poor: after over four decades of tar sands developments, Alberta Environment issued its first reclamation certificate in March 2008 to Syncrude Canada for reclaiming one square kilometre of land.[12] As one participant in the oil sands consultations noted, 'Development is going along at hyperspeed but reclamation is going along at geological speed' (quoted in Alberta Energy 2007, 18).[13]

Finally, criticism is now frequently directed towards ineffectual public consultation (or, where consultation is adequately conducted, unheeded public consultation) on tar sands–related projects. During the land rights issuance process there is no opportunity for public input and perhaps even little public notice.[14] In Wenig's (2004) appraisal, the CMDRC's work is a 'black box': there is no public involvement at this stage and very little public information about what the committee does.[15] At the level of the ERCB, public involvement may occur, but only if the proposed project is brought to a hearing, and a hearing will be triggered only if people protest that they will be 'directly and adversely affected' by an ERCB decision. If there are no private landowners or occupants of the land in question, a hearing may not be triggered.

Public consultation has occurred more frequently through Alberta Environment than through Alberta Energy, but it is severely constrained in both the EIA and licensing processes. During the EIA process, public involvement is permitted in a limited way, but only in the later stages. There is some room for input from directly affected individuals during

Alberta Environment's licence-issuing processes for tar sands projects, and these licences may be appealed through the Environmental Appeals Board (EAB), the quasi-judicial tribunal of Alberta Environment. But EAB decisions, like EIA reports, are non-binding on the minister of the environment (Environmental Law Centre 2006a, 2006b, 2006c, 2006d). Consultations such as the 2006–07 oil sands consultation led by the Multi-Stakeholder Committee and the 2007 Royalty Review Panel Consultation were more inclusive, but these consultations produced only recommendations. If consultations have no legal standing, the government is not required to enact recommendations (Vlavianos 2007c, 64). Therefore, as Ricardo Acuña, executive director of the Parkland Institute, noted at the time, 'although the actual consultation process is an improvement over the window-dressing consultations of the Klein years, it would appear that the outcome will be no different – a government with no interest in actually acting on what Albertans are recommending' (2007).

For these reasons, ENGOs frequently worry that government-initiated consultations provide a mere illusion of participation, while diverting the energies of activists.[16] The longstanding problems of governmental accountability and public access to information from Alberta Energy have been repeatedly documented (Alberta Royalty Review Panel 2007; Dunn 2007; Valentine 2008).

Analytical Gap: Cumulative Impacts

A final problematic regulatory trend of note – one taken up in greater detail in chapter 4 – is the limited scope of environmental analysis, which focuses only on the immediate impact of specific projects as opposed to long-term and cumulative effects. As industrialization expands, a regulatory process examining *individual* permits or projects without an analysis of regional, cumulative impacts is ever more inadequate – hence Timoney and Lee's urgent call for 'comprehensive, peer-reviewed assessments of the cumulative impacts of tar sands development' (2009, 65).

Since the late 1990s there have been multiple institutional attempts to overcome the fragmented nature of decision making on tar sands development. Key examples include the Regional Sustainable Development Strategy (RSDS) for the Athabasca Oil Sands Area and the CEMA (Kennett 2007). These initiatives have been marked by continual delays, primarily due to difficulties (genuine or contrived) in reaching a consensus. So far, there have been no tangible recommendations on development trade-offs and no clear framework for departments to address cumulative effects.[17] Overall, therefore, these policy integration attempts seem to be 'parking lots' for complex issues with no significant impact on the development process. Worse, Hoberg and Phillips (2011) argue that multi-stakeholder consultations and bodies like CEMA are defensive strategies of government and industry to 'bolster the legitimacy of the policy process while maintaining control over decision-making, rules and venues'; they can be strategies of 'cooptation or manipulation' (509–10). In 2015, following the election of the NDP government, both the industry funders and the provincial government withdrew their support from CEMA. The new government has indicated interest in developing a new government-led multi-stakeholder organization to address the cumulative environmental effects in the province (McDermott 2015). Cumulative effects monitoring may also be undertaken through the Alberta Environmental Monitoring, Evaluation and Reporting Agency, the Joint Canada-Alberta Implementation Plan for Oil Sands Monitoring, or the province's climate change office (announced in early February 2016).

As discussions on cumulative impacts slowly proceed and research accumulates, tar sands projects advance toward an anticipated fivefold expansion (*CBC News* 2007; Oil Sands Experts Group Workshop 2006). Hence Wenig et al.'s (2006) criticism (relating to freshwater extraction) of 'regulatory foot dragging' within the Alberta government. As these authors note, while 'bemoaning' the lack of a cumulative effects plan, 'the province's Energy and Utilities Board has continued approving, and Alberta Environment has continued issuing new water licences for, successive oil sands operations' (2006, 24). This pattern continued via the ERCB (since replaced by the Alberta Energy Regulator), as noted above, regarding its weak implementation of tailings pond regulations.

In response to growing public concern with the lack of cumulative effects assessment and inadequate environmental monitoring, the Government of Alberta established the Alberta Environmental Monitoring, Evaluation and Reporting Agency (AEMERA) in 2014 to monitor all air, land, and water quality, as well as biodiversity, with a specific focus on oil sands effects. The agency was created as a body independent of government, one committed to comprehensive monitoring of the cumulative effects of economic development in the province and to full public transparency of data and reports. Yet there are already clear signals that the agency falls short of the goal of political independence, in great part due to the government-appointed board (Read 2013).

Taken together, these trends – consideration of environmental impacts that is poorly timed and weakly integrated into the decision-making process on tar sands projects, alongside important regulatory and analytical gaps – constitute a fragile system of environmental regulation that has served the interests of short-term profit maximization for the energy corporations. The role of these corporations in shaping Alberta's oil-dependent political ecology is discussed below.

The Industry Lobby

A number of chapters in this volume discuss the long-term, concentrated government support for the tar sands via subsidies, low taxes, and royalties, as well as public relations campaigns to protect and support the industry. Building on these accounts, I examine how this government support has been inextricably intertwined with and reinforced by the oil and gas industry lobby, which attempts to influence the provincial government and the public to ensure its investments and profits in tar sands developments are protected.

Data on corporate spending in the tar sands show that from 1997 to 2010, capital, operating, and royalty expenditures for in situ and open-pit mining as well as upgrading operations totalled $220.8 billion dollars, peaking in 2010 at $34.2 billion dollars (CAPP 2011, table 4.16b). These investments have generated impressive earnings and profits. Tar sands producers' annual sales increased dramatically from $4 billion in 1997 to $37.8 billion in 2008 (CAPP 2011, table 4.19b). Even during the post-2008 recession period, significant investments have continued. Comprehensive information on net profits is unavailable; however, data are available on net income for individual companies, providing a sense of the value of these operations. In a recent industry comparison of the net income of publicly traded oil and gas companies in Canada in 2009, the top five net earners, all active in tar sands developments (Encana Corp., Canadian Natural Resources Ltd., Imperial Oil Ltd., Husky Energy Ltd., and Suncor Energy Ltd.), had an average net income of $1.5 billion ('The Top 100 Canadian' 2010). The top five net income earners in 2013 (Suncor, Imperial Oil, Husky Energy, CNRL, Canadian Oil Sands Ltd.) averaged $2.4 billion ('The 100 Largest' 2014).[18]

These are high stakes, and industry players have extensive means to protect their access to lucrative resources. Coordinated and led by the Canadian Association of Petroleum Producers (CAPP), they have entered the environmental regulation debate, lobbying the provincial and federal legislatures. Available data demonstrate this well: for example, data from the Office of the Commissioner of Lobbying of Canada show that in 2009 the environment minister, Jim Prentice, had the highest number of contacts with lobbyists, and the majority of these were with fossil fuel industry representatives, including major players in the tar sands (McGregor 2010). Communicating with political leaders, the fossil fuel industry repeatedly raises concerns about costs and delays associated with environmental assessments and regulations. For example, in May 2009 the CAPP, along with the Canadian Association of Oilwell Drilling Contractors and the Small Explorers and Producers Association of Canada, presented arguments to the House of Commons Standing Committee on Industry, Science and Technology for loosening environmental regulations generally, and for clarifying uncertainty around GHG regulations. They argued that these regulations have added to operating costs in Canada to the point that, according to the executive director of the Small Explorers and Producers Association of Canada, Gary Leach, 'Canada provides among the lowest rates of return on investment in the world' (quoted in Akin 2009). The Harper governments delivered policy in line with this corporate pressure. Using omnibus budget bills, the Conservative Party removed or weakened environmental regulation and assessment requirements around oil and gas development. Robert MacNeil (2014) describes these changes to environmental policy as part of the federal government's ongoing "unabashed assault on environmental regulations" (175).

The fossil fuel industry in Alberta has lobbied hard against emissions reductions. Since the mid-1990s, campaigns to avoid emissions reductions requirements have been led by the Canadian Council of Chief Executives, CAPP, and specific companies operating in the tar sands. Industry advocated for voluntary emissions reductions or intensity-based emissions targets (as opposed to targets that would ensure an absolute decline in emissions), a low per-tonne price for carbon emissions, and tax breaks for spending on reducing emissions. Canadian political scientist Douglas Macdonald describes the anti-Kyoto lobby led by the oil and gas industry and the province of Alberta in 2002 as 'the largest public campaign ever seen in Canadian environmental politics' (2010, 16; see also Urquhart 2005).

While lobbying key government ministries, tar sands companies also fund political campaigns and parties to encourage policy development amenable to oil interests. This is an under-researched subject; however, 2005 research by Trevor Harrison reported a close correlation between corporate donations by oil and gas companies (among the major funders of the Conservatives) to political parties and the fossil-friendly policies of those parties (2005, 100–1). Conversely, oil companies withdrew funding after Alberta Premier Stelmach's government indicated, in late 2007, that the royalty regime might be altered to capture more revenues: oil industry donations to the Alberta Progressive Conservatives declined by 41 per cent prior to the 2008 provincial election (Romanowska 2009).[19]

These political manoeuvres are supplemented by media campaigns at key moments in the debate on the tar sands, for example, CAPP's fall 2006 media campaign to counter the charge that tar sands operations use a significant amount of water. In the winter of 2008–09, CAPP members also funded an intensive advertising campaign on Alberta television and radio networks to argue against the Royalty Review Panel's recommendations that royalty rates be increased. This was followed by CAPP's launch of a website in winter 2009 ('Canada's Oil Sands: A Different Conversation') to refute high-profile media criticisms of the environmental impacts of tar sands developments. Similarly, in June 2009 executives of the Canadian Heavy Oil Association (CHOA) stated their intention to work – in coordination with other associations, companies, and government agencies – to improve the 'perception of the oil sands' until the message reaches a 'critical mass' (Tracy Grills, CHOA vice president, as quoted in Collison 2009). In 2010, Enbridge began running ads on commercial radio stations throughout British Columbia and Alberta to sell its Northern Gateway pipeline project to northern communities. Then as the debate on the pipeline intensified in 2014 in anticipation of the federal decision on the project, CAPP launched a social media campaign to emphasize the employment and economic benefits of resource extraction. CAPP's 'Canada's Energy Citizens' website and Facebook page invited those who identify as 'energy citizens' to 'join the team' and become an 'industry advocate.'[20]

Industry 'grassroots' community 'engagement' projects are also central in the lobbying effort, for example, CAPP's Energy in Action program (Energy in Action 2009), or Synergy Alberta, a non-profit organization it co-founded with a mission of 'fostering and supporting mutually satisfactory outcomes in Alberta communities' (Synergy Alberta n.d.). Understood more critically, Synergy Alberta is interpreted as a 'civil peacekeeping organization' that measures success by pipelines developed or wells dug and jobs and profits created (Jaremko 2006). Likewise, Enbridge created the Northern Gateway Alliance (a pro-development coalition of local councillors and business people) and a website to create support for its pipeline proposal.

To ensure continued access to tar sands resources and legitimacy to extract within an undemanding environmental regulatory system, the tar sands industry combines government lobbying and political financing with strategic media and community relations and philanthropy. Major players make targeted, high-profile donations to community infrastructure such as recreation, health care, and the performing arts. For example, Suncor has committed millions to develop the Northern Lights Regional Health Foundation's programs (including provision of medical equipment), the Suncor Energy Centre for the Performing Arts, and the Suncor Community Leisure Centre at MacDonald Island in Fort McMurray, which includes CNRL arenas, the Total Fitness Centre, Shell Place (a performance and sport stadium), and the Syncrude Aquatic Centre.

Even more significant is the corporate funding to postsecondary educational institutions. A key element of this strategy is the energy industry's funding of targeted research programs or facilities in colleges and universities, which help to ensure that companies have the labour force and research they require, all heavily subsidized by general tax revenue. An industry publication reported in 2004 that, since the 1970s, approximately $3 billion had been invested by industry and government to support research on augmenting fossil fuel development in the province (Polczer 2004). One key joint initiative is the University of Calgary's Institute for Sustainable Energy, Environment and Economy (ISEEE), founded in 2003 by fossil fuel industry leaders to increase conventional and unconventional oil recovery rates. The institute is aligned with the Alberta government–funded Alberta Energy Research Institute (AERI) (recently rebranded as Alberta Innovates – Energy and Environment Solutions), which was formerly the Alberta Oil Sands Technology and Research Authority (AOSTRA) and has close ties with Alberta's fossil fuel industry. Another obvious example of the oil industry's reach into the university system is Imperial Oil Limited's $10 million commitment to the University of Alberta in 2004 for its Imperial Oil Centre for Oil Sands Innovation. At the time, this was the 'largest investment ever made by Imperial in a university, and the largest single corporate cash commitment ever received by the university's Faculty of Engineering' (University of Alberta 2006). Colleges have also received investments by industry, such as Ledcor Group's (a tar sands construction company) $1.5 million investment in the Northern Alberta Institute of Technology for an applied research chair in oil sands environmental sustainability, and $250,000 in scholarship funds (Healing 2010). Syncrude donated $5 million to Fort McMurray's Keyano College (Syncrude Canada Ltd. 2009, 36).

The tar sands industry works to protect billions of dollars of investments and profits via political lobbying, political funding, public relations, and local 'engagement' activities, including co-opting the public education system. Meanwhile, influenced by these industry strategies and also motivated by revenues from oil development, Alberta's governments have legitimized and protected tar sands projects. The environmental regulatory regime resulting from the overlap of these powerful interests is, unsurprisingly, ineffective. But is Alberta's condition an anomaly among first world cases, or part of a broader trend across oil-dependent 'subnational' cases in North America?

Comparing Environmental Policy in 'First-World' Petro-polities

This chapter has argued that the political economics of the petro-state, primarily marked by the close symmetry of government and industry interests in developing remaining oil reserves, has resulted in the ineffectual environmental policy regime surrounding tar sands developments in Alberta. Here we see public policy as 'the embodiment' of the tensions and interests in the state (Brynt 1992, 18). Alberta is not alone in exhibiting this pattern; it can easily be compared to other first world petro-polities exhibiting parallel environmental policy trends such as Saskatchewan, Newfoundland and Labrador, Alaska, and Wyoming. I outline the main tensions in each below, drawing upon ongoing research on environmental policy in oil-dependent cases in the USA and Canada.

Saskatchewan is on the cusp of copying and perhaps surpassing Alberta's environmental mismanagement of the tar sands (Prebble et al. 2009). The development of the province's tar sands has been overseen by the conservative Saskatchewan Party government. This government has sought to under-bid Alberta for investment, by offering even lower royalty rates and less rigorous environmental processes and standards. Premier Wall's government has frequently been congratulated by the North American business community for its welcoming business climate (Enterprise Saskatchewan 2010). The environmental impacts of this approach may soon be tested in Saskatchewan's tar sands: exploration has occurred and the first tar sands development licence was requested in 2010 (Oilsands Quest Inc. 2010).

Further east, in Newfoundland and Labrador, oil developers are pushing the limits of offshore oil development in ultra-deep wells – even deeper than BP's Deepwater Horizon oil rig – in 'one [of] the most productive marine areas in the world' (Wiese and Ryan 2003, 1091). The environmental risk is potentially great but so are the oil revenues: royalties and other funds from oil in 2009 amounted to nearly 40 per cent of total provincial revenues. This trend will become more pronounced as the province captures more revenue through equity stakes, initially 4.9 per cent, in the new Hebron project (a share hard won from Exxon Mobil by former premier Danny Williams in 2008). Environmental policy trends surrounding oil in this case echo problems in Alberta and exemplify the pro-oil bent of the provincial government: the province has limited marine areas protected from potential oil development; there are no regulations on carbon emissions; standards for routine discharges of oil waste into the ocean are problematic (Fraser, Russell, and von Zharen 2006); and there is a lack of post-development monitoring (Fraser and Ellis 2008) as well as longstanding problems with transparency. It has been difficult to obtain even basic data on oil spills offshore (Fraser, Ellis, and Hussain 2007). Broader still are the questions that have been raised about the potential conflict of interests within the primary environmental regulator, the Canada-Newfoundland and Labrador Offshore Petroleum Board (McCarthy 2010) and the government in general.

Looking across the border, Alaska also struggles with containing the environmental impacts of oil development in the fragile Arctic North Slope fields, developments that have long been documented as causing 'inevitable accumulated undesirable effects' (Committee on Cumulative Environmental Effects of Oil and Gas Activities on Alaska's North Slope 2003, 11). Yet with oil revenues contributing as much as 40 per cent of total state revenue, the sway of the oil industry in Alaska is great, as is the pressure to open new areas to oil drilling, such as the Arctic National Wildlife Refuge (see, for example, Haycox 2002; Ott 2005; Standlea 2006; Strohmeyer 2003). Environmental organizations actively block this expansion in part because of the poor record of environmental protection in northern Alaska. Numerous environmental regulation problems plague oil development in this state, similar to the situation in Alberta. In discussions of oil development, environmental concerns are continuously trumped by development interests due to the primary position the Department of Natural Resources occupies relative to the chronically underfunded and constrained Department of Environmental Conservation and Department of Fish and Game. Multiple, fast-paced public consultation processes overwhelm community capacity to participate effectively. This problem is compounded by the need for a comprehensive environmental impact analysis to ascertain the full range of the developments' effects (hence the National Research Council's emphasis in 2003, via the Committee on Cumulative Environmental Effects of Oil and Gas, on the need for a regional, cumulative, independent analysis; the council noted the lack of basic data on Arctic ecosystems). Another problem concerns the lack of monitoring once projects commence – an issue most evident with regard to air pollution and pipeline spills. Reclamation standards are ambiguous and so rates of land recovery have been slow.

Finally, in Wyoming, rapidly expanding oil and gas production threatens to turn fragile sagebrush habitats for vulnerable and endangered species into densely developed industrial areas. Contaminated water tables, polluted air, and other environmental impacts are increasingly noted near development sites. But in Wyoming, with nearly two-thirds of total state revenue coming from mineral severance taxes that are predominantly based on oil and gas developments (Liu 2008, 24–5), oil and gas production continues to expand. Since the 2001 American National Energy Policy, the pace of oil and gas development has been startling: a decade ago, as much as one-fourth of the state was already leased to oil and gas companies – a move facilitated by the strategic fast-tracking of leasing and streamlining of environmental regulations. The environmental policy trends are similar here to other high oil-dependency cases: plans for land development are either non-existent or not respected, and while leasing expands, there has been no parallel increase in monitoring and enforcement capacity. Regulators at the state and federal level cannot match the pace of development, and this void is filled by companies self-monitoring and self-reporting. Buffer zones and winter stipulations created to protect key species, particularly sage grouse and mule deer, have long been acknowledged as inadequate. Further, reclamation either does not occur (wells are understood to operate indefinitely) or reclamation standards are weak. Non-governmental groups criticizing the standards argue that operators are required to simply 're-seed and walk away.' As in other cases, public consultation processes are frequently criticized as exclusionary or, where the public is actively involved, the timelines and sheer size and complexity of documents make participation difficult. Then public concerns are frequently unheeded in the decision-making process. Finally, here too there is a need for human health and environmental cumulative impact analysis.

Crippled environmental policy in response to the political economy of oil dependence is, upon initial analysis, widespread in these petro-polities. Alberta's condition is not a singular one within the broader North American context. Of course, that these difficult circumstances are shared across jurisdictions is no great comfort. How can Alberta, and other jurisdictions like it, overcome the negative impacts of great resource wealth? The election of a social democratic party to government in 2015 may constitute a challenge to the entrenched interests that have dominated environmental policy making in the past, although it is still too early to tell how profound this challenge will be.

Conclusion: Turning a Resource Curse into a Blessing?

> Does this place called Alberta really exist, or did you make it up?
>
> Cover of *Alberta Views*, December 2010

Alberta exhibits multiple key elements of the resource curse: over the first decade of the twenty-first century, non-renewable resource revenues, driven by energy exports, accounted for as much as half of total provincial revenue,[21] and the direct and indirect GDP impact of the oil and gas industry from 1971 to 2004 has been estimated at 42 per cent of provincial GDP (Mansell and Schlenker 2006, ii–iii). The province is exposed to the negative economic impacts of high dependence on oil, such as non-diversification and revenue unpredictability. Alberta also experiences negative political impacts associated with the resource curse, which have worrisome implications for Alberta's democracy. The province's tax system, which could insulate Alberta's public services from the vicissitudes of oil prices and eventual oil reserve declines, has been eroded; governments have engaged in wasteful spending to shore up their legitimacy (for example, public relations campaigns 'green-washing' the tar sands); citizen engagement with the state has waned (notable in the historically low 40.6 per cent voter turnout in the March 2008 provincial election – the lowest turnout in the last half-century of Canadian provincial elections ['Low Voter Turnout' 2008]); and one party ruled the province from 1971 to 2015. Alongside these negative economic and political impacts is the trend of weakened environmental policy systems and, therefore, the serious negative environmental impacts outlined in this chapter.

Undoubtedly, as demonstrated by the resource curse literature, dependence on natural resources poses particular political-economic challenges to states, as well as environmental challenges. But does oil *determine* these outcomes? Is the 'curse' inevitable? In some cases, like Norway – long held as an example of successful management of resource dependence – governments have overcome the challenges of resource dependence to reap long-term economic and political benefits with comparatively better environmental records. Hence, in Auty's words, the resource curse is 'not an iron law, rather it is a strong recurrent tendency' (1994, 12).

Given this, researchers are now elaborating on how state policy can turn a potential oil curse into a blessing (Bridge 2008; Humphreys, Sachs, and Stiglitz 2007; Rosser 2006; Weinthal and Jones Luong 2006), and these are policy recommendations that would benefit Alberta. First, governments can overcome resource curse outcomes by maintaining public control over the oil industry or, at least, by negotiating contracts with private developers that guarantee that the owners of the resource receive the majority of the benefits. Second, governments should ensure transparency of information concerning oil developments and the impact they have on government revenue and corporate profits. Government and corporate accountability tools advocated by international NGOs such as 'PWYP' (publish what you pay) provide citizens with information on profits from public resources and on how governments spend the resource funds. To this I would add the need for clear and publicly accessible information on the full range of the environmental impacts of the tar sands.[22] Third, governments are advised to manage oil wealth wisely: rather than allow royalties and other funds from oil to enter the government budget as general revenue, governments must save or invest oil revenues as non-renewable assets. Alberta's woefully underfunded Heritage Fund needs to be rejuvenated and restructured to channel oil revenues towards productive, long-term public investments directed by an arm's length body that can protect the fund from wasteful spending. Alberta's oil fund, established by former premier Peter Lougheed in 1976, has been undercut since the Klein government's dramatic reduction of fund payments in the late 1980s. In 2013, the government reported the value of the AHST Fund as $16.4 billion – a meager amount compared to the cumulative value of Alberta's oil production and oil revenues, and in sharp contrast to Norway's $664 billion dollar petroleum fund (Campbell 2013). Funds from a revitalized Albertan sovereign wealth fund could be invested in transitioning the province to a renewable energy economy. Together, these changes would protect Alberta's economy from the volatility of oil prices.

Do Humphreys et al. (2007), in proposing these policy recommendations, assume a long-term-thinking and public-interest-oriented government? Is it naive to imagine an Albertan government implementing these policies in light of the tremendous vested interests in oil development? Is such radical policy change possible in a province where oil has long been viewed as being fundamental to all else? For, as Laird notes, 'Energy built modern Alberta' (2005, 146). Goldberg et al. (2008) argue that the kind of governance in place at the moment when oil dollars reach the state tends to reproduce itself; oil wealth is used to maintain political power and status quo political-economic systems. Can this cycle be broken in Alberta?

Reorienting the Albertan petro-state will be no easy task. But opportunities do exist, particularly through re-articulating and re-engaging with Alberta's diverse political history and culture. These are elaborated, for example, by Doreen Barrie, who retrieves the 'other strands in the ideological tapestry' that have been smothered by a seemingly 'monochromatic' conservatism (2006, xi). Lois Harder undertakes a similar project to provide an analysis of feminism in Alberta that 'disrupts prevailing views of the constancy of the province's conservatism' (2003, 1). Contrary to the one-dimensional stereotype of conservative Alberta is the province's history of early progress towards workers' rights (such as the formation of the One Big Union in Calgary in 1919) and women's rights (for example, the 'Famous Five' Albertan women who fought to have women defined as persons and thereby secured women's right to be appointed to the Canadian Senate in 1929). Alberta was also the birthplace of the Co-operative Commonwealth Federation (CCF) (the precursor to the New Democratic Party), established in Calgary in 1932. It is on this other stream of politics that contemporary progressive civil society organizations draw. Even within conservative politics there is useful debate and diversity of opinion. A notable example is, of course, former Alberta premier Peter Lougheed's advice to the PC party, in 2009, to reassert the province's position as the resource owner, rather than see the resource 'subjugated to the wishes of the petroleum industry, who are basically lessees' (quoted in Radwanski 2009), to control the pace of development to allow time to deal with environmental impacts, and to diversify the economy beyond oil. Hope for reforming the Albertan petro-state is to be found in the possibility of strengthening these debates and transforming them into an effective political movement that re-engages voters to build a new Alberta with a diversified, renewable economy, a sustainable environment, and a re-invigorated democracy.

NOTES

1 This is not to say that government and industry do not face opposition, or that environmental policy does not also incorporate some of the gains of past social struggles by environmentalists and other actors. Other chapters in this volume contribute such additional elements of the broader picture of the terrain of contestation.

2 The common threshold of 'high dependence' is when oil represents one-third of exports, GDP, or government revenues (Atkinson and Hamilton 2003; Ross 2001; Sachs and Warner 1995, 1999, 2001; Stevens 2003; Weinthal and Jones Luong 2006).

3 To date, Syncrude Canada Ltd., Suncor Energy Inc., and Shell Albian Sands Energy are the three major bitumen producers using primarily surface mining methods with in situ methods on the rise. Syncrude is a joint venture among Canadian Oil Sands Limited (36.74 per cent), Imperial Oil Resources (25 per cent), Suncor Energy Oil and Gas Partnership (12 per cent), Sinopec Oil Sands Partnership (9.03 per cent), and smaller shares to Nexen Oil Sands Partnership, Mocal Energy Limited, and Murphy Oil Company Ltd. Shell Albian Sands is operated by Shell Canada Energy, a wholly-owned subsidiary of Royal Dutch Shell.

4 The scientific literature on the environmental impacts of the tar sands, albeit limited, is summarized in Timoney and Lee (2009) with an important new addition provided by Kelly et al. (2009). The Pembina Institute has also provided independent, publicly oriented analysis of the environmental impacts of tar sands developments.

5 Beyond scholarly literature and more publicly oriented research reports, the primary sources for this section include interviews conducted since spring 2007 with policy makers in key Albertan government departments and agencies (Alberta Environment, Alberta Energy, and the Energy and Utilities Board), as well as elected officials and representatives of involved environmental NGOs, social justice NGOs, research and law institutes, and independent researchers. In accordance with an ethics clearance from

the Interdisciplinary Committee on Ethics in Human Research at Memorial University of Newfoundland, these interviewees will remain anonymous. Where direct quoting is necessary, I refer to the interviewee's position in his or her organization.

6 In January 2008 the EUB was divided into the ERCB, mandated to regulate oil, natural gas, tar sands, and coal as well as pipeline developments, and the Alberta Utilities Commission (AUC), regulating the utilities sector (electricity and natural gas markets) (Low 2009).

7 Alberta Energy policy maker, interview with author, 27 April 2007.

8 EUB policy maker (business operations and development), interview with author, 24 April 2007. Note that this political pressure can be easily applied given the EUB staff: at the time of research on this issue in 2007 and 2008, the board was a highly politicized entity with nine board members appointed by cabinet through non-debated orders-in-council.

9 Alberta Environment policy makers (oil and gas policy sector and electricity/minerals sector), interview with author, 23 April 2007. On this point, also see Vlavianos's analysis of the relationship between the EUB and Alberta Environment (2007c, 58–9).

10 Alberta Environment policy maker (oil and gas policy sector and electricity/minerals sector), interview with author, 23 April 2007.

11 Alberta Environment policy maker (Environmental Policy Branch), interview with author, 23 April 23, 2007.

12 For an analysis and criticism of reclamation efforts to date, see Grant et al. (2008).

13 Note that after months of negative press following the oiling of ducks in Syncrude's Aurora tailings pond in April 2008, the Alberta government developed (but has not yet effectively implemented) Directive 074 in February 2009, and it continues to subsidize research in this area. Examples include the 2002 provincial and federal investments in the University of Alberta's Oil Sands Tailing Research Facility, in partnership with tar sands companies, a total of $2.3 million dollar capital investment (Oil Sands Tailing Research Facility n.d.), the 2007 formation of the Imperial Oil-Alberta Ingenuity Centre for Oil Sands Innovation (a partnership between the University of Alberta's Faculty of Engineering and Imperial Oil with funding from the Alberta Government's Alberta Ingenuity Fund), and the Government of Alberta $3 million grant to the University of Alberta's School of Energy and the Environment through the Energy Innovation Fund.

14 As Alberta's Environmental Law Centre Fact Sheet on Oil and Gas Developments and Surface Rights (2006d) explains, even for 'potentially affected surface owners or occupiers,' there is 'no direct notice' when

rights to the land are offered for auction and leased. See also Vlavianos (2007b).

15 See also Vlavianos's (2007a) comments on the 'complete lack of public participation' at crucial stages of the tar sands decision-making process. This issue is explored in detail in Vlavianos (2007b).

16 Fluet and Krogman (2009) studied one public consultation process, the Northern East Slopes Sustainable Resource and Environmental Management Strategy. They found it to be marked by industry capture, ENGO exclusion, and overly narrow terms of debate that excluded the option of non-use or the recognition of intrinsic values. They observed that these consultations 'may be represented as democratic but simultaneously maintain the power relations that produce the current model of economic development' (138). McInnis and Urquhart made a similar point much earlier, describing public participation processes as 'Symbolic, Manipulative Politics' (McInnis and Urquhart 1995, 247).

17 The 2008 Land-Use Framework held out some hope that cumulative effects would be integrated into land-use planning, but the first regional land-use plan to be developed under this framework (the Lower Athabasca Regional Plan) was criticized for privileging energy sector growth over ecological criteria. See Parkins (2011).

18 With Canadian Oil Sands Ltd. coming a distant fifth, however, to the top four, it is worth noting that the average net income of the top four companies in 2013 was $2.7 billion.

19 The Conservative Government was initially pressured to move in this direction by the recommendations of an independent panel (a rare phenomenon in Alberta) that had been appointed by the government to review the province's oil and gas royalties regime, and that had urged the government to increase the royalty rates.

20 As of June 2014, CAPP's web campaign was accessible at www.energycitizens.ca and www.facebook.com/CanadasEnergyCitizens.

21 Data on Alberta's total provincial revenues are from Statistics Canada's table 385-0002 (available at http://www5.statcan.gc.ca/cansim/a26?lang=eng&id=3850002); data on non-renewable resource revenue were provided by a director of policy, planning and external relations with Alberta Energy (e-mail to author, 5 November 2007).

22 Ecologist and statistician Kevin Timoney suggests the government has attempted to cover up health and environmental impacts of the tar sands, calling information control in Alberta 'world class' (CBC 2007). Another example is the Alberta government's reluctance to release updated estimates of the number of migratory birds killed in April 2008 in Syncrude's

tailings pond – which showed that 1,606 had died as opposed to the originally reported 500. Syncrude reported increased numbers in the summer of 2008, but the more alarming count was not released by the government until April 2009. Similarly, the government has been reluctant to admit the seriousness of toxic waste leakages from tailings ponds, even as evidence of this problem mounts (Kelly et al. 2009; Price 2008).

REFERENCES

Acuña, Ricardo. 2007. 'Don't Ask Questions If You Don't Want Answers.' *Vue Weekly*, 10 April. http://vueweekly.com/front/story/dont_ask_questions_if_you_dont_want_answers/ (accessed 5 May 2010).

'Agencies May Clear Oil Sands under Energy Law.' 2008. *Pipeline & Gas Journal*, 1 April, 6.

Akin, David. 2009. 'Ottawa Urged to Ease Regulations.' *Calgary Herald*, 6 May, D4.

Alberta Energy. 2006. *Alberta Oil Sands Royalty Guidelines: Principles and Procedures*. Edmonton.

Alberta Energy. 2007. *Oil Sands Consultations: Multistakeholder Committee Final Report*. Edmonton.

Alberta Energy. 2008. *Alberta's Oil Sands*. http://www.energy.gov.ab.ca/OurBusiness/oilsands.asp (accessed 10 October 2008).

Alberta Environment and Fisheries and Oceans Canada. 2007. *Water Management Framework: Instream Flow Needs and Water Management System for the Lower Athabasca*. Edmonton.

Alberta Royalty Review Panel (William Hunter, Evan Chrapko, Judith Dwarkin, Ken McKenzie, André Plourde, and Sam Spanglet). 2007. *Our Fair Share: Report of the Alberta Royalty Review Panel*. Alberta Royalty Review Panel. Edmonton.

Atkinson, Giles, and Kirk Hamilton. 2003. 'Savings, Growth and the Resource Curse Hypothesis.' *World Development* 31 (11): 1793–1807.

Auty, Richard. 1993. *Sustaining Development in Mineral Economies: The Resource Curse Thesis*. London: Routledge.

Auty, Richard. 1994. 'Industrial Policy Reform in Six Newly Industrializing Countries: The Resource Curse Thesis.' *World Development* 22 (1): 11–26.

Barrie, Doreen. 2006. *The Other Alberta: Decoding a Political Enigma*. Regina: University of Regina.

Bramley, Matthew, Derek Neabel, and Dan Woynillowicz. 2005. 'The Climate Implications of Canada's Oil Sands Development.' Calgary: Pembina Institute.

Bridge, Gavin. 2008. 'Global Production Networks and the Extractive Sector: Governing Resource-Based Development.' *Journal of Economic Geography* 8: 389–419.

Brynt, Raymond. 1992. 'Political Ecology: An Emerging Research Agenda in Third-World Studies.' *Political Geography* 11 (1): 12–36.

Bytnerowicz, Andrzej, Witold Fraczek, Susan Schilling, and Diane Alexander. 2010. 'Spatial and Temporal Distribution of Ambient Nitric Acid and Ammonia in the Athabasca Oil Sands Region.' *Alberta Journal of Limnology* 69 (1): 11–21.

Campbell, Bruce. 2013. 'The Petro-Path Not Taken: Comparing Norway with Canada and Alberta's Management of Petroleum Wealth.' Ottawa: Canadian Centre for Policy Alternatives.

CAPP (Canadian Association of Petroleum Producers). 2008. *What We're Doing: Social Impacts*. 2008. http://www.canadasoilsands.ca/en/what-were-doing/social-impacts.aspx (accessed 25 June 2009).

CAPP. 2009. *Energy in Action*. http://www.capp.ca/aboutUs/events/EnergyInAction/Pages/default.aspx (accessed 21 July 2009).

CAPP. 2011. *Statistical Handbook for Canada's Upstream Petroleum Industry*. http://www.capp.ca/GetDoc.aspx?DocId=184463&DT=NTV (accessed 12 February 2012).

Carter, Angela. 2011. 'Environmental Policy in a Petro-State: The Resource Curse and Political Ecology in Canada's Oil Frontier.' Doctoral diss., Cornell University.

CBC. 2007. *Crude Awakening*. Documentary aired on CBC Television's *The National*, 13 December.

CBC News. 2007. 'U.S. Urges "Fivefold Expansion" in Alberta Oilsands Production.' 18 January. http://www.cbc.ca/news/canada/u-s-urges-fivefold-expansion-in-alberta-oilsands-production-1.638883 (accessed 21 March 2016).

Chaudry, Kiren Aziz. 1997. *The Price of Wealth: Economies and Institutions in the Middle East*. Ithaca: Cornell University Press.

Collison, Melanie. 2009. 'From Education and Networking to Advocacy.' *Oilsands Review* 2009 (June): 48–50.

Committee on Cumulative Environmental Effects of Oil and Gas Activities on Alaska's North Slope. 2003. *Cumulative Environmental Effects of Oil and Gas Activities on Alaska's North Slope*. Washington, DC: National Academies Press.

Cooper, David. 2009. 'Shell Wins ERCB Decision; Environmentalists Lose Bid for New Hearing on Mine Expansions.' *Edmonton Journal*, 13 June, E1.

Cross, P. 2008. 'The Role of Natural Resources in Canada's Economy.' *Canadian Economic Observer* (Statistics Canada), November, 3.1–3.10.
Dunn, Fred. 2007. *Annual Report of the Auditor General of Alberta, 2006–2007.* Edmonton: Auditor General of Alberta.
Dyer, Simon. 2010a. 'Oil Sands Regulator Does Not Enforce Toxic Tailings Rules.' Pembina Institute press release, 23 April 23. http://www.oilsandswatch.org/media-release/2004 (accessed 26 April 2010).
Dyer, Simon, Jeremy Moorhouse, Katie Laufenberg, and Rob Powell. 2008. *Under-Mining the Environment: The Oil Sands Report Card.* Drayton Valley: Pembina Institute/WWF.
Ecojustice. 2013. 'Legal Backgrounder: Bill 2: Responsible Energy Development Act.' http://www.ecojustice.ca/files/reda-backgrounder-may-2013/at_download/file (accessed 20 June 2014).
Ecojustice, Toxics Watch Society, and the Pembina Institute. 2009. 'Backgrounder: Court Ruling Demonstrates Oil Sands Review Process Broken.' 28 August. http://pubs.pembina.org/reports/bg-leave-denied-oil-sands-process-broken.pdf (accessed 21 February 2012).
Energy Resources Conservation Board. 2010. *ERCB Approves Fort Hills and Syncrude Tailings Pond Plans with Conditions.* https://www.aer.ca/documents/news-releases/NR2010-05.pdf (accessed 21 March 2016).
Enterprise Saskatchewan. 2010. *What the World Is Saying about Saskatchewan.* http://www.enterprisesaskatchewan.ca/quotes (accessed 14 July 2010).
Environment Canada. 2010. *Key Data Tables: Table 3: Summary of GHG Emissions by Facility (2008).* http://www.ec.gc.ca/ges-ghg/default.asp?lang=En&n=DF08C7BA-1 (accessed 8 July 2010).
Environmental Defence, Équiterre, and Pembina Institute. 2010. *Duty Calls: Federal Responsibility in Canada's Oil Sands.* http://pubs.pembina.org/reports/ed-fedpolicy-report-oct2010-web-redo.pdf (accessed 12 February 2012).
Environmental Law Centre. 2007a. *Fact Sheet: Environmental Appeals Boards (EAB).* Environmental Law Centre. https://www.elc.ab.ca/pdffiles/TheEnvironmentalAppealsBoard_EAB.pdf (accessed 20 April 2007).
Environmental Law Centre. 2007b. *Fact Sheet: Environmental Approvals and Licences.* Environmental Law Centre. https://www.elc.ab.ca/pdffiles/EnvironmentalApprovalsandLicences.pdf (accessed 20 April 2007).
Environmental Law Centre. 2007c. *Fact Sheet: Environmental Assessment.* Environmental Law Centre. https://www.elc.ab.ca/pdffiles/Envrionmental_Assessments.pdf (accessed 20 April 2007).
Environmental Law Centre. 2007d. *Fact Sheet: Oil and Gas Developments and Surface Rights.* Environmental Law Centre. https://www.elc.ab.ca/

pdffiles/OilandGasDevelopmentandSurfaceRights.pdf (accessed 20 April 2007).

Fluet, Colette, and Naomi Krogman. 2009. 'The Limits of Integrated Resource Management in Alberta for Aboriginal and Environmental Groups: The Northern East Slopes Sustainable Resource and Environmental Management Strategy.' In *Environmental Conflicts and Democracy in Canada*, ed. Laurie Adkin, 140–56. Vancouver: UBC Press.

Fraser, Gail, and Joanne Ellis. 2008. 'Offshore Hydrocarbon and Synthetic Hydrocarbon Spills in Eastern Canada: The Issue of Follow-Up and Experience.' *Journal of Environmental Assessment Policy and Management* (10): 173–87.

Fraser, Gail, Joanne Ellis, and Leena Hussain. 2007. 'An International Survey of Governmental Disclosure of Hydrocarbon Spills from Offshore Oil and Gas Installations.' *Marine Pollution Bulletin* (56): 9–13.

Fraser, Gail, Janet Russell, and Wyndylyn von Zharen. 2006. 'Produced Water from Offshore Oil and Gas Installations on the Grand Banks, Newfoundland and Labrador: Are the Potential Effects to Seabirds Sufficiently Known?' *Marine Ornithology* (34): 147–56.

Gale, Fred. 2000. 'Regulating Accumulation, Guarding the Web: A Role for (Global) Civil Society?' In *Nature, Production, Power: Towards an Ecological Political Economy*, ed. Fred Gale and R. Michael M'Gonigle, 195–214. Cheltenham: Edward Elgar.

Gelb, Alan. 1988. *Oil Windfalls: Blessing or Curse?* New York: Oxford University Press.

Goldberg, Ellis, Erik Wibbels, and Eric Mvukiyehe. 2008. 'Lessons from Strange Cases: Democracy, Development, and the Resource Curse in the U.S. States.' *Comparative Political Studies* 41 (4/5): 477–514.

Government of Alberta. 2006. *Investing in Our Future: Responding to the Rapid Growth of Oil Sands Development.* Oil Sands Ministerial Strategy Committee final report, 29 December. Edmonton.

Government of Alberta. 2008. *Alberta's 2008 Climate Change Strategy.* Edmonton.

Greenpeace Canada. 2009. 'Greenpeace Strikes Again: Activists Occupy Shell Upgrader Expansion Site in Fort Saskatchewan.' 2 October. http://www.greenpeace.org/canada/en/recent/stopstarsands3_action/ (accessed 12 February 2012).

Griffiths, Mary, Amy Taylor, and Dan Woynillowicz. 2006. *Troubled Waters, Troubling Trends: Technology and Policy Options to Reduce Water Use in Oil and Oil Sands Development in Alberta.* Drayton Valley, AB: Pembina Institute.

Griffiths, Mary, and Dan Woynillowicz. 2009. *Heating Up in Alberta: Climate Change, Energy Development and Water*. Drayton Valley, AB: Pembina Institute.

Harder, Lois. 2003. *State of Struggle: Feminism and Politics in Alberta*. Edmonton: University of Alberta Press.

Harrison, Lynda. 2013. 'Environmental Lawyers Prepare For Single Energy Regulator.' *Daily Oil Bulletin*, 13 June.

Harrison, Trevor. 2005. 'The Best Government Money Can Buy? Political Contributions in Alberta.' In *The Return of the Trojan Horse: Alberta and the New World (Dis)Order*, ed. Trevor Harrison, 95–112. Montreal: Black Rose.

Haycox, Stephen. 2002. *Frigid Embrace: Politics, Economics, and the Environment in Alaska*. Corvallis: Oregon State University Press.

Healing, Dan. 2010. 'Ledcor, NAIT Announce Oilsands Research Initiative.' *Calgary Herald*, 23 March.

Hoberg, George, and Jeffrey Phillips. 2011. 'Playing Defence: Early Responses to Conflict Expansion in the Oil Sands Policy Subsystem.'*Canadian Journal of Political Science* 44 (3): 507–27.

Holroyd, Peggy, Simon Dyer, and Dan Woynillowicz. 2007. *Haste Makes Waste: The Need for a New Oil Sands Tenure Regime*. Drayton Valley, AB: Pembina Institute.

Humphreys, Macartan, Jeffrey Sachs, and Joseph Stiglitz, eds. 2007. *Escaping the Resource Curse*. New York: Columbia University Press

Irving, Jacob. 2008. 'Oil Sands: An Industry Overview.' Paper presented at the North American Centre for Transborder Studies Tour conference, Fort McMurray, 3 September. http://www.google.ca/url?sa=t&rct=j&q=jacob%20irving%

Jaremko, Gordon. 2006. 'Disaster Relief Now Means Healing Relations.' *Edmonton Journal*, 20 November, A16.

Jeffries, Dean, Raymond Semkin, and John Gibson. 2010. 'Recently Surveyed Lakes in Northern Manitoba and Saskatchewan, Canada: Characteristics and Critical Loads Of Acidity.' *Journal of Limnology* 69 (1): 45–55.

Jensen, Nathan, and Leonard Wantchekon. 2004. 'Resource Wealth and Political Regimes in Africa.' *Comparative Political Studies* 37 (7): 816–41.

Karl, Terry. 1997. *The Paradox of Plenty: Oil Booms and Petro-States*. Berkeley: University of California Press.

Kelly, Erin, Jeffrey Short, David Schindler, Peter Hodson, Mingsheng Ma, Alvin Kwan, and Barbra Fortin. 2009. 'Oil Sands Development Contributes Polycyclic Aromatic Compounds to the Athabasca River and Its Tributaries.' *Proceedings of the National Academy of Sciences of the United States of America* 106 (52): 22346–51.

Kennett, Steven. 2007. *Closing the Performance Gap: The Challenge for Cumulative Effects Management in Alberta's Athabasca Oil Sands Region*. Calgary: Canadian Institute of Resources Law.

Laird, Gordon. 2005. 'Spent Energy: Re-fueling the Alberta Advantage.' In *The Return of the Trojan Horse: Alberta and the New World (Dis)Order*, ed. Trevor Harrison, 156–72. Montreal: Black Rose.

Le Billon, Philippe, and Angela Carter. 2012. 'Securing Alberta's Tar Sands: Resistance and Criminalization on a New Energy Frontier.' In *Natural Resources and Social Conflict: Towards Critical Environmental Security*, ed. Matthew A. Schnurr and Larry A. Swatuk, 170–92. London: Palgrave Macmillan.

Liu, Wenlin. 2008. *Wyoming Economic Insight and Outlook*. Cheyenne: Economic Analysis Division, State of Wyoming.

'Low Voter Turnout in Alberta Election Being Questioned.' 2008. *CBC News*, March 5. http://www.cbc.ca/canada/edmonton/story/2008/03/05/edm-turnout.html (accessed 5 March 2008).

Macdonald, Douglas. 2010. 'Factors Influencing the Ability of the Oil and Gas Industry to Influence Government of Canada Climate Change Policy.' Paper presented at conference of the Canadian Environmental Studies Association, Concordia University, Montreal, 1 June 1.

MacNeil, Robert. 2014. 'Canadian Environmental Policy Under Conservative Majority Rule.' *Environmental Politics* 23(1): 174–8.

Mansell, Robert, and Ron Schlenker. 2006. 'Energy and the Alberta Economy: Past and Future Impacts and Implications.' *Alberta Energy Future Project*. Calgary: Institute for Sustainable Energy, Environment and Economy, University of Calgary. http://www.iseee.ca/media/uploads/documents/AB%20Energy%20Futures/policypapers/1-Energy%20and%20the%20Alberta%20Economy_%20Past%20and%20Future%20Impacts%20and%20Implications.pdf (accessed 12 February 2012).

McCarthy, Shawn. 2010. 'Newfoundland Rejects Calls to Change Oil Industry Regulator.' *Globe and Mail*, 11 May, B3.

McDermott, Vincent. 2015. 'With No Financial Support for 2016, CEMA Prepares to Shut Down.' *Fort McMurray Today*, 10 December. http://www.fortmcmurraytoday.com/2015/12/10/with-no-financial-support-for-2016-cema-prepares-to-shut-down (accessed 21 March 2016).

McGregor, Glen. 2010. 'Prentice Leads Way in Cabinet Contacts with Lobbyists: Analysis.' *Canwest News Service*, 9 January. http://www.canada.com/news/Prentice+leads+cabinet+contacts+with+lobbyists+analysis/2425057/story.html (accessed 9 January 2010).

McInnis, John, and Ian Urquhart. 1995. 'Protecting Mother Earth or Business?: Environmental Politics in Alberta.' In *The Trojan Horse: Alberta and the Future of Canada*, ed. Gordon Laxer and Trevor Harrison, 239–53. Montreal: Black Rose.

Mehlum, Halvor, Karl Moene, and Ragnar Torvik. 2006. 'Institutions and the Resource Curse.' *Economic Journal* 116: 1–20.

Miller, Byron. 2007. 'Green Cities Are Great Cities: Making Alberta's Cities Global Leaders in the Fight against Climate Change.' In *Alberta's Energy Legacy: Ideas for the Future*, ed. Robert Roach, 133–54. Calgary: Canada West Foundation.

Mittelstaedt, Martin. 2008. 'Polluted Tar Sands Ponds Leaking, Report Indicates.' *Globe and Mail*, 8 December, A6.

Nankani, Gobind T. 1979. *Development Problems of Mineral Exporting Countries*. World Bank staff working paper 354. Washington, DC: World Bank.

Neumayer, Eric. 2004. 'Does the 'Resource Curse' Hold for Growth in Genuine Income as Well?' *World Development* 32 (10): 1627–40.

Obad, Joe. 2010. 'Where Will the ERCB Draw the Line on Tailings Ponds?' *Water Matters*, 28 June. http://www.water-matters.org/story/384 (accessed 15 July 2010).

Oil Sands Experts Group Workshop: Security and Prosperity Partnership of North America. 2006. *Oil Sands Workshop SPP Report*. Houston: Natural Resources Canada/U.S. Department of Energy.

Oilsands Quest Inc. 2010. 'Oilsands Quest Files for Environmental Approval of Axe Lake SAGD Project.' Press release, 17 May. http://www.oilsandsquest.com/pdf/BQI_Axe_Lake_filings_May.pdf (accessed 21 February 2012).

Oil Sands Tailing Research Facility. 2010. *Funding*. http://www.ostrf.com/funding (accessed 15 July 2010).

'The 100 Largest Oil and Gas Producers in Canada: Alberta Oil Magazine's 100 Ranking for 2014.' 2014. *Alberta Oil*, 26 May. http://www.albertaoilmagazine.com/2014/05/the200-100-largest-oil-gas-producers-canada/ (accessed 16 May 2015).

Ott, Riki. 2005. *Sound Truth and Corporate Myths: The Legacy of the Exxon Valdez Oil Spill*. Cordova: Dragonfly Sisters.

Parkins, John R. 2011. 'Deliberative Democracy, Institution Building, and the Pragmatics of Cumulative Effects Assessment.' *Ecology and Society* 16 (3): 20.

Pembina Institute. 2007 'Government Protects Oil Sands Industry, Fails to Protect Athabasca River.' 2 March. http://www.pembina.org/media-release/1384 (accessed 9 October 2008).

Pembina Institute. 2010. *Alberta's Oil Sands*. http://www.oilsandswatch.org/os101/alberta (accessed 1 May 2010).

Plourde, André. 2010. 'Oil and Gas in the Canadian Federation.' Paper presented at the conference Canadian-United States Energy Issues after Copenhagen: Oil Sands and Energy Interdependence, Buffett Center for International and Comparative Studies, Northwestern University, Evanston, IL, 28 May.

Polczer, Shaun. 2004. 'Alberta Government Grants $1 Million To U of C Research Institute.' *Daily Oil Bulletin*, 13 August.

Prebble, Peter, Ann Coxworth, Terra Simieritsch, Simon Dyer, Marc Huot, and Helene Walsh. 2009. *Carbon Copy: Preventing Oil Sands Fever in Saskatchewan*. Drayton Valley, AB: Pembina Institute/Saskatchewan Environmental Society/Canadian Parks and Wilderness Society.

'Prentice Defends Oilsands Following National Geographic Article.' 2009. *CBC News*. 25 February. http://www.cbc.ca/canada/story/2009/02/25/oilsands-articles.html (accessed 1 June 2009).

Price, Matt. 2008. *The Tar Sands' Leaking Legacy*. Toronto: Environmental Defence.

Radwanski, Adam. 2009. 'The Premier Who Helped Kick-Start Alberta's Oil Sands Development Has Emerged as a Forceful Critic of How It Has Proceeded.' *Globe and Mail*, 8 June, B2.

Richardson, Lee. 2007. *The Oil Sands: Toward Sustainable Development*. Ottawa: House of Commons Standing Committee on Natural Resources.

Read, Andrew. 2013. 'Alberta's New Monitoring Bill Mixes Science with Politics.' Pembina Institute. 15 November. http://www.pembina.org/blog/764 (accessed 20 June 2014).

Romanowska, Patrycja. 2009. 'Political Penalty: A Year after a Contentious Royalty Hike, Alberta Tories Saw Corporate Campaign Donations Nearly Cut in Half.' *Alberta Oil*, 2 February. http://www.albertaoilmagazine.com/?p=631#more-631 (accessed 10 October 2010).

Ross, Michael. 2001. 'Does Oil Hinder Democracy?' *World Politics* 53 (3): 325–61.

Rosser, Andrew. 2006. 'The Political Economy of the Resource Curse: A Literature Survey.' Brighton: Institute of Development Studies working paper 268, University of Sussex.

Sachs, Jeffrey, and Andrew Warner. 1995. 'Natural Resource Abundance and Economic Growth.' Working paper no. 5398. Cambridge, MA: National Bureau of Economic Research.

Sachs, Jeffrey, and Andrew Warner. 1999. 'The Big Push, Natural Resource Booms and Growth.' *Journal of Development Economics* 59: 43–76.

Sachs, Jeffrey, and Andrew Warner. 2001. 'The Curse of Natural Resources.' *European Economic Review* 45: 827–38.

Schneider, Richard, and Simon Dyer. 2006. *Death by a Thousand Cuts: Impacts of In Situ Oil Sands Development on Alberta's Boreal Forest*. Drayton Valley, AB: Pembina Institute/Canadian Parks and Wilderness Society.

'Scientist Apologizes to Oilsands Researchers.' 2010. *CBC News*, 21 June 21. http://www.cbc.ca/technology/story/2010/06/21/edmonton-mceachern-defamatory-apology.html (accessed 21 June 2010).

Standlea, David. 2006. *Oil, Globalization, and the War for the Arctic Refuge*. Albany: State University of New York Press.

Stelmach, Ed (premier). 2007. *Premier Ed Stelmach: Speeches: Canadian Association of Petroleum Producers (CAPP) Oil and Gas Investment Symposium*. http://www.premier.alberta.ca/speeches/speeches-2007-june-19-CAPP.cfm (accessed 21 July 2009; the link has since been removed from the premier's website).

Stevens, Paul. 2003. *Resource Impact: Curse or Blessing? A Literature Survey*. Dundee: IPIECA.

Strohmeyer, John. 2003. *Extreme Conditions: Big Oil and the Transformation of Alaska*. Anchorage: Cascade.

Syncrude Canada Ltd. 2009. *2008/09 Sustainability Report*. http://www.syncrude.ca/assets/pdf/sustainability-reports/SyncrudeSD2008-2009.pdf (accessed 21 March 2016).

Synergy Alberta. N.d. 'Vision and Mission." http://www.synergyalberta.ca/vision-and-mission (accessed 12 February 2012).

Thomson, Graham. 2009. *Burying Carbon Dioxide in Underground Saline Aquifers: Political Folly or Climate Change Fix?* Toronto: Munk Centre for International Studies, University of Toronto.

Timoney, Kevin. 2007. *A Study of Water and Sediment Quality as Related to Public Health Issues, Fort Chipewyan, Alberta*. Prepared on behalf of the Nunee Health Board Society, Fort Chipewyan, Alberta, 11 November, http://www.borealbirds.org/resources/timoney-fortchipwater-111107.pdf.

Timoney, Kevin, and Peter Lee. 2009. 'Does the Alberta Tar Sands Industry Pollute? The Scientific Evidence.' *Open Conservation Biology Journal* 3: 65–81.

'The Top 100 Canadian Publicly Traded Oil & Gas Companies.' 2010. *Alberta Oil*, 7 June. http://www.albertaoilmagazine.com/2010/06/top-100-canadian-oil-and-gas-producers/?sort=net (accessed 3 August 2011).

University of Alberta, Faculty of Engineering. 2006. 'Imperial Oil Contributes $10 Million for Oil Sands Research.' April 17. Accessed February 22, 2012. http://www.uofaweb.ualberta.ca/jmshaw/news.cfm?story=45280.

Urquhart, Ian. 2005. 'Alberta's Land, Water, and Air: Any Reasons Not to Despair?' In *The Return of the Trojan Horse: Alberta and the New World (Dis)Order*, ed. Trevor Harrison, 136–55. Montreal: Black Rose.

Valentine, Peter. 2008. *Building Confidence: Improving Accountability and Transparency in Alberta's Royalty System*. http://www.energy.alberta.ca/Org/pdfs/Valentine_ABRoyalty.pdf (accessed 1 November 2010).

Vlavianos, Nickie. 2006. *The Potential Application of Human Rights Law to Oil and Gas Development in Alberta: A Synopsis*. Calgary: Alberta Civil Liberties Research Centre, Canadian Institute of Resources Law.

Vlavianos, Nickie. 2007a. 'Key Shortcomings in Alberta's Regulatory Framework for Oil Sands Development.' *Resources* 100: 1–6.

Vlavianos, Nickie. 2007b. 'Public Participation and the Disposition of Oil and Gas Rights in Alberta.' *Journal of Environmental Law and Practice* 17 (3): 205–33.

Vlavianos, Nickie. 2007c. *The Legislative and Regulatory Framework for Oil Sands Development in Alberta: A Detailed Review and Analysis*. Calgary: Canadian Institute of Resources Law.

Weinthal, Erika, and Pauline Jones Luong. 2006. 'Combating the Resource Curse: An Alternative Solution to Managing Mineral Wealth.' *Perspectives on Politics* 4 (1): 35–53.

Wenig, Michael. 2004. 'Who Really Owns Alberta's Natural Resources?' *LawNow* (December 2003/January 2004).

Wenig, Michael, Arlene Kwasniak, and Michael Quinn. 2006. 'Water under the Bridge? The Role of Instream Flow Needs (IFNs) in Federal and Interjurisdictional Management of Alberta's Rivers.' Paper presented at the Water: Science and Politics conference of the Alberta Society of Professional Biologists, Calgary, 25–28 March.

Wheeler, David. 1984. 'Sources of Stagnation in Sub-Saharan Africa.' *World Development* 12 (1): 1–23.

Wiese, Francis, and Pierre Ryan. 2003. 'The Extent of Chronic Marine Oil Pollution in Southeastern Newfoundland Waters Assessed Through Beached Bird Surveys 1984–1999.' *Marine Pollution Bulletin* (46): 1090–1101.

Woynillowicz, Dan. 2006. 'Presentation to the Oil Sands Multi-Stakeholder Committee.' Fort McMurray, 19 September. http://www.google.com/url?sa=t&rct=j&q=dan%20woynillowicz%20presentation%20to%20the%20oil%20sands%20multi-stakeholder%20committee&source=web&cd=1&ved=0CCIQFjAA&url=http%3A%2F%2Fpubs.pembina.org%2Freports%2FOilsands_PCP_Dan_Ft.McMurray.pdf&ei=F1U4T4RWia-DB8WojOgF&usg=AFQjCNHtfpSrMBpDAnNlzxq32voTVP-Cpw&cad=rja (accessed 12 February 2012).

6 Turning Up the Heat: Hegemonic Politics in a First World Petro-State

LAURIE E. ADKIN AND BRITTANY J. STARES

There's always going to be individuals or organizations who want to rid the world of fossil fuel reliance, and let's be clear – that is what's behind most of these campaigns. That's simply not going to happen.

(Alberta Energy Minister Ron Liepert, quoted in Krugel 2010)

This chapter offers a close examination of the hegemonic politics of a first world petro-state confronted with growing opposition to fossil capitalism from local and globally networked actors. Shoring up consent 'domestically' has entailed re-suturing the ruptures opened up in multiple places, such as a housing crisis in Fort McMurray, a highly skewed provincial labour market, antagonized farmers and ranchers, or the budgetary effects of wildly fluctuating provincial revenues. Increasingly, Progressive Conservative[1] governments have been obliged to counter pressures for change emanating from opposition beyond the province's borders (publics, NGOs, governments, and international bodies). We cannot, of course, examine every aspect of hegemonic politics in this context; our focus is, rather, on the Stelmach governments' responses to environmental criticisms of the oil sands.[2] These responses include the restructuring of government institutions to attempt to better manage the multiple contradictions thrown up by the hydrocarbon-extractive political ecology; policy initiatives and decisions; political marketing or branding campaigns; and the government's use of sustainable development and 'nativist' discourses. While spending on environmental regulatory capacity has declined in recent decades, and investment in 'green transition' has not taken off, spending on government 'communication strategy,' branding exercises, international promotion of the government's 'clean energy' brand for the province, and subsidization of research aimed at extending the life of fossil capitalism have grown.

We identify the criticisms or campaigns that the Stelmach government and its successors have perceived to be most threatening, and document a clear shift from concern with 'local' legitimation of the development model to a concern with international legitimation, manifested by actions taken since 2007. This shift reflects the spreading environmental campaigns against the oil sands and the coalescing of networks and coalitions that link local points in a global web of fossil fuel production, refining, and transportation. In Canada, too, linkages are being made among climate change (especially since the intense Kyoto debate in 2002), the local environmental and social harms associated with fossil fuel extraction (oil sands mining, coal mining, coalbed methane, sour gas, off-shore deep-water drilling – the last under intensive scrutiny following the BP well blowout in the Gulf of Mexico in 2010), fuel transportation and storage (pipelines, shipping by sea, LNG terminals), upgrading, and refining. Communities across North America are engaged in struggles against the environmental and health risks and social justice issues associated with each link in these chains.

While the organizational and discursive-strategic transformations of opposition to fossil capitalism are not explored in depth in this chapter (see Haluza-DeLay and Carter, this volume), such opposition constitutes the 'other part' of the dialectic that shapes hegemonic politics. The government's 'clean energy' and 'responsible environmental management' branding exercises are clearly responses to the 'dirty oil' discourse of environmental organizations targeting US and European publics. This chapter documents the increasing activity and resources devoted to such legitimation efforts – efforts that support the characterization of the Alberta state as a first world petro-state.[3] Our analysis, however, raises another question: Why has the Alberta state been so poorly equipped, or so unwilling, to respond to the multiple crises of fossil capitalism by investing in a transitional (post-carbon) development strategy? What does its intransigence tell us not only about the nature of the petro-state, but also about the global political economy in which Alberta's political ecology is imbedded? What roles have civil society actors (including the forces that shape public opinion and conceptions of citizenship) played in shifting the grounds of these hegemonic struggles? While much more work needs to be done to answer these questions, we hope that this and other studies in this volume will

Figure 6.1. Billboard sponsored by Corporate Ethics International, June 2010 campaign called 'Rethink Alberta.'

help to develop a more comprehensive picture of the obstacles to, and potential for, a greener future for Alberta.

A Note on Research

To document the government's strategy since 2006 in response to the critics of fossil capitalism, we undertook several analyses. One was of governmental discourse itself, as manifested in media statements, speeches, blog entries, ministerial reports, or any other government documents that addressed a public audience and that pertained to the environmental management of the oil sands, or the role of the oil sands in the provincial economy.[4] We sought to identify specific patterns and recurring elements that constituted a governmental discourse about the role of the oil sands in the provincial economy versus other possible paths of development, the environmental sustainability of the oil sands, and the depiction of critics of this model.

Like Maartin Hajer (1995, 44), we view a discourse as 'a specific ensemble of ideas, concepts, and categorizations ... through which meaning is given to physical and social realities.' Rooted in assumptions about human nature and non-human nature as well as social relations, 'discourses construct meanings and relationships, helping to define common sense and legitimate knowledge' (Dryzek 2005, 9). Not only do discourses aim to influence public perceptions and opinions about issues (such as the necessity of exploiting non-conventional fossil fuel reserves), and hence the range of policy options that may be 'legitimately' pursued, but they do so in relation to other, contemporaneous discourses, seeking to counter conflicting interests, beliefs, or types of knowledge. Recognizing that discourses can only be comprehended in the broader social context in which they operate, key target audiences (provincial, Canadian, and international publics) and salient events in the multi-scalar environmental campaigns were considered in reconstructing the discourse of the provincial government since 2006.

Second, we attempted to identify all areas of government spending related to the defence of oil sands exploitation in provincial, national, and international contexts, and to map these, roughly, against developments in the environmental campaigns. It soon became clear that isolating promotional/communication expenses related to the oil sands would not be a straightforward task. Numerous ministries are involved in the promotion of investment in the energy sector (Alberta Energy, Alberta Environment, Alberta Treasury Board, Sustainable

Resource Development,[5] the Ministry of International and Intergovernmental Relations, and the Executive Council, which includes the Public Affairs Bureau). Other ministries are also involved in consultations, policy development, or public relations work related to the oil sands, such as the ministries of Aboriginal Relations or Infrastructure. Ministries have their own communications staffs, who do not record the time they devote to particular projects. Government budget documents use broad and often uninformative categories; spending on energy public relations is often bundled into broader initiatives. Clearly itemized expenditure data for the Alberta International Offices were unavailable. Most useful were publicly available figures for the international travel of government representatives, which proved key in identifying an increased effort from 2008 onward to defend the oil sands to international investors, governments, and publics. These findings are presented in figures 6.5 and 6.6.

Third, it was necessary to verify the government's claims regarding its expenditures on measures to reduce greenhouse gases or to promote renewable sources of energy or energy conservation, and to investigate at least cursorily the sums being channelled to corporate, higher education, or multi-partner agencies to conduct such research. Here, our aim was to get an overall picture of how the direction of resources by the government corresponds to its stated claims and relates to its discursive strategy to assuage critics of oil sands development.

Political-Economic Context

Oil sands mining operations have expanded dramatically since 1999, as indicated by the figures on oil sands expenditures in table 6.1 and illustrated in figure 6.2. Corporate investment in oil sands mining (measured by capital and operating costs combined, as reported by members of the Canadian Association of Petroleum Producers, or CAPP) grew steadily from approximately $3.6 billion in 1997 to $10.5 billion in 2004. Then in 2005 investment jumped significantly to $16.7 billion, reaching $47.3 billion in 2012. New capital investment in the oil sands outstripped capital investment in the conventional oil sector in Alberta for the first time in 2008, and did so again in each of 2011, 2012, and 2013 (Masson 2014, 7).

The significant increase in the price of crude oil in 1999–2000 (illustrated in figure 6.3) contributed to making investment in the oil sands more attractive; this trend continued to 2008, when crude oil prices reached a high of $138/barrel before a worldwide recession temporarily dampened oil and gas prices.

In 2008, for the first time, corporate revenues from the oil sands surpassed revenues from natural gas. Table 6.1 and figure 6.4 also show a hefty increase in royalty revenues from $819 million in 2005 to $3.5 billion in 2008.

Along with the oil sands–driven economic boom in the provincial economy came the less desirable effects documented in other chapters: rapid inflation, housing and labour shortages, lagging infrastructure, and increased drug use and prostitution servicing predominantly male labour camps, as well as international attention to the negative consequences of the oil sands for climate change and for First Nations communities living near to and downstream of the oil sands.

For these reasons, the Conservative governments led by Premier Ed Stelmach (who became party leader and premier in December 2006, and whose government was subsequently re-elected in March 2008) faced greater challenges of legitimation than earlier Conservative governments led by Premier Ralph Klein (from 1993 to 2006).[6] When Stelmach became premier, there was evidence of widespread concern among Albertans that oil sands development was out of control.[7] In addition, there was a growing sentiment that the citizens of the province were not getting a fair rent from the exploitation of oil and gas resources.[8] One of Stelmach's first acts as premier was to appoint a panel of experts to review the provincial royalty regime and make recommendations for changes (discussed further below). Another development faced by the Stelmach government was the linkage of a previously marginalized (if persistent) local environmental opposition to global counterparts, particularly those based in the USA (the major market for Alberta's oil exports) and Europe (where publics have been pushing for governmental action to stop global warming).

Institutional Responses to Rising Opposition to the Oil Sands

Provincially, by the time Ed Stelmach became premier in 2006, environmental organizations were pushing within multi-stakeholder bodies for restraints on oil sands expansion. The Pembina Institute was publishing one report after another on the environmental consequences of oil sands emissions and water use. Polling data were showing rising levels of concern among Albertans about the negative environmental and social effects of 'oil sands fever' (Pembina 2007). These develop-

Table 6.1 Oil Sands Investment, Revenues, and Crude Oil Prices, 1986–2012

Year	Oil sands expenditures ($ millions)			Average yearly crude oil prices $CAD/bbl (highest crude oil price during the year in parentheses)[b]	Value of Alberta producers' sales ($ millions)		
	Total capital expenditures[a]	Total operating costs	Royalties paid for in situ and mining production		Crude oil and condensate	Oil sands[c]	Natural gas
1986				20.19	6268	1710	5049
1987				24.00	7847	2204	4195
1988				18.32	6088	1696	4374
1989				22.18	6897	2161	4572
1990				27.64	8210	2799	4667
1991				23.39	6633	2255	4344
1992				23.52	6835	2455	4812
1993				21.78	6428	2376	6648
1994				21.87	6868	2672	8091
1995				24.05	7579	3257	6200
1996				29.23	9124	4033	7760
1997	1915	1665	270	27.64	7995	4037	9455
1998	1543	1654	67	20.08	5289	3076	9587
1999	2372	1784	269	27.35 (38.31)	7062	4887	12507
2000	4223	2289	816	44.33 (52.35)	10757	8044	23091
2001	5907	2753	265	39.13 (44.28)	8388	6864	27763
2002	6751	2557	182	39.91 (45.19)	8455	9258	19176
2003	5048	3794	274	43.14 (53.71)	8599	11159	30264
2004	6183	4341	769	52.54 (64.52)	9858	14940	31082
2005	10437	6305	819	68.72 (78.73)	11885	17742	41198
2006	14337	8051	2187	72.77 (85.62)	12248	23259	32820

Table 6.1 (*Concluded*)

Year	Oil sands expenditures ($ millions)			Average yearly crude oil prices $CAD/bbl (highest crude oil price during the year in parentheses)[b]	Value of Alberta producers' sales ($ millions)		
	Total capital expenditures[a]	Total operating costs	Royalties paid for in situ and mining production		Crude oil and condensate	Oil sands[c]	Natural gas
2007	18065	8135	2716	76.35 (89.63)	12644	24059	31987
2008	18113	11105	3545	102.16 (138.04)	17481	37770	36321
2009	11227	11781	2110	65.9 (78.30)	10441	28187	16911
2010	17195	13275	3747	77.50 (85.57)	12302	36690	15565
2011	22688	18183	4467	95.03 (110.72)	15835	49325	13160
2012	27199	20089	3683	86.13 (96.63)	16378	49031	8663
2013	N/A	N/A	N/A	92.9 (105.71)	N/A	N/A	N/A

Note: Figures have been rounded to nearest million.

[a] Capital expenditures and operating costs include costs for in situ, mining, and upgraders.

[b] Calendar average of refiners' postings for Canadian Par at Edmonton ($Cdn/bbl) 825/0.5%s.

[c] Figures reported in this column include bitumen and do not represent the true value of all synthetic crude oil.

Sources: Canadian Association of Petroleum Producers, *Statistical Handbook*, 2013, tables 4.16a, 4.16b ('Canada Oil Sands Expenditures'); table 4.19b ('Value of Producers' Sales [Alberta] 1986–2012'); tables 05-05a to 05-05g ('Reference Crude Oil Prices and Foreign Exchange Rates, 1986–2012'). Data accessible at http://www.capp.ca/library/statistics/handbook/Pages/statisticalTables.aspx.

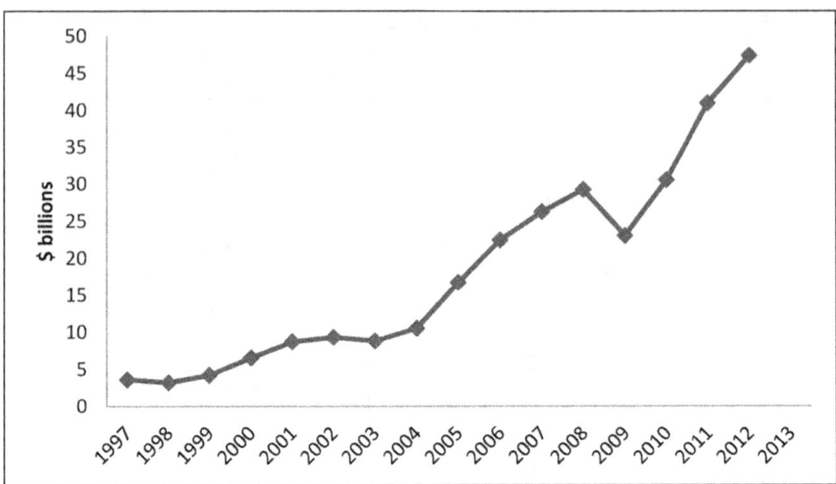

Figure 6.2 Expenditures in oil sands production ($ billions). (*Source*: Data in table 6.1. These are the calendar averages of refiners' postings for Canadian Par at Edmonton [$Cdn/Bbl].)

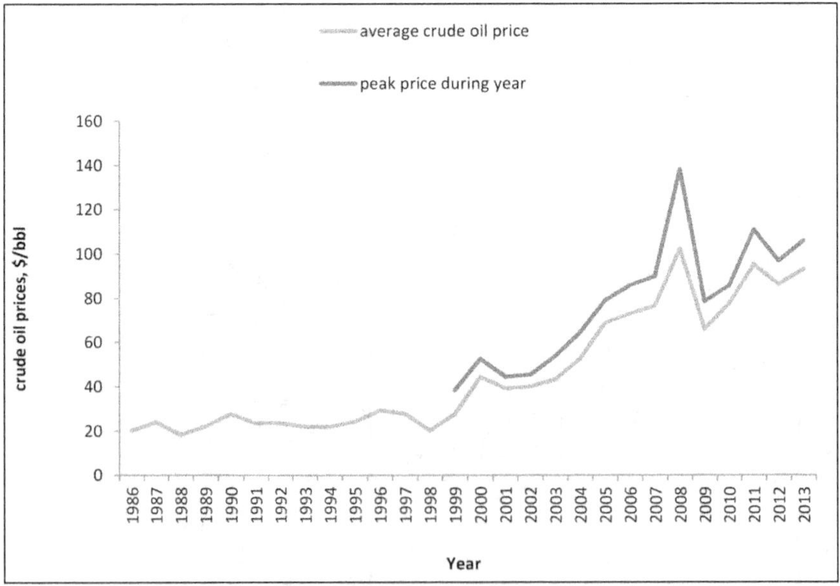

Figure 6.3. Average yearly crude oil prices and peak prices in each year, 1986–2013. (*Source*: Data in table 6.1. These are the calendar averages of refiners' postings for Canadian Par at Edmonton [$Cdn/Bbl].)

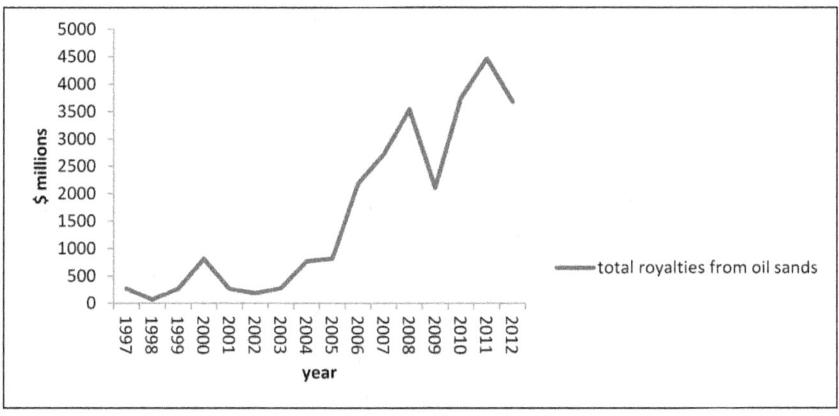

Figure 6.4 Royalties from oil sands production, 1997–2012. (*Source*: Data in table 6.1.)

ments underpinned the creation, in 2007, of three new branches within government tasked with environmental management of the oil sands.[9] The Oil Sands Environmental Division was created in January of that year within Alberta Environment, following a worldwide personnel recruitment campaign entitled 'Take on the World' (Alberta Environment 2007). An Oil Sands Branch was created within Sustainable Resource Development, to be involved in the environmental impact assessment process of oil sands projects, cumulative effects management, reclamation, biodiversity, monitoring initiatives, and compliance enforcement (Alberta, Ministry of Sustainable Resource Development 2009). In the summer of 2007, an Oil Sands Sustainable Development Secretariat was developed within the Alberta Treasury Board, to address issues accompanying the rapid development of the oil sands region (Alberta Treasury Board n.d.). The budget for the latter increased from $1.1 million in 2007–08 to $3 million in 2010–11.[10] By 2013 the Oil Sands Sustainable Development Secretariat was a program administered by the Ministry of Energy, and its budget estimate for 2014–15 was $3.2 million.[11]

As the oil sands continued to drive the provincial economy, other institutional changes occurred. The Alberta Energy Regulator (AER), created in 2013 in response to growing pressures from multiple directions to streamline applications for project approvals, and to improve oversight of environmental regulation (see chapter 3), reports through the minister of energy to the cabinet. The AER is now 'the single regulator of energy development in Alberta – from application and exploration, to construction and development, to abandonment, reclamation, and remediation.'[12] The AER's mandate is to ensure 'the safe, efficient, orderly, and environmentally responsible development of hydrocarbon resources over their entire life cycle. This includes allocating and conserving water resources, managing public lands, and protecting the environment while providing economic benefits for all Albertans.' Interestingly, the AER took over regulatory functions that had previously been the purview of the Ministry of Environment[13] and of the Energy Resources Conservation Board (ERCB), relocating and combining these under the authority of the minister of energy. The AER is intended to be '100 per cent funded by industry' via an administrative fee levied on oil and gas wells, oil sands mines, and coal mines. The budget of the Ministry of Energy for 2014–15 shows that revenue raised in this way amounted to approximately $202 million.

The Alberta Petroleum Marketing Commission (APMC), created in 1974 as a crown corporation to help market conventional crude oil, has, since 1996, been brought increasingly under the management of the Ministry of Energy. Its mandate was expanded in 2012 to include efforts to remove the blockages to the export of non-conventional crude from the oil sands. It is also responsible for marketing the bitumen that is transferred to the province as royalty 'in kind' (BRIK) from the oil sands producers. Legislation proclaimed in January 2014 amended the Petroleum Marketing Act to transfer the APMC's authority to determine 'public interest' to the minister of energy, and placed the agency under closer direction by the minister.

The BRIK arrangements are, nevertheless, agreed upon between the government and the private-sector oil producers, as is virtually everything having to do with the province's energy sector. The new Petrinex system, for example, created to exchange 'key volumetric, royalty and commercial information associated with the upstream petroleum sector' among private and government actors, is described as 'a joint strategic organization supporting Canada's upstream oil and gas industry ... represented by Government (Alberta Department of Energy (DOE), the Alberta Energy Regulator (AER) and the Saskatchewan Ministry of the Economy (ECON)), and Industry (represented by the Canadian Association of Petroleum Producers (CAPP) and the Explorers and Producers Association of Canada (EPAC)).'[14]

In general, the Energy Ministry dominates the provincial cabinet, accounting for an enormous share of provincial revenue (thanks to royalties and land lease sales), administering a large budget, and exercising a major influence in related policy areas such as land use, greenhouse gas emissions, and water regulation. The multiplication of branches, secretariats, and agencies to administer the royalties system, coordinate responses to infrastructure shortfalls, and wrestle for social licence to operate, collect data, monitor, and regulate in areas related to energy production has occurred in tandem with the expansion of investment in the oil sands since the mid-2000s (see table 6.1).

Political Marketing of the Oil Sands

Notably, 2007 was a 'quiet' year in terms of government activism on the legitimation front. We surmise that the new premier and his cabinet may have been preoccupied with the review of the royalty regime (discussed below), and with the problems thrown up by the economic boom (labour and housing shortages, infrastructure problems, inflation), as well as with setting policy directions in other areas (especially health). Local criticism of the oil sands in the context of the massive influx of foreign investment was likely not considered a serious threat. Moreover, the federal Conservative government (elected in 2006) was fending off international pressures on Canada to implement measures to reduce GHG emissions that might have reduced corporate profits in the Alberta oil patch.

It was not until 2008 that the Stelmach government felt the sting of the international ENGO campaigns branding the province's oil exports as 'dirty.'[15] It is in this year that we see a large increase in the public resources dedicated to defending the oil sands from environmental critics, such as the allocation of $25 million to the Public Affairs Bureau to implement the 'Alberta Brand' (discussed below), and a significant jump in Ministry expenditures on international travel (figure 6.5) and in the number of such trips (figure 6.6). In 2008 we also see a new central theme in these promotional missions: that of rebranding Alberta's oil as 'clean.'

Figure 6.5 illustrates the trend in expenditure on international travel by government representatives charged with 'telling the story' of Alberta's environmentally sustainable and responsibly managed oil production. (Summaries of the objectives of these 'missions' and expense reports are made available on government websites.) The ministries involved in such communications work include Alberta Energy, Alberta Environment, Sustainable Resource Development,[16] the Ministry of International and Intergovernmental Relations (formerly the Ministry

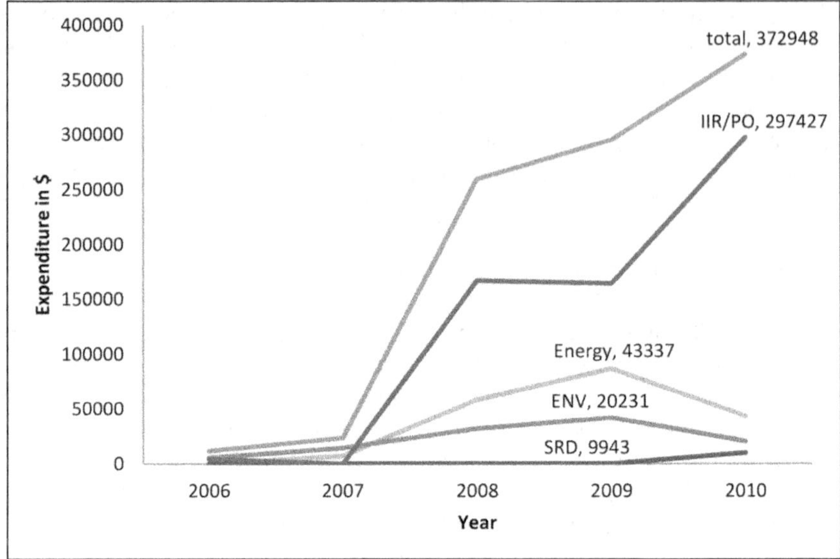

Figure 6.5 Ministry expenditure on international travel related to environmental criticism of exploitation of the province's bituminous sands, 2006–2010. (*Source*: Government of Alberta, http://alberta.ca/home/international_travel_expenses.cfm?.)

Note: We included in our calculations for figures 6.2 and 6.3 only those trips for which a significant 'environmental story' component was mentioned by the Ministry's summary of the mission's objectives or in accompanying media releases. One mission included in 2010 for the IIR/PO was not a trip, but the paid advertisement in the *Washington Post* (10 July) that took the form of an open letter from Premier Stelmach defending the oil sands and the Keystone XL pipeline project. The cost of this advertisement was $55,800.

of International, Intergovernmental and Aboriginal Relations), and the Office of the Premier. Prior to 2008, the primary objective of government missions related to the energy sector was to promote investment (especially in the oil sands) and to expand Alberta producers' shares of international markets (especially in the USA). However, in 2008 the theme of promoting Alberta as an 'environmentally responsible' producer of 'clean energy' emerged, indicative of a direct attempt to counter critics. Mission objectives increasingly included statements such as:

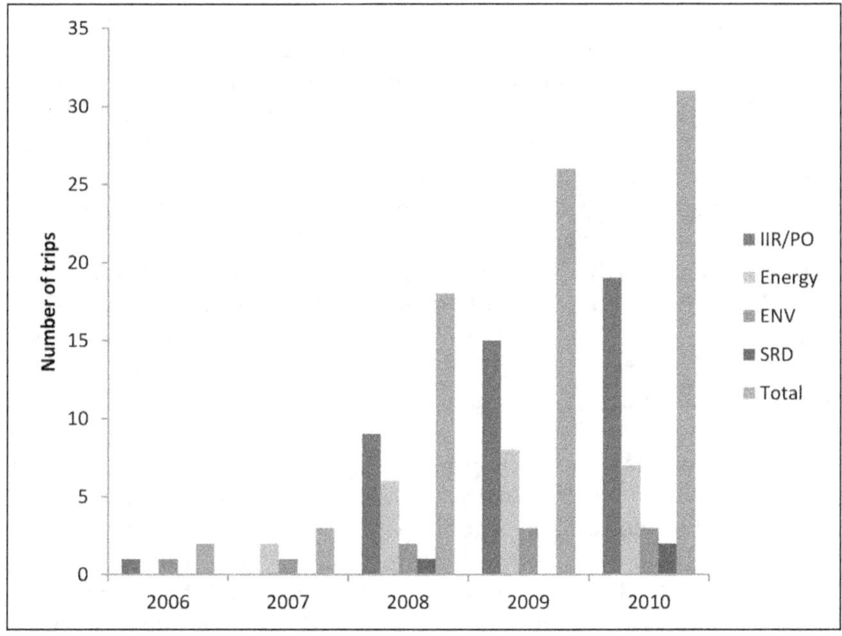

Figure 6.6. International missions by ministry, 2006–2010. (*Source*: Government of Alberta, http://alberta.ca/home/international_travel_expenses.cfm?.)

'demonstrate Alberta's commitment to environmental responsibility in the development of its resources' (Ministry of International, Intergovernmental and Aboriginal Relations 2008b), 'ensure U.S. leaders know the full Alberta energy story' (Alberta Energy 2009a), and 'dispel myths about the environmental impacts of oil sands production' (Ministry of International and Intergovernmental Relations 2008).

Particularly significant is the role assigned to the Ministry of the Environment in performing this legitimation work. Prior to the end of 2007, the Ministry of Environment conducted primarily fact-finding missions and trips to United Nations conferences, the latter being the only forum where promoting the province's clean energy development took place. Beginning in December 2007, 'telling the story' of Alberta's environmental achievements in energy development to a much wider audience became the central theme of international travel undertaken by the Ministry. In travels to Norway, the UK, Germany, Poland, the

Netherlands, Denmark, the United States, and elsewhere, successive environment ministers have met with policy makers and prospective investors. The minister of environment's speeches abroad consistently highlight the province's investment in carbon capture and storage (CCS) technology, its emission 'caps' for large CO_2 emitters, and the oil sands industry's water conservation efforts. This trend is mirrored in the increasing output of Environment Ministry publications such as 'Alberta's Clean Energy Story.'[17]

Our research further revealed an increasingly prominent role for the Ministry of International and Intergovernmental Affairs (IIR), created in March 2008 when the former Ministry of International, Intergovernmental, and Aboriginal Affairs was split into two ministries. Responsibility for 'investment attraction' was moved to IIR from the Ministry of Employment, Immigration, and Industry (Ministry of International, Intergovernmental, and Aboriginal Relations 2008a, 104). Its first minister, Ron Stevens, also held the position of deputy premier. Then in May 2009, Premier Stelmach took over responsibility for IIR in addition to the Executive Council, indicating the priority attached to IIR's functions. In 2008–09 the premier himself led five international missions intended to – as the department's 2008–2009 annual report put it – 'reinforce Alberta's commitment to environmentally responsible development of its energy resources' (Ministry of International and Intergovernmental Relations 2009a, sec. 1:10). In 2008–09 the IIR was authorized a budget for approximately $29 million, with $19.3 million of that allocated to international relations. In 2009–10 the authorized budget for IIR was $26.3 million, with $9.8 million allocated to international relations and $6.9 million to the international offices. When we take into account these developments, along with the IIR's roles in organizing meetings with investors and government officials abroad and hosting delegations to Alberta, it appears that the IIR, in conjunction with the Executive Council, had become the lead department in organizing the government's responses to international environmental campaigns. (The predominant role of IIR and the Premier's Office is illustrated in figures 6.5 and 6.6.) This shift in government resources to PR efforts outside of Canada coincides with the government's heightened concern with international ENGO campaigns, beginning in 2008.

There has been particular emphasis on influencing US policy makers, since the US remains, for the time being, the main market for Alberta energy exports. In 2008, for instance, the Alberta Washington Office organized a visit to the oil sands by the chair of the United States

House of Representatives' Subcommittee on Energy and Air Quality, to generate 'increased awareness of the environmental requirements in place for [oil sands] development' (Ministry of International and Intergovernmental Relations 2009b, 27). Prior to the Copenhagen climate change talks in 2009, the Alberta government was reported to be spending $40,000 a month on a team of consultants, led by former Michigan governor James Blanchard and former Canadian diplomat Paul Fraser, 'to improve the province's environmental image in Washington' (*CBC News* 2009). An article in the magazine *Alberta Oil* reported in May 2010 on the work of Alberta's representative in Washington, Gary Mar, to build the province's 'energy brand' and to counter 'dirty oil' publicity by 'hard-green factions' (Jaremko 2010, 14). The Alberta government's 'international reputation management effort,' the author wrote, 'focuses on authorities and economic sectors with roles or connections in the multibillion-dollar energy trade. Current ERCB chairman Dan McFayden, for instance, talks to the U.S. Federal Energy Regulatory Commission. In Washington, provincial representatives meet energy and state department officials, at times reaching inner sanctums of the executive branch such as the National Security Council' (14). He continued, noting evidence of success:

> The effort to brand Alberta production as clean and responsible energy spans the continent ... Mar roves across the U.S. He has personally delivered reminders about intertwined livelihoods and supplies of essential products to governors, legislators, employers, and workers in more than 20 states from California to Massachusetts.
>
> Special attention is currently given to jurisdictions working on introducing clean fuel standards that are liable – sometimes inadvertently, for instance by favoring homegrown fuels – to penalize oil sands exports. Wisconsin, always a keen supporter of fuels distilled from its farm crops, is reminded that the Alberta bitumen belt is one of the biggest customers for Milwaukee heavy equipment assembly plants. Illinois, Indiana and South Carolina factories make parts for jumbo oil sands mining machinery
>
> The oil industry and Alberta government sees [*sic*] the recent White House decision to approve the Enbridge pipeline from Alberta to the US, through Montana (the Alberta Clipper), as evidence that their efforts to re-educate the Obama administration have been successful. The State Department's statement said: 'Concerns have been raised about higher-than-average levels of greenhouse gas emissions associated with oil sands crude. The administration has considered these concerns and considers

> that on balance they do not outweigh the benefits to the national interest.' (14)

Responding to growing opposition in the United States to the Keystone XL pipeline expansion[18] (including a call by fifty congressional representatives for the project to be delayed, and a vote by Bellingham, Washington, councillors to boycott the use of oil sands petroleum in city vehicles), the Alberta government paid $55,780 to place a letter from the premier in the *Washington Post* (2 July 2010, A4). The ad represented Alberta as 'a stable, responsible and reliable source of energy' for the USA, and defended the environmental record of the province's oil sands. Recent regulatory battles surrounding the Northern Gateway, Alberta Clipper, Keystone XL, and other pipeline projects have placed growing demands on the IIR Ministry (along with the Public Affairs Bureau) to engage in lobbying efforts with publics and governments beyond provincial boundaries. Other chapters in this volume describe the mobilizations by communities and First Nations across North America against megaload hauling, pipelines, tanker shipping proposals, and new refineries linked to Alberta crude. Alberta's oil and gas exports are linked to long and complex chains of effects, both economic and ecological. The conflicts arising from these effects, and the ever-shifting strategies of civil society actors, impel continual strategic, personnel, budgetary, and organizational changes within the provincial state.

Branding and Blogging

A government PR initiative that has attracted significant criticism from environmentalists is the 'Alberta Brand,' referred to in the *Alberta Oil* article quoted above. Initiated amidst peaking oil sands criticism outside the province, the development of the Alberta Brand demonstrates sensitivity to the province's international image (Public Affairs Bureau 2008, n.d.).[19] Although the branding exercise was also linked in government statements to tourism, its central thrust was the attraction of investment capital, entailing minimization of threats of environmental regulation or boycott campaigns, as well as the depiction of the province as having a strong technology infrastructure and skilled workforce (Public Affairs Bureau 2008, n.d.). While energy is not the only sector to which the government hopes to attract investors, it is the sector targeted by environmental campaigns, and the one the government has laboured to defend. In response to criticisms that the $25 million would be better spent actually remediating environmental sites such as the tailings ponds, Premier Ed Stelmach argued:

> Twenty-five million dollars is well spent in ensuring that we protect the integrity of this province not only within Canada but within North America and around the world. We will continue to do that, and here's why: 30 years ago Syncrude pioneered the bird aversion strategy, the research. For 30 years things went well. One year we have one incident [reference to the death of 1,600 waterfowl in a Syncrude tailings pond in northern Alberta in April 2008]. That is what's being used by that party [the Liberals] to try and damage the reputation not only of the Department of Environment and of a company but also of this Legislature, because they're using this Legislature to communicate that misinformation.[20]

As well as managing the Brand, the Public Affairs Bureau (PAB) – which is overseen by the premier (the minister responsible for the Executive Council) – has been kept busy responding to criticisms of the oil sands from all directions and on multiple fronts. The Stelmach government created, for example, an oil sands website linked to its home page, which provided short videos, fact sheets, links to reports, and other materials all making the case for the 'responsible' nature of the government's environmental management of the oil sands. *The Alberta Blog*, managed by PAB, responded to critics of provincial energy development and provided journal-like accounts of the premier's travels or those of other ministers, always highlighting the same message. In the wake of a renewed branding and (tourism) boycott campaign by a coalition of ENGOs in the USA, timed to influence the State Department's decision on the approval of the Keystone XL Pipeline, the Stelmach government announced that it would spend another $200,000 on a campaign to 'educate' Albertans, other Canadians, and Americans about the oil sands (Brooymans 2010).[21]

The Sustainable Development Discourse

While neither the Alberta nor the federal Conservative government implemented effective measures to reduce industrial greenhouse gas emissions, both deemed it necessary to defend their ongoing subsidies for hydrocarbon extraction and their comparative failure to invest in green alternatives.[22] Both levels of government adopted a sustainable development discourse that includes (1) an emphasis on technological solutions (no need to change the structure of the economy or consumption norms or to limit economic growth); (2) minimization of the problem[23] with regard to GHGs, water use, boreal forest loss, or toxic contamination; (3) reassurance that 'responsible management' will succeed in 'balancing' (or dissolving) the social and environmental harms of hydrocarbon extraction and processing while retaining the economic benefits thereof. We look at these elements of the Alberta government's sustainable development discourse in more detail below, before turning to another key element of the discursive strategy in the following section.

Technology Is the Answer

Discourse is rooted in worldviews as well as being strategic; the role of technology in the government's response to environmental criticism reflects both its Promethean beliefs and assumptions about human ingenuity, nature, and progress, and its alignment with regard to existing capitalist and colonial power relations. In a word, technology is the key means by which fossil fuel extraction is made environmentally 'sustainable.' This sustainable development discourse – promulgated also by large business associations and by the federal government – reassures the public that no deep or radical changes to the economy or to consumption norms are necessary in order to prevent ecological decline. The faith placed in technology by the Alberta government – given the enormous costs and high risk of failure – is quite extraordinary. Stares (2010, 51–2), for instance, notes that technology is presented as the solution to virtually every environmental consequence of oil sands exploitation, including water use, climate change, devastated landscapes, toxic tailings lakes, and air pollution. Moreover, technology holds out the promise of reducing costs of production and providing a more reliable source of fuel to Canada and the USA. In a speech given in both New York and Washington, DC in March 2009, Environment Minister Rob Renner stated: 'In Alberta, our greatest opportunity to truly reduce emissions and continue as a global energy producer will come from innovation and technology. By supporting innovation and advancing technology, we are increasing our potential ... to improve efficiency, reduce or even stop emissions and ultimately improve environmental performance' (Alberta Environment 2009, 10).

Table 6.2. Research Projects or Institutes Receiving Funding from Government of Alberta for Research into Fossil-Fuels-Related Technologies (Environmental Remediation, GHG Emission Reduction, Enhanced Oil Recovery, CCS, 'Clean Coal,' etc.), 2004–2011

Alberta Ingenuity Centre for In Situ Energy, University of Calgary
Alberta Innovates – Technology Futures
Bitumen and Heavy Oil Research Program
The Heartland Area Redwater Project
Materials and Reliability in Oil Sands
Alberta Innovates – Energy and Environmental Solutions
Alberta Energy Research Institute
• supported 39 projects and 6 research chairs at 3 universities
Canadian Centre for Clean Coal/Carbon and Mineral Processing Technologies
Carbon Management Canada, University of Calgary
Centre for Intelligent Mining Systems, University of Alberta
Centre for Oil Sands Innovation, University of Alberta
Helmholtz/Alberta Initiative, University of Alberta
Institute for Sustainable Energy, Environment, and Economy, University of Calgary
Metagenomics for Greener Production and Extraction of Hydrocarbon Energy, University of Calgary
National Centre for Upgrading Technology, Devon, Alberta
Oil Sands Tailing Research Facility, University of Alberta
School of Energy and the Environment, University of Alberta

In highlighting the importance of technology for Alberta's environmental sustainability, the government often points to 'clean' energy development, which, along with carbon capture and storage (CCS) and occasional references to energy conservation or efficiency gains, constitutes Alberta's approach to reducing greenhouse gas emissions (Alberta Environment 2008a, 7). 'Clean' energy development entails the use of technology to reduce the environmental impacts of extraction and production. While the province has funded research into biofuel technologies (see figure 6.7), no secret is made of the fact that 'greening' energy development refers first and foremost to fossil fuel production: 'the development of clean hydrocarbons is essential to Alberta's energy future ... this must be Alberta's responsibility and focus. Alternative and renewable energy sources will play a growing role in Alberta energy's future, but they cannot match the importance to Alberta of "clean" fossil fuels' (Alberta Energy 2008, 20).

Indeed, the province has made multiple investments in 'clean' fossil fuel technology research. Table 6.2 provides a list of projects and institutes that have received provincial government funding for research into fossil-fuels-related technologies since 2004. It is difficult to ascertain the full amount spent on this in recent years: funding is disbursed by a variety of agencies and programs located in different ministries, announcements of grants do not always clearly identify their budgetary sources, and funding information is not always made available. Confusing matters further is the major reorganization of the research infrastructure that took place in 2009, merging units from various agencies into new ones.[24] On the basis of the data available to us, we estimate that provincial government funding for the agencies listed in table 6.2 totalled more than $430.4 million between 2004 and 2010. (About $80 million of this came from the Energy Innovation Fund, discussed below.)

There are several primary sources of funding for 'clean' energy research and technology. The Energy Innovation Fund was announced in 2006 as a $200 million fund to be spent over three years on research, technologies, and projects 'focusing on energy supply and protection of the environment' (Alberta Environment n.d.[a]). According to Alberta Energy (email communication to authors, 15 June 2010) at least $135.5 million had been spent from this fund on various projects as of June 2010. The Climate Change and Emissions Management (CCEM) Fund was launched in 2009 as part of the 2007 Climate Change and Emissions Management Amendment Act and its accompanying Specified Gas Emitters Regulation. Under this legislation, large emitters were required to reduce the 'emissions intensity' of production, or to pay $15 into the CCEM fund per one tonne of emissions that exceed their reduction target.[25] (Alternative options are to purchase Alberta-based offset credits or purchase 'Emission Performance Credits.') The fund is managed by the 'arm's length' Climate Change and Emissions Management Corporation (CCEMC), whose directors disburse grants to projects in seven categories (CCEMC 2014): cleaner energy production from fossil fuels, carbon capture and storage, adaptation to climate change, renewable energy, energy efficiency, biological energy technologies, and carbon uses.

As of May 2014, $404 million had been paid into the CCEM Fund since its creation in 2009, comprising half of large emitters' compliance with the Specified Gas Emitters Regulation (Read 2014, 4). This fund, in particular, has been highlighted in government speeches abroad, for example, to international audiences in Washington, DC, New York, and the UK (Alberta Environment 2008b, 2009). Of the ninety projects that the CCEMC lists (in May 2014) as having received funding since 2009, only five support the application of wind or solar technologies, and one supports the development of net zero housing. The total amount granted for these six projects is $30,724,673, or 7.6 per cent of the fund's total.

Investments in 'clean' energy research and technology have come also from Alberta's share of the federal ecoTrust (Alberta 2009a) and from Memorandums of Understanding between companies and Alberta Advanced Education and Technology (Alberta 2008a). Stares (2010) notes that the province refers to these grants in its discourse as proof of its environmental commitment.

While investments in technology undoubtedly have some practical benefit for the environment, their effectiveness is limited as, overwhelmingly, they are not intended to reduce dependence on hydrocarbons. An analysis of government budgets from 2006 to 2014 shows that spending by the Ministry of Energy on energy conservation and efficiency incentives or R&D occurred in only one year (2008–09), and amounted to approximately $21.8 million, or 4.6 per cent of the total spent on all 'clean energy' initiatives funded by the Ministry since 2006 (see figure 6.7). Several provincial 'clean' energy initiatives serve an environmental PR function as much as an environmental improvement function. Research investments into oil sands technologies, for instance, essentially subsidize and support an emissions-intensive industry. Nowhere is this more evident than in the disbursements of the CCEMC, whose board of directors retains close links to the industries paying into it (see CCEMC 2010c, and www.ccemc.ca). Suncor, for example, has been granted approximately $21.8 million to date for projects based in its oil sands operations (CCEMC 2014). Moreover, the low sum required from large emitters paying into the fund ($15 per tonne over target) provides little incentive for companies to make costly internal changes to increase the efficiency of operations and reduce emissions; the Pembina Institute (2010) reports that, for oil sands producers, 'this works out to be at most about 18 cents per barrel of oil – too little to have any material effect on firms' production decisions.' In the absence of an effective climate change policy that advances a transition to a low-carbon economy, the province's greenhouse gas emissions continue to rise (Environment Canada 2013, 35).

'Clean energy' research investments further serve a legitimation function through the partnerships created with higher education institutions (indicated in table 6.2) – bodies that should be drivers of change. The lure of government-funded grants or new research chairs effectively serves two purposes relevant to maintaining the petro-state: first, it silences a potential source of opposition from these institutions, and second, it helps to further legitimize rapid development of hydrocarbon resources. As officials proclaimed in 2011, the University of Alberta alone houses 'close to 50 oilsands research programs.'[26] The University of Alberta's president drew attention, in an April 2013 fund-raising letter to alumni, to the 'approximately 800 industry and university researchers providing the research and innovation needed to power Alberta's largest industry [the oil sands].'[27] In an advertisement for the Canadian Association of Petroleum Producers, available on the association's website, the dean of engineering at the University of Alberta states:

> People are quite energized when I describe what's happening at the University of Alberta. Over a thousand researchers collaborate here on this one major topic: the responsible development of our oil sands. We do two types of research. One would be improving existing processes. The other type would be looking for breakthroughs. How do we make it better? Less impact on the air. Less impact on the water. Less impact on the land.[28]

In its efforts to legitimize and prolong the extraction of fossil fuels, the provincial and federal governments have engaged in a restructuring of the funding available to researchers in the post-secondary sector. These priorities have far-reaching implications for the production of knowledge and for the resources available to contest the existing model of development and to advance alternatives. To date, the effects of the reorganization of the research infrastructure in the province in accordance with petro-state priorities have been under-documented.[29] We are, however, experiencing pressures very much like what Bret Gustafson calls 'university capture' by the fossil fuel industry (2012, 322).

Counting on CCS

Perhaps nowhere is a legitimation function more clearly served than in the promises regarding carbon capture and sequestration (CCS) technology, labelled by the government as an 'environmental' initiative in which carbon dioxide from large emitters is captured and stored indefinitely underground. In 2008, at the moment when international campaigns against the oil sands were really heating up, the Alberta government pledged $2 billion to support research and development of CCS, boasting that such projects would 'reduce greenhouse gas emissions by five million tonnes annually beginning in 2015' (Alberta Energy 2010a). No initiative has received more attention in provincial discourse on climate change: 'It's a significant commitment to make a substantial reduction in global greenhouse gas emissions' (Alberta 2009b, 10), and, in a classic example of sustainable development discourse: 'Carbon capture and storage provides the province with the greatest potential to substantially reduce greenhouse gas emissions, while at the same time maintaining Albertans' quality of life' (Alberta 2008b).

Despite this Promethean faith in the capacity of technology to progressively dissolve limits to capitalist economic growth, CCS is an extremely high-risk strategy for preventing irreversible climate change. Critics of CCS have argued that the technology is unproven, too long term, and exorbitantly expensive, and that the money could be better spent on other strategies to reduce GHGs (see, especially, Thomson 2009). Moreover, 'carbon capture and storage' as discussed in the Alberta context is misleading, as most of the 'CCS' projects funded so far are not primarily concerned with the reduction of GHGs, but rather, are about 'enhanced oil recovery' (EOR) (the use of carbon dioxide to extract oil from depleted oil fields). In this process, CO_2 may remain beneath the surface for some time, but not necessarily indefinitely. It is additionally misleading to represent EOR projects as a solution to global warming because the oil extracted from the depleted conventional fields is eventually burned, thus contributing to global emissions of GHGs. In other words, Thomson concludes, 'as far as the global climate is concerned almost no carbon dioxide would have been removed from the atmosphere' from such projects (2009, 23).

CCS funding was initially set aside for four projects, two of which focused explicitly on EOR: $495 million was pledged to Enhance Energy Inc. and North West Upgrading's Alberta Carbon Trunk Line (ACTL) – a 240-kilometre pipeline that will transport CO_2 to depleting conventional oil fields for EOR (Alberta Energy 2010b); $436 million was earmarked for TransAlta Corporation and partners for 'Project Pioneer,' which was also intended to capture CO_2 for EOR in nearby oil fields (Alberta Energy 2010b; TransAlta 2009). A third project also had an EOR component: Swan Hills Synfuels was granted funds to create wells that would access coal seams traditionally too deep to mine, and convert the coal into a 'clean synthetic gas known as syngas' for high-efficiency power generation. According to project descriptions, the CO_2

created in the process would also be captured and used in EOR (Alberta Energy 2010b). The only project receiving funding from the province for CCS that appears unconditionally to attempt to capture and store CO_2 is known as Quest, a partnership between Shell Canada Energy and the Athabasca Oil Sands Project. If successful, this project will 'capture and store 1.2 million tonnes of carbon dioxide annually beginning in 2015 from Shell's Scotford upgrader and expansion, near Fort Saskatchewan' (Alberta Energy 2010b). This constitutes a capture of 35 per cent of the upgrader's CO_2 emissions (MIT 2013a).

Notwithstanding the central place of CCS R&D in the government's international campaigning from 2008 to 2012, very little of the originally heralded $2 billion was spent during this period (see figure 6.7). Despite the availability of substantial government subsidies, a number of the corporate partners decided not to go ahead with the projects. In the case of Swan Hills Synfuels, the price of natural gas remained too cheap to make investment in synthetic gas production profitable (Alberta Energy 2013). Project Pioneer was shelved in 2012 following a technical and economic feasibility study that found that although the 'technology works,' the 'market for carbon sales and the price of emissions reductions were insufficient to allow the project to proceed' (Project Pioneer 2012). Translated by journalist Carrie Tait in the *Globe and Mail* (26 April 2012), this meant: 'Even though TransAlta's project was backed by $778.8-million in provincial and federal funding, the company said it could not justify spending the millions of dollars necessary, arguing that Canada's weak regulations for carbon pricing made it unattractive.' The ACTL project is still in the planning stages, with pipeline construction 'intended to take place in 2014' (Enhance Energy 2012, 2; see also MIT 2013b). It was reported in 2013 that $495 million had been committed by the provincial government to the ACTL, expected to start operating in 2015 (Cooper 2013, D7). As for Quest, the investors (Shell Canada, Chevron Canada, Marathon Oil Sands) announced in April 2013 that work had begun in fall 2012, and that the injection of CO_2 should begin sometime in 2015.[30] The provincial government has committed $745 million to the Quest project, in addition to $120 million from the federal government (MIT 2013a). These subsidies for the ACTL and Quest projects add up to $1.2 billion, close to the amount that the provincial government was heralding in 2014 as its commitment to CCS.[31]

Table 6.3 and figure 6.7 tell a somewhat different story, however. According to the Ministry of Energy's budget documents from 2006–07 to 2012–13, much less has actually been *spent* by the department on carbon

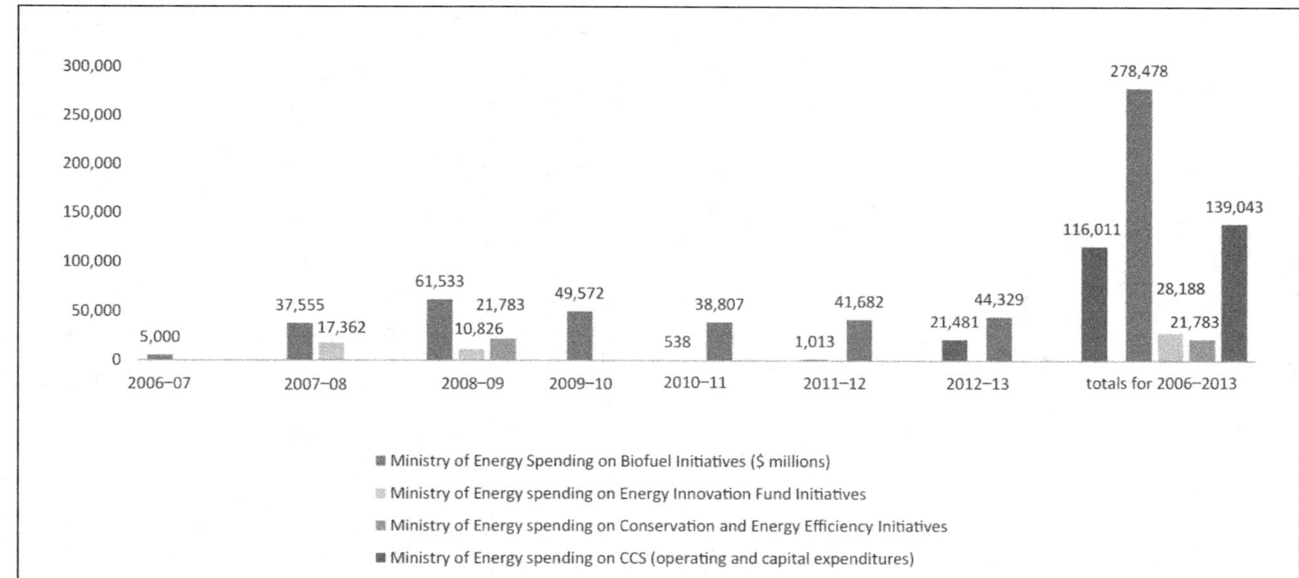

Figure 6.7. Ministry of Energy spending on 'clean energy' initiatives, 2006–2012 ($ millions). (*Source*: Government of Alberta budge estimates for Ministry of Energy from 2006–07 to 2014–15.)

Table 6.3. Ministry of Energy Commitments to 'Clean Energy' Initiatives, Aggregated, 2007–2014

Area of spending	Amount spent (millions $)	Amount budgeted (millions $)
Biofuels	278	525
Carbon Capture and Storage	139	742
'Energy Innovation Fund Initiatives'	28	36
'Conservation and Energy Efficiency Initiatives'	22	30

Source: Government of Alberta budget documents for Department of Energy, 2006–07 to 2014–15

capture and storage initiatives. These data show that only $139 million had been spent as of budget year 2012–13. Indeed, the Ministry of Energy's budget figures show that it has spent twice as much on biofuel initiatives as it has on CCS. Notwithstanding the very limited contribution of CCS to the reduction of GHG emissions to date, representatives of the Alberta government continue to refer to the CCS investment to persuade international audiences that Alberta is a leader in action on climate change.[32]

In July 2014 the province's auditor general (AG) reported to the legislature on the performance of the Ministry of Environment and Sustainable Resource Development in implementing its 2008 greenhouse gas emissions reduction strategy, which, as mentioned above, claimed that CCS technology would permit the province to reduce its GHG emissions by 5 million tonnes/year beginning in 2015. In addition, CCS was to account for 70 per cent of Alberta's 2020 and 2050 targeted emission reductions. The AG found that, 'with only two carbon capture storage projects planned, the total emissions reductions are expected to be less than 10% of what was originally anticipated' (Alberta Auditor General 2014, 44). With regard to the government's 2008 'climate change plan,' the AG reported that the government had not made a plan to implement it until 2012; indeed, the audit 'found no evidence that the department regularly monitored performance between 2008 and 2012 against the 2008 strategy targets' (39). The plan developed in 2012, the AG concluded, 'lacked the information necessary to monitor performance of actions and the government's overall progress with reducing greenhouse gas emissions and meeting its strategic targets ... The department has not yet developed criteria for selecting new climate change actions and evaluating existing ones' (39–40).

As if to confirm the role of provincial climate change policy documents and statements in securing international 'social licence to operate' in the oil sands, in July 2014 the leading candidate for leadership of the Conservative Party, commenting on the shortcomings of the Environment Ministry in the wake of the auditor general's report, said: 'You can't advance an environmental agenda but more to the point, you can't defend yourself either domestically or internationally. That frankly has been the position Alberta has been in for many years ... We've not had the strength inside Alberta Environment, the scientific strength, the monitoring capacity or the regulatory regime to defend ourselves internationally as protecting the environment' (Jim Prentice, quoted in Wood 2014).

Minimization of Environmental Risks and Harms

Key to the government's response to environmental criticism is its framing of the environmental impacts of hydrocarbon extraction. In her analysis of the government's discourse on the environmental management of the oil sands, Stares (2010) documents the use of rhetorical devices that serve to minimize the negative environmental consequences. Although that study focused on discourse directed to a provincial audience, similar patterns are observed in the government's discourse addressed to national and international audiences. Such devices include an anthropocentric view of nature, the depiction of the environmental impacts of development as marginal and/or manageable, undermining of the credibility of environmental critics, and positing of some level of environmental harm as necessary to the development that brings about the prosperity of Albertans. Together, these devices function to alleviate concern and deflect criticism.

Stares (2010) documents numerous strategies employed to minimize environmental risks and harms, including representing the negative effects of exploitation as temporary, negligible, localized, and/or reducible (often with technology), the use of success stories, and the omission of information. One tactic, for instance, involves using broad comparisons to downplay environmental effects – a strategy often described as 'putting things into perspective' (e.g., Alberta 2009b, 7). In his address to the Ninth Annual Arctic Gas Symposium in March 2009, for instance, Premier Ed Stelmach said: 'In more than four decades of oil sands mining, the industry has disturbed about one-hundredth of one percent of Canada's boreal forest. The oil sands account for less than one tenth of one percent of the world's greenhouse gases, or to put it another way about half the greenhouse gases produced by the City of Hong Kong each year. And US coal-fired electricity production accounts for about sixty times the annual emissions from the oil sands' (Office of the Premier 2009).

Rhetorically minimizing the environmental impacts of oil sands development also involves challenging claims to the contrary. ENGO claims – selectively recognized in government discourse – are often portrayed as being hyperbolic, if not blatantly false. Those spreading such 'myths' (Alberta Environment 2008b, 5) may be identified vaguely in government discourse as 'critics,' but specific organizations are also identified, such as Greenpeace (*Edmonton Journal* 2010b), the Natural Resources Defense Council (Public Affairs Bureau 2009), or National Geographic (Office of the Premier 2009). When not implying that ENGOs exaggerate the environmental effects of development, the government directly accuses them of misrepresenting impacts by not telling the full story. In some cases, albeit less frequently, critics are depicted as irrational, foolish, 'scare-mongering,' or manipulative, seeking to deliberately mislead the public (Stares 2010, 39). On the particularly hot issue of characterizing oil sands crude as being more greenhouse gas intensive than other sources of oil, for instance, government officials wrote (on *Alberta Blog*) of critics: '[The environmentalists] purposely make a complex topic all the more confusing ... [they] take the most damaging snapshot of oil sands GHG emissions, delete the rest of the lifecycle where the emissions begin to balance with those of other crude oils, and pretend that part of the lifecycle doesn't exist' (Public Affairs Bureau 2009).

In contrast, information reported by the government is presented as factual and corrective to this 'misinformation' (Office of the Premier 2009). For instance, at a speech given in London in 2008, Environment Minister Rob Renner stated:

> We have all heard that there are claims made that the development comes without environmental controls. That the extraction methods are harmful to the environment. That we leave huge scars on the face of the earth as a result of the activity, we're responsible for tremendous greenhouse gas emissions and that we're damaging the air and the water. And so if I can, I just want to address those issues – *set the record straight* from my position as the Minister of Environment. (Alberta Environment 2008b, 5, italics added)

In attacking critics, however, the government rarely tells the full story itself. On the same issue, drawing repeatedly on two reports commissioned by the Alberta Energy Research Institute from US energy consulting firms (Jacobs Consultancy and TIAX LLC) and released to the public by the government in July 2009, the government claims that the 'life cycle' emissions associated with oil sands crude are in the same range as those of other crudes refined in the USA (Public Affairs Bureau 2009).[33] The methodologies of these studies, however, as well as their conclusions, have come under strong criticism by university and ENGO-based scientists.[34] A second example stems from the government's identification of international concern about climate change as a major threat to Alberta's crude oil exports. Despite the fact that Alberta's existing legislation requires only reductions of emissions *intensity* per barrel of oil (allowing total emissions to rise), the government presents its GHG management system as one that substantially reduces emissions, and refers misleadingly to its 'emissions caps.' One political commentator called Alberta's climate change policy 'a license to pollute dressed up in rhetorical petticoats' (Simpson 2008).

A final aspect of the province's discursive strategy in deterring environmental criticism is that it constitutes what John Dryzek (2005, 157) calls a 'rhetoric of reassurance': 'We *can* have it all: economic growth, environmental conservation, social justice, and not just for the moment, but in perpetuity.' The public is repeatedly assured that through the use of technology and careful, 'responsible' management, every challenge of oil sands exploitation can be mitigated, and every benefit reaped. In constructing this reassurance, the government explicitly denies that hydrocarbon development is taking place at the expense of the environment. Attributes of competence and leadership are used in the government's description of itself (see Stares 2010). Frequently, the government compares itself favourably to others: 'Alberta's reduction commitment [on greenhouse gas emissions is] the largest identified and published by any province in Canada,' and 'while others are talking, Alberta is acting' (Alberta 2009b, 10). The argument is made, for example, that of all the sources of oil imports to the United States, only Canada has in place measures to reduce greenhouse gas emissions. In this argument, the 'cleanness' of Canada's oil is compared with that of Saudi Arabia, Mexico, Venezuela, and Nigeria.[35]

Nativist Neoliberalism

Another key discourse utilized by governments of Alberta since the 1970s to mobilize consent for its *laissez faire* approach to the exploitation of the province's mineral wealth (i.e., its open-door policy toward, and massive subsidization of, private-sector exploitation of these resources) is – unlike the sustainable development discourse – directed almost exclusively to the provincial population. Drawing upon historical grievances about Alberta's peripheral position in confederation (see chapters 1, 17), Conservative governments of Alberta have for decades constructed, discursively, a combative, defensive Albertan identity, unified in opposition to 'central Canadian' interests. The narrative is one of a besieged people that must repeatedly defend its greatest source of wealth (oil and gas revenues) from the grasping hands of governments representing the larger, urban populations of eastern Canada.[36] While this is not a 'nativist' discourse in the typical sense of a discourse that depicts a 'native' population threatened by new 'foreign' arrivals who are taking their jobs or draining their social services,[37] it does portray Albertans as a group with settler roots and credentials whose ownership of provincial resources and the rights to benefit from them are constantly threatened by envious or hostile outsiders. The irony of this discourse – in light of the dispossession of Aboriginal peoples as a consequence of settler exploitation of oil and other natural resources – is lost on the Conservatives.

Political economist John Richards observed that by the time the Province of Alberta was created, in 1905, settlers in Saskatchewan and Alberta already perceived Ottawa as 'an imperial government, a complex of institutions organized by central Canadian elites for the purpose of dominating and plundering the hinterlands. The provincial administration, whatever its political colouration, became the indispensable agent for attacking political colonialism and bargaining with external economic interests' (1979, 17). Historian Nelson Wiseman reminds us that the federal government did not relinquish to Alberta control over its natural resources until 1930; thereafter, Ottawa continued policies that privileged the interests of central-Canada-based financial and industrial interests over those of the petty commodity producers of the prairies (2013, viii). The Social Credit government led by William Aberhart had 'an extremely acrimonious relationship with its federal counterpart' (Barrie 2006, 118; see also Mallory 1954). As Wiseman summarizes:

> Standing up to Ottawa has been politically profitable in Alberta; it was so when the UFA and Social Credit governed and continued to be so through most of the century ... Social Credit's Progressive Conservative successors

> also used Ottawa as an expedient punching bag in their provincial election campaigns; their 1975 election slogan, for example, "Vote for Alberta" ... conveyed the idea that a vote for an opposition party was a vote against Alberta itself. (2013 ix)

Political economist Larry Pratt observed that the 'resource nationalism' discourse of the Lougheed (Conservative) government, driven by a growing middle class with interests in economic diversification, constructed the 'Toronto-Montreal establishment' as the greatest obstacle to Albertans' economic development (1984, 195).

Since the 1970s, various parties or associations espousing provincial autonomy (even independence, or merger with the United States) have arisen in Alberta, including the Western Canada Concept Party, Western Independence Party, Citizens Centre for Freedom and Democracy (which launched the Wildrose Alliance Party in 2007), the Alberta Independence Party, and the Separation Party of Alberta. Parties with provincial sovereigntist agendas and federal electoral ambitions include Reform, Canadian Alliance, and the post-merger Conservative Party of Canada – all originally based in Alberta.

It has not been difficult for the Conservative governments that have been in office in Alberta since the early 1990s to slot the issue of climate change policy into the same framework, constructing federal proposals for carbon pricing or cap-and-trade systems (however tepid) as yet another attempt to tax away the hard-earned prosperity of Albertans, or casting environmentalists in the role of threatening outsiders. For example, in a 4 March 2009 speech to the Ninth Annual Arctic Gas Symposium, Premier Stelmach described part of his job in 'telling Alberta's story' as 'combating the misinformation and distortion peddled by those who would damage the future prosperity of our province.'[38]

An additional assumption comes into play in the government's response to critics of its development model, that is, its intransigent insistence on the primary and non-substitutable role of intensive hydrocarbon extraction and export in the provincial and national economies. From the perspective of the ruling Progressive Conservative governments, all threats to this source of wealth had to be vigorously deflected, and no alternatives – while the reserves last – could seriously be considered. While the statement of the energy minister that leads off this chapter captures such beliefs, other examples abound. In a speech to investors at a TD Securities meeting in July 2008, Premier Stelmach said: 'There is opposition to the development of this global resource –

here in Canada, and in the U.S. and elsewhere. As a province and as a country, that's something we need to take seriously. As an energy-producing nation, Canadians must have a clear understanding of these issues – our future prosperity depends on it.'[39] Figures on industry investment in the sector are frequently cited by both government officials and industry representatives; references are made repeatedly to the numbers of jobs created and to the general prosperity enjoyed by Albertans as a result of the province's resource endowment.

In this discourse, the interests of Albertans are consistently homogenized and identified with those of the oil companies in maximizing exports and securing markets in the USA (against threats of boycotts or low-carbon fuel standards).[40] The Stelmach government chose to maximize the expansion of oil sands operations and the growth in crude oil exports in the short term, rather than to 'touch the brakes,' or to redirect significant public investment towards renewable and alternative energy sectors and conservation.[41] The Progressive Conservative governments that succeeded that of Premier Stelmach did not deviate from this path.

Paradoxically, while Alberta governments have defended this extractive economy in the name of maximizing the benefits of resource ownership for Albertans, they simultaneously have rejected calls to increase the economic rents the province obtains for these resources. The recommendations of the Royalty Review Panel appointed by the Stelmach government in 2006 – including that the province modernize its royalty regime by raising rates to levels comparable to those of other jurisdictions[42] – were only partially, and temporarily, accepted.[43] Following threats from oil and gas companies to move investment to BC, Saskatchewan, or elsewhere, the Stelmach government announced in February 2009 that it would undertake a 'competitiveness review' comparing Alberta's regulatory requirements, land practices, and royalty rates for conventional oil and gas producers with those in other jurisdictions.[44] This time, the review was led by the Ministry of Energy, in close consultation with oil and gas industry representatives. In March 2010, following the 2009 decline in world crude oil price and a slump in investment in the sector, the Stelmach government decided to reverse increases that had only just come into effect (January 2009), thereby restoring a regime more favourable to the oil and gas companies.[45] The government reduced the maximum royalty rate on conventional oil wells from the (January 2009) level of 50 per cent to 40 per cent, and for gas wells from 50 per cent to 36 per cent. In addition, the government made permanent an incentive that was introduced in 2009 to reduce royalty rates for the first year of a well's life to 5 per cent.[46] A specialist in energy royalties who examined the results of the government's 2009 adjustments to the royalty regime concluded that 'the big change was in gas royalties, which dropped by just over $5 billion per year following the 2009 change. Gas went from providing two-thirds of total royalties to providing only one-sixth of total royalties' (Roy 2015, 3). Royalty rates for oil sands production were not altered. However, according Roy, the combination of lower gas royalties and status quo (comparatively low) oil royalties cost the province $10 billion in foregone revenue between 2009 and 2014 (1).[47] All the same, further accommodating the interests of the oil and gas industry, the government announced in March 2010 that it was launching a cross-ministry task force to study support for 'innovative technologies' and ways to 'streamline' the regulatory system.[48]

While oil and gas companies expressed satisfaction with these changes, energy economists did not view the royalty cuts as justified by economic conditions. Pedro van Meurs, quoted in a *Calgary Herald* article as 'a world-renowned oil and gas economist,' said 'I think Alberta is competitive the way it is. I don't see any reason for change.'[49] André Plourde, an energy economist and a member of the Royalty Review Panel of 2006–07, responded in stronger terms.

> The province has failed to act in the best interest of Albertans – the owners of the resource. The government is more interested in generating short-term activity than snaring the most royalties possible over the longer term, he said, and has failed to adequately consult Albertans on the changes. "What concerns me in all of this is who's speaking for the owners? It's clear the government is not and hasn't been for some time ... You sit here as an owner and say it's a shameful performance."[50]

What is more, recent research has confirmed that the lion's share of resource wealth has, over a long period of time, been channelled not to major improvements in, or expansions of, public goods in the province, but rather to private corporations. Three economists at the University of Alberta reported in May 2010 that 'from 1989 to 2008, provincial spending per person on health care rose 37 per cent; spending on schools rose six per cent; spending on social services fell 25 per cent; personal incomes rose 39 per cent; and corporate profits climbed 314 per cent.'[51]

In this context, it is difficult to argue that a key priority of the Conservative governments of Alberta was to maximize hydrocarbon extraction revenues for residents of the province. Rather, the state's over-riding objective – as André Plourde argued –was to maximize the *rate of exploitation* of the resources, even if the economically rational strategy was to profit from higher long-term prices (related to eventually declining global oil supply) and to obtain *a greater share of rent* from the producers.[52] Even the province's former Conservative premier, Peter Lougheed, argued for a slower rate of exploitation. His successors, more tightly wrapped in neoliberal ideology, have claimed that it is up to market forces, not government, to determine the rate of exploitation (Gray 2008). (This stance was contradicted by their *governmental* decision to provide an incomparably favourable investment environment for oil and gas companies.) Meanwhile, questions about the *ecologically and socially* desirable pace of oil and gas extraction were not included in the Ministry of Energy's consideration of the 'best interest' of Albertans.

When such contradictions have threatened to destabilize the legitimacy of the government/corporate drive to extract every last drop of oil from Alberta's land as quickly as possible, conservative governments have turned to nativist politics. The hostile others who conspire to steal 'Albertans'' wealth/prosperity/future have taken the form of central Canadians, Liberals, Liberal federal governments, the provincial or federal New Democratic Party (NDP) leadership, environmentalists (local or foreign), or misinformed European publics. Ranged on the side of 'Albertans' are the corporations that have chosen to 'develop' the province, and the government that aggressively and tirelessly defends Albertans' right to exploit their resources as 'they' see fit.[53] This siege discourse conceals both the elite interests driving the economic strategy and the existence of real conflicts of interest among differently situated groups of Albertans. Most notably, the assertion of provincial sovereignty and the construction of a generic, 'true' Albertan completely erase the colonial nature of resource exploitation, its inherent environmental racism, and the racial-, class-, and gender-based distribution of oil and gas revenue.

Conclusion

In this chapter we have examined the institutional adaptations, use of energy policy resources, and discursive strategies that have constituted elements of Alberta governments' hegemonic politics. What we see, with regard to institutional restructuring, is a growing bureaucracy (dominated by the Energy Ministry) dedicated to managing the local economic and environmental effects of oil sands exploitation, along with a substantially expanding allocation of resources to the political legitimation of the province's reliance on hydrocarbon extraction and export. From 2008 onward, these legitimation efforts were concentrated on containing and countering the 'dirty oil' campaigns of ENGOs, indigenous organizations, and other social movement actors that target publics and investors in the United States and Europe. Less visible, but of ongoing importance, have been negotiations with the federal government and other provinces regarding climate change policy. For these reasons, legitimation functions have fallen increasingly – since 2008 – to the Ministry of International and Intergovernmental Relations and to the Public Affairs Bureau, strongly supported by the ministries of Energy, Sustainable Resource Development, and Environment.

Energy and environmental policy are closely entwined with these legitimation efforts, and these departments play important roles in making the case that the province is committed to the sustainable development of the oil sands and other resources, and is taking action to reduce greenhouse gas emissions. On closer inspection, however, the available evidence suggests that most of the public funding going to 'environmental' research and development has been channelled to projects aiming to enhance oil recovery or reduce the environmental impacts of the oil sands and coal-fired electricity, rather than to projects advancing a transition to renewables and energy conservation. Moreover, the centrepiece of the province's climate change policy under its conservative governments – investment in carbon capture and storage – is considered by many experts to be too much of a long shot to be a realistic solution to the problem. As of 2015 it was evident that CCS would not achieve even the modest emission reduction targets that the province had set for 2020. In this light, CCS may be little more than a highly costly public relations exercise.

The hegemonic discourse of Alberta's petro-state has evolved in important ways since the 'resource nationalism' of the early 1970s (Pratt 1976). Under the Klein governments from 1993 to 2006, neoliberalism became deeply entrenched in government discourse and policy. The blending of 'nativism' and neoliberalism differentiates the Alberta Conservatives' right-wing populist discourse from that of far-right populist parties in Europe. The latter have shifted, since the 1980s, from neoliberal to anti-globalization stances in order to attach the support of the social strata most adversely affected by unemployment and other forms of insecurity. In Alberta, on the other hand, the geology of mineral wealth and the geography of proximity to the oil-guzzling American market have made possible a political economy that has sustained (with important exceptions) economic growth and relatively high per capita incomes (albeit with growing income inequality). The party that represents the primary beneficiaries of hydrocarbon extraction has had no need of anti-globalization discourse. Indeed, the only protectionism of interest to the Alberta Conservatives is the protection of the right to continue to mine and export resources without interference from federal Liberal governments or international environmental campaigns for the boycotting of 'dirty oil.'

Since the mid-2000s, however, Alberta governments have felt increasingly besieged by climate change science and global calls for action to reduce GHGs, as well as by Canadian and international environmental campaigns against the oil sands. To the neoliberal and nativist discourses described above, then, has been added a discourse of sustainable development. While sustainable development has been interpreted in such a way as to encompass greater expansion of oil sands exploitation (mainly by the Promethean promise of technological solutions), environmental critics have been included in the motley crew of outsiders who threaten 'Albertans'' prosperity. This discursive construction of environmentalists has been put forward with some caution to differentiate between well-meaning but misinformed environmentalists and the more 'extreme' kind, like Greenpeace, who cross the line of legitimate dissent with their direct actions in the oil sands and elsewhere. For the conservative parties, the former must be met with 'truth,' 'facts,' and the dispelling of 'myths' about the oil sands – and the latter with criminal prosecution and association with terrorism.

Albertans themselves seem to be more inclined to grant credibility to environmentalists' claims regarding the harms associated with the oil sands, while at the same time recognizing the dependence of the local economy on massive investment in resource extraction. In addition to the spring 2007 poll conducted for the Pembina Institute, mentioned above (see note 7), subsequent polls have found that Albertans have a high level of environmental concern. A survey of 900 Albertans commissioned by the *Edmonton Journal* and *Calgary Herald,* and conducted in February 2008, found that almost two-thirds of the respondents agreed the government should 'limit greenhouse gas emissions produced by oilsands development, even if it means some projects would be delayed or cancelled' (Henton 2008). A poll conducted in July 2008 by the government-contracted Innovative Research Group found that only 16 per cent of Albertans thought the province was doing a good job of protecting the environment, and that only 3 per cent of non-resident Canadians thought the same.[54] At the same time, 47 per cent of Albertans 'strongly agreed' that oil revenues are necessary to provide 'proper funding for social programs like health and education,' compared to 31 per cent of non-Albertan Canadians who 'strongly agreed.' The pollsters found it paradoxical, however, that 51 per cent of Albertans doubted that they benefited personally from 'oil sands profits.' (Given the findings of Taft et al., reported above, they were very likely correct.) These results – along with the multiple forms of resistance described in other chapters in this volume – manifest the numerous fractures in the consensus, or 'consent,' that the provincial government and its industrial partners continually seek to patch together. This chapter has attempted to make the workings of this hegemonic project more transparent, and to enlarge the space for asserting alternatives to fossil capitalism.

NOTES

1 Although the name of the conservative party that ruled Alberta from 1971 to 2015 is officially the Progressive Conservative Party, in this chapter we use 'Conservative' and 'Progressive Conservative' interchangeably. Note that the federal Conservative Party has dropped the qualifier 'progressive.'

2 That this model has also generated enormous wealth for a corporate elite, along with high-paying jobs for a large workforce, thereby creating deeply entrenched and well-resourced interests, is also of critical importance to an analysis of hegemonic politics in the province. While we do not undertake an analysis of *corporate* strategies of legitimation with regard to the oil sands in this chapter, it should be noted that these often resemble closely those adopted by the provincial government. More detail is offered in Carter's chapter in this volume.

3 Note that we do not give equal attention in this chapter to what might be considered the repressive or coercive aspects of hegemony as exercised by the provincial and federal states, such as surveillance, criminalization, and prosecution of environmental, First Nation, and other opponents of oil and gas extraction. In the post–September 11, 2001 environment, and with the

arrival in Alberta of Greenpeace, there was a growing tendency in governmental discourse (Albertan and federal) to associate environmental actions with terrorism. See Arsenault (2009); Canadian Press (2009); Le Billon and Carter (2010).

4 Fifty-seven such documents were systematically analysed to identify how the Stelmach government has represented the environmental implications of oil sands development to the Alberta public, and what discursive strategies, such as word choice, use of metaphor and imagery, or omission of information, have been used to achieve this. This research is reported in depth in Stares (2010).

5 As of 2014 this ministry had been folded into the Ministry of Environment, creating a renamed Ministry of Environment and Sustainable Resource Development.

6 Although the Klein government mobilized to prevent ratification of the Kyoto Protocol by the federal Liberal Government in 2002, governments under Klein were predominantly concerned with provincial and national opposition to the growth of the oil sands. The Klein government also promoted investment in, and US markets for, oil sands crude, in conjunction with the companies working in the sector. See, for example, Gauthier (2008, 2009).

7 A poll commissioned by the Pembina Institute in spring 2007 found that 71 per cent of Albertans believed there should be a halt to new oil sands project approvals until infrastructural and environmental issues had been addressed; 74 per cent thought that the rate of development should be managed by the government, compared to 20 per cent who favoured leaving this up to market forces. The poll results may be found on the website of the Pembina Institute: http://www.pembina.org/pub/1446.

8 According to the same Pembina poll referred to in note 7, 56 per cent of those surveyed were of this opinion.

9 In 2006 an Oil Sands Ministerial Strategy Committee commissioned the Radke Report, which provided recommendations to better manage the effects of oil sands development. According to provincial government websites, this report contributed to the creation of the Oil Sands Environmental Division within Alberta Environment and the Oil Sands Branch within Sustainable Resource Development (Alberta, Ministry of Sustainable Resource Development 2009).

10 Figures provided by the Minister of the Alberta Treasury Board, written communication to author, 10 June 2010.

11 Government of Alberta, *2014–15 Government and Legislative Assembly Estimates*, Ministry of Energy, p. 66, http://finance.alberta.ca/publications/budget/estimates/est2014/energy.pdf.

12 A description of the AER's mandate and functions is provided on its website: http://www.aer.ca/about-aer/who-we-are.
13 "Ministry of Environment" is used here, although the actual name of the ministry charged with environmental regulation changes periodically, and in 2013 was the Ministry of Environment and Sustainable Resource Development.
14 Petrinex's mandate and governance structure are described on its website, http://www.petrinex.ca/54.asp.
15 In 2008 Greenpeace set up an office in Alberta as well as a website on the oil sands. A coalition of ENGOs lobbied the governors of the western states in the USA to stop importing Alberta's 'dirty oil,' and campaigns were launched in Europe directed at corporate shareholders and investors.
16 The ministries of Environment and Sustainable Resource Development were combined after our research on department expenditures and travel had been completed.
17 This brochure was available on the Ministry of Environment's website in June 2010, at http://environment.alberta.ca/documents/AB-Clean-Energy-Story.pdf. A similar brochure is posted on the government's oil sands website: http://www.oilsands.alberta.ca/documents/FS-CleanEnergy Story.pdf.
18 TransCanada hopes to expand its Keystone pipeline system to ship 510,000 barrels of crude oil a day to Texas from Alberta, including a leg that would take crude from the glutted Cushing, Oklahoma, storage hub to Gulf of Mexico refiners. The resulting network will stretch more than 1,600 miles, from Alberta down through six American states.
19 Research conducted during the development phase of the Brand initiative evidences this concern. Albertan, Canadian, and American participants in the surveys conducted by the contracted PR firm, Harris/Decima, were specifically asked about their perceptions of the oil sands and the sands' environmental costs (Harris/Decima 2009, 14–15). One of the findings of the firm's report was that the government was perceived as being indifferent towards the environment (8).
20 Premier Stelmach, statement in provincial legislature, 30 April 2008, recorded in Hansard, 27th Legislature of Alberta, Session 1, 30 April 2008 (afternoon), http://www.assembly.ab.ca/Documents/-c62f24ac7d18/1/doc/.
21 See Jason Fekete, 'Gov't to Spend $200,000 Refining Energy Message,' *Calgary Herald*, 19 June 2010. The ENGO campaign, led by Corporate Ethics International, included billboards in four US cities that compare the pollution from the oil sands to the BP Deep Horizon well blowout in the Gulf of Mexico. The Government of Alberta's counter-campaign was to have been

managed by the PAB. The Alberta Ministry of Energy, in response to a request from the authors for an account of how the government's campaign money was spent, stated in an email of 19 May 2011 that the campaign 'did not go ahead.'

22 The 2008–2009 annual report of the Alberta Energy Research Institute (AERI) illustrates the heavy weighting of funding by the provincial government towards research into technologies aimed at reducing the carbon footprint of fossil fuels, improving oil recovery, bitumen upgrading, and so on; renewable energy projects received only 9 per cent of AERI's funding in that year.

23 Stares (2010) notes that environmental consequences of development are rarely presented as 'problems' in Alberta government discourse on the oil sands.

24 The Alberta Ingenuity Fund, for example, established with an endowment of $800 million in 2000, was merged into Alberta Innovates – Technology Futures in 2009 (along with units of the Alberta Research Council and other research bodies).

25 Actual reductions of total emissions were not mandated in this legislation.

26 University of Alberta, *Express News*, 8 April 2011, http://www.expressnews.ualberta.ca/NewsArticles/2011/04/HelmholtzAlbertapartnershiptakesnextstep.aspx.

27 Indira Samarasekera, president, University of Alberta, letter to alumni, 25 April 2013, 1.

28 See http://www.capp.ca/canadaIndustry/oilSands/Innovation/media/Pages/David.aspx, last accessed 19 July 2014.

29 The restructuring of post-secondary education to accommodate the imperatives of the petro-state is the subject of research currently in progress under the direction of Laurie Adkin. Some of the preliminary results are reported in this chapter.

30 David Howell, 'Construction Hits Midway Point on Shell's Quest Carbon-Capture Project,' *Edmonton Journal*, 17 April 2014. http://www.edmontonjournal.com/Construction+hits+midway+point+Shell+Quest+carbon+capture+project+with+video/9750554/story.html (accessed 14 July 2014).

31 Government of Alberta, 'Budget 2014 Advances Alberta's Carbon Capture Projects,' media release, 17 April 2014, http://alberta.ca/release.cfm?xID=36237692013C2-9656-5F58-20508CCA717E2819.

32 There are many examples, but one is the presentation made by Alberta's representative in Washington, DC, David Manning (a former president of the Canadian Association of Petroleum Producers), to a public hearing in South Portland, Maine, on 9 July 2014. A video of the presentation is avail-

able at http://daveberta.ca/2014/07/canada-oil-sands-south-portland-maine/ (accessed 21 July 2014). In this presentation, Manning also referred to the $400 million CCEM Fund, without mentioning that the $15/tonne price on GHG emissions is ineffective in actually reducing emissions.

33 The two reports may be found on the website of Alberta Innovates (formerly the Alberta Research Centre): http://www.albertainnovates.ca/energy/major-initiatives/lca.

34 See, for example, the 'stakeholder comments,' including responses by the Pembina Institute, scientists at the universities of Calgary and Toronto, and the senior climate policy analyst at the California Energy Commission, available at http://albertainnovates.ca/media/15768/post%20workshop%20stakeholder%20input.pdf. The US-based National Resources Defence Council also published criticisms of the reports on its blog: http://switchboard.nrdc.org/blogs/sclefkowitz/studies_confirm_tar_sands_dirt.html.

35 The CAPP takes the same line; see the CAPP page on climate change: http://www.capp.ca/environmentCommunity/Climate/Pages/Circumstance.aspx#9dVozgShpJ5M (retrieved 10 June 2010). A brochure produced for the government's oil sands website argues that 'full life cycle GHG emissions' for oil sands crude oil do not differ substantially from those of crude oils imported from other US suppliers (http://www.oilsands.alberta.ca/documents/FS-OilSands-GHG.pdf). See also http://www.oilsands.alberta.ca/documents/GHG_oil_sands.pdf. In the conflict over the Keystone XL pipeline, the premier also argued that 'when considered in the context of other leading suppliers of crude, including Saudi Arabia, Venezuela, Nigeria, Iraq, Angola and Algeria, the energy security benefits of oil from Alberta are clear' (Stelmach, letter to *Washington Post*, 2 July 2010, http://www.edmontonjournal.com/pdf/washingtonpostletter.pdf [accessed 22 July 2010]).

36 For the history of oil and gas royalty conflict between Alberta and the federal government, see Richards and Pratt (1979), esp. chap. 9.

37 Indeed, the Alberta Labour Federation has so far been very careful to avoid the construction of an opposition of this kind between Albertan workers and temporary foreign workers, notwithstanding the scale of the TFW program approvals by the federal government. (As of summer 2014 there were an estimated 85,000–87,000 TFWs in Alberta.) The AFL has called instead for a moratorium on the program, and the integration of the TFWs already in Canada through immigration channels.

38 The text of this speech is available at http://www.premier.alberta.c.a/speeches/2009_25404Addresstothe9thAnnualArcticGasSymposium.cfm.

39 The text of this speech may be found at http://www.premier.alberta.ca/speeches/speeches-2008-July-9-TD_Oil_Sands.cfm. In March 2009, following *National Geographic's* story on the oil sands, and in a speech introducing the Alberta Brand, the premier stated the government's intention to 'keep promoting Alberta in the global marketplace, to preserve our export markets and our future prosperity, 'http://www.premier.alberta.ca/speeches/2009_25568ClosingAddresstotheAAMDCSpring2009Convention.cfm. Conservative prime minister Steven Harper likewise stated in 2007 that increasing oil exports to the USA would be one of his top priorities. See http://www.cbc.ca/canada/story/2007/01/17/oil-sands.html.

40 This concern is stated, for example, in the government's February 2010 Speech from the Throne: 'While we must develop new opportunities to participate in markets like China and India, our economy will be seriously harmed if access to the US energy market is impaired. Alberta fought hard for free trade, which has proven a boon for our people. We cannot lose those hard-fought advantages, and must secure access to the emerging clean energy market south of the border' (http://alberta.ca/home/1240.cfm?).

41 In his speech to the Canadian Association of Petroleum Producers (CAPP) Oil and Gas Investment Symposium, Calgary, 19 June 2007, Premier Ed Stelmach said: 'The response of some has been to demand that we "touch the brake." That approach has been rejected by my government. It's my belief that when government attempts to manipulate the free market – bad things happen.'

42 The Alberta Royalties Review Final Report (2007), entitled *Our Fair Share*, was submitted to the Minister of Finance on 18 September 2007, and is available at http://www.albertaroyaltyreview.ca/panel/final_report.pdf.

43 'Alberta Relaxes Royalty Increase Plans,' Oil and Gas Insight series, *Business Monitor International*, November 2008, http://www.oilandgasinsight.com/file/71074/alberta-relaxes-royalty-increase-plans.html (accessed 7 June 2010). In November 2008 the government exempted unconventional oil and natural gas wells drilled at depths of 1,000–3,500 metres from the new royalty regime (until 2013).

44 For information about the review and its report, *Energizing Investment*, see the website of the Ministry of Energy, Government of Alberta: http://www.energy.alberta.ca/About_Us/Royalty.asp. This is not the first time the oil and gas corporations have used disinvestment threats to turn back royalty increases in Alberta or federally. See Ed Shaffer's description of the conflict of 1979–81 (1984, 188–91).

45 Indeed, in May 2009 it was revealed that the provincial government had,

effectively, not collected royalties from oil sands producers in the first quarter of 2009. See http://www.upstreamonline.com/live/article178363.ece.

46 'Alberta Cuts Conventional Royalty Rates,' *Business Monitor International*, Oil and Gas Insight series, March 2010, http://www.oilandgasinsight.com/file/87063/alberta-cuts-conventional-royalty-rates.html (accessed 7 June 2010).

47 Jason Fekete, 'Alberta's Energy Industry Urges New Royalty Review,' *Calgary Herald*, 28 December 2009, http://www.nationalpost.com/Alberta+energy+industry+urges+royalty+review/2389146/story.html.

48 The Regulatory Enhancement Task Force included representatives of the ministries of Energy, Environment, and Sustainable Resource Development, and was charged with making recommendations by 31 December 2010 for 'a renewed and integrated policy assurance system that will contribute to Alberta's overall competitiveness while protecting the environment and ensuring public safety and conservation of resources.' See the Ministry of Energy website: http://www.energy.alberta.ca/Initiatives/RegulatoryEnhancement.asp.

49 Quoted in Jason Fekete, 'Alberta to Slash Royalties,' *Calgary Herald*, 11 March 2010, http://www.calgaryherald.com/news/Royalties+revamp+expected+review/2669326/story.html.

50 Quoted in Jason Fekete, 'Alberta to Slash Royalties.' Elsewhere, Plourde commented: 'The government is really interested in getting a lot of activity in the very short term. You want to encourage investment, get drilling done, get a lot of production done ... We're trading revenues down the road for a bit more activity in the present' (quoted in Archie McLean, 'Oil, Gas Royalties Cut,' *Edmonton Journal*, 12 March 2010, http://www.edmontonjournal.com/news/royalty+rates/2673614/story.html#ixzz0qU7X5fk4.

51 Kevin Taft, 'Public Spending Stayed Flat as Alberta Economy Grew,' *Edmonton Journal*, 1 May 2010, reporting on a study carried out by Kevin Taft, Mel McMillan, and Junaid Jahangir (2012). The rate of increase of corporate profit reflects not only the royalty regime, of course, but the overall corporate tax regime in the province.

52 The priority of attracting investment in the province's resource sectors continues to be manifested in such initiatives as the Alberta Competitiveness Act, the Alberta Competitiveness Council, and the Alberta Competitiveness Forum (2 June 2010). See the 'competitiveness' page on the website of the Ministry for Finance and Enterprise: http://finance.alberta.ca/economic-development/competitiveness/index.html.

53 Once again, this alliance predates the Stelmach government, as documented by Leadbeater (1984) and Richards and Pratt (1979), who describe the marriage between the provincial elite and multinational oil companies in the 1970s and1980s.

54 The results of the poll were made available on the Government of Alberta website: http://alberta.ca/home/documents/2009BrandResearch Summary.pdf.

REFERENCES

Alberta. 2008a. 'Multi-million Dollar Research Agreement Aims to Improve Water Management and Use in Alberta.' http://www.alberta.ca/acn/200810/24510E2341D13-B789-7532-BC2178A424FE799F.html (accessed 19 June 2010).

Alberta. 2008b. 'Council to Help Develop Alberta's Path Forward on Carbon Capture.' http://alberta.ca/ACN/200804/23372811D6E8B-ECBD-6A20-0AB89FD6BAF75321.html (accessed 22 June 2010).

Alberta. 2009a. 'Alberta Invests in U of A Clean Energy Research Partnership.' http://alberta.ca/home/NewsFrame.cfm?ReleaseID=/acn/200912/2743650139EF0-087D-55E3-166A569A7CAA3156.html (accessed 17 June 2010).

Alberta. 2009b. 'Alberta's Oil Sands: Resourceful. Responsible.' http://environment.gov.ab.ca/info/library/7925.pdf (accessed 22 June 2010).

Alberta Energy. 2008. 'Provincial Energy Strategy.' http://www.energy.alberta.ca/Initiatives/strategy.asp (accessed 17 June 2010).

Alberta Energy. 2009a. 'Final Mission Report: Travel to Great Falls, Montana, May 13–14, 2009.' http://www.energy.gov.ab.ca/Travel/2009_05_Mitzel_MontanaMissionReport.pdf (accessed 2 June 2010).

Alberta Energy. 2009b. 'Business Plan (2009–12).' http://www.finance.alberta.ca/publications/budget/budget2009/energy.pdf (accessed 2 June 2010).

Alberta Energy. 2010a. 'Facts and Statistics.' http://www.energy.alberta.ca/OilSands/791.asp (accessed 20 June 2010).

Alberta Energy. 2010b. 'Carbon Capture and Storage.' http://www.energy.alberta.ca/Initiatives/1438.asp (accessed 20 June 2010).

Alberta Energy. 2013. 'Carbon Capture Funding Agreement Cancelled.' News release. 25 February. http://alberta.ca/acn/201302/33717121A0157-D5E3-839F-750E02B41D039F98.html.

Alberta. Environment. 2007. 'FAQ Oil Sands.' http://environment.gov.ab.ca/info/faqs/faq5-oil_sands.asp (accessed 14 June 2010).

Alberta Environment. 2008a. 'Alberta's 2008 Climate Change Strategy.' http://environment.gov.ab.ca/info/library/7894.pdf (accessed 22 June 2010).
Alberta Environment. 2008b. 'London School of Economics.' Speech given by Environment Minister Rob Renner, 22 October 2008. http://environment.alberta.ca/ documents/London_School_of_Economics_Oct22-08.pdf (accessed 22 June 2010).
Alberta Environment. 2009. 'Climate Change.' Speech given by environment minister, 30 and 31 March. http://environment.alberta.ca/documents/Climate-Change-Was-NY-speech-Mar-30-31-2009.pdf (accessed 12 June 2010).
Alberta Environment. n.d.(a). 'Energy Innovation Fund.' http://environment.alberta.ca/02014.html (accessed 18 June 2010).
Alberta Environment. n.d.(b). 'Climate Change and Emissions Management Fund.'http://www.environment.alberta.ca/02486.html (accessed 18 June 2010).
Alberta Environment. n.d.(c). 'Greenhouse Gas Reduction Program.' http://environment.alberta.ca/01838.html (accessed 30 June 2010).
Alberta Finance. 2010. *Budget of Alberta 2010*. http://www.finance.alberta.ca/publications/budget/budget2010/energy.pdf (accessed 30 June 2010).
Alberta, Ministry of International and Intergovernmental Relations. 2008. 'Final Report. Premier's Mission to Western Governors' Association, Jackson Hole, Wyoming, June 28–July 1, 2008.' http://www.international.alberta.ca/documents/International/ Mission_report-WGA-June08.pdf (accessed 16 June 2010).
Alberta, Ministry of International and Intergovernmental Relations. 2009a. *Annual Report 2008–2009*. http://www.international.alberta.ca/documents/IIRAnnualReport_2008-2009.pdf (accessed 30 June 2010).
Alberta, Ministry of International and Intergovernmental Relations. 2009b. *International Offices Activity Report 08.09*. http://international.alberta.ca/documents/2008-2009_InternationalOfficesActivityReport.pdf (accessed 18 June 2010).
Alberta, Ministry of International, Intergovernmental and Aboriginal Relations. 2008a. *Annual Report 2007–2008*. http://www.international.alberta.ca/documents/About_Us/IIARAnnualReport_2007-2008.pdf .
Alberta, Ministry of International, Intergovernmental and Aboriginal Relations. 2008b. 'Final Report. Mission to Washington D.C., January 15–18, 2008.' http://www.international.alberta.ca/documents/International/Premier_Mission_to_WashingtonDC.pdf (accessed 6 June 2010).
Alberta, Office of the Premier. 2009. 'Address to the 9th Annual Arctic Gas

Symposium' 4 March. http://www.premier.alberta.ca/speeches/2009_25404Addresstothe9thAnnualArcticGasSymposium.cfm (accessed 2 June 2010).

Alberta, Ministry of Sustainable Resource Development. 2009. 'Oil Sands.' http://www.srd.alberta.ca/ManagingPrograms/OilSands/Default.aspx (accessed 15 June 2010).

Alberta, Treasury Board. n.d. 'Oil Sands Sustainable Development Secretariat.' http://www.treasuryboard.gov.ab.ca/OilSandsSecretariat.cfm (accessed 13 June 2010).

Arsenault, Chris. 2009. 'Canada; Govt Threatens Tar Sands Activists with Anti-Terror Laws.' Tar Sands Solutions Network, 20 October, http://tarsandssolutions.org/in-the-media/canada-govt-threatens-tar-sands-activists-with-anti-terror-laws.

Barrie, Doreen. 2006. *The Other Alberta: Decoding a Political Enigma*. Regina: Canadian Plains Research Centre.

Brooymans, Hanneke. 2010. 'Proposed Rules for Oilsands Water-Removal Inadequate: First Nations (Wants Athabasca Protected During Low Flow),' *Edmonton Journal*, 23 July.

Canadian Press. 2009. 'Top Lawyer Defends Oilsands Activists, Chides Alta Gov't for Terrorism Remark,' 6 October. http://www.canadaeast.com/rss/article/815129.

CBC News. 2009. 'Alberta Hires Consultants to Lobby Washington,' 3 April. http://www.cbc.ca/canada/calgary/story/2009/04/03/cgy-lobby-alberta-washington.html (accessed 2 June 2010).

CCEMC (Climate Change and Emissions Management Corporation). 2010a. 'Climate Change and Emissions Management (CCEMC) Corporation Announces $37.5 Million in Funding for Renewable Energy Projects.' Media release, 16 June. http:// ccemc.ca/_uploads/June-16-Media-Release.pdf (accessed 18 June 2010).

CCEMC. 2010b. 'Climate Change and Emissions Management (CCEMC) Corporation Announces More than $5.7 Million in Funding for Six Energy Efficiency Projects.' Media release, 23 June. http://ccemc.ca/_uploads/CCEMC-Media-Release-June23.pdf (accessed 23 June 2010).

CCEMC. 2010c. 'Board Members and Executive.' http://ccemc.ca/about-us/board-members-and-executive (accessed 18 June 2010).

CCEMC. 2014. 'The CCEMC Projects.' www.ccemc.ca (accessed 17 July).

Cooper, Dave. 2013. 'Carbon Capture and Storage Project Goes Modular.' *Edmonton Journal*, 5 April.

Dryzek, John S. 2005. *The Politics of the Earth: Environmental Discourses*. 2nd ed. New York: Oxford University Press.

Edmonton Journal. 2010a. 'Climate Change Projects Move to Slow Lane,' 10 February. http://www.edmontonjournal.com/technology/Alberta+slows+climate+change+spending/2542319/story.html#ixzz0qH2n0XR4 (accessed 20 May 2010).

Edmonton Journal. 2010b. 'Premier Should Know When to Clam Up,' 25 March. http://www.edmontonjournal.com/opinion/venting/Venting/2572869/Video+Squirrel+monkey+born/2698729/Graham+Thomson+Premier+should+know+when+clam/2723597/story.html?id=2723597 (accessed 8 June 2010).

Enhance Energy. 2012. Newsletter, issue 6 (November), http://www.enhanceenergy.com/pdf/News/newsletters/Enhance_Newsletter_November_2012_Issue_6.pdf.

Environment Canada. 2013. 'Canada's Emissions Trends." October.

Fluet, Colette, and Naomi Krogman. 2009. 'The Limits of Integrated Resource Management for Aboriginal and Environmental Groups: The Northern East Slopes Sustainable Resource and Environmental Management Strategy.' In *Environmental Conflict and Democracy in Canada*, ed. Laurie E. Adkin, 123–39. Vancouver: UBC Press.

Gauthier, Jennifer L. 2008. 'Whose Mission Accomplished? Alberta at the 2006 Smithsonian Folklife Festival.' *American Review of Canadian Studies* 38 (4): 451–72.

Gauthier, Jennifer L. 2009. 'Selling Alberta at the Mall: The Representation of a Canadian Province at the 2006 Smithsonian Folklife Festival.' *International Journal of Cultural Studies* 12, no. (6): 639–59.

Gray, John. 2008. 'The Second Coming of Peter Lougheed.' *Globe and Mail, Report on Business,* September, 72–80.

Gustafason, Bret. 2012. 'Fossil Knowledge Networks: Industry Strategy, Public Culture and the Challenge for Critical Research." In *Flammable Societies: Studies on the Socio-economics of Oil and Gas,* ed. John-Andrew McNeish and Owen Logan, 311–34. London: Pluto Press.

Hajer, Maartin A. 1995. *The Politics of Environmental Discourse: Ecological Modernization and the Policy Process*. Oxford: Oxford University Press.

Harris/Decima. 2009. 'Research Summary: Branding Alberta Initiative.' http://alberta.ca/home/documents/2009BrandResearchSummary.pdf (accessed 27 May 2010).

Henton, Darcy. 2008. 'Environment Trumps Oilsands: Poll.' *Edmonton Journal,* 28 February, http://stoptarsands.wordpress.com/2008/02/28/environment-trumps-oilsands-poll/.

Howell, Trevor Scott. 2008. 'Rebranding Alberta Contract Awarded to Tory-Friendly Ad Firm.' http://www.ffwdweekly.com/article/news-views/

news/rebranding-albertacontract-awarded-tory-friendly-/ (accessed 7 June 2010).

Jaremko, Gordon. 2010. 'Ethical Energy.' *Alberta Oil*, vol. 5, issue 7 (May), 14.

Krugel, Lauren. 2010. 'U.S. Ad Campaign Seeks to Tar Alberta's Image,' Canadian Press, 14 July. http://travel.ca.msn.com/canadianpress-article.aspx?cp-documentid=24874378 (accessed 23 July 2010).

Leadbeater, David, ed. 1984. *Essays on the Political Economy of Alberta*. Toronto: New Hogtown Press.

Le Billon, Philippe and Angela Carter. 2010. 'Dirty Security? Tar Sands, Energy Security, and Environmental Violence.' In *Critical Environmental Security: Rethinking the Links Between Natural Resources and Political Violence*, ed. Matthew A. Schnurr and Larry A. Swatuk. Halifax: Centre for Foreign Policy Studies, Dalhousie University, 1-18. http://www.dal.ca/content/dam/dalhousie/pdf/cfps/pubs/critical-environmental-security/chapter9.pdf.

Mallory, J.R. 1954. *Social Credit and the Federal Power in Canada*. Toronto: University of Toronto Press.

McLean, Archie. 2010. 'Stelmach Seeks Clinton's Support for Oil Sands Pipeline to U.S.' *Edmonton Journal*, edmontonjournal.com, 8 July 2010.

MIT. 2013a. 'Quest Fact Sheet: Carbon Dioxide Capture and Storage Project.' Carbon Capture & Sequestration Technologies Program. CCS Project Database. http://sequestration.mit.edu/tools/projects/quest.html. Modified 28 March.

MIT. 2013b. 'Alberta Carbon Trunk Line Fact Sheet: Carbon Dioxide Capture and Storage Project.' Carbon Capture & Sequestration Technologies Program. CCS Project Database. http://sequestration.mit.edu/tools/projects/quest.html. Modified March 28.

Pembina Institute. 2007. 'Albertans' Perceptions of Oil Sands Development: Poll.' Backgrounder, 8 May. http://pubs.pembina.org/reports/Poll_Env_mediaBG_Final.pdf (accessed 22 July 2010).

Pembina Institute. 2010. 'False Claims Highlight Alberta's Inadequate GHG Regulations.' http://www.aenweb.ca/media/false-claims-highlight-albertas-inadequate-ghg-regulations (accessed 19 June 2010).

Pratt, Larry. 1984. 'The Political Economy of Province-Building: Alberta's Development Strategy, 1971–1981.' In *Essays on the Political Economy of Alberta*, ed. David Leadbeater, 194–222. Toronto: New Hogtown Press.

Project Pioneer. 2012. Press release, Calgary, 26 April. http://www.projectpioneer.ca/ (accessed 6 April 2013).

Public Affairs Bureau. Executive Council. Government of Alberta. 2009. 'Natural Resources Defense Council Could Follow Its Own Advice.' *Alberta*

Blog. 9 September 2009. http://alberta.ca/blog/home. cfm/Energy?&startRow=16 (accessed 25 June 2010).

Public Affairs Bureau. Executive Council. Government of Alberta. 2008. 'Facts about Alberta's Brand Campaign.' http://www.publicaffairs.alberta.ca/pab_images/Brand_Fact_Sheet.pdf (accessed May 2010).

Public Affairs Bureau. Executive Council. Government of Alberta. n.d. 'Alberta Brand Initiative: Overview.' http://publicaffairs.alberta.ca/pab_images/Brand_Overview.pdf (accessed 7 May 2010).

Read, Andrew. 2014. 'Climate Change Policy in Alberta.' Pembina Institute. Backgrounder, 14 July. http://www.pembina.org/pub/climate-change-policy-in-alberta (accessed 20 March 2016).

Richards, John, and Larry Pratt. 1979. *Prairie Capitalism: Power and Influence in the New West*. Toronto: McClelland and Stewart.

Roy, Jim. 2015. 'Billions Forgone:The Decline in Alberta Oil and Gas Royalties.' Fact sheet. Edmonton: Parkland Institute, 23 April. http://s3-us-west-2.amazonaws.com/parkland-research-pdfs/billionsforgone.pdf (accessed 17 May 2015).

Shaffer, Ed. 1984. 'The Political Economy of Oil in Alberta.' in *Essays on the Political Economy of Alberta*, ed. David Leadbeater, 174–93. Toronto: New Hogtown Press.

Simpson, Jeffrey. 2008. 'A Licence to Pollute Dressed Up in Rhetorical Petticoats,' *Globe and Mail*, January 26.

Stares, Brittany. 2010. 'A Way with Words: Analyzing Provincial Government Discourse of the Alberta Oil Sands.' Honours thesis. Department of Political Science, University of Alberta.

Taft, Kevin, with Mel McMillan and Junaid Jahangir. 2012. *Follow the Money*. Calgary: Detselig Enterprises Ltd.

Tait, Carrie. 2012. 'Alberta's Carbon Capture Efforts Set Back.' *Globe and Mail*, 26 April 26, updated 6 September, http://www.theglobeandmail.com/report-on-business/industry-news/energy-and-resources/albertas-carbon-capture-efforts-set-back/article4103684/.

Thomson, Graham. 2009. 'Burying Carbon Dioxide in Underground Saline Aquifers: Political Folly or Climate Change Fix?' Paper for the Program on Water Issues, Munk Centre for International Studies. University of Toronto, September. http://beta.images.theglobeandmail.com/archive/00242/Munk_Centre_Paper_242701a.pdf.

TransAlta. 2009. 'TransAlta's Project Pioneer Awarded Government Funding to Proceed.' News release, 14 October. http://www.transalta.com/newsroom/news-releases/2009-10-14/governments-canada-and-alberta-have-partnered-transalta-build-one- (accessed 6 April 2013.

Wiseman, Nelson. 2013. Introduction to C.B. Macpherson, *Democracy in Alberta: Social Credit and the Party System*, 3rd ed., vii–xxx. Toronto: University of Toronto Press.

Wood, James. 2014. 'Prentice Dismisses Carbon Capture as 'Science Experiment.' *Calgary Herald*, 10 July. http://www.calgaryherald.com/news/politics/Prentice+dismisses+carbon+capture+science+experiment/10016830/story.html.

7 The Alberta Oil/Tar Sands and Mainstream Media Framings between Globalization and Polarization

CONNY DAVIDSEN

The public debate around oil dependency, climate change, and the local-global environmental impacts of oil extraction is characterized by multiple, highly politicized discourses. These discourses make reference to the complex spatial and temporal scales of geopolitical resource interests, and seek to establish the socio-environmental costs and socio-economic benefits of fossil fuel reliance, corporate-governmental accountability for these, and the political responsibilities of individual citizens. Discursive framings of what is at stake in oil sands exploitation include 'energy security,' 'hydrocarbon dependency,' local 'ecological degradation' 'environmental justice,' 'climate justice,' 'economic prosperity,' Canada as an 'energy superpower,' and much more. Such discourses appeal for urgent action throughout the entire commodity and emission chain, from hydrocarbon extraction to energy consumption worldwide. In sum, the oil/tar sands constitute what we call 'a wicked problem.'

This chapter uses a political ecology perspective to analyse mainstream framings of the wicked problem that surrounds the Alberta oil/tar sands, and explores the concept of structural literacy as a way to improve our culture of critique and contestation for a better public understanding of what environmental problems entail. By 'mainstream' framings I mean the particularly powerful and visible stream of mediated reporting within the entirety of the broader universe of discourses surrounding the Alberta oil/tar sands.

While the multitude of policy actors involved in the Alberta oil/tar sands, such as environmental and social movements, contribute to distinct streams of the discourse realms of the topic, the persistent 'mainstream' of the discourse is predominantly defined by the industry, government, and mass media, including large-circulation newspapers that may hold a strong position of influence in the region.

Environmental problems are difficult to understand, let alone define. Environment is both a vertical issue that touches upon all scales from the global to the very local, and a horizontal issue that touches upon all areas of life from livelihoods to resource industries, economies, health and agriculture, land base, and biodiversity. The more complexity and transdisciplinarity is involved in its overlapping scales of ecosystemic, economic, political, health-related, and other issues, the more an environmental issue becomes a wicked problem. These are particularly difficult to deal with in public spheres and by non-experts, as the complex issue needs to be divided, interpreted, simplified, prioritized, framed, and inevitably communicated in selective pieces as opposed to the whole.

The findings of this research suggest that although the debate surrounding the Canadian tar sands is becoming much more active ('mobilized') with regard to frequency and public attention, mainstream discourses surrounding the topic show signs of a selective reduction of the choices presented to the public, thus immobilizing the public from a comprehensive rethinking of the problems that surround the issue. The chapter focuses on two driving forces to explain this process: globalization and polarization.

The chapter then focuses on media dynamics to ask how the framing of what is at stake might be transformed in ways that permit a deeper understanding of the root causes of the environmental crisis and possible alternative modes of development. It highlights ways in which wicked problems can be dealt with more critically, effectively, and comprehensively by promoting a new level of literacy for environmentally and democratically engaged citizens: environmental literacy, political literacy, and media literacy. These are suggested as new spaces of democratic engagement, political critique, and a more pluralist discourse that may counteract current dynamics of simplification, polarization, and immobilization of choices in the oil/tar sands debate as it tends to be presented in mainstream media.

The chapter is structured into six parts. The following section introduces concepts of 'mainstream' media and problem 'framings' in the oil/tar sands. Section 3 discusses media pressures that limit the discursive spectrum, and section 4 examines created language categories of 'oil sands,' 'tar sands,' and 'dirty oil,' and the polarizations they represent. Section 5 analyses the interface of mainstream media discourses and public engagement, and more specifically the roles of discourse, action, and silence. Finally, section 6 recaptures the dynamics and limitations of mainstream media discourses and identifies potential spheres of transformation through a combination of different types of (environmental, political, media) literacy, called structural literacy.

'Mainstream' Media and the Framing of 'Problems' in the Age of Globalization

Environmental problems are framed based on a particular concern that typically emerges because of a new scarcity, degradation, or other clash with established patterns of our human-environment system. They are selective observations out of a vast and complex array of multidimensional connections, overlaps, and dynamic changes. These environmental issues are difficult for humans to grasp in their larger extent and scope, let alone understand them sufficiently to develop an informed opinion. Media and communication technologies have a vital role in exchanging information, positions and arguments that ideally provide a discursive arena for differences and consensus as public opinions shape and change over time. Discourses, ultimately, affect environmental decision-making processes at their core, because they shape the ontology of environmental issues by creating categories, concepts, and context. They can be understood as a 'shared way of apprehending the world,' as John Dryzek (2005, 9) suggests. Discourses provide interpretations of meanings and connections, define common and legitimate knowledge, and constitute a shared ground on which ideas can be discussed, exchanged, and challenged. This includes an acceptance of foundational social norms, environmental values, and assumed rights and responsibilities in human-environment relationships. Language, for example, often goes unnoticed but represents a powerful influence on the construction of framings and meanings, for which discourse analysis may provide a deeper insight into the strategic uses of terms and rhetorical devices by different actors.

Mass media deeply affect the political economy and ecology of communication in the digital age, and have arguably become one of the most dominant influences on public debates, from their ontological framing of 'issues' for the general public to the public attention and relevance these receive. The globalization of mass media has profoundly changed the social and political environment of discourses for the past decades, especially the arrival of the digital age with new information and communication technologies (ICTs). On the one hand, the internet's expanding breadth of information sources has rendered portrayals of environmental issues less monolithic because alternative framings of issues are finding more access to the public. Many powerful and diverse discourse arenas are available for voices outside the 'mainstreamed' realms of the mass media, which suggests that media spectacle could be more open to contestation and democratic mechanisms of control and accountability.

At the same time, however, the globalization of new 'mainstream' mass media technologies has caused a fragmentation of information that discourages complexity. In-depth stories, multi-level argument structures, and a critical review of existing categories are often incompatible with the needs of short television clips or newspaper articles, and are replaced by short punchlines and simple arguments that skip over the context of background and reasoning (Chomsky and Barsamian 1992). Instead of a few consciously selected news sources a day, individuals gather information from selective bits that are increasingly served from a plethora of outlets that reach the consumer ubiquitously and constantly. This intensifies the competitive pressure to increase pace, brevity, and particularly the spectacularity of environmental representations to gain the attention of the consumer. Images and stories of ecological disasters, activist campaigns, and public reactions circulate the world within a day, often mainstreamed into a few highly selective headlines and pictures that catch just enough public attention to survive amidst the daily overabundance of news reports. With emerging narratives of global human suffering and civilization at the brink of survival, environmental issues are gaining an unprecedented sense of urgency in news reporting, intensified by the fast-growing scales of impact. Global environmental issues seem to have become the major issue of conflict of the twenty-first century. From funding-dependent non-governmental organizations to political election campaigns, reality TV entertainment and environmental infotainment, ecotourism and nature theme parks, media-marketed environmental activism and advertising campaigns of the local zoo, the interested and engaged public is increasingly exposed to Disneyfied narratives of conservation and ecological destruction that focus on emotionally gripping but selective and limited storylines and cater to the demands of the news industry.

This suggests a deeper layer in which the globalization of new media has not only changed the hardware and interaction dynamics of environmental communication and discourse, but also changed the environmental experience and media literacy of the individual on a deeper level. An increasing disconnect of modern society's personal experience with the physical realities of the environment has created media consumers who are unable to evaluate the selective representations of an environmental problem from their own personal range of experience, and are increasingly unable to tell fiction from reality. Modern lifestyles with high technology dependency have enabled such environmental performances or spectacles to become more pervasive and misleading because people are more and more likely to interact with a digital interface instead of a real human being or material interaction with nature. High-profile media portrayals with advanced technology thus affect our cultural sense of the aesthetics of nature, such as BBC's 'Planet Earth' series or IMAX nature documentaries, as they create a different, spectacularized version of reality. They influence our understanding of human-environment relationships. Steve Irwin's television show 'Crocodile Hunter,' for example, became a part of popular culture for its thrill-seeking but affectionate attitude towards dangerous animals. Filmmaker Rob Stewart's feature-length documentary 'Sharkwater' that challenged our socially framed fears of sharks is an educational piece on human-animal interaction and marine conservation while it uses visual spectacularity to capture the audience and make them care. While the media can be a highly effective element in environmental education, environmental spectacles bear several risks for informed environmental decision making and public understanding of environmental problems. The spectacular portrayals of nature, of its ecological destruction and human roles in this, have become a central influence on mainstream perceptions of environmental problems and human responsibilities. This is of particular concern for highly urbanized societies in which environmental conceptualizations are even more dependent on media representations as opposed to a material, individual interaction with nature.

This accelerated, disconnected portrayal of realities has led to what Bauman (2000) called 'liquid modernity,' a world of perpetual change in which the seemingly infinite breadth of possibilities and risks creates an intimidating level of individuation and lack of guidance. Ultimately, mass media spectacles become more attractive than the mundane and less spectacular world of life actually lived. Accordingly, Tsing argues that global markets have turned into a 'global economy of appearance' in which 'dramatic performance has become an essential prerequisite of economic performance' (2005, 57). (An example of this may be the dramatic volatility of stock markets in the past years.) While the material realities of extractive industries continue to be important, economic growth increasingly depends on a dramatic performance to sell everything from corporate investment chains to stock market strategies, leading to a choreographed 'spectacular accumulation' that may long precede the material reality of these economic gains on paper.

In order to participate in competitive investment markets, countries, provinces, and cities are also enticed into dramatizing their local potential for investors (Tsing 2005). Environmental media portrayals have become part of the larger political economy in which environmental organizations, governments, corporate actors, and others compete for public attention and its related support and revenue. Alberta's public relations efforts towards positive energy-related imagery clearly follow these patterns of choreographed economies of appearance in pursuit of a positive public image. In 2008, for example, the Alberta government launched a $25-million public relations campaign to re-brand Alberta as a place to invest, with environmental aspects as a substantial motivation of the larger campaign. Adkin and Stares (this volume) show how the Stelmach government employed new public relations strategies to 'clean' or green its energy policy. In summer 2010, the Alberta government announced another pro-oil-sands public relations campaign targeted at its own voters. An intense two-week PR blitz, estimated to cost CAD $268,000 (Cryderman and D'Aliesio 2010), aimed at countering environmental criticisms. The campaign was to include large advertisements in Alberta daily and weekly newspapers as well as radio spots. A few weeks earlier, the Alberta government had spent $58,000 on a half-page advertisement in the *Washington Post* that declared: 'A good neighbour lends you a cup of sugar. A great neighbour supplies you with 1.4 million barrels of oil per day.' The timing of the advertisement was tied to the approval process in the US for the proposed Keystone XL oil sands pipeline that had become highly controversial.

These PR trends, together with the weak environmental regulation described by Angela Carter (this volume), suggest a strategic shift of the government away from managing realities to managing opinions. In fact, the director of Fort McMurray's Oil Sands Developers Group, Don Thompson, made clear in an interview that the biggest strategic focus of the oil/tar sands operations is not to improve its environmental performance, but rather to strengthen its public relations, in order to legitimate what he called a 'social license' [to operate] (interview with the author, 8 June 2010; also see Zalik, this volume).

Delimiting the Discursive Spectrum

Public legitimization of the tar sands depends on public knowledge of available alternatives. Noam Chomsky argued that 'the smart way to keep people passive and obedient is to strictly limit the spectrum of acceptable opinion, but allow very lively debate within that spectrum' (Chomsky and Barsamian 2003, 43). Following the unifying force of popular culture, globalized environmental media portrayals lead towards a mainstreaming of environmental experience and perceived ranges of options. Furthermore, homogenized representations of diverse global realities grant legitimacy and coherency to the dominant worldview of an already powerful historic bloc, as Gramsci (2000) suggests. This worldview limits the thinkable options to those available within a neoliberal capitalist political economy. The solutions that dominate are thus primarily limited to a Promethean perspective and some aspects of ecological modernization. There are few critical voices presented in public media whose focus is not solely on the immediate environmental impact of unconventional oil production, but extends to the underlying political economy, power structure, and discursive spaces of the tar sands debate (a gap this book may help to fill).

The limitation and polarization of discourses in mainstream media coverage of the oil/tar sands debate is difficult to counteract because it is reinforced by inequalities of power and influence. Dominant worldviews and interests are not only closely tied into the mainstream discourse, but they actively reproduce and buttress the legitimation of the ruling elite's worldview over time (Brockington, Duffy, and Igoe 2010). Low-income groups, groups that are socially marginalized on multiple grounds (such as recent, racialized immigrants, temporary foreign workers, or members of First Nations), students, and radical intellectuals may gain power by uniting, but often form a scattered and marginalized periphery in media coverage of perspectives and alternative visions.

Consumers and citizens are confronted with such a profusion of environmentally relevant information that careful and intuitive selection has become a critical skill. One of the key obstacles to a more comprehensive, differentiated, and effective use of the media as a discursive tool for democratic debate is the media's tendency to simplify complex issues and to focus on dynamic, spectacular, short-term events. Media are pressed to commodify information into news 'products,' which have to withstand competition in a fast-paced news industry.

This makes it difficult to translate complex policy issues into pieces of information that are accessible to the public both in news size and required level of background knowledge, while still embedding the interesting news bite into a relatively uneventful overall picture. Nuanced, thoughtful approaches, long-term perspectives, and careful complexity in environmental debates are seldom heard in the existing media landscape, as they compete with other, more spectacular(ized) versions. As a result, media representations may distort or exaggerate the actions and actual change that are occurring on the ground.

Mainstream media coverage can also shift the public's focus towards more easily communicable aspects of the problem. For example, tar sands critics have established a strong moral narrative surrounding the 'dirty' extraction and processing of bitumen, which is certainly a true part of the story. At the same time, however, this narrative focus has somewhat diverted attention from the ramifications of carbon-intense political economies, for example, the fact that the burning of fossil fuel by the end consumer alone produces up to 80 per cent of the oil's overall carbon footprint. The media (and environmentalist?) focus on the extraction end of the commodity chain is driven by several dynamics surrounding mass media reporting. First, directing the critique at an object or process away from the consumer means that the audience can support the issue without having to accept its own failures, the material or moral responsibilities of overconsumption, or arising obligations for drastic lifestyle changes. Second, this framing of the issue offers a strategically easier storyline that is more likely to spark public interest and fundraising because it offers environmental campaign supporters a positive affiliation for a struggle against a tangible and purportedly 'evil' target. Yet such framings may suggest that more rigorous government regulation or the application of new technologies constitute adequate solutions to the problem, leaving untouched the more troubling conclusion that our whole energy system of petroleum dependency and high energy consumption needs to be remodelled. Narrow critiques solely focused on the realms of oil production distract from this important point. Thus one might see such criticisms as falling within a 'mainstream' framing of the problem associated with the oil/tar sands – one that is ultimately amenable to managerial and technical solutions because they create new opportunities for economic growth. This narrow range of mainstream options creates the illusion of environmental progress, but ultimately suppresses momentum for real energy transitions and social change. Media representations of environmental issues, in summary, not only create illusions of the environmental problem at hand, but also create a carefully framed perception of the socio-political reality surrounding it. These representations shape the underlying assumptions of the debate about the tar sands.

Even if the media were to present a comprehensive in-depth discussion of the issues, lack of scientific literacy in public communication also limits the communication of environmental problems. Science and politics are driven by different objectives, values, codes of interaction, and approaches for resolving conflict (Parson 2000). This causes misunderstanding, ineffectiveness, and incompatibilities between the two different domains that cause a communicative disconnect in their discourses. Complex arguments that are embedded in a plethora of existing knowledge are difficult to translate into media-friendly, bite-sized chunks of information, and will almost inevitably be misunderstood, misinterpreted, and used against their sources, either strategically or accidentally.

Moreover, competition for public attention is stronger than ever. Environment has become an increasingly attractive theme for the media industry over the last decades, but primarily when processed into a mediated version that follows the demands of the market. Infotainment and market competition between news services encourage a spectacularization of news in order to gain market shares. This leads to an unbalanced portrayal of the issue that may be presenting only the extremes, may exaggerate the extremes, or may take the message out of context altogether. Media consumers are overwhelmed with a plethora of options, and are inclined to limit their daily news selection based on the most efficient – to-the-point, accessible, and entertaining – use of their time. More complex arguments that go beyond existing categories, in contrast, would require carefully collected background knowledge, attention, and space, all three of which are often incompatible with changing media pressures and demands (Herman and Chomsky 1988).

Oil Sands, Tar Sands, and Dirty Oil: Labels and Implications

One example of the underlying polarization of the discourse is the subtle dichotomy of terms used for description. 'Oil sands' and 'tar sands' refer to the same hydrocarbon resource but represent politically opposite terms that are associated with the two different political views of oil sands supporters and tar sands critics. Their conscious usage in the current debate functions as a shibboleth (Gwendolyn Blue in Davidsen and Wells 2010), a 'word or pronunciation that distinguishes people of one group from another' (Merriam Webster 2010). The technical term is 'bituminous sand,' which is less of a catchphrase and did not gain traction in the debate. Language is powerful, and the choice of language is a highly strategic tool in creating imagery that supports political goals (Myerson and Rydin 1996) and holds 'all kinds of consequences' (Dryzek 2005, 10). Environmental terms and phrases evolve and change over time as their ideas become contested, revised, popularized, or defeated. The term 'oil sands' shifts the emphasis to the final product, moves away from the difficulties of the extraction process, and enforces the imagery of a rich natural composite in which the oil is already there. Andrew Nikiforuk has argued in his book *Tar Sands* that 'if that lazy reasoning made sense, Canadians would call every tomato *ketchup*' (2008, 12). The term 'tar sands,' on the other hand, emphasizes the black and sticky characteristics of the soil, which is closer to its visual appearance. It also recognizes the natural occurrence of the sticky bitumen in the ground, as opposed to the term 'oil sands,' which refers to an end product (oil) that depends on technological processing and human ingenuity to extract. However, more recent uses of the term 'tar sands' have turned the term's descriptive natural characteristics into a political connotation of being sticky, hideous, and 'dirty.'

Both terms have experienced differences in their public acceptance over time. Figure 7.1 shows an analysis of keyword hits from twelve large Canadian newspapers across the country, conducted by Daniel Graham (2010) at the University of Calgary. It shows that the usages of 'oil sands' and 'tar sands' have traversed different trajectories over the past decade. While the term 'tar sands' has only slightly increased in usage, the usage of the term 'oil sands' has grown tremendously in the same years.

The dramatic increase in usage of one term over the other indicates two things. First, the increased usage of 'oil sands' suggests the growing salience of the debate over the oil/tar sands. Second, 'oil sands' seems to be more accepted by the newspapers. To determine whether this is true of all the newspapers surveyed, figures 7.2 and 7.3 track the relevant keyword hits of each individual newspaper over time. They show that the term 'oil sands' was used to different extents by various newspapers, but only a few newspapers, for a relatively short period of time, used the term 'tar sands.' The overlay of the two figures indicates that none of the newspapers consistently adopted the term 'tar sands' over 'oil sands.'

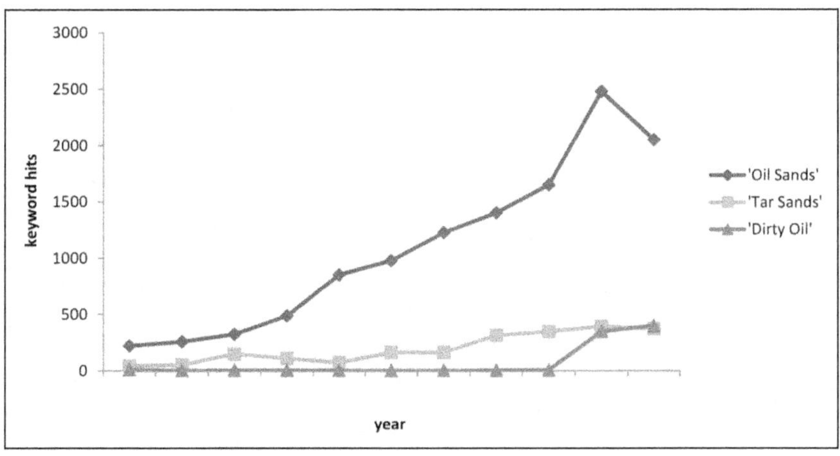

Figure 7.1. Oil/tar sands language in Canadian newsprint media 1999–2009. (*Source*: Graham 2010.)

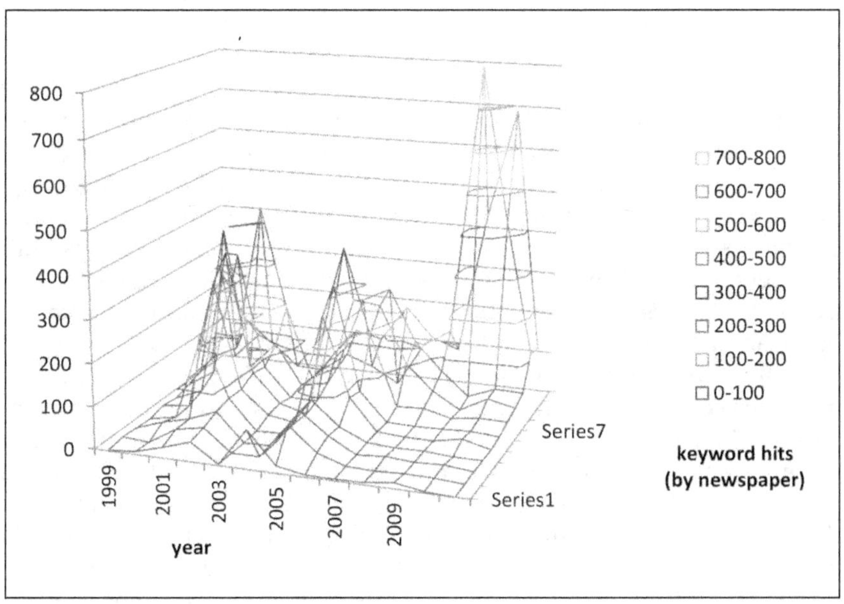

Figure 7.2. Usage of term 'oil sands' in selected Canadian newsprint media, 1999–2009. (*Source*: Graham 2010.)

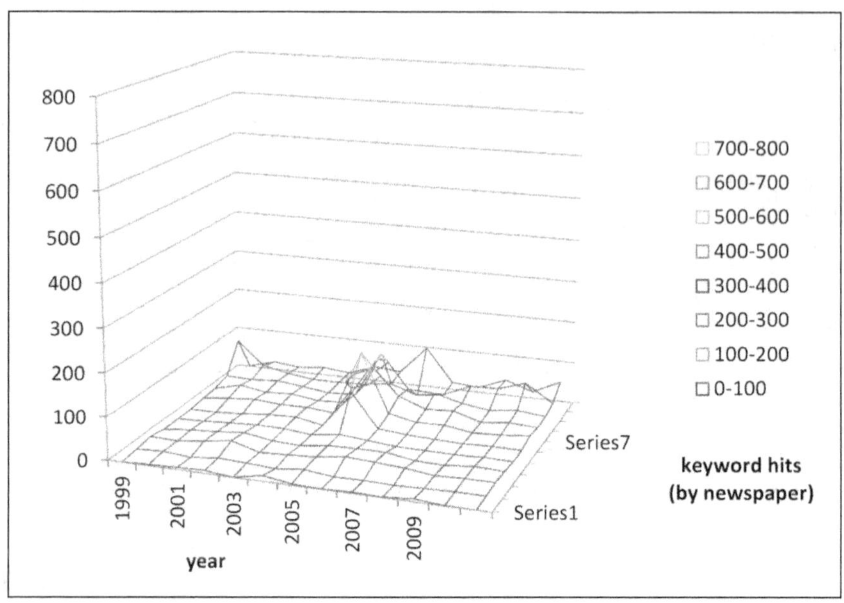

Figure 7.3. Usage of term 'tar sands' in selected Canadian newspapers, 1999–2009. (*Source*: Graham 2010.)

Interviews among key policy actors and the general public on the oil/tar sands issue, carried out by the research and film team of the 2010 documentary *Oil Literacy* (Davidsen and Wells 2010), allowed interviewers to identify categories of public responses. The interview responses suggested the existence of at least five different groups with different reasons for their choices of vocabulary ('oil sands' or 'tar sands'). The first group includes those relatively unfamiliar with the issue and its context. Street interviews with these individuals indicated a certain level of insecurity about which term to use. Some also thought that the terms referred to different things. The second group are members of the general public who know *of* the issue, may have friends or family working in the industry, and stay at a careful distance from the issue while adopting the dominant language. They tend to use the term 'oil sands' and explain that it is the term they hear the most in the media and among friends and relatives. The third group are environmentally engaged individuals who, either as professionals or as citizens, want to express a certain level of criticism; these use the term 'tar sands.' The fourth group are professional members of the industry or government, highly integrated into the political economy surrounding the oil/tar sands. They use 'oil sands' because it reflects and supports their pro-development perspective. The fifth group the interviews identified included professional members of environmental organizations and the media who are trying to reach and engage the public and key actors, in order to bring as many actors as possible to the table. They, too, use 'oil sands' in order to keep the more critical views visible in the 'mainstream' discourse by at least sharing the same language, in an attempt to reverse the problems that a polarization of language entails. While the fifth group has a valuable argument for a reconciliation that may reopen closed doors in the debate, I have deliberately used 'tar sands' in this chapter in order to emphasize the need for a deeper social and environmental critique that does not give in to, but distances itself from, the hegemonic (governmental-corporate) discourse.

Even stronger political reactions and economic implications surround the term 'dirty oil.' Petroleum derived from the oil/tar sands has been labeled 'dirty' because of severe environmental impact throughout the production process, in particular higher greenhouse gas emissions but also the tailings ponds, water pollution, air pollution, or destruction of boreal forest. The framing of 'dirty oil' campaigns caused powerful waves of public attention and environmental criticism in the United States and Europe. The term managed to create a strikingly simplistic and media-effective label that epitomized the tar sands' mainstream environmental concerns, provided an easy catchphrase for social condemnation even for those unfamiliar with the issue, and most importantly, created a language of stigma that helped mobilize consumer boycotts of Alberta oil exports.

Getting the Public Involved: Discourse, Action, and Silence

The oil/tar sands are operating in an increasing void between the status quo of hydrocarbon resource extraction economies on one hand and growing public criticism on the other. There seems to be broad consensus among involved policy actors and the general public that something needs to change, but how, exactly, is unclear. According to the 2010 Global Thought Leader Survey on Sustainability in which McAllister Opinion Research (2010) questioned five thousand experts and government officials on behalf of the Pembina Institute, approximately 75 per cent of the respondents rated Canada's efforts at addressing climate change as poor or very poor. Sixty-eight per cent also rated Canada's efforts in stimulating economic reorientation towards renewable energies as poor or very poor. The same percentage identified federal leadership as the most important factor affecting the implementation of sustainable energy solutions. Another 58 per cent found subsidies for petroleum production to be the policy that needs to change the most. As Pembina's executive director, Marlo Raynolds, sums up, leaders within government, academia, and the private sector 'have a high agreement on how best to tackle' the current problems, but nonetheless Canada 'lacks the political commitment to take sufficient action' (Pembina 2010).

These trends of consensus and differentiated solution approaches among the policy actors continue to be juxtaposed with simplified media reports that either emphasize the conflict between the cores of each view or favour the strongest and most conventional political view entirely. This suggests some parallels to the discourse history of climate change, as Oreskes and Conway summarize the case of the US: 'Until recently the mass media presented global warming as a raging debate – twelve years after President George H.W. Bush had signed the U.N. Framework Convention on Climate Change, and *twenty-five years* after the U.S. National Academy of Sciences first announced that there was no reason to doubt that global warming would occur from man's use of fossil fuels' (2009, 242–3, italics in original). Mass media sources not only prolonged but aggravated the momentum of disagreement based on false or oversimplified reports that impeded discussion of acceptable approaches and solutions, and ultimately delayed urgent action against climate change by causing further confusion over the substance of claims rather than facilitating a content-driven democracy.

This again emphasizes the importance of media literacy and the difficult roles of the media in providing a robust and reliable communication forum towards democratic discourse and decision making. One important step towards this would be to strengthen the spheres of contestation between the state, economy, and citizens in order to encourage an ongoing social renegotiation of priorities, concerns, and critique (Gottweis 2009). The media hold a central position in facilitating critical feedback between the actors and interests of society, which would enable a more direct inclusion of the public in key discussions. They can enable a stronger sphere of contestation if they manage not only to create spheres of exchange and comparison between actors, but also keep track of past dialogue with its disbanded arguments and reached consensus. Accordingly, transparency and literacy are vital building steps towards an informed, consequential and progressing dialogue between citizens and governmental agencies, industry actors and consumers, governmental planning and research/development sectors, and non-governmental groups in continuous exchange with government and industry actors. Such mediated spheres of contestation could foster the communication between different epistemological communities or 'semiotic universes' and allow space for their differing presentations of realities towards policy decisions (Beck, Giddens, and Lash 1994; Hajer 1995).

Discourse transparency, however, has limitations on many levels. During the filming of the documentary *Oil Literacy*, our team observed substantial obstacles to discursive participation based on notions of personal obligation: a considerable percentage of the population declined to be interviewed and argued that they felt obliged to remain silent because their socio-economic status depends entirely or in part on the tar sands. The film team found that street interviews were particularly difficult in Alberta. On some days, more than thirty per cent of the people approached in the streets declined an interview on-camera based on the same argument: they had a family member working in the industry and would rather keep quiet in order to avoid negative reactions from employers, colleagues, or offended family members. Social control thus plays a role in discouraging citizen, employee, and consumer engagement in a system that is small, tightly woven, and regionally fixated on the benefit of one industry.

Oil sands exploitation, it seems, continues in part because of an immense layer of non-action silently agreed upon by those who think they benefit from it, as Theresa Garvin (this volume) also finds in her study of public participation in the regulation of sour gas production. This silence is an enticing one in that it is easier to refrain from action and not risk controversy, or risk exposure for whatever reason. It is easy to get away with this because there is no pronounced moral discourse about citizenship rights and responsibilities in relation to the public good. Also, depending on the social values of peers, there is still very little reward mechanism in place for those who do change their lifestyles and manage to reduce their consumption. One problem is that consumers are too easily absorbed into the anonymity of mass consumption of fossil fuels because these are currently not marketed as diversified or distinguishable commodities with labels identifying sources and production processes (eco-, 'ethical,' or fair trade) that would be able to guide consumer choices. Overall, transparent consumer information and peer control are more difficult to implement for crude oil sales than for certified organic food or fair trade clothing, although this may change in the future (similar to recent market changes surrounding 'green' electricity). At present, the marketing options for labelling and retail channels are narrower; the physical product is less tangible and has no easily marketable 'face.'

The silence also draws strength from traditional industrialist roles in which consumers were protected from criticism as long as consumption was understood as a good thing and more consumption as an even better thing, following the post-war paradigm in which consumer demand served economic growth and prosperity. The US and Canadian governments actively employ narratives in which maintained or increased levels of consumption are a citizen responsibility to support the national economy. A historical sense of entitlement continues to surface where such discourses in the global north assume a high consumption standard as an indisputable privilege. This continues in international arguments of the US and Canadian governments surrounding climate change action where the CO_2 impact of North America is being deflected by finger-pointing at China and India. This strategic disregard of historical responsibilities was also deployed to avoid the emission reduction commitment of the Kyoto Protocol; the Alberta and federal Canadian governments have taken an active role in discouraging a politically literate reflection of its citizens on their carbon-intensive lifestyles, the robustness and future of the regional environment and economy, and democratic decision making surrounding the oil/tar sands issue. Instead, their rationale of global competition among national economies has reinforced the growth narrative, ignoring alternative measurements of development that would promote a sense of global environmental citizenship.

A second theme of mainstream discourse – one that permits a certain spectrum of discussion while ultimately limiting the full range of choices – is ecological modernization. This offers a controllable range of debate with technological solutions and advertises environmental progress under the noble guise of sustainability efforts. Such claims create an illusion of change while maintaining the structural status quo of energy policies and the surrounding political economy. Davidson and Mackendrick 2003) show how the Alberta government has used the narrative of ecological modernization for its public relations and media strategies – advertising to the public that environmental performance regulations of natural resource extraction are an important goal of the government – while very few changes have been implemented on the ground, and new approvals for expanding extraction operations continue at an accelerated rate. Adkin and Stares (this volume) analyse in detail the provincial government's strategy to legitimate and sustain the current focus on petroleum extraction in what have essentially become petro-capitalist power structures. Using an ecological modernization narrative creates a more proactive environmental image at the same time as it buttresses the status quo, deflects criticism of key issues, and shifts these storylines to more palatable and marketable activities such as research and development in 'green technologies.' Environmental studies, however, have increasingly suggested that a shift is necessary towards solution approaches based on a more comprehensive and transdisciplinary understanding of socio-environmental problems, that value social ingenuity as well as scientific ingenuity, and that move us beyond a carbon-intensive society (Hawken 2007).

In summary, current Canadian 'mainstream' discourses regarding the oil/tar sands can be described as somewhat immobilized by a simplification and polarization of choices, paired with a limited presentation of problem framings. Consumption-related critiques, for example, are seldom heard in the mainstream framing of the 'problem' surrounding the oil/tar sands issue.

Lessons, Transformative Shifts, and Structural Literacy

The Alberta oil/tar sands have become the centre of an increasingly active, politicized, and polarized public relations battle to frame what is at stake in the province's continuing reliance on hydrocarbon extraction and export. The oil/tar sands have severe environmental implications of which a democratic society needs to be fully aware, but the public framing of the 'problem' is distorting the range and proportionality of choices. The breadth of the discourse has become fairly delimited, evolving into a discussion of the tar sands as a smaller issue of environmental performance with stricter regulation, 'greening' the productive process of petroleum production, and a gradual inclusion of renewable energy. The spectacularized local frame of 'dirty' operations of the Alberta tar sands has created an absorbing public framing of the problem within which the debate can be heated and active, but it does not jeopardize the power base of the petro-capitalist economic framework. The discursive universe rarely embraces a larger debate about the political-economic causes of global environmental degradation and climate change.

In this chapter I have discussed the effects of two key forces on the framing and distortions of the oil/tar sands issue: globalization and polarization. An increasing globalization of media communication accelerates the already existing oversimplification of environmental representations, while an increasing disconnect between the media consumer and the environmental issue limits our ability to assess these representations and encourages us to accept them without the control to verify their content. Polarization tendencies across public representations of the oil/tar sands debate across Canadian mass media have formed imagined narratives of two not only opposing views but mutually exclusive options – a dichotomy that also becomes evident at deeper layers of language in the divergence of the politicized terms 'oil sands' versus 'tar sands,' or the contentious use of the term 'environmentalists' vis-à-vis those supporting Albertan economic growth. Another consequence is the current communication paradigm that delineates somewhat polarized identities shaped around 'us' versus 'them' on various levels, adopted by many actors in the energy debate who compete for the loyalties of citizens or consumers. The simplification or spectacularization into antagonistic positions is paired with an omission of choices in between and beyond, which has immobilized the range and momentum of public debate.

The Alberta oil/tar sands and their 'mainstream' framings of the problem highlight an urgent need for a more comprehensive, deeper, and open understanding of a broader range of societal choices – without constructed loyalty ties to limited narratives. The discussion of possible solutions from here needs to address all short- and long-term consequences of action and non-action without the limitations of existing 'mainstream' frames. Citizens require stronger literacy skills with regard to media, scientific-environmental, economic, and political issues, creating a strong, comprehensive, and ultimately 'wicked' tool to approach complex problems with the necessary level of curiosity, critique, and creativity. This comprehensive ability is what I refer to as structural literacy.

The Canadian oil/tar sands are arguably the biggest environmental and economic challenge of modern Canadian history, and need to be urgently addressed with the necessary level of education, context, and critique. As Albert Einstein once suggested, in order to solve a problem our thinking must evolve beyond our level of thinking when we created it. A new level of thinking and engagement is needed that enables a more critical literacy in order to counteract the problems of media consumption in times of globalization and polarization. This includes a stronger political culture of well-informed and transparent critique, comprehensive arguments, and democratic discourse. Only this kind of media consumer (or media citizen) can question the presented assumptions of what 'environmental problems,' 'economic risks,' and 'growth' actually mean, or what else they could possibly be beyond the offered framings. This comprehensive understanding of literacy also entails a stronger structural critique of the surrounding context, including the society that has enabled the emergence of these environmental problems. We must come to acknowledge that the causes are rooted in the consumption of a petroleum-dependent society, and that the oil/tar sands are not so much the source of the problem as the symptom of a global affliction – a problem that runs much deeper.

The Canadian oil/tar sands are part of a much larger wicked problem, and we need to heighten our level of discursive complexity in order to meet the complexity of the issue. The media are a central and crucial tool in enabling democratic spheres of dialogue and contestation, but their service and mission reflect the literacy levels of the public. Enabling the media as transformative spaces of dialogue (both in content and interaction) requires life-long public education, more emphasis on the connections of science, politics, and media in the educational curriculum for all ages, and more openness beyond a 'mainstream' of mass media representations. Structural literacy is critically needed to review existing categories used in mainstream discourses, to allow citizens to look beyond problem framings offered by the media, and to reinvigorate democratic spaces of engagement in the debate. Ultimately, a stronger emphasis on structural literacy – in our educational policies, in our political culture, in our media environment – would allow us to reshape our discursive culture so that it becomes more democratic, more environmentally accountable, and more capable of informing environmental citizenship.

REFERENCES

Appenzeller, Tim. 2004. 'The End of Cheap Oil.' *National Geographic Magazine,* June. http://ngm.nationalgeographic.com/ngm/0406/feature5/index.html.

Beck, Ulrich, Anthony Giddens, and Scott Lash. 1994. *Reflexive Modernization: Politics, Tradition and Aesthetics in the Modern Social Order*. Cambridge, MA: Polity.

Brockington, D., R. Duffy, and J. Igoe. (2008.) *Nature Unbound: Conservation, Capitalism, and the Future of Protected Areas*. London: Earthscan.

Buckminster Fuller, Richard. 2008 [1969]. *Utopia or Oblivion: The Prospects for Humanity*. Switzerland: Lars Muller.

Chomsky, Noam, and David Barsamian. 1992. *Chronicles of Dissent*. Stirling: AK Press.

Chomsky, Noam, and David Barsamian. 2003. *The Common Good*. Tucson: Odonian.

Cryderman, Kelly, and Renata D'Aliesio. 2010. 'PR Blitz Rolls Out as Oilsands Protests Grow – Tories Spending $268,000 to Silence Critics.' *Calgary Herald*, 4 August.

Davidsen, Conny, and Robert Marshall Wells (executive producers). 2010. 'Oil Literacy.' Documentary. MediaLab, Pacific Lutheran University, Tacoma, WA, and University of Calgary.

Davidson, Debra, and Norah MacKendrick. 2003. 'All Dressed Up with Nowhere to Go: The Discourse of Ecological Modernization in Alberta, Canada.' *Canadian Review of Sociology and Anthropology* 41: 1 47–65.

Dowling, Robyn. 2010. 'Geographies of Identity: Climate Change, Governmentality and Activism.' *Progress in Human Geography* 34 (4): 488–95.

Dryzek, John. S. 2005. *The Politics of the Earth: Environmental Discourses*. 2nd ed. New York: Oxford University Press.

Foucault, Michel. 1972. *The Archaeology of Knowledge and the Discourse of Language*. London: Tavistock.

Gottweis, Herbert. 2009. 'Theoretical Strategies of Poststructuralist Policy Analysis: Towards an Analytics of Government.' In *Deliberative Policy Analysis: Understanding Governance in the Network Society*, ed. Maarten Hajer and Hendrik Wagenaar, 247–65. Cambridge: Cambridge University Press.

Graham, Daniel. 2010. *Canadian Climate Change Discourse in Canadian Newsprint Media*. MGIS research paper, Department of Geography, University of Calgary.

Gramsci, Antonio. 2000. 'War of Position and War of Maneuver.' In *The Antonio Gramsci Reader*, ed. D. Forgacs, 192–228. New York: New York University Press.

Hajer, Maarten. 1995. *The Politics of Environmental Discourse: Ecological Modernization and the Policy Process*. Oxford: Oxford University Press.

Hardin, Garrett. 1968. 'The Tragedy of the Commons.' *Science* 162 (3859): 1243–8.

Hawken, Paul. 2007. *Blessed Unrest*. New York: Penguin.

Herman, Edward S., and Noam Chomsky. 1988. *Manufacturing Consent: The Political Economy of the Mass Media*. New York: Pantheon.

Igoe, Jim, Katja Neves, and Dan Brockington. 2010. 'A Spectacular Eco-Tour around the Historic Bloc: Theorising the Convergence of Biodiversity Conservation and Capitalist Expansion.' *Antipode* 42 (3): 486–512.

Illich, Ivan. 1974. *Energy and Equity*. London: Calder and Boyars.

IPCC (Intergovernmental Panel on Climate Change). 2007. *Climate Change 2007: Synthesis Report*. Contribution of working groups I, II and III to the Fourth Assessment Report of the IPCC. Ed. Core Writing Team, R.K. Pachauri, and A. Reisinger. Geneva: IPCC. http://www.ipcc.ch/publications_and_data/ar4/syr/en/contents.html.

Kohler, Nicholas. 2009. 'We're Not So Dirty.' *Maclean's*, 16 March, 44.

Krauss, Clifford. 2005. 'In Canada's Wilderness, Measuring the Costs of Oil Profits.' *New York Times*, 9 October.

Kunzig, Robert (author), and Peter Essick (photographer). 2009. 'Scraping Bottom: The Canadian Oil Boom.' *National Geographic Magazine*, March. http://ngm.nationalgeographic.com/print/2009/03/canadian-oil-sands/kunzig-text.

Macdonald, Douglas. 2009. 'The Government of Canada's Search for Environmental Legitimacy: 1971–2008.' *International Journal of Canadian Studies/Revue internationale d'études canadiennes* 39–40: 191–210.

McAllister Opinion Research. 2010. *The 2010 Global Thought Leader Survey on Sustainability: Climate Change, Sustainable Energy and Oil Sands*. Survey commissioned by Pembina Institute, May 2010.

McDonough, William, and Michael Braungart. 2002. *Cradle to Cradle*. New York: North Point.

Minio-Paluello, Mika. 2010. 'Rock That Burns: Dirty Oil Deposits around the World.' *New Internationalist* 431: 16–17.

Moorhouse, Jeremy, Marc Huot, and Simon Dyer. 2010. *Drilling Deeper: The In-Situ Oil Sands Report Card*. Drayton Valley, AB: Pembina Institute. http://pubs.pembina.org/reports/in-situ-report-card.pdf.

Motta, Sara C. 2009. 'Old Tools and New Movements in Latin America: Political Science as Gatekeeper or Intellectual Illuminator?' *Latin American Politics and Society* 51 (1) 31–56.

Myerson, George, and Rydin, Y. 1996. *The Language of Environment: A New Rhetoric*. London: UCL.

Nikiforuk, Andrew. 2008. *Tar Sands: Dirty Oil and the Future of a Continent*. Toronto: Greystone.

Parson, Edward A. 2000. 'Environmental Trends and Environmental Governance in Canada.' *Canadian Public Policy* 26 (2): 123–43.

Prudham, Scott. 2010. 'Pimping Climate Change: Richard Branson, Global

Warming, and the Performance of Green Capitalism.' *Environment and Planning A* 41 (7): 1594–1613.

Redclift, Michael. 2005. 'Sustainable Development (1987–2005): An Oxymoron Comes of Age.' *Critical Perspectives on Sustainable Development* 13 (4): 212–27.

Sabatier, Paul A., and Hank C. Jenkins-Smith., 1993. 'The Advocacy Coalition Framework: Assessment, Revisions, and Implications for Scholars and Practitioners.' In *Policy Change and Learning: An Advocacy Coalition Approach*, ed. P.A. Sabatier and H.C. Jenkins-Smith, 211–36. Boulder.: Westview.

Sachs, Wolfgang. 1999. 'Sustainable Development and the Crisis of Nature: On the Political Anatomy of an Oxymoron.' In *Living with Nature: Environmental Politics as Cultural Discourse*, ed. F. Fischer and M. Haajer, 23–42. Oxford: Oxford University Press.

Tsing, Anna. 2005. *Friction: An Ethnography of Global Connection*. Princeton: Princeton University Press.

8 The Gendered and Racialized Subjects of Alberta's Oil Boomtown

SARA O'SHAUGHNESSY AND GÖZE DOĞU

So much has been written, opined, and investigated about the Athabasca oil sands on everything from royalty rates to worker shortages and statistics on emissions, barrels produced per day and required pipeline capacity, that it is easy to miss the fact that at the heart of it all are people whose lives are being fundamentally altered as a result of oil sands development. In a booming and busting resource economy where a singular product overwhelmingly dominates the political, economic, physical, and social landscapes, widespread marginalization often accompanies the mass of opportunities created. Generally, the inequalities and hardships generated by this type of development are overshadowed by the dominant narrative of job generation, growth, and prosperity. That these inequalities conform to markedly gendered and racialized patterns goes virtually unmentioned in mainstream accounts of oil sands development. Yet a feminist political ecology approach focuses our attention on these often neglected relationships. Research on the gendered and racialized dimensions of the world's largest mega-development project is only in its infancy, and this chapter seeks to make a contribution to their unravelling. Specifically, we explore the ways in which shifting gender practices and relations alter, and are altered by, the wider social, economic, and environmental changes in the continually booming and busting oil town of Fort McMurray.

Northern Canadian resource communities are often perceived as having strongly gendered divisions of labour where men and wom-

The authors thank Dr Laurie Adkin for her helpful contributions to this chapter.

en occupy polarized social and economic roles. At first, Fort McMurray seems like a resource community where gender roles are stable. Yet these gender roles are often more fluid and flexible than they outwardly appear, and are shaped by community restructuring associated with resource development. The unprecedented scale and pace of development in the Athabasca oil sands is altering, for some individuals, the ways in which gender roles are performed in northern Canadian resource communities. For others, the transformations associated with the oil sands are reinforcing traditional gender practices. The demand for labour at the height of the recent boom (2006–2008), for example, created employment opportunities for women in male-dominated fields like engineering, skilled trades, and local administration. At the same time, many occupational fields continue to reflect a traditional sexual division of labour. Women are over-represented in lower-paying occupations that replicate reproductive labour at the wider community level, such as in-home childcare, industrial cleaning services, and the non-profit sector; while oil industry workers are overwhelmingly men. In this chapter, we explore the ways in which gender practices unfold in a booming and busting resource community deeply entrenched in the national and global political economy, and how these practices are negotiated within the often contradictory discourses of frontier masculinity and traditional family values. We focus specifically on the localized identities and practices at the site of production – the oil sands – and at the site of reproduction both at home and at the community level.

Making sense of these heterogeneous and occasionally contradictory gendered experiences requires an understanding of the gendered discourses that permeate the most intimate aspects of individuals' lives within the community while ideologically supporting an extractive-based economy. We begin by situating the particular context of Fort McMurray and highlighting its unique challenges as the staging area for changing (and also reinforcing) gendered practices. Next, we describe the ways in which gender practices and relations are constructed through discourses of frontier masculinity and traditional family values. We then explore what we call the 'maxims' commonly heard among residents of Fort McMurray that speak to perceptions of gender, ethnic, or racial identity. A maxim is a common way of talking about oneself or one's group, and reveals an aspect of identity. We were attentive to the ways in which the women we interviewed referred to themselves, as their language provides insight into how they perceive their identities.

Theory and Method

Feminist political ecology (FPE) is a critical theoretical lens for viewing human-environmental relations that integrates insight from a wide body of thought, including political ecology, feminist political economy, post-structuralism, post-colonialism, and ecofeminism (Rocheleau, Thomas-Slayter, and Wangari 1996). FPE stresses that power circulates through localized gendered practices in ways that support particular human-environmental relations, and that power is fundamentally tied to political, economic, and gendered regimes (Rocheleau et al. 1996; Salleh 2009). Feminist political ecologists foreground both the material *and* discursive practices that reinforce gendered inequalities in decision-making processes, policies, resource allocations, and so on. They also understand that individuals have multiple subjectivities and that there exist complex interactions among racial, class, gendered, sexual, and other subject positions (Bavington, Grzetic and Neis 2004; Hovorka 2006). FPE has primarily been used to investigate women's experiences of environmental change and resource management and in rural 'third world' contexts. Recently, a growing number of voices have advocated the expansion of feminist political ecology's scope to include urban areas (Hovorka 2006) and the 'first world' context (Reed and Christie 2009; Reed and Mitchell 2003). Our research contributes to these emerging areas of feminist political ecology by exploring the shifting gender identities, relations, and practices that are associated with the exploitation of the Canadian tar sands. Our analysis is focused on a particular place – a town located in the tar sands mining zone and that has experienced the economic and social extremes of the boom-and-bust cycle of the global oil economy.

Fort McMurray is located at the extraction point in the commodity chain for crude oil, and its fortunes have indeed been chained to the vagaries of international oil prices and the investment decisions of multinational corporations. In boom times it draws in workers from other parts of Canada and from foreign countries, and these migrants often seek to make a life for themselves in this northern, resource-based town. The outcome is a growing, increasingly multi-ethnic mix of inhabitants who are quite differently situated in regard to the benefits of the oil economy, and in regard to Canadian citizenship. But within this mix we discern gendered and racialized patterns that constitute important parts of the explanation for the socio-economic inequalities that characterize this microcosm of the petro-state.

We draw on qualitative data from two research projects to inform our analysis: the first one, undertaken by O'Shaughnessy, consisted of semi-structured interviews done in 2009 with twenty female residents in Fort McMurray who are employed in the oil sands, social services, and non-profit sectors.[1] The second project was a two-year ethnographical study of the social impacts of oil industry development in Fort McMurray, undertaken by a team of researchers between 2007 and 2009. Doğu and the other members of the research team interviewed a variety of people (over 150) living and working in Fort McMurray.[2] We also draw upon other accounts of the discursive constructions of gender relations in Fort McMurray to enrich our data and analysis. The oil sands – and Fort McMurray – receive a huge amount of attention, mostly from the media. Though much of this coverage is focused on the economic and environmental dimensions of production, dominant conceptions of masculinity and femininity are often tacitly embedded and reproduced in such accounts.

Fort McMurray: The Staging Area

Located in the heart of the Athabasca oil sands, Fort McMurray is frequently described as a place of contradictions – a small town with big city problems. In some ways, Fort McMurray represents the archetypical Canadian boomtown. Yet in other ways it defies predictions and expectations. From rapid population growth, soaring income inequality and cost of living, to social disruption, Fort McMurray has not only hit all the markers of classic boomtowns, it has taken them to the extreme. The happenings of this relatively small community command the world's attention. Media outlets report on the numerous feats of engineering, economic growth, population mobility, and environmental destruction accomplished in and around Fort McMurray on a near-daily basis.

Fort McMurray's population almost doubled from 2003 to 2008 (Regional Municipality of Wood Buffalo [RMWB] 2008). Much of this population growth has been driven by the migration of people from other Canadian communities that have suffered stagnation, if not collapse, locally. Oil sands exploitation has altered the political economy of the entire country, as Angela Carter and Anna Zalik also point out in their chapter in this volume. There has been a massive migration of unemployed – mostly male – workers from the depressed Atlantic provinces to the Alberta oil sands, as described in this *Chatelaine* article:

> According to Statistics Canada, more than 33,000 Atlantic Canadians moved to Alberta between 2001 and 2006. Not included in that number are the region's thousands of invisible oil-patch commuters, men who live in Alberta work camps but have permanent homes in Atlantic Canada. An Alberta report estimates that more than half of the Fort McMurray area's 25,000 migrant workers are from Atlantic Canada. But even without doing the math, you'd be hard pressed to find a family in Sydney or St. John's or Glace Bay that doesn't have a male relative working in the oil sands. And there's more evidence this trend is on the rise: Between 2005 and 2007, passenger traffic between the Halifax and Edmonton airports increased by 250 percent. In the past two years, Air Canada and WestJet have added four new weekly flights from St. John's to Fort McMurray, estimating the grand total of those travelling to Alberta each year at 51,000. (Beaton 2008)

This migration has had a huge impact on families and communities in the Atlantic provinces. In particular, women in this region have been left to care for children and manage households on their own.

Other oil sands workers come from even further away. Fort McMurray's growth is also attributable to international immigrants and temporary foreign workers (TFWs) who have been encouraged to fill the perceived labour shortage in Alberta during the height of the recent boom. Reflecting the effects on migration of the global boom-and-bust cycle in oil prices and investment, the number of temporary foreign workers who came to work in Alberta fell 26.8 per cent in 2009 – from 39,088 in 2008 to 28,610 in 2009 (Government of Alberta 2011, 6). As of September 2010, about 58,000 TFWs were estimated to be living in Alberta (Alberta Federation of Labour 2011). The increase in population exacerbated the pressures Fort McMurray experienced during the boom. Turnover rates were significantly high across a variety of sectors. The social and health services have been particularly affected and stressed, with high turnover rates amongst teachers, health workers, and municipal employees (Archibald 2006).

The high-income-earning possibilities in Fort McMurray's 'oil patch' are well known. Indeed, Statistics Canada reported that the median family income in 2005 in Fort McMurray was $124,592 (2006), compared to $60,600 for the rest of Canada (2011).[3] Due largely to the high wages in the resource sector, the cost of living in Fort McMurray has skyrocketed. In 2009, Fort McMurray surpassed Edmonton and Calgary as the most expensive housing market in the province of Alberta (Toneguzzi

2009). A *Fortune* magazine story reported the average one-bedroom apartment rent to be $2,200 at the height of the recent boom, in 2008 (Heinrich 2008). As a result, there are many working poor, a category defined as those earning less than $40,000 per annum (Canadian Press 2006). Moreover, media commentaries on Fort McMurray's high wages or housing problems typically neglect to mention that social conditions and job opportunities in the region are markedly gendered. Men in Fort McMurray, on average, earned $76,645 CAD in 2005, while women earned only $24,452 (Statistics Canada, 2006). Women employed in the retail and social services sector account for a significant percentage of the working poor population.

Constructing Men and Women in Fort McMurray

Frontier Masculinity

According to Rebecca Scott, 'working-class communities are typically identified according to what the men do: a steel town, a mining town ... Female-coded types of employment do not identify communities' (2007, 493). The preeminence of masculinity – specifically frontier masculinity – and male-coded work has been commonly associated with northern resource-dependent regions and extractive industries (Anahita and Mix 2006; Dunk 2003; Hogan and Pursell 2008; Miller 2004; Scott 2007). Frontier masculinity is premised on a vision of the masculine as strong, rugged, self-sufficient – conquering the dangerous wilderness in the hope of striking it rich (Anahita and Mix 2006; Miller 2004). This particular form of hegemonic masculinity is imbued with nostalgia for (real and fictive) accounts of a historical era where independent white men were iconized as heroes. Women were only minimally present and committed to serving the needs of men, while the natural environment was viewed as a boundless source of riches. In the context of Canadian settler society, frontier masculinity is a working-class notion of masculinity that defines itself against all that is considered feminine, non-white, and urban (Hogan and Pursell 2008; Scott 2007); in other words, it is defined against all that is deemed incapable of enduring the tough conditions of the frontier. As Raewyn Connell (1993) argues, hegemonic masculinities, including frontier masculinity, do not simply provide the cues through which men construct their gendered identities; masculinities also shape social relations and political and economic regimes. (See also Anahita and Mix 2006; Scott 2007.)

Nowhere is frontier masculinity more 'out there' in Canada than in Fort McMurray. The perception of the Athabasca oil sands as a modern-day northern frontier has been featured in academic accounts, media reports, and in the words of local residents. The isolation of Fort McMurray and the surrounding work camps, the dangers associated with oil extraction and the main highway leading to the work camps, the sheer wealth being generated, and the 'rough and tumble' (Fillion 2006) nature of resource workers are all key fixtures in nearly every media account of the oil sands development, as the following quotations demonstrate:

> Just as the California gold rush came to define the American dream, so Fort McMurray defines a particular kind of Canadian dream. *Go west*. Fort McMurray can change your life. But first, perhaps, consider what price you are willing to pay. (Edemariam 2007)

> Fort McMurray, for example, has an overall crime rate five times the provincial average. Some experts chalk up the problem to a '*wild west frontier mentality*' that sees people embrace a work hard, play harder outlook. (Ferguson 2007, italics added)

Yet, in spite of the dangers, the isolation, and the frigid weather associated with northern regions, the perceived larger-than-life opportunities in 'Fort McMoney' are drawing thousands of would-be oil workers trying to 'make it' – much like the northern Canadian gold rushes of days past. However, in the Fort McMurray oil patch, 'making it' is measured by the number of zeros at the end of one's pay-cheques. The prospect of making it rich through hard physical labour is also a fundamental mechanism through which men can assert their masculinity.

While frontier masculinity is a discourse with evident influence on men's identities, it also delineates certain subject positions that are available to women. As Stoeltje (1975) points out, popular stories and myths of frontier times rarely feature women in a prominent or positive way. Representations of women in the frontier are limited to the 'good helpmate' or 'bad woman.' The helpmate is a woman who, unlike refined women in urban centres, is tough enough to withstand the harsh conditions of the frontier. The helpmate, according to Stoeltje, is the glue of the household and the community, defined by 'their ability to fulfill their duties which enable their men to succeed, and to handle crises with competence and without complaint' (1975, 32). Whereas the helpmate is frequently represented in desexualized ways, the 'bad woman' – the polar opposite of the helpmate – is necessarily defined by her sexual behaviour. Whether portrayed as whores or racialized 'others' who fulfilled the exotic sexual fantasies of frontier men, these women are represented as the antithesis of the women who maintain community on the frontier.

Such extreme representations of frontier masculinity and femininity continue to be imposed on men and women and on gendered relations in Fort McMurray, particularly in the popular media and in the perceptions of residents. For instance, in a report on impaired driving and on drinking behaviour in Alberta, Rothe suggests:

> Bar fights may be a result of demographics – an imbalanced male-female ratio found in Fort McMurray. As a kind of *frontier one-industry city*, oil industry-related workers are dominantly males. They are highly over represented in the local population. Hence there is a keen competition for women, a factor which several interviewees suggested leads to fights. (Rothe 2005, 19, italics added)

An interesting, and somewhat contrasting, perspective is offered by a controversial article that appeared during the height of the boom in *Chatelaine*, a popular Canadian women's magazine: "There are guys in them thar' hills of this booming frontier town, but you'll need to step over the gold diggers, divorcées and escorts to get 'em" (Preville 2006, 119). The widely perceived demographic disparity between men and women in Fort McMurray is a classic trait of frontier boomtowns. As the quotations above suggest, it can engender a circumstance – real or imagined – where relationships between women and men are highly competitive, sexualized, and even acrimonious. As one of our interviewees described:

> You'll notice that when you're trying to meet people … it's a whole different dynamic. Because they're used to the whole 'gold digger' … or the animosity between the men and women about 'girls up here are like this,' and 'guys up here are like this' and most of the women who think that the guys up here are all creepy, and then all guys up here think that all women are gold diggers.

The notion of frontier masculinity resonates beyond gender relations in the community. It also circulates broadly in local and regional economic and political structures, which then circulates back into the community. The urgency and frenetic pace of oil sands development is premised on an elision of productivity with wealth (and of wealth with well-being), which necessarily privileges a very masculine form of labour and a masculine relationship of domination of nature that harkens back to colonial settlement. Every year, thousands of new and often long-distance commuting (Canadian) or temporary foreign workers are lured to the oil sands, separated from existing social support networks, and lodged in work camps. Many of the costs of housing and servicing this workforce have been socialized, for example, by provincial government subsidies for housing, spending on new hospital or other facilities, and so on. The Alberta Federation of Labour (2010, 2011) argues that the prolific applications of employers in the province to bring in TFWs constitute a strategy of keeping wage rates from rising in what would otherwise be a very tight labour market.

Traditional Family Values

Conforming to the dualistic image of women in frontier times as either 'helpmates' to men or, conversely, as 'bad women,' a discourse of traditional family values pervades Fort McMurray. What constitutes 'traditional' behaviour and values in extractive communities, of course, varies through time and space. However, as Walsh (1985) notes, there are some underlying descriptors based on commonly held beliefs about gender roles within family structures that guide women's behaviour. Being a traditional woman in an oil boomtown means deriving one's 'formal' identity through men (husbands, fathers); women publicly maintain a subordinate position to men and publicly acknowledge a man's primacy in the home and in household affairs. Finally, when possible, 'traditional' women make a career out of taking care of home and family. Although they willingly work outside the home when the economic need or opportunity arises, 'making a living' is not their primary responsibility unless they are single and/or heads of households.

At the heart of the 'traditional family values' discourse is a highly normative conception of the 'family' as white, heterosexual, and hard working, where men are the primary breadwinners and women are responsible for reproduction, regardless of whether they are formally employed (Harder 2002; see also Fraser 1989; Stacey 1998). This particular notion of the family is generally promoted by prevailing political discourse, public policy, and economic institutions. However, it also implies a set of social and moral responsibilities that influence relationships and interactions with others. Often, women's social role within the family as an unpaid caregiver in the household translates into a responsibility for social reproduction at the wider community level, which facilitates an economic system in which (primarily male) labourers can devote themselves almost exclusively to production. Mellor refers to this economic arrangement as a 'masculine experience economy, a ME economy that has cut itself free from the ecological and social framework of human *being* in its widest sense. Its ideal is "economic man," who may also be female. Economic man is fit, mobile, able-bodied, unencumbered by domestic or other responsibilities' (2009, 254). The ME economy to which Mellor refers is clearly seen in the mobile, work-camp lifestyle of many workers in the region, where food, lodging, cleaning, and other services are provided. However, this sexual division of labour is also replicated in the broader community. Resource workers who do reside in the community rather than work camps are often exempted from both household and community responsibilities and serviced by female labour. In fact, in the early history of northern resource communities, companies realized early on that 'the presence of women and children and the formation of stable family household units would provide a social basis for the growth of a stable workforce' (Luxton 1980, 26). In extractive industries, encouraging the settlement of families is a cost- effective way of feeding, sheltering, and entertaining workers without employers having to directly incur these costs themselves or compromise productivity.

As one female resident of Fort McMurray suggested, women in the community disproportionately take on household and community responsibilities, even when they interfere with their own workplace responsibilities:

> I have noticed is that if there is something that needs to be taken care of, like a bank appointment or the kids are sick at school, in my office I've noticed it's us ... You know all of the home-front issues are being dealt with by us. It is due partly to the fact that our husbands are working out on site and we're working right downtown and the bank's just down the road. But I think that it just becomes an expectation for that ... we're very accommodating of ... life needing to take precedence over work sometimes.

As this woman points out, a noticeable consequence of the tendency for women to take on the majority of household and community responsibilities is that this particular sexual division of labour becomes ingrained as an expectation. Furthermore, this widely held expectation tacitly contributes to a devaluation of the non-extractive work in the community – such as social and health care work – that is more commonly undertaken by women.

The local municipal government has drawn on the discourse of traditional family values to promote Fort McMurray as a family-friendly place and to counteract the negative media attention the town has incurred as a rough-and-tumble place. It is the lack of support for traditional families that is blamed for many of the social ills in Fort McMurray. This opinion has also been echoed by local residents, who draw on a notion of family to define the boundaries of desirable community in Fort McMurray:

> I think it's better since the construction's stopped, because people that are here are the families that are left, and we don't see so much [of Fort McMurray's negative side] ... we like it because it's turned into more of a family community again like it used to be and, you know, you don't hear about the drugs, you don't hear about the violence that's happening.

In fact, local residents have suggested that many opportunities for social integration and community participation revolve around family activities. Those who do not have a 'traditional family,' however, report barriers to integrating in the community:

> So many things to do that you hear about are going on, are like, 'oh it's family oriented' – even going garage sale-ing ... we started going to garage sales here 'cause we were like, 'I wonder what it's like when people have so much disposable income, I wonder what the garage sales are like?' And we sorta stopped going because it's like, all kids' books or all kids' toys. These are the most boring garage sales ever. Everyone has kids' stuff.

The literature on gender and resource development suggests that the discourse of family values combined with frontier masculinity shapes women's lives in ways that may not be immediately evident. Meg Luxton's (1980) study of women in Flin Flon, Manitoba, a post-boom mining community facing economic bust, exposed the extent to which women's well-being and daily lives can become determined by the schedules and desires of the men within their households, leaving little opportunity to pursue their own economic and social interests and well-being. Early psychological and health studies of northern resource communities also suggested that women in resource communities face higher levels of depression and other mental health issues than their male counterparts (Evans and Cooperstock 1983; Moen 1981). These threats to well-being are often attributed to the considerable isolation and the difficulty of forming new social connections in communities where class divisions and social distinctions between newcomers and long-term residents are highly pronounced (Walsh and Simonelli 1986).

However, the portrayal in some of the literature on women in resource communities as agency-less victims of resource development has been challenged in other accounts. For instance, Freudenburg (1981, 223) has suggested that in rural areas where traditional sex roles may be more deeply socially entrenched, it is possible that 'energy boom development could potentially have a liberating and beneficial effect upon rural women by disrupting traditional patterns.' As some primary resource industries become increasingly complex and differentiated and others collapse, it is important to recognize that employment opportunities are likely to become increasingly heterogeneous, with women occupying an increasing variety of professional roles in the secondary and tertiary sectors (McLeod and Hovorka 2008). In Fort McMurray, considerable employment opportunities have opened up for women in the non-profit sector. Yet despite these material changes in resource communities, the ideological legacy of the traditional sexual division of labour remains influential (Reed 2003).

Between Frontier Masculinity and Family Values: Shifts in Gender Relations

Feminist political ecologists remind us that in uncovering questions of social justice involving the distribution of, or access to, resources, participation in decision making, or environmental harms, we must be attentive to gendered and racial as well as class-based relationships of power. Accounts of the political ecology of a place like Fort McMurray, then – or indeed that of the province as a whole – should take care to 'look inside' the category of 'the political economy' or the 'human' relationship to nature and to ask how hierarchies of power and privilege may be gendered and racialized. Reflecting on our extensive body of interview work with women in Fort McMurray, we see how the intersections of class, race, and gender position women in different ways with regard to the social effects of the local economy. We also see how the global capitalist economy produces migrant-sending and -receiving sites, and in the process constructs gendered and racialized hierarchies within local labour forces.

Temporary Foreign Workers in the Oil Economy

Most of the TFW women we interviewed were from the Philippines and were overwhelmingly employed in housekeeping and childcare. There were two streams of workers: those who were admitted to Canada under the Live-in Caregiver Program (LCP) and those who were admitted under the Temporary Foreign Worker Program (TFWP). Members of the first group are required to work as live-in caregivers for two full years before they are permitted to apply for permanent residency, at which point they are released from the obligation to do live-in domestic work. These women described their two-year period as 'doing time.' The promise of potential citizenship motivates caregivers to endure the hardships of their work for two years. 'Camille,'[4] who had an education degree and who had been working in Fort McMurray for a little over a year when we interviewed her, said:

> I have a goal. I want to become an immigrant and bring my family here. If this means that I have to stay on the job for two years and be patient, then, I will do that. I have to be tolerant ... At times, it is kind of difficult to put up with certain demands of my employer but I don't want any problems and just get this over with.

The caregivers in Fort McMurray had various different (and most of the time, professional) backgrounds. Though their personal histories were diverse, they generally conformed to the pattern of de-skilling through immigration that has been well documented in the literature (Bonifac 2008; Pratt 2008). The research also shows that de-skilling is usually followed by ghettoization within low-wage occupations and low monetary returns on educational investments for immigrant women, and especially for Filipinas (Gardiner Barber 2008; Hiebert 1999; Pratt 2004; Stasiulus and Bakan 2005). For live-in caregivers, the experience of coming to Canada as a nanny evidently narrows occupational opportunities long after the requirements of the program have been fulfilled. Two former live-in caregivers, both with professional degrees and now permanent residents, commented on the difficulties they had finding a job other than housekeeping and care work after the two-year period was over. One, who was 'luckier,' landed a clerical/administrator job in a daycare centre (again a highly gendered occupation), while the other, after having tried for so long to get out of the 'cycle,' gave up and took another nanny job because she could not afford not to have a job for more than a month in Fort McMurray. She is now taking industrial training at the local college with an eye on jobs in the oil sector, thinking her ship is yet to arrive.

Such experiences are not exceptions among the other immigrant women we have interviewed. For most of them, one of the most promising ways of securing immigrant status is to upgrade their education and apply through the skilled worker program. However, getting educational training or gaining Canadian work experience in one's own profession as a temporary foreign worker is almost impossible under the current federal law. This was a common concern among the TFW women with whom we spoke.

Although the federal government refers to the TFWP as a 'labour mobility program,' it is precisely the lack of labour mobility that puts these workers in a vulnerable position. Unlike other workers, they cannot move to more attractive work sites; leaving their jobs – or being fired – almost always means deportation. As citizens of countries in which poverty and unemployment remain pervasive, and like their counterparts in other cities, most TFWs in Fort McMurray are careful to protect their migratory livelihoods by acceding to employer demands, even when tired or injured. Although TFWs in Fort McMurray are not necessarily a cheap form of labour compared with their counterparts in the US, they gladly take jobs with tasks and hours seen as undesirable to Canadians with other employment options.

Employers of temporary foreign workers can choose both the country that will supply them with labour and the gender of their workers. The ability to choose the origin country, for example, means that labour-sending countries (and workers themselves) compete for work placements in Canada. Employers play countries against one another in order to achieve the flexibility demanded by a competitive global marketplace. Adding gender to the mix, employers' ability to choose workers on this basis means they can divide their labour force by hiring men to do the 'men's work' – for example maintenance – and women to do 'women's work.' Gender plays a key role in the implementation of foreign worker programs in high-income countries, as well as in the everyday experiences of migrant workers and their families. Geraldine

Pratt says, 'there is an efficient way of rendering the immigrant stories; this is that Canada as a society exploits women as a low-cost labor force' (2004, 39).

Bringing in foreign workers mostly from the Philippines is the choice of employers in Fort McMurray. Although there are no official statistics on the ratio of Filipina women to Filipino men working in Fort McMurray, it is our impression that Filipina women outnumber men in the service sector and hospitality jobs. But no Filipina's position is secure. Rendered as, and expected to be, more docile, Filipina women report that, when they question their working or living conditions, employers and/or supervisors threaten that they could easily be replaced with men.

Most migrants come to Canada as a strategy to provide for their families, particularly their children. However, the experiences of men and women, especially women who are single-household heads, are rather different. For example, as two single mothers reported, they often suffer the anxiety of leaving their children behind in their home countries. While men (and some women) come to Canada knowing their children are being cared for by a parent, most women must rely on kin, neighbours, or even older children during their migratory absences. Their stress is compounded by the fact that while migration for men involves fulfilling their primary gender role, of breadwinner, for women it means abandoning theirs – that is, motherhood (at least as it has been defined traditionally).

Racialized immigrant women commonly experience a mix of racial and gender stigmatization. Both the TFWs and live-in caregivers mentioned that they had to be resilient to all sorts of degradations. For example, migrant women report that they receive a lot of unwanted male attention, mainly from co-workers but also from supervisors and employers, and that they are expected to be open to advances. This hurts the women, especially the ones who are self-reportedly religious (Catholics and Baptists). Secondly, white Canadians commonly assume that Filipina women are 'low-skill' caregivers when in fact they hold a wide range of skills and occupations.

Geraldine Pratt (2004) has identified the same kind of stigmatization in her extensive research within the Filipina community in Vancouver. She sees this stereotyping revolving around a distinction between immigrant and nanny, and involving important gendered and class associations. Although, admittedly, the live-in caregiver program has been one of the most accessible routes for many Filipina women who wish to initiate immigration to Canada, there are other settlement patterns among Filipino men and women. The term 'immigrant' is used to refer to individuals who entered Canada through the conventional immigration system: the point-based system or business-class programs. Hardly anybody assumes that Filipino men are caregivers – or nannies, to use the popular term; rather, they are assumed to be immigrants. A Filipina engineer we met in Fort McMurray shared her experiences of being 'mistaken' as a nanny all the time:

> I am an engineer, working for one of the biggest oil companies in Fort McMurray, but people don't know that. When I chat with strangers, they always assume I am a nanny and get surprised when I say I am an engineer ... I guess it is all right to be mistaken like that. There are indeed a lot of nannies around.

This woman may be comfortable with being mistaken for a nanny, but some actual nannies with whom we spoke thought that being a nanny carried a stigma.

> Sundays are my days off. I meet with my friends, mostly other nannies, to go to church and spend the day with them and in a few occasions when other people saw us together, they said 'oh, there go nannies' and laughed. I am not ashamed of being a nanny; I am not doing anything bad, I just take of the children, cook and clean and that's it.

'Not doing anything bad" refers to prostitution and the myth that Filipina women steal other women's husbands. We did not actively pursue these stigmas in Doğu's ethnographic interviews with Filipina women, but they came up in various other contexts once the topic of Filipinas came up. For instance, in a conversation with some Aboriginal women, when we mentioned the Filipinas living in a hamlet close to Fort McMurray, both our interviewees started laughing and shaking. Having noticed our blank faces, they spilled the question: 'You know what they call Filipina women around here? Husband stealers.' One of them then joked that when she sees a Filipina coming she covers her husband's eyes. A community member we spoke to reported rumours that Filipina women marry Canadian citizens, then bring other relatives or friends to Canada to do the same, and that this constitutes a form of prostitution. Employers of nannies are aware of rumours that Filipinas are prostitutes, saying things like:

> I don't care and I don't want to know what she [referring to her nanny] does on her days off. You know, you hear things like some girls venture on other, more lucrative ways of earning quick money and if she does things like that, I am better off not knowing about them.

The repeated devaluation of racialized women, especially those working in unskilled jobs in various businesses across town as temporary foreign workers, is significant. Employers we interviewed drew on a discourse of sexual difference to produce men as trainable skilled workers and women as untrainable. They spoke about men as quick learners, strong, and dependable. Women were characterized as quitters and emotional, hence not as easily trainable. Women's 'untrainability' and intrinsic suitability to unskilled labour is a staple stereotype.

Recent Immigrants

Our interviews revealed that the decision to migrate to an extractive boomtown is based on a number of calculations that range from employment opportunities to the desire to be in an exciting place. Far from random, unknowing transience, immigrants who move to boom areas do so – if not because they are sent by a company for which they already work – because of advice from friends or from family already in the region. Pauline Gardiner Barber (2008) found in her research in the Philippines that Filipinos are sophisticated assessors of immigration possibilities in other countries. Of course, in Fort McMurray, the extensive media coverage and aggressive provincial and municipal advertising that 'we have the energy' also helped 'spread the word' in and outside of Canada. Yet informal networks of friends and relatives extend across the region and relate information about job availability and living conditions.

We were able to interview a number of Somali women who migrated to Fort McMurray in the mid-2000s. Our conversations revolved around issues of race, culture, religion, (lack of) employment opportunities, and housing challenges in Fort McMurray. Experiences of racism and discrimination, especially in employment practices, were frequently recounted. A young Somali who was raised in Canada from the age of three told us: 'As you can see, there is a lot of racism here. I am looking for work. Although they [employers] say they are hiring they are not hiring us [Somalis]. On the phone, it is okay. But when they see me, the way I dress and my skin color, then there is a wall. They don't even want to talk.' The way Somali women dress has been 'an issue' in some companies that offer cleaning services – the most common line of work that Somali women are hired for. Several women were let go, all at the same time, because they were not following company safety policy regarding work outfits, which requires them to wear pants instead of long skirts. 'They are taking our skirts off; next day it is going to come to our headscarves. I can't allow this; I want to live as my culture dictates,' said a woman who had been fired.

'Living as my culture dictates' was a common theme. In fact, one of the recurring themes of the interviews was how the Somali immigrant women experienced living between Somali and Canadian cultures. The recounted their challenges in trying to maintain a balance between cultures, and the methods they used to cope with the issues that come up in their new lives for themselves, their children, and other family members in Fort McMurray. The women described an experience of living in an in-between, or liminal, state. They talked about the tension in living in this in-between space – of being pulled back towards their traditional ways (including their gender roles) and at the same time being pulled into a new Canadian life. It is the role of women to maintain the cultural traditions within the family and to pass them onto their children. The traditions of communality, the bonds of family kinship, respect for elders, and a strong Muslim lifestyle are some of the most important traditions that the Somali women strive to maintain. The women we spoke to talked about the pressures from their family members and the Somali community to continue these cultural traditions after their move to Fort McMurray. One woman related: 'there is a lot of pressure on women who come here. You are carrying so much, you are carrying the culture of Somalia.' The women have the social responsibility to maintain cultural traditions and the family values in Fort McMurray; however, many of these values have to be renegotiated and redefined by the immigrant women.

The Somali community has a strong sense of group solidarity that provides an important buffer for immigrant women in Fort McMurray. This communality was most evident in their visits to the Markaz-al-Islam (The Islamic Centre), which serves as the religious and social centre for the Somali women and also as a local cafe, owned and operated by Somalis. The women are able to continue the Somali traditions of language, dress, food, and music in Fort McMurray through their association with the Markaz. These venues allow them to have a sense of belonging and familiarity and a religious and social network that is a positive and significant influence in their lives in Fort McMurray. On the other hand, the Somali women remain isolated from contact with others outside of their ethnic/religious community. The women we spoke to did not have much social contact with other communities and individuals in Fort McMurray. Notably, racialized communities are spatially segregated, as Somali families and other racialized minorities live mostly in apartment complexes in the downtown area, whereas white middle-class Canadians are more likely to live in single-family homes in other neighbourhoods. One woman reflected on the isolation she felt and regretted that she did not have any Canadian friends:

> I don't know anywhere here [to go], I just go to work and come back home. I don't even have any friends, just my kids and me ... Some people are lucky; they come here and get Canadian friends and maybe you go out and hang out with them and your kids could play with each other.

Such isolation from the larger community also has repercussions for 'problem solving.' The women often rely upon privatized solutions to familial and social problems. Unless absolutely necessary, they hardly ever seek the assistance of formal agencies such as the police or social workers.

Given the high cost of living in Fort McMurray and the volatility of family income for most immigrant families, more Somali women are working outside the home. Employment creates many new challenges for the Somali women and their families. Even the notion of working outside of the home is a step towards renegotiating the gender roles within the household. But it does not stop there. Taking a job means the women have to learn to speak English to their co-workers; they have to find transportation to work; and they have to learn a new set of rules and social norms of the workplace.

Another persistent theme in the interviews with immigrant women was the effects of shift work on their family relationships and gender roles. For those who are married to oil sands workers, the prolonged absences of their husbands have resulted in some changes in gender roles. A few women wanted especially to talk about how they and their friends have learned to become more independent – shouldering responsibilities while their husbands are at work for ten or twenty consecutive days. The experience of the following interviewee and the way she reflects on her own experience is quite interesting. She is from India, as are most of the people in her circle. She says that most of her female friends in Fort McMurray have learned to manage finances, deal with mechanics, electricians, and plumbers, change light bulbs, pay bills, negotiate with bank managers, and generally undertake a whole range of traditionally masculine roles. While some of her friends made efforts to revert to their 'feminine' roles when their husbands are at home, she explained that she chose not to do so. She felt that she was capable of doing 'anything' now that she had learned to manage independently during her husband's absences. She had a 'newfound' empowerment. 'I compare myself to my family – my sisters, that they still depend on their husband. Which I don't, I just can get out, do anything on my own. So, it makes a real difference.' Nevertheless, she said that she remains conscious of the continued pressure from her community to conform to traditional gender roles. Another woman from the Indo-Canadian community in Fort McMurray told us:

> By now, I have got so used to doing things on my own, that sometimes I forget that he is around and that he can do it. Don't wait for him. He tells me 'You don't need me at home at all!' We are at the bank, because I am used to doing everything, I just go ahead to the counter do whatever had to be done. He is with me but he is just standing. And sometimes if he is angry or just teasing me, or speaking to my mum, he says 'Your daughter doesn't need me actually; I am just here. She can handle all things on her own'.

As opposed to the wives of oil sands workers, who tend to be better off in terms of family income, some of the women we spoke to had become the sole breadwinners of the family after their husbands lost their jobs during the economic downturn of 2009. These women were not so privileged to reflect on their feelings of independence or power, despite the fact that they were the sole breadwinners of their households. In almost every case, these immigrant women either had no marketable skills or were unable to find a job for which they were trained (because of the language barrier and non-recognition of their occupational credentials in Canada). They worked at minimum-wage service sector jobs, babysat at home, or did precarious jobs here and there. Basically, they all talked about having tried to do everything they could to take care of their families. Whether traditional or not, practicality is sometimes the single most important guiding principle in these women's lives. This was not always true for men, especially for immigrant men who had had jobs in the oil field prior to the bust. They would be more likely to not work and receive employment insurance than to accept lower-paying jobs. Often they made as much, if not more, money collecting employment insurance than their wives earned at minimum wage.

Capitalizers

This maxim is perhaps the most widely held and negative stereotypical image of newcomers to Fort McMurray. It characterizes the gendered practices of people who are viewed as transients or 'down-on-their luck' drifters who gravitate to the town. The town's boom-and-bust finances parallel the global oil economy and it was during the time of the boom (2006–08) that 'capitalizers,' like many others, were drawn to Fort McMurray. They are there for other kinds of opportunities a boomtown would offer; at times these involve more 'shady' opportunities, such as prostitution or drug dealing.

A considerable amount of Fort McMurray's media attention has been on the sex, drugs, and alcohol trio. As oil sands expansions were taking off in the mid-2000s, Criminal Intelligence Service Alberta reported an influx of organized crime groups linked to the drug trade in the region (Tetley 2005). The Alberta Alcohol and Drug Abuse Commission reported in 2005 that Fort McMurray ranked number one in the province for drug abuse. As many as 16.82 per 1,000 persons aged fifteen and older were said to be using illicit drugs, compared to a rate of 2.64 per cent in the southern city of Lethbridge and 2.41 per cent in the city of Calgary (Tetley 2005). According to the Drug Rehab Services (2009), marijuana, crack cocaine, and crystal meth are all widely used in Fort McMurray. Close to 40 per cent of the workers in Fort McMurray test positive for cocaine and marijuana, and the province's comparatively high industrial accident rate is thought to be related to drug use. Big money makes for big problems in a city where the average age is 31.8 (according to Fort McMurray's mayor [Fillion 2006]). A lot of people attribute the rise in drug and alcohol use to the increasing cash in town and also to the perception that there is 'nothing to do.' 'There is nothing to do here but to sit around and drink,' a pipe fitter told us in a bar while she was sipping her third beer. In April 2010 the provincial solicitor general assigned twenty-one additional Royal Mounted Police officers to a new 'elite integrated crime unit' in Fort McMurray to help deal with 'the booming drug trade' and associated crime (Alberta Solicitor General 2010; 'Fort McMurray Gets Boost in Police Ranks to Fight Drugs, Organized Crime,' *Globe and Mail*, 20 April 2010).

Whereas men are often portrayed as capitalizers in regard to their temporary attachment to the community – they are only there to make a lot of money in a short time – women capitalizers are seen as taking advantage of their sexuality and the perceived male-to-female imbalance to earn their living. The October 2006 *Chatelaine* article, mentioned above, and other references to 'gold diggers' are examples of what we call the 'capitalizer' maxim. In the *Chatelaine* article (Preville 2006), prostitutes, gold diggers, and divorcées are disturbingly lumped together and blamed for 'tempting' the lonely, male, mobile workforce into committing its social evils. In a town with too many bloated paycheques and a high cost of living, the *Chatelaine* article posits, sex is used to barter for basic needs such as groceries or rent, or for sustaining a certain lifestyle. The reporter states that 'every last guy in Fort McMurray claims to have been through a similar experience' of women bypassing the getting-to-know-you stage and heading straight for the wallet. The article attracted a lot of criticism through its portrayal of Fort McMurray's single women as gold diggers who will not date or bed a man until they know the size of his salary.

The evidence of a widespread class of gold diggers pointed to by local residents is, at best, anecdotal, yet it is an important gender stereotype that colours the ways in which many women define themselves (as the 'other' to the gold-digging woman) and affects the interactions between men and women. The widely held belief about women 'choosing' to 'take advantage' of men in Fort McMurray masks the very real vulnerability of women who are in the community with few resources to support themselves – often because of ending relationships with men upon whom they were financially dependent. For instance, we spoke with a young woman in her early twenties who had migrated from the East Coast as a teenager with her much older boyfriend. Although she is currently apprenticing as an electrician and working a second retail job, since ending her relationship she has struggled to make a sufficient income to cover her living expenses and house herself. She reported living in eighteen different homes during her three years in Fort McMurray – entirely dependent on the willingness of casual acquaintances to put her up and always at risk of being asked to leave at a moment's notice. Although, in this woman's case, she had not resorted to prostitution to ensure a roof over her head, she spoke at length about the sexual overtures by her male co-workers as a constant pressure.

The lack of affordable housing in Fort McMurray greatly increases the vulnerability of women to prostitution. A report by Wood Buffalo's

Homelessness Initiatives Steering Committee in 2007 said sixty-five young people between the ages of eleven and seventeen were without a permanent home in Fort McMurray; most of them said they slept at friends' homes or in parks, and some were trading sexual favours for a place to stay (*CBC News*, 27 September 2007). Aboriginal women are highly represented in street prostitution. The Prostitution Awareness and Action Foundation of Edmonton reports that 55 to 65 per cent of women working in the sex trade are Aboriginal (Narine 2010, 4). The Edmonton Urban Aboriginal Affairs Committee estimates, similarly, that more than half of Alberta's prostitutes are Aboriginal (Mahaffey 2007, 37).[5] Finally, there are numerous anecdotal reports of high rates of female prostitution in Fort McMurray and other northern resource towns involving women who constitute another kind of 'transient' workforce. These sex trade workers come to Fort McMurray from other parts of the province, the country, or even from other countries to service the large population of single males, and their numbers fluctuate according to the booms and busts of the local economy (see Mahaffy 2007).[6]

Stabilizers

This maxim refers to women in the community who embrace the maternal, helpmate roles assigned to them in the discourse of frontier masculinity. Stabilizers tend to be white, long-time residents and/or women employed in the helping professions. Although stabilizers represent one of the most widely acknowledged images of women in resource communities, the context of Fort McMurray's booms and busts shapes these women's practices considerably. Whereas employment opportunities for women in many smaller resource communities are notoriously limited, the huge demand for labour in the lucrative oil sector during the boom created a space for women to take on a considerable proportion of responsibility for maintaining and stabilizing community in public forums. Not surprisingly, women greatly outnumber men in the health and social services sector in Fort McMurray (Statistics Canada 2006).

Working in the helping professions is not the only occupation of 'stabilizers.' Numerous women spoke of their volunteer work and participation on various community boards as part of their investment in the community, or positioned themselves as 'defenders' of the community. The primary motivation for stabilizers is not financial, but is rather framed as a moral commitment to the wider community. It is a form of maternal politics, described by Murphy as 'the transference of women's traditional maternal, nurturing, and caretaking role to the political sphere' (1995, 166). Some women working in this sector, however, do express alternative motivations in their decision to enter the helping professions, such as the ability to quickly rise up to managerial and director positions. While this is a source of pride for some, others described their situations as 'putting in their time up north' in order to secure better positions upon return to more desirable communities. Those who considered themselves stabilizers were somewhat disdainful of women whose motivations for working in the helping professions were self- rather than community-minded.

Parallel with the official discourses around family values, creating a family-friendly community and making Fort McMurray a good place to live are critical for retaining long-term residents (Dorow and Do u, 2011). The relevance of the discourse of family values for stabilizers is quite apparent here, as much of the community work performed by stabilizers is geared towards creating a family-friendly community. As we have already mentioned, the literature on women in extractive communities conceptualizes women's roles essentially as stabilizers of the workforce, as wives and mothers, contributing to the well-being of the community through their reproductive and community work (Bron 2001). Some women we spoke to indeed identified with the gendered roles pertinent to family values. The following quote from a mother of two young children, who works part time in the NGO sector, is a good instance of this:

> I am satisfied with my life. I never really wanted to be a career person. Of course, when the boys are gone I probably would want to work more than I do now. I think about going to full time work now and then I think, 'no, because then I couldn't go and do things with the kids.' And my husband works very hard and he doesn't have time to take care of things. But when the boys are gone, things that I might want to do are get more education, maybe, and do something. But right now, I just do not have the time for it. 'Cause to me family comes first.

However, this division of labour requires something approaching a family wage for one spouse, and can be translated into reality by only a limited portion of the population in Fort McMurray. A number of women working in the non-profit or volunteer sector bluntly stated that even working in the helping professions would not provide sufficient income to support a family. Their volunteer work was predicated upon the financial support of a partner employed in a more lucrative oil-related job. Although this maternal (rather than primary breadwinner) role is a source of pride for women, it also creates a tension by reinforcing dependency on men. Additionally, many women would need nannies in order to work full time, given the unavailability of affordable childcare spaces in Fort McMurray. As we have seen, staying at home is not an option for most of the immigrant women we spoke to. In fact, for many people from different walks of life, women's pay-cheques are crucial family income. This reality means that traditional gender roles are subject to renegotiation. Indeed, the economic benefit of the female partner's having a steady income – particularly in the public sector – during the recent slowdown in the oil sector is disrupting the primacy of male breadwinners in many households. Whether this disruption is temporary or will have more lasting implications, however, has yet to be seen.

Women Oil Industry Workers

This maxim covers women working in trades and professional positions as well as those living in work camps. Women's employment opportunities in resource sectors, as previously mentioned, are notoriously limited. Due largely to the size of operations and the scope of labour demands, however, the oil sands near Fort McMurray are a curious case. Women are now one of the targeted populations that have been sought to fill the labour shortage in the trades in Alberta (CWDFC 2004). In fact, women are increasingly entering into nearly all facets of the oil industry. Among the occupational fields available to women, the oil patch is the most likely sector in which they can earn a living. Yet, regardless of whether they are employed as labourers, clerical workers, or engineers, women in this sector face considerable challenges. Crossing the gender boundaries into male-dominated fields results in a contradictory situation for women, where they are simultaneously marginalized because of their gender and rejected if they attempt to act like 'one of the boys.'

Bonnie Watt-Malcolm observes that women in male-dominated trades occupations must 'ignore, accept, or challenge' (2008, 224) existing gender practices. Ignoring and accepting the masculine workplace culture were the most widespread tactics adopted by the women we spoke with. For example, the necessity of minimizing one's femininity in order to fit in – whether by altering their appearance, their personalities, or their behaviour – is a common theme among women in Alberta's oil industry (see also Bron 2001; Miller 2004; Watt-Malcolm 2008). As one female professional we spoke with explained, 'in the field, I mean, there's no other way to behave but, you know, very strong willed and clearly spoken and very assertive. So there was just no room for my usual personality out there.'

Many women we interviewed spoke at length about the 'improper' behaviour of other women they worked with who did challenge the masculine workplace culture. For instance, one female construction crew supervisor and self-identified 'tomboy' bemoaned the 'Tiffanies' she worked with who had 'giggly-girly attitudes, wearing make-up, tank tops, g-strings etc. to work and chatting up guys.' These women were looked down upon by other women and seen as not tough enough to last in their fields.

At the same time, women who do not fit into the image of what is considered appropriately feminine are often stigmatized for transgressing gender boundaries. One oil patch worker described two types of women in her field: 'bitches' and 'normal girls' – terms that she herself found perturbing. 'Bitches,' she explained, are women who have an aggressive, unfriendly demeanour towards others, including other women. 'Normal girls,' conversely, are those who have a more friendly and cordial disposition. Although the women she described as being in the 'bitches' category may be more successful on the worksite, they were generally disliked and excluded from on- and off-site socializing. Another woman we met resided in a work camp. She had insisted on staying in the men's trailer because of the threat of men who are drunk or high on crack who would attempt to break into the women's trailer. The men in her crew referred to her by a derogatory term (a shortened version of the word 'transvestite') because of her 'mannish' behaviour.

Regardless of the ways in which female oil workers represent themselves and perform their femininity, there are tangible forms of discrimination and marginalization at the worksite. Lower pay rates than male counterparts are commonly reported. As one former tradeswoman, now employed in a work camp, states: 'if you are a woman trying to make a living in trades and there is a man, nine times out of ten a man will get the better [paid] job and you are going to do the same job as he does only at a lesser wage.' Many of the women who report that pay discrimination is common explain that they were originally grateful to be employed, and were unaware that they could negotiate higher salaries.

Training opportunities for women in the resource sector to advance to higher-paying positions are limited (Tallichet 2000). Especially during boom times when jobs are plentiful and there are not enough experienced workers available, most training is on the job. Because of existing prejudice against hiring women in traditionally male-dominated trades, and the unstructured nature of on-the-job-training, few are actually able to acquire the skills needed to compete for high-paying, energy-related positions. Verbal and physical sexual harassment is also widely normalized on the job. Although there are increasing efforts to address sexual harassment on worksites and in work camps – such as barring men from entering women's trailers – these regulations often have the secondary effect of further ostracizing women as outsiders.

Conclusion

Using a feminist political ecology lens, we can clearly see that oil extraction is not only a physical and economic process; it is also a gendered one. In various ways, the oil companies benefit from long-rooted discourses of frontier masculinity and family values that help to mould workers and their dependants to the requirements of the labour process. On the other hand, the great demand for labour during growth periods may destabilize traditional gender roles as women are recruited into non-conventional occupations in the oil sands. This trend, in turn, may generate demand for lower-wage domestic caregivers from the global south, typically under Canada's live-in caregiver program. The influx of temporary foreign workers and immigrants from many countries has turned Fort McMurray into a much more multi-ethnic place than it was twenty years ago, in the process creating a racially differentiated and hierarchized workforce. Gender roles in immigrant families may also be destabilized as women are drawn into paid labour. In times of economic downturn, when men lose their higher-paying jobs in the oil sands operations, women may become the sole household wage-earners – a situation that may further destabilize traditional gender roles.

The predominant sexual division of labour remains one in which men hold the high-paying jobs in construction and oil sands operations, while women predominate in lower-paid service jobs and in the human service professions. But the particular character of the male workforce – the thousands of men living in work camps and the migrations back and forth across the country of thousands of men from the Atlantic provinces, the youth of these men, their high disposable incomes, and so on – creates the opportunity for a flourishing sex trade that is also based, to some extent, on migratory workers.

Given these complexities it is not surprising that we see multiple representations of masculinity and femininity, such as the division between 'good' women who help their men and care for their households and communities and 'bad' women who wield their sexuality for personal gain. These representations, or 'maxims,' attach to racialized subjects, often compounding the marginalization or racism experienced by women occupying the lower rungs of the local labour market. The booms and busts of the oil economy are constantly remaking the social and economic spaces in which gender identities and roles are negotiated. Thus we see that gender identities and practices are neither uniform nor stable; a strong degree of fluidity exists between each of the maxims we identify.

Whether or not they think of it this way, the daily lives of women and men in Fort McMurray are very significantly organized around sustaining oil production. For the most part, women and men are positioned within this local economy in quite different ways, with men benefiting most in terms of employment and income. Our account attempts to show that this hierarchy is further complicated by racialized differences, in which the lowest-paid service sector jobs are held predominantly by immigrant women. An interesting continuity with earlier incarnations of the extractive boomtown is the opportunity opened up by the particular characteristics of the (largely migrant) male workforce (the gold miners) for entrepreneurial sex trade workers (the gold 'diggers'). Oil has transformed the life chances of all these individuals in ways that are sometimes welcomed (e.g., greater autonomy for some groups of women, income for migrant men to remit to families in distant places, opportunities for immigration to Canada) and sometimes tragic (drug addiction, trafficking of Aboriginal women by gangs, industrial accidents, separation from distant families, vulnerability to domestic violence, homelessness). It has also transformed the natural environment of the community and the quality of life available to the inhabitants of Fort McMurray in both positive and negative ways.

The sociological study of individual self-perceptions and world views is crucial to identifying the hopes, fears, or visions that may weave together movements for social change. Our research sketches a preliminary, and by no means comprehensive, picture of the lived experiences of the gendered and racialized subjects located in the heart of Alberta's oil economy. Our interviewees demonstrated awareness of many of the problems associated with this model of development and its implications for their own lives. Notably, largely missing from their stories is a sense of how they, as citizens, might – or should – be able to influence the developments taking place around them. Where, in the daily struggles to find housing, pay the rent, manage the household, navigate the dangerous highways, protect oneself from crime, or deal with the multitude of other issues thrown up by life in Fort McMurray, are the opportunities to reflect upon the necessity of, or the alternatives to, the oil-driven economy, its colonial foundations, or its hegemonic masculinity? In the end, our research points to the absence of such spaces and practices of citizenship.

NOTES

1 This SSHRC-funded project was headed by Dr Naomi Krogman, professor of rural sociology in the Department of Rural Economy, University of Alberta.

2 This SSHRC-funded project was headed by Dr Sara Dorow, professor of sociology at the University of Alberta.

3 *Fortune* magazine put the average salary in Fort McMurray in 2008 at $110,000 CAD per year (Heinrich 2008).

4 We use pseudonyms to protect the anonymity of our interviewees.

5 Gonorrhea rates doubled across Alberta between 2000 and 2005, and were highest among First Nations people (261 per 100,000 compared to 37.4 per 100,000 for the rest of the population) (Mahaffy 2007, 37).

6 There are also stories about high-end prostitution and not necessarily only in the form of street prostitution, which is the common depiction in the media, but also in other spaces such as bars, clubs, and massage parlours in town. But even more fascinating is the purchasing of sexual service at a distance. We have heard about some out-of-town, and even out-of-country, escorts. They come up on the bus/plane to 'do a shift' and then go back to wherever their home bases are. The 'client' base of escorts is said to be mostly oil sands workers who live in work camps outside of Fort McMurray. There has been a mention of prostitutes from Houston flying up for the weekend, taking cabs to the work camp and working full-time over the weekend and then flying back home.

REFERENCES

Alberta Federation of Labour. 2010. *Report on Temporary Foreign Workers in Alberta*. 10 December. http://www.afl.org/index.php/View-document/267-2010-Dec-16-Report-on-Temporary-Foreign-Workers-in-Alberta.html (accessed 14 August 2011).

Alberta Federation of Labour. 2011. 'Alberta's Use of Temporary Foreign Workers 'Shameful,' Says Labour Group.' 2 *August*. http://www.afl.org/index.php/AFL-in-the-News/albertas-use-of-temporary-foreign-workers-shameful-says-labour-group-many-employers-said-to-use-guest-worker-program-as-first-choice-not-last-resort.html (*accessed 13 August 2011).*

Alberta Solicitor General. 2010. 'ALERT Opens Elite, Integrated Crime Unit in Fort McMurray.' Media release, 20 April. http://www.alberta.ca/acn/201004/281691BC63930-F309-73E2-AC32B91D05329C67.html (accessed 14 August 2011).

Anahita, Sine, and Tamara L. Mix. 2006. 'Retrofitting Frontier Masculinity for Alaska's War against Wolves.' *Gender & Society* 20 (3): 332–53.

Archibald, Gerald. 2006. *Fort McMurray Quality of Life and Social Indicator Review*. http://www.woodbuffalo.ab.ca/municipal_government/municipal_departments/pdf/AlbianEUB/report1_archibald.pdf (accessed 20 September 2009).

Bavington, Dean, Brenda Grzetic, and Barbara Neis. 2004. 'The Feminist Political Ecology of Fishing Down: Reflections from Newfoundland and Labrador.' *Studies in Political Economy* 73: 159–82.

Beaton, Eleanor. 2008. 'Oil Rig Wives: Concerns for When the Men Are Away.' *Chatelaine*, 5 June. http://www.chatelaine.com/living/oil-rig-wives-concerns-for-when-the-men-are-away/.

Bonifac, Glenda Lynna Anne Tibe. 2008. 'I Care for You, Who Cares for Me? Transitional Services of Filipino Live-in Caregivers in Canada.' *Asian Women* 24 (1): 25.

Bron, Ingrid. 2001. 'Finding Their Place: Women's Employment Experience in Trades, Technology and Operations.' Master's thesis, Queen's University.

Connell, Robert W. 1993. 'The Big Picture: Masculinities in Recent World History.' *Theory and Society* 22 (5): 597–623.

CWDFC (Construction Workforce Development Forecasting Committee). 2004. *Alberta Construction Workforce Supply Demand Forecast 2004–2008*. http://www.coaa.ab.ca/LinkClick.aspx?link=pdfs/forecast1.pdf&tabid=67 (accessed 20 September 2009).

Dorow, Sara K., and Göze Doğu. 2011. 'The Spatial Distribution of Hope in and beyond Fort McMurray.' In *Ecologies of Affect: Placing Nostalgia, Desire, and Hope*, ed. Tonya K. Davidson, Ondine Park, and Rob Shields, 271–92. Waterloo: Wilfred Laurier University Press.

Drug Rehab Services (Canada). 2009. 'Drugs, Sex, and Parties – Fort McMurray Alberta Has It All.' Media release, 8 July. http://express-press-release.net/63/Drugs,%20Sex%20and%20Parties%20Fort%20McMurray%20Alberta%20Has%20It%20All.php (accessed 13 August 2011).

Dunk, Thomas W. 2003. *It's a Working Man's Town*. Montreal: McGill-Queen's University Press.

Edemariam, Aida. 2007. 'Mud, Sweat and Tears.' *Guardian*, 30 October. http://www.guardian.co.uk/environment/2007/oct/30/energy.oilandpetrol (accessed 20 September 2009).

Evans, Jennifer, and Ruth Cooperstock. 1983. 'Psycho-Social Problems of Women in Primary Resource Communities.' *Canadian Journal of Community Mental Health* 2 (1): 59–66.

Ferguson, Amanda. 2007. 'Cocaine Easier to Buy than Pizza.' *Edmonton Journal*, 26 August. http://www.canada.com/edmontonjournal/news/story.html?id=cd92c68f-21d9-4009-986b-03ecc3d00144&p=1 (accessed 20 September 2009).

Fillion, Kate. 2006. Interview with Melissa Blake, mayor, Fort McMurray. *Maclean's*, 119, no. 48 (4 December), 16–19.

'Fort McMurray Teens Swapping Sexual Favours for Shelter: Report.' 2007. *CBC News*, 27 September. http://www.cbc.ca/canada/edmonton/story/2007/09/27/fortmc-favours.html (accessed 20 September 2009).

Fraser, Nancy. 1989. *Unruly Practices: Power, Discourse, and Gender in Contemporary Social Theory*. Minneapolis: University of Minnesota Press.

Freudenburg, William R. 1981. 'Women and Men in an Energy Boomtown: Adjustment, Alienation and Adaptation.' *Rural Sociology* 46 (2): 220–44.

Gardiner Barber, Pauline. 2008. 'The Ideal Immigrant? Gendered Class Subjects in Philippine-Canada Migration.' *Third World Quarterly* 29 (7): 1265–85.

Government of Alberta. 2011. *2010 Annual Alberta Labour Market Review*. Edmonton: Government of Alberta. http://www.employment.alberta.ca/documents/LMI/LMI-LFS-labour-market-review.pdf (accessed 13 August 2011).

Harder, Lois. 2003. *State of Struggle: Feminism and Politics in Alberta*. Edmonton: University of Alberta Press.

Heinrich, Erik. 2008. 'Canada's Boomtown.' *Fortune*, 157, no. 7 (14 April), 30.

Hiebert, Daniel. 1999. 'Local Geographies of Labor Market Segmentation: Montreal, Toronto, Vancouver, 1991.' *Economic Geography* 75: 339–69.

Hogan, Maureen P., and Timothy Pursell. 2008. 'The "Real Alaskan": Nostalgia and Rural Masculinity in the "Last Frontier."' *Men and Masculinities* 11 (1): 63–85.

Hovorka, Alice J. 2006. 'The No. 1 Ladies' Poultry Farm: A Feminist Political Ecology of Urban Agriculture in Botswana.' *Gender, Place and Culture* 13 (3): 207–25.

Liepe-Levinson, Katherine. 2002. *Strip Show: Performances of Gender and Desire.* London: Routledge.

Luxton, Meg. 1980. *More Than a Labour of Love: Three Generations of Women's Work in the Home.* Toronto: Women's Educational Press.

Mahaffy, Cheryl. 2007. 'The Hidden Face of Prosperity.' *Alberta Views,* November, 36–40.

McLeod, Colleen M., and Alice J. Hovorka. 2008. 'Women in a Transitioning Canadian Resource Town.' *Journal of Rural and Community Development* 3 (1): 78-92.

Mellor, Mary. 2009. 'Ecofeminist Political Economy and the Politics of Money.' In *Eco-Sufficiency & Global Justice: Women Write Political Ecology,* ed. Ariel Salleh, 251–67. London: Pluto.

Miller, Gloria E. 2004. 'Frontier Masculinity in the Oil Industry: The Experience of Women Engineers.' *Gender, Work and Organization* 11 (1): 47–73.

Moen, Elizabeth. 1981. 'Women in Energy Boom Towns.' *Psychology of Women Quarterly* 6 (1): 99–112.

Murphy, Kathleen. 1995. 'Women's Political Strategies in a Logging Town.' In *Living on the Edge: The Great Northern Peninsula of Newfoundland,* ed. Lawrence F. Felt and Peter R. Sinclair, 164–84. St John's: Institute of Social and Economic Research.

Narine, Shari. 2010. 'The Unofficial Olympic Sport to Attract Women in the Trade.' *Alberta Sweetgrass,* 1 January, 4.

'Oilsands Growth beyond Regulator's Reach: Analysts.' 2006. Canadian Press, 28 December. http://www.ctv.ca/servlet/ArticleNews/story/CTVNews/20061228/oilsands_alberta_061228/20061228?hub=Canada (accessed 20 September 2009).

Pratt, Geraldine. 2004. *Working Feminism.* Philadelphia: Temple University Press.

Pratt, Geraldine. 2008. 'From Registered Nurse to Registered Nanny: Discursive Geographies of Filipina Domestic Workers in Vancouver, BC.' In *Reading Economic Geography,* ed. Trevor J. Barnes, Jamie Peck, Eric Sheppard, and Adam Tickell, 375–88. On-line Blackwell Publishing. http://www3.interscience,wiley.com/cgi-bin/bookhome/117881217/ (accessed 10 August 2011).

Preville, Philip. 2006. 'Down & Dirty in Fort McMurray.' *Chatelaine*, October, 118–31.

Reed, Maureen G. 2003. *Taking Stands: Gender and the Sustainability of Rural Communities*. Vancouver: UBC Press.

Reed, Maureen G., and Shannon Christie. 2009. 'Environmental Geography: We're Not Quite Home – Reviewing the Gender Gap.' *Progress in Human Geography* 33 (2): 246–55.

Reed, Maureen G., and Bruce Mitchell. 2003. 'Gendering Environmental Geography.' *The Canadian Geographer* 47 (3): 318–37.

Regional Municipality of Wood Buffalo. 2008. *The Regional Municipality of Wood Buffalo Municipal Census 2008*. http://www.woodbuffalo.ab.ca/business/demographics/pdf/2008_municipal_census.pdf (accessed 20 September 2009).

Rocheleau, Dianne, Barbara Thomas-Slayter, and Esther Wangari. 1996. 'Gender and Environment: A Feminist Political Ecology Perspective'. In *Feminist Political Ecology: Global Issues and Local Experiences*, ed. D. Rocheleau, B. Thomas-Slayter, and E. Wangari, 3–23. London: Routledge.

Rothe, J. Peter. 2005. *Impaired Driving as Lifestyle for 18-29-Year-Old Alberta Drivers: Focus Group Analysis*. Alberta Centre for Injury Control and Research, 12 October.

Salleh, Ariel. 2009. 'Ecological Debt: Embodied Debt.' In *Eco-Sufficiency and Global Justice: Women Write Political Ecology*, ed. A. Salleh, 1–40. London: Pluto.

Scott, Rebecca R. 2007. 'Dependent Masculinity and Political Culture in Pro-Mountaintop Removal Discourse: Or, How I Learned to Stop Worrying and Love the Dragline.' *Feminist Studies* 33 (3): 484–509.

Stacey, Judith. 1998. *Brave New Families: Stories of Domestic Upheaval in Late-Twentieth-Century America*. Berkeley: University of California Press.

Stasiulis, Daiva, and Abigail Bakan. 2005. *Negotiating Citizenship: Migrant Women in Canada and the Global System*. Toronto: University of Toronto Press.

Statistics Canada. 2006. *Community Profile: Wood Buffalo*. http://www12.statcan.ca/census-recensement/2006/dp-pd/prof/92-591/details/page.cfm?Lang=E&Geo1=CSD&Code1=4816037&Geo2=PR&Code2=48&Data=Count&SearchText=wood%20buffalo&SearchType=Begins&SearchPR=01&B1=All&Custom= (accessed 20 September 2009).

Statistics Canada. 2011. 'Median Total Income, by Family Type, by Province and Territory (All Census Families).' CANSIM, table 111-0009. http://www40.statcan.ca/cbin/fl/cstprintflag.cgi. Last modified 28 June 2011.

Stoeltje, Beverly J. 1975. '"A Helpmate for Man Indeed": The Image of the Frontier Woman.' *The Journal of American Folklore* 88 (347): 25–41.

Tallichet, Suzanne E. 2000. 'Barriers to Women's Advancement in Underground Coal Mining.' *Rural Sociology* 65 (2): 234–52.

Tetley, Deborah. 2005. 'Sex, Drugs and Alcohol Stalk the Streets of Fort McMurray.' *Calgary Herald*, 22 October.

Thomas-Slayter, Barbara, Esther Wangari, and Dianne Rocheleau. 1996. 'Feminist Political Ecology: Crosscutting Themes, Theoretical Insights, Policy Implications.' In *Feminist Political Ecology: Global Issues and Local Experiences*, ed. D. Rocheleau, B. Thomas-Slayter, and E. Wangari, 287–307. London: Routledge.

Toneguzzi, Mario. 2009. 'Fort McMurray Surpasses Calgary, Edmonton as Most Expensive Housing Market.' *Calgary Herald*, 22 September. http://www.calgaryherald.com/business/Fort+McMurray+surpasses+Calgary+Edmonton+most+expensive+housing+market/2023809/story.html (accessed 10 October 2009).

Walsh, Anna C. 1985. 'Effects of a Boom-Bust Economy: Women in the Oil Field.' *Free Inquiry in Creative Sociology* 13 (2): 133–6.

Walsh, Anna C., and Jeanne Simonelli. 1986. 'Migrant Women in the Oil Field: The Functions of Social Networks.' *Human Organization* 45: 43–52.

Watt-Malcolm, Bonnie J. 2008. 'Women in Trades: Policy in Theory and Policy in Practice.' Doctoral diss., University of Alberta.

9 Constructing Participation in the Regulation of Alberta's Sour Gas

THERESA GARVIN

Around the world, communities are struggling to deal with the local impacts of global forces related to late-industrial capitalism. In Alberta this struggle is being played out in the arena of the rapidly expanding petroleum industry. In addition to its basis in oil, however, this economy also includes energy-based resource developments around natural gas. This distinction is particularly important in Alberta, where natural gas revenues make up a substantial component of the province's annual budget. From 1993 until 2009, natural gas was the Alberta government's largest single source of revenue from energy resources, reaching peaks of $7.2 billion in 2000/01 and $8.4 billion in 2005/06. Natural gas was overtaken by oil sands royalties revenue only in 2009–10 (Alberta Energy n.d.[a]). According to Alberta's Ministry of Energy, Alberta's total marketable natural gas production, including coal-bed methane, was approximately 5 trillion cubic feet of natural gas in 2012/13, falling to 3.2 trillion cubic feet in 2013/14 (Alberta Energy n.d.[b]). About 70 per cent of this gas is transported by pipelines to other provinces and to the United States. Due to low market prices for natural gas and the government's 2009 changes to the royalty regime (see chapter 6), revenue from natural gas royalties fell to just under $1 billion in 2012/13, or 12.5 per cent of non-renewable resource revenue for the provincial government in that year (Alberta Energy n.d.[a]). However, natural gas plays a crucial role in fuelling oil sands extraction and other industry in the province, as well as heating homes and generating electricity. As natural gas reserves become depleted, industry has adopted horizontal drilling, deep drilling, hydraulic fracturing (fracking), and new processing techniques in order to tap into resources that were previously considered inaccessible or insufficiently profitable. Consequently, as figures 4.2 and

4.3 in Quinn et al. (this volume) show, greater numbers of Albertans are seeing oil and gas developments in and around their communities.

This chapter draws upon a five-year research project that examined public participation in the regulation of sour gas in Alberta via policy analysis and a set of community-based focus groups located throughout the province. Results from the policy analysis revealed the changing ways in which regulatory agencies have structured citizen participation, thereby shaping the exercise of citizens' rights and the role of the citizen in governance. Importantly, this citizen participation takes place some*where* (Massey 2004) but is not simply limited to what happens at the local level (what Amin [2004] calls 'politics of propinquity'); rather, it is increasingly tied into global processes and into a 'politics of connectivity' (Amin 2004). In the case of sour gas in Alberta, the *where* includes local communities. These communities, when faced with proposed sour gas facilities, must engage in a construction of policy and political discourse that produces an inequitable scalar distribution of costs and benefits (Paulson, Gezon, and Watts 2005). In other words, communities located near sour gas must bear the risks, while provincial and national governments, as well as multinational corporations, reap the benefits of resource exploitation.

Focus groups allowed us to document citizens' resistance to the ways in which their participation has been structured, as well as their changing perceptions of government and the legitimacy of regulatory processes. Most notably, citizens' previous understandings of the oil and gas industry have been called into question, as has the notion of this industry being central to what has been termed the 'Alberta Advantage' (see Adkin and Phillips, this volume). Citizens have become more aware of the risks and disadvantages of sour gas development, and more critical of the dominant construction of 'economic development.' The experiences and knowledge of citizens have collided with the representations of sour gas development put forth by government and corporate actors, and as a result, citizen trust in governance processes has been eroded (Evans and Garvin 2009). Further, the weakening of citizen trust has been accompanied by questions about government agencies' capability to act in the public interest (Hierlmeier 2008; Paulson, Geezon, and Watts 2005). However, despite the growing recognition that there are tensions between sour gas development and the autonomy of local areas, citizens are uneasy with challenging the underlying ethos that sour gas development – and oil and gas development more generally – is in the public interest.

Overall, the work reported here suggests that there are shifting sands of policy legitimation in relation to the regulation of sour gas. Over the course of their engagement in public consultation processes, citizens concerned with the risks of sour gas began to question the ways in which this resource is being governed – particularly in regard to the degree and nature of their own inclusion in the decision-making process (see, for example, Masuda, McGee, and Garvin 2008).

Background

In Alberta, the development of energy resources is rife with tension as citizens feel constrained in their ability to contest oil and gas development (Evans and Garvin 2009; Fluet and Krogman 2008). This is due in part to the government's overarching representation of resource development as a public good. In 2010/11 the oil and gas industry contributed $8.4 billion (or 22 per cent) to Alberta's government revenues, with natural gas accounting for 4 percent, or 1.42 billion (Government of Alberta 2011). As Carter and Zalik explain in their discussion of Alberta as a 'rentier state' (this volume), royalty revenue supports provincial programs and services such as health and education. Without a provincial sales tax, these resource revenues are Alberta's main source of income,[1] creating a resource-based economic hegemony. The lesser-known story of this model of development is the growing public antipathy towards an expanding petroleum industry and increasing concern over the negative externalities associated with the extraction and processing of oil and gas. These concerns are particularly salient with regard to sour gas wells (Evans and Garvin 2009).

Sour gas is natural gas that contains a hydrogen sulphide (H_2S) content of 1 per cent or higher. H_2S is poisonous, invisible, and highly toxic (Guidotti 1996). Acute exposures to low levels of H_2S can result in unconsciousness, pulmonary edema, and, if untreated, death (Alberta Energy and Utilities Board Gas Department [AEUB] 1990; Roth and Goodwin 2003). However, comparatively little is known about long-term, low-level exposure (Roth and Goodwin 2003; Scott 1998). In Canada, about 30 per cent of natural gas is considered 'sour' and over 85 per cent of that is in Alberta (Petroleum Communication Foundation 2000). The past decade has seen a rise in the complexity and uncertainty of managing the risks of sour gas development. As public awareness and concern have grown, there has been increased conflict between local communities, gas companies, and regulators about what is considered an acceptable risk for communities who have nearby sour gas wells (Marr-Laing and Severson-Baker 1999).

Conflict over sour gas is rooted in the inherent tension in the scalar distributions of risks and benefits related to extracting and processing this resource. In essence, communities (or localities) are being asked to shoulder the health and environmental risks for sour gas production, while the benefits accrue primarily to private companies and the provincial government. While the oil and gas sector employs about 139,000 persons,[2] questions remain about the distribution of economic benefits when profits are paid out to corporate shareholders who have little connection to the communities where oil and gas facilities are located. Though the state has the authority to permit development, and the local landowner receives individual payments, the bulk of profits from oil and gas wells (in particular) accrue not at the local level, but at the provincial, national, and even international level in the form of royalties and corporate profits.

Natural gas, in its preprocessed state, is either 'sweet' (with little or no H_2S), or 'sour' (with more than 5,000 ppm of H_2S). Although found in many parts of the world, sour gas is heavily concentrated in the oil and gas fields of Alberta, where there is a naturally high concentration of sulphur (Guidotti 1996). In particular, sour gas is found in deep, high-pressure deposits such as those in the foothills of the Rocky Mountains (Government of Alberta 2009b). Approximately 30–40 per cent of the gas extracted in Alberta is considered sour (Canadian Association of Petroleum Producers 2008). Once a deposit is recognized as sour, it is ranked by concentration of H_2S, and the well is labeled as level I (low risk) through level IV (high risk). Because these deposits are found in the foothills of the Rockies, they are on both private and crown lands. Once collected at the well site, sour gas travels via pipeline to a processing plant. Before it can be distributed via the main natural gas distribution system, sour gas must be 'sweetened' through a cleaning process that removes the H_2S and results in commercially viable sulphur and local emissions of sulphur dioxide (SO_2). Public health concerns about sour gas stem from potential exposure to both raw sour gas (H_2S) through leakage or blowout events, and to the SO_2 emissions from the sweetening process. Exposure can result from any of the over 6,000 sour gas wells, 18,000 kilometres of sour gas pipelines, or 250 sour gas processing plants in Alberta (Government of Alberta 2009b).

As mentioned above, the H_2S found in sour gas is highly toxic, and human health effects vary by exposure (Guidotti 1996). Most research on human health effects is found in occupational health studies that focus on acute exposure events rather than on the low-level, long-term exposure that may occur in communities located near sour gas processing facilities (Kilburn 1997). Occupational health studies show that acute exposure to mild concentrations of H_2S is associated with headaches, nausea, and tearing of the eyes. However, exposure to even moderate concentrations of H_2S can paralyse the olfactory senses and prevent subjects from noticing the gas; therefore there is little warning of overexposure (Guidotti 1996; Roth and Goodwin 2003). A number of serious problems are related to acute exposure to moderate and high concentrations of H_2S, including unconsciousness (called 'knockdown' in the petroleum industry), amnesia, and pulmonary edema. While very little work has been done on community health effects of low-level, ambient exposure (Roth and Goodwin 2003), one small-scale study from Texas (Kilburn 1997) found that even moderate, downwind exposure to H_2S may lead to long-term and permanent neurological impairment. By comparison, a Nebraska study (Inserra et al. 2004) found no neurological effects related to long-term, low-level H_2S exposure. Research in Alberta has also concluded that there is little-to-no effect of short-term exposure on either humans (AHW 2006) or animals (WISSA 2006). Despite these findings, there is ongoing and rising concern over the community health effects of sour gas and its emissions, as well as anecdotal evidence supporting public concern (for a discussion, see Evans and Garvin 2009).

Even without H_2S exposure through potential leakage or other acute events, processing plants emit SO_2 as an output of the sweetening process. These emissions are regulated and controlled by a government agency that reports that total SO_2 emissions from sour gas processing plants in Alberta have declined since 1972. In addition, despite increases in the number of plants and the volumes of sour gas processed, sour gas emissions began flattening out in 1982 (AEUB 1995). Nevertheless, these assurances have done little to calm public fears, which have on the contrary been reinforced by the designation of SO_2 as a group 1 carcinogen (one of the strongest) by the International Agency for Research on Cancer (IARC 1992).

Public fears exploded in the late 1990s, concurrent with a series of bombings that took place at sour gas facilities in Hythe and Beaverlodge, communities located in northwestern Alberta. After two local farmers were charged and then convicted of this 'ecoterrorism,' many questions remained about the safety of sour gas in general, and the lack of evidence regarding human and animal health effects.[3] Another string of bombings at sour gas pipelines across the provincial border in Dawson Creek and Tomslake, British Columbia, in 2010 – characterized by some as 'ecoterrorism' (O'Donnell 2010) – once more focused attention on public health concerns associated with sour gas. In June 2013 rising floodwaters ruptured a sour gas pipeline in Turner Valley (southwest of Calgary) and led to the evacuation of over fifty homes (*CBC News* 2013). While media attention ebbs and flows regarding the dangers associated with sour gas extraction, distribution and processing, these dangers are ever-present.

The Regulatory Process: Policy, Framing, and the Construction of Participation

Public policies, particularly those that appear in the form of reports and written policy statements, are increasingly recognized as important vehicles for communicating sets of values and morals (Yanow 1996). These documents have embedded statements that reflect underlying motivations (Ianuntuono and Eyles 1997) and convey particular narratives about what is good, what is right, and what is possible (Therborn 1978). In short, reports and policy statements are reflections of ideology (Yanow 1996).

There have been three main policy documents produced in Alberta concerning sour gas and public participation over the past two decades (Environmental Resources Conservation Board [ERCB] 1994; Energy Utilities Board [EUB] 2000, 2007). The following discussion briefly summarizes the regulatory context for sour gas in Alberta and then critically evaluates each of the three key policy documents. Each of these documents reconstituted the notion of citizen participation, while at the same time defining who could speak and what was permitted to be discussed. As a result, these policy documents represent boundary-setting exercises defining the allowable participation and discourse in the regulation of sour gas development.

Regulatory Context of Sour Gas in Alberta

Since the creation of the Petroleum and Natural Gas Conservation Board in 1938, the provincial government has assigned the authority to regulate oil and gas development (including the siting of sour gas wells and facilities) to a series of agencies. While the name and mandate of energy

Table 9.1 Alberta Resource Regulatory Agency Titles over Time

1938	The Alberta government creates the Petroleum and Natural Gas Board (PNGB) to ensure orderly development of the province's energy resources.
1971	The Petroleum and Natural Gas Conservation Board becomes the Energy Resources Conservation Board (ERCB).
1995	The Alberta Public Utilities Board and the Energy Resources Conservation Board are combined to form the Energy and Utilities Board (EUB).
2008	On 1 January 2008 the Alberta Utilities Commission Act split the EUB back into two new regulatory bodies, the Energy Resources Conservation Board (ERCB) and the Alberta Utilities Commission (AUC). The ERCB is made responsible for the development of Alberta's oil and gas resources, and the AUC is made responsible for the distribution and sale of electricity and natural gas to Alberta consumers.
2012	The Alberta Energy Regulator (AER) replaces the ERCB and takes on regulatory functions related to energy development that were previously located in the Alberta Environment(or the Alberta Environment and Sustainable Resource Development) ministries. The AER becomes the single regulator of energy development in Alberta and oversees applications, exploration, construction, development, abandonment, reclamation, and remediation. While this transition began in 2012, it was not fully complete until 2014.

Adapted from AER 2014; Government of Alberta 2009b; Jaremko 2013

regulation agencies have changed over time (see table 9.1), their primary function has been to manage the development of the province's energy resources (Government of Alberta 2009a). At the time of writing, the latest restructuring of energy development regulation in the province, which took place between 2012 and 2014, has produced the Alberta Energy Regulator (AER). Prior to the establishment of the AER, responsibilities for energy discovery, management, development, delivery, environmental regulation, and reclamation of the oil sands and drilling sites were divided among the Environment Ministry (which itself has been restructured in various ways over the years, as explained in chapter 3); the Ministry of Energy, and the Energy Resources Conservation Board (ERCB).The AER brings together all energy-related regulatory issues into a 'lifetime' approach and oversees company applications for exploration, development, and construction as well as the management of site abandonment, reclamation, and remediation (Jaremko 2013). The AER has also been charged with overseeing all public consultation regarding oil and gas development, including sour gas.

Keeling (2001) identifies the uneasy co-existence that has developed over time between Alberta's agricultural sector and the oil and gas industry. Ongoing confrontations between rural residents, industry, and regulators have forced continual re-evaluation of provincial development and review processes. For sour gas, specifically, there have been three critical reframings of public debate and discussion. Reports produced in 1994, 2000, and 2007 have reflected the changing government notion of the role of the public and of public participation in the siting process and governance of sour gas in Alberta.

1994 Report: Public Safety and Sour Gas

In the early 1990s, as concern over sour gas in Alberta was nascent, responsible government agencies focused on the highly technical aspect of control and management of oil and gas. Environmental engineering reports, risk assessments, drilling plans, emergency evacuation plans, and impact assessments all employed highly technical language related to specialized evaluations of proposed projects. Written in expert language, for expert audiences, this information was far out of reach of the average Albertan trying to understand the implications of sour gas development in his or her community. Douglas Torgerson (1986) refers to this kind of policy management as technocratic, valuing expert assessments over lay experience and science over local knowledge. Likewise, Mary Richardson, Michael Gismondi, and Joan Sherman (1995), Theresa Garvin (2001), and Jeff Masuda, Tara McGee, and Theresa Garvin (2008) equally attest to the imbalance of power when technocratic approaches are used to downplay the importance of local experiences.

Indeed, the 1994 report, entitled *Public Safety & Sour Gas* (ERCB 1994), reflected this technocratic management approach by focusing on issues related to risk assessments, setbacks, and emergency response. The report goes on to suggest that the public needs to better understand what has been done in terms of prevention and mitigation, as well as to appreciate the concept of 'balanced development.' As early as this 1994 report, there is recognition on the part of the regulatory agency that 'conflicts between the sour gas industry and surface landowners and residents may occur because people feel that they take the risks but do not receive economic benefits in return' (ERCB 1994, iii). However, little follow-up occurred in 1994 in response to the growing level of public concern. Likewise, the growing distrust of the public towards regulators and industry was generally dismissed as attributable to the public's lack of information and knowledge about the scientific evidence.

The 1994 report concludes that the ERCB should better educate the public on the technical complexities behind sour gas development and risk. The focus at this time was clearly on how to improve the public's understanding of the 'science' around the management of sour gas. By concluding that government agencies need to engage in 'better communication' by educating the public, the approach in the early 1990s was embedded in traditional knowledge transfer systems that see the public as passive receivers of expert knowledge (Leiss 2001; Torgerson 1986), waiting to receive and 'learn' from those with greater knowledge and expertise (Garvin and Lee 2003). The role of the public, as 'participants' in the policy-making process, was to learn about the science and evidence behind the sour gas regulatory process. By contrast, the role of the regulators and industry was to provide better communication and to better educate the public. In other words, it was up to the experts to educate lay people; once the public better understood the technical risk-management techniques in effect, the public would better trust the regulators and companies in managing the risks related to sour gas (ERCB 1994). This is a classic response from agencies that manage risk, from the perspective of experts needing to inform the public about what is safe and what is dangerous (Freudenburg and Pastor 1992) and from policy makers who fail to understand public perceptions of risk (Garvin 2001; Leiss 2001).

2000 Report: Public Safety and Sour Gas

Despite the findings of the 1994 report, public concern over sour gas continued to grow in Alberta. By the late 1990s, some residents reported ongoing distrust of the regulatory commission and of the government's ability to address the needs of local citizens (EUB 2000). The high-profile case of northern Alberta rancher Weibo Ludwig, who was convicted of bombing a sour gas well adjacent to his property, brought national attention to the issue (Nikiforuk 2001). In response, the EUB requested another report – commissioned in January 2000 and completed in December 2000 – on how to address the rising concern about public safety and sour gas in Alberta (EUB 2000).

The primary finding from the 2000 report was that communication and education were not sufficient to address public concern (EUB 2000). Rather, a more consultative approach was recommended in order to address growing public resistance. The 2000 report found that there was insufficient evidence regarding the environmental and health effects of sour gas, and that a better overall understanding of sour gas effects was needed not only for the public but for industry and regulators as well. Equally, this report identified problems with the regulatory system itself, including an unclear role for communities in the regulatory process, lack of accepted guidelines for risk and hazard assessments, and procedures that resulted in inequitable access to knowledge and resources. Overall, the 2000 report concluded that there was a need to reduce the health and safety impacts of sour gas facilities and that there was a strong need for an improved consultation process around sour gas in Alberta (EUB 2000).

By focusing on consultation, as opposed to public education, the 2000 report was a marked departure from the 1994 report's technical policy frame. Where the 1994 report recommended a unidirectional communication and education process emanating from experts and downloaded to lay people, the 2000 report specifically identified a more relational, communicative approach to community interactions (EUB 2000). This new approach advocated the generation of more multidisciplinary and integrated knowledge transfer, a greater respect and understanding of multiple viewpoints, and the recognition of the situated knowledges available in the various stakeholder groups. Because of this more relational approach, the 2000 report falls within a more consultative policy frame, as compared to the political and technical management approach used earlier (Torgerson 1986). Finally, focusing on a more process-oriented framework, the 2000 report concluded with a set of eighty-seven recommendations in five categories, with a clear request that the regulatory agency 'develop a plan to implement all of its recommendations, to make the plan public, and to publicly report progress toward implementation, each calendar year quarter, until all of the recommendations have been addressed' (EUB 2000, vi).

By focusing on process, this document clearly reflects the broader shift taking place in the fields of community engagement (Rowe and Frewer 2005) and public participation (Hendriks 2005). In environmental planning and policy, engagement and participation are increasingly proposed as means to address citizens' concerns about risk (Draper and Mitchell 2001; Taylor 2002). This approach to engagement is at the intersection of participation, citizenship, knowledge, and power (see, for example, Marinetto 2003) and suggests that an increased level of public engagement can be equated to 'better' or 'more effective' public inclusion in governmentality. However, such attempts at engagement and participation face many challenges (Adkin 2009). Just because a citizen is 'active' and 'engaged' does not necessarily mean that citizen's opinions, values, and wants influence outcomes. Indeed, some propose that this notion of 'active citizenship' may serve to legitimize existing power and decision-making structures that effectively concentrate decision making in the hands of bureaucrats, institutions, and governments (Callaghan and Wistow 2006; Masuda et al. 2008). In other words, rather than being truly inclusive, processes such as those advocated in the 2000 report are criticized as providing only a token form of consultation intended to legitimate decisions that have already been made in behind-the-scene meetings (Masuda et al. 2008), and thus represent only symbolic participation and knowledge sharing (Bickerstaff and Walker 2005; Delli Carpini et al. 2007). In response to the 2000 report, the Energy and Utilities Board committed to a follow-up process to address the document's eighty-seven recommendations. Seven years later, the EUB produced its final report on those follow-up activities (EUB 2007). The following section analyses the 2007 EUB report from the perspective of its risk management and consultative approaches.

2007: Public Safety and Sour Gas: Final Report

The EUB's final report classified the earlier set of recommendations into the five key categories used to track progress: health effects and sour gas research; sour gas development planning and approval; sour gas operations; emergency preparedness; and information, communication, and consultation.

In a marked divergence from its predecessors, the 2007 report clearly identifies that there has been a lack of knowledge and considerable scientific uncertainty around many of the health and environmental effects of sour gas. This report also acknowledges that, regardless of what experts might say, there remains considerable public concern around sour gas; it states much more clearly than the 2000 report the problems faced by the public in accessing sour gas information due to complex regulatory structures, inability to access information through the EUB, and an unclear set of Industry Recommended Practices (IRPs).

Structuring Participants through the Regulatory Process

Following the seven-year process of consultation and the generation of responses to the eighty-seven recommendations in the 2000 report, the construction of participation presented in the 2007 report differs substantially from its predecessors. Recall that the 1994 report structured participants as unknowledgeable and passive receivers of information. The role of the regulatory agency was to 'better educate' the public, and the assumption behind this report was that once the public better understood the science behind sour gas regulation and management, public concern would abate. By contrast, the 2000 report recognized that public concern centred less on 'knowledge' and more on access to information, input into decision-making processes, and lack of trust among communities, the regulatory agencies, and industry. The 2007 report went even further, acknowledging that missing scientific information (or equivocal and conflicting scientific information) drove uncertainty and raised public anxiety. The problem, according to the 2000 and 2007 reports, was not that the public did not have the knowledge, but that the public did not have access to the knowledge that was available, and that the regulatory system was disorganized and inconsistent. The narrative of the 2007 report structured the public neither as adversaries nor as consultants, but as collaborators in the regulatory process. As collaborators the public and communities had both a right to and a need for information, as well as a legitimate set of concerns that must be included in the decision-making process. If, as Bickerstaff and Walker (2005) and Delli Carpini et al. (2007) suggest, 'good' participation models pay attention to the location and circulation of knowledge, the 2007 report appears to move towards more integrative and inclusive forms of governance.

Public Trust, Access to Information, and Citizen Resistance

In 2002–2003 ten focus groups were held in four different communities throughout Alberta (Bonnyville, Pincher Creek, Medicine Hat, and Grande Prairie).[4] Focus group members were recruited randomly from community telephone books and approached first by telephone to seek approval to participate, followed by an information package in the mail. Participants were told that the research focused on issues related to sour gas, and that the purpose of the research was to understand what 'average citizens' felt on the topic. Groups ranged in size from six to twelve participants from eighteen to seventy years of age, with an even mix of men and women. The group discussion focused first on what it was like to live in the local community (participants' sense of place), followed by their understandings of sour gas, the sour gas regulatory process, and what kind of information they needed to bet-

Table 9.2 Sour Gas Regulatory Documents and Policy Framing

Document	Key Findings	Conclusions	Policy Frame	Construction of Participant
1994	• Improve the public's understanding of prevention and mitigation measures • Improve public understandings of the balance between risks and benefits • Improve the delivery of consistent and sound public information programs by government regulators and the petroleum industry	Better education required	Technical	Students requiring education
2000	• Generate a better understanding of sour gas • Improve the sour gas regulatory system • Improve impacts on public health and safety • Improve public consultation	Better interaction required	Consultation	Consultants whose opinions need to be acquired
2007	• Greater transparency and access to information has been provided • Public concern is based on lack of sound evidence • Regulatory system is now better coordinated • Greater access to information and resources has been given to the public	Better cooperation required	Response	Collaborators whose concerns need to be addressed

ter understand sour gas. These focus groups identified three key issues related to the regulation of sour gas: trustworthiness of sources, access to local information about sour gas, and the importance of resistors in the regulatory process.

Public Trust and Trustworthiness of Sources

When it came to who would be trusted to act in the public interest, most participants indicated a greater sense of trust in local actors, as opposed to those from farther away. Here, for example, are some of the statements made by participants in the focus groups:[5]

> M11: But I think I agree with [name] in one sense, that you tend to trust people that you know, so most local politicians are not connected with a political party. They're there because you know them.
>
> M12: Nine times out of ten, I'll trust the Alberta government, and zero times out of ten, I'll trust the federal government. I have absolutely no faith in the federal government whatsoever. But yet, for some stupid reason, I trust the provincial government, and for even some dumber reason, I trust our municipal government, even maybe more. I guess the higher up the hierarchy they come, the less I trust them.
>
> M09: With politicians, sometimes you feel that, whether they believe it or not, if it's the party line, they'll give it to you. Alberta is so tied up with the oil and gas industry that I think I would be [pause] not doubtful of things coming from the provincial government, but certainly sceptical of believing everything that they said.
>
> F12: When you're born and raised on the farm, you don't trust government [general laughter].

Participants clearly articulated that their trust in government regulators was connected to a perceived political power shift from rural to urban areas in Alberta. Given that more than half of the province's population lives in the cities of Edmonton and Calgary, approximately half of the seats in the legislature come from these municipal areas. Indeed, these cities have historically been under-represented in the Legislature due to previous electoral boundaries. In its 2009–10 report the Alberta Electoral Boundaries Commission (2010) recommended an increase in legislature seats for the cities of Edmonton (1) and Calgary (2), while creating one other new riding and reorganizing boundaries in the rest of the province. Yet the urbanization of the province generates a perception among rural residents that they inhabit a hinterland whose resources are exploited primarily to the benefit of outsiders.

> M02: So that's what I mean, is you've got your population centres in the south, but your revenue's generated up here in [local area]. And that's what's going to happen is, politically, if you've got most of your politicians representing other areas, all of a sudden, that twinning of some highway through Calgary gets priority over the one to [local areas].

> M05: What I'm hearing you talk about is something that's been bugging me for a long time, and that's – and this is coming from a kid who grew up in the city ... I'm talking about people who live outside the metropolis areas. Right now in Alberta, I'm talking about the Edmonton-Calgary corridor. If you take that capsule-shaped unit out of the centre, that's two-thirds of the population of Alberta. The rest of the province, which supports, the natural resources that support that core, are outside of there. Once the balance of power to the legislature shifts to the urban side – and it is, it's shifting now – it's going to [be] harder and harder to convince provincial politicians to put the money into rural areas ... because they won't get elected on the rural vote. Rural Alberta used to be able to bring in the Conservative government, it used to [pause] a couple of premiers ago, if you owned rural Alberta, you owned Alberta. Now if you can sway Edmonton-Calgary your way, you could be premier and never set foot outside the city. That frightens me when I look in my crystal ball, because as long as they keep the ugly things outside of the cities, they can still garner enough votes inside to stay in power. And politicians are all about staying in power.

Finally, there were clear indications of the contradictory experiences of communities surrounded by the petrochemical industry, including by sour gas wells. Participants recognized that their own livelihoods were bound up in a global economy dependent on petrochemicals. They equally recognized that the very standard of living that they enjoy is presently dependent on *someone* taking on the risks of extractive activities. As a result, they often reluctantly trusted that the best was being done to protect their communities, as shown by the three women in the following interaction:

> F01: It's like a[n] evil necessity that we have to put up with to get the things that we want in life. To have the life that we have now, we have to put up with certain things. And the companies have major control over these, so the little guy doesn't have much say about where a well's going to go.
>
> F02: I heard lots from farmers about sour gas, or sulphur dioxide, I guess, with the barbed wire rusting and cows aborting. Or is it? Or is it the pesticide they use?
>
> F01: It is a catch-22 situation, isn't it?
>
> F03: Our economy grows with that though!
>
> F01: And are we ready to do without electricity and natural gas as a people? Or the coal? Are we ready to do without that?
>
> F03: I think there's a standard of living that goes along with it and that's hard to give up.
>
> F01: We all want to live comfortably.

These women recognize the trade-offs that they feel must be made in order to sustain their standard of living. Interestingly, they fail to see alternatives outside the 'Alberta Advantage' approach of ongoing oil and gas development, and tend to speak to an 'all or nothing' set of options for change in energy use. These participants appeared to feel trapped in a set of irreconcilable challenges (for similar discussions, see Adkin 1998 and Luginaah et al. 2002).

Access to Information

Considerable time was spent discussing both how to find appropriate information about sour gas and what information, once found, could be considered trustworthy. In general, participants indicated that access to information about sour gas was quite difficult, confirming what was concluded in the 2007 report (EUB 2007):

> M07: There's a lot [of] questions that are raised. And you can get all the information you want on crude separation, you can get all the information

you want on natural gas. Sour gas, generally, the only information you get is the emergency precautions and the side effects of sour gas. That's pretty much where all the research points. It doesn't tell you where they're located, nor does it tell you where the pipelines are. When you say, 'where are they located?' 'Don't know. Can't find a map.'

The three main potential sources of information identified included the internet, companies, and government sources. Of these, the internet was treated with both caution and respect because of the availability of conflicting information and viewpoints:

M18: You can get on the internet and find out almost anything you want today. Consequently, I think the population is more educated to what's going on out there. It only takes one or two people to bring up a topic like sour gas, and everybody'll be heading to their computers to see what's on there and see more about it.

M14: The problem with the internet is that any fool can put up a web page and make it look really fancy and make it look really officious ... so it's easy to put up misinformation on the web, and people will buy it.

M15: This day and age, information's very easy to come by. You can request it from the government, you can request it from any company, you can go on the internet. Of course, you only believe half of what you read on the internet.

The benefit, however, was that the breadth of information allowed each individual to make his or her own informed decision:

F02: I think they all tend to give you the story they want you to believe, and it's up to you, as a thinking person, to evaluate what you hear.

M06: See, that's the keyword right there, is – thinking person.

F02: But it's good to hear both sides of the story ... and then you go through and you decide.

Interestingly, some of the least credible information was seen as that coming from industry, primarily because it was seen as a biased source. Whereas, in the past, there was more public confidence in corporate information, that situation has changed, in part as a result of greater experience with sour gas events:

> M07: I think if it was something to do with oil and sour gas, I probably wouldn't go first of all to an oil company to get the lowdown, basically because I think it would not, perhaps, be in their self- interest to tell me the bad points.
>
> F03: If you got information from the company, years ago, that might have been okay. As long as the jobs followed and they kind of did what else they said, then the risk factor for the community, my impression would be that it was less a factor. If they said it wasn't going to hurt them, then they were more likely to be believed, and I don't think that's the case now.
>
> M03: Why I don't trust their information? I don't know. Maybe it's just precaution, very much so. 'Cause I've been in the patch for too many years to see what's going on and when push comes to shove, you see the executives landing in helicopters, you see every big shot coming in the helicopters, and all of a sudden – boom – there's a blanket over the whole thing, and 'You guys got to go over there.' 'You got to sit in this room for two hours.' 'What for?' 'Confidentiality. Not a word of this can leak out.'

By contrast, there was some disagreement about the degree to which government information could be trusted. Some pointed to past experiences of questionable information and considered government sources to be as biased as corporate ones:

> M01: I think there's a bit of the Charlie Brown and Lucy football-kicking thing happening. Lots of times, we have faith that the government is doing the kind of testing they should be, giving us the information we need, and every now and then we run into an incident that suggests to us that's not so.
>
> M04: Is it complete, is it a hundred percent? Are they telling you everything, right? Then you're into the ethical side ...
>
> F06: So you're talking about the companies?
>
> M04: Companies, yeah, and even government agencies to a certain extent, because they're making money off this, too, and they're promoting

growth, and they see it helping the economy and everything. So if they paint too ugly of a picture, then it's not going to happen, right?

On the other hand, some participants suggested that the public needs to trust in government agencies to work in the public interest. Most groups also expressed a desire to trust, but where some participants reported being readily able to trust, others expressed scepticism about the validity of government information:

F14: I think there's a lot more public faith in the regulatory community that things are tested and sampled, than actually is the case. I mean, we have all kinds of things where you're going, 'How do we know that's safe?' 'Well, we believe it is because the government wouldn't allow it to happen if it wasn't.'

F17: I think it would be nice to be able to trust, to believe the information, but I think you have to be self-informed and cautious.

M13: Very cautious; I just think that'd be better.

M17: I think I'd tend to put that safety factor into the hands of the government. They look at ... all kinds of stuff like that, anyway. They've got to spend all kinds of money and research on making sure this gas industry is safe, the same as they make sure restaurants are safe, the same way they make the cars safe. They're making sure bridges are safe that I drive across; they're making sure the water I drink is safe. They're looking after all the other safety, so why would I put more onus on sour gas than I would making sure the restaurant I just ate at was tested according to the schedule that they're supposed to be? Like, who cares?

Overall, participants tied trust to the source of information, stressing the importance of multiple viewpoints in developing one's own opinion:

M09: Each entity is only going to give you the information they want you to have ...

F05: There's always more information, but you're not going to get it all ...

F07: And there's always ways of interpreting the information you get, too, and each interpretation can put a different slant on it, a different view to it.

M09: You determine the motives of the people that provide that information ...

F01: Like, what are they about to gain ...

F05: That's a good clue as to where you put your trust.

Since 1994, the public consultation and regulatory process around the siting of sour gas facilities has claimed to provide clear and open discussion in the public interest based on the 'values, choices and priorities' of Alberta (Hierlmeier 2008, 311). However, focus group participants clearly indicated that a lack of trust meant that information was conceptualized as a commodity in the regulatory process. Access to information and knowledge about sour gas and regulatory processes were forms of power-brokering in the public participation process (Masuda et al. 2008). Participants also reported varying degrees of trust in regulators, suggesting the potential for recreancy (Freudenburg 1993), or lack of faith in the regulators to do their jobs.

Resistance

Nested within discussions of trust and information were indications of how participants viewed resistance to the petrochemical industry in general, and sour gas exploitation, in particular. Because of the high-profile nature of the bombings of sour gas facilities in previous years, the case of Weibo Ludwig was discussed in every group and almost all expressed a sympathetic view of both the man and his situation. While they agreed that his concerns were legitimate, they did question his methods:

F11: We can look at Weibo and go, 'He was an extremist' – and he was. But I'm not sure he's totally off-base; in fact, I'm pretty sure he's not totally off-base. Just a disagreement in technique [laughs] more than what he was concerned about.

M09: I totally agree with what he was going to do, but I disagree with his methods.

> M107: That's what I disagree with. If he would have went about it a different way, you bet, I would have been up there and supported him. I would have even given him a paycheque or two to help him out. I totally agree with the man. I feel sour gas is deadly, period.

> F11: It's sad that that's the way he handled it, because you don't win friends – well, you don't win friends, period, when you do that – but still, he maybe had a legitimate concern. But you don't go around doing the sorts of things that happened there, because you turn people off ... I think that's what he was complaining about, I believe.

Resistors, in general, were viewed as playing an important role in public discussions about sour gas and other petrochemical developments. And while some in the groups revealed that they thought environmentalists were akin to rabble-rousers, others acknowledged the important role that resistors played in ensuring alternative viewpoints are expressed:

> F11: There's always two points of view, very strongly. But you know, sometimes, too, I don't know how well they come together, but whatever project happens, sometimes, thanks a lot to these watchdogs, a lot extra happens because those people are bold enough to stand up and say [pause] 'It better be done right, it better be this, it better be that.'

> M12: They're not the only ones out there [referring to oil and gas companies]. But I'll tell you, they know those people [resistors] are coming to those meetings and they're prepared. They have the information, they have – I mean, I come away from those meetings learning a whole lot more than I ever thought I would have because – did I ask the questions? No. I hadn't even thought of a lot of them. But because of them, I've gained.

Importantly, many participants in the focus groups identified the challenges to engaging in resistance. These challenges related, primarily, to the social, economic, and family costs involved in taking part in public processes around siting of petrochemical facilities, as well as a certain fatalism that their voices would not be heard:

> F16: I think people do care. I think a lot of people are [too] wrapped up in their own lives right now to want to do anything about it until it personally affects them.

> M18: I've heard of places where they had it [sour gas], and they were really against it, but they couldn't stop it because once the oil companies have the right to enter land, they can't stop it.

> M15: I'd like them to listen to people for a change, and they don't and that's why I wonder why sometimes I even participate again. Not that – my time is as valuable as anybody else's in my opinion. I think it counts a little bit, too. But when you're talking and they don't hear you, why talk?

Summary

The issues of trust, access to information, and resistance overlapped in the minds of focus group participants, creating a picture of a public that is deeply mistrustful of the petrochemical industry. Focus group members also portrayed Alberta's government agencies as somewhat complicit in development, and in supporting the petrochemical industry at the expense of local communities' concerns and anxieties. In contrast to previous work that found rural Albertans' evaluations of resistance to be largely negative (Evans and Garvin 2009; Masuda and Garvin 2008), focus group members reported that resistors play an important role in the regulatory process by ensuring that alternative viewpoints and information are included in public discussions.

Discussion

Results from the policy review indicated that there has been a shift over time in how government regulators, via the public participation process, sequentially constructed the 'appropriate' participant first as a passive receiver of information, then as one of many stakeholders suitable to be involved in consultation, and finally as collaborators whose opinions and concerns need to be included in the process. From the focus groups we see a public that is deeply mistrustful of those same government agencies as well as sceptical about information sources overall.

The petrochemical complex is dependent upon naturally occurring resources that are place dependent. Oil and gas are found only in certain geological formations, and Alberta is rife with these resources. Therefore such resources can only be developed and processed in certain communities, using readily available technology. However, demand for these resources is 'out there' and, most often, in locales far

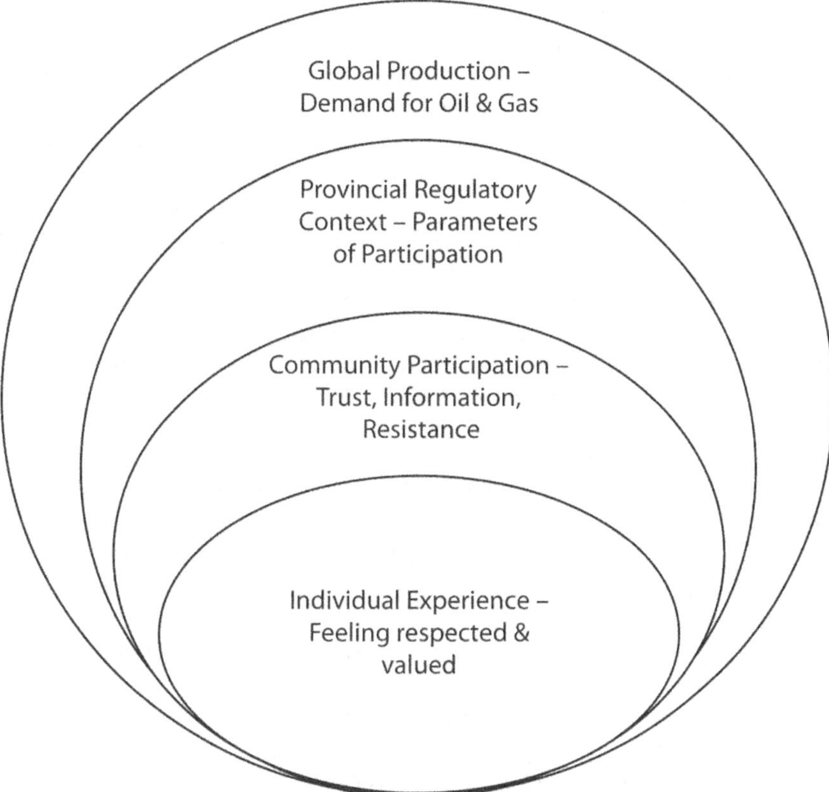

Figure 9.1. Nested relations of sour gas.

beyond the region of extraction (in particular, the United States). As a result, the structuring of global economic relations drives demand for petrochemicals, including sour gas, which in turn influences demand for production in Alberta's rural areas. As shown in figure 9.1, the individual experience of feeling valued and respected, and that the physical surroundings of one's home are valued and respected, is shaped by the public participation experience. The siting of petrochemical facilities at the local level is nested within regional and provincial parameters that define participants and the notion of participation. At the same time those provincial issues are themselves nested within a global structuring of the petrochemical complex and international demand for oil and gas that tends to assume all places are negotiable for development and rewrites spaces into primarily commodity zones (Zaup 2008).

Traditional political ecology approaches focus on the dialectical nature of poverty and environmental degradation (Blaikie and Brookfield 1987). They also suggest that the risks and benefits of resource use are distributed inequitably (Robbins 2004). This is certainly the case for sour gas in Alberta where the financial benefits (via royalties) accrue at the provincial and national levels, while the environmental and health 'disbenefits' are felt primarily at the local level. On the other hand, focus group participants tend to believe that their 'good life' (Bellah et al. 1991) is underpinned by industrial development and the consumption of fossil fuels. While they generally laughed off such contradictions, their ontological insecurity (Giddens 1986) was evident and can be seen as an example of the reflexivity of late-modern societies (Beck1992) with regard to the claims of technology and science. In other words, participants accepted that they had to live with mistrust in exchange for the personal and community-based economic benefits related to oil and gas production. At the same time, participants considered how power and knowledge operate within discourses about risk, industrial development, and citizen participation (Masuda et al. 2008).

Analysis of the three policy documents identified how the regulatory system that controls oil and gas development frames particular forms of public participation that allow (or disallow) particular voices to be heard. Over time, the regulatory agency changed the identity of the participant from being a receiver of information, to being a person to be consulted, to being a participant integral to (though not completely 'integrated in') the decision-making process. Certainly the documents examined here reflect the regulatory agency's power to define how participation is produced (Robbins 2004). By setting the parameters for participation, these policy documents discursively construct the policy and politics of sour gas exploitation (Neumann 1998; Paulson, Gezon and Watts 2005; Pulido 1996; Watts 2009). The documents suggest that behaving in the approved manner means being included in the public exercise. By contrast, they also imply that there are costs to participants in not behaving in the regulated manner, including being excluded (for examples of this, see Masuda et al. 2008). In other words, those not adhering to the behaviour of a constructed participant (or those identified as 'resistors' in the focus groups) run the risk of being either discredited or left out of the discussion. While this might not always happen, the 'costs' of resisting are high and can include stigmatization and becoming a social outcast at the community level (a detailed explanation of this is provided in Evans and Garvin 2009). This participant role constructed by the policy documents 'bumps up' against the set of expectations identified by the focus group participants, who want to trust decision-makers but feel that approval decisions are really a foregone conclusion and that dissenting voices, though important, are dismissed as not reflecting concerns held by the general public.

Further, the analysis of policy documents finds that by controlling who has a voice in deliberative processes, as well as by delineating what is or is not permitted to be discussed,[6] the regulator controls flows of power and the meaning of citizenship (Masuda and Garvin 2006). The regulator also assumes the existence of a ready-made 'citizen' waiting to receive power in order to exercise his or her rights, and relies upon something like Dean's (2004) notion of the heroic citizen acting to fulfil a civic duty. However, this is far from the reality conveyed in focus groups, where participants saw citizenship as a relational assemblage of negotiated co-constituencies that were simultaneously coercive and voluntary (Cruikshank 1999). As summarized by Masuda et al. (2008, 362): 'Discourses of participation and empowerment are the means to structure the possible "fields of action" for citizenship ..., which implies acting upon citizens' interests and desires in order to influence their action in a way that helps to legitimize particular goals in, for example, environmental planning.' Meanwhile, the role of resistors, and of resistance in general, was seen by some focus group members as a critical component of the public participation process. In their view, individuals and groups who push back against the participant role as constructed by regulatory agencies play an important part in ensuring that multiple voices and perspectives surface.

The focus groups also exposed the depth of public anxieties, mistrust, and scepticism. As the number and extent of public participation and consultation activities increase, the role of resistors will likely become critical for ensuring that alternative views are engaged. For example, the Land Use Framework (Government of Alberta 2010) was predicated on extensive public input. However, the underlying value-basis of the framework was one of growth and expansion, immediately constraining policy options and defining a 'good' citizen as one who participates in making mild recommendations around mitigation, thereby implicitly agreeing with industrial growth and economic development. In other words, good participants can 'tweak around the edges' but not fundamentally question the underlying assumption that unbridled growth and development is good for Alberta. The development discourse that dominates in Alberta produces a particular form of regulated citizen who nominally 'participates' in governance activities that both promote and protect industrial development (Robbins 2004). The 2008–09 budget goal of Alberta Environment was to 'inform' Albertans about the government's leadership on environmental stewardship. However, this 'inform and educate' approach is precisely the one that was rejected in the EUB's 2000 report, which advocated a more consultative process.

At the federal level, the Harper government's passage in June 2012 of Bill C-38, the Jobs, Growth and Long-term Prosperity Act, further excluded citizens from important decisions regarding oil and gas development. The act's amendments to a wide swathe of environmental legislation seriously curtailed the opportunities for intervention in and environmental assessment of proposed facilities by limiting the definition of an 'environmental effect' (in the Canadian Environmental Assessment Act) and shortening the timeline for deciding if an assessment is required. Further, the act eliminated the requirement for Canadian Environmental Assessment Agency review of projects involving federal money and allowed the federal cabinet to override decisions of the National Energy Board. These are precisely the moves towards authoritarianism and away from public participation that various authors have pointed to as evidence of Canada's shift to petro-state behaviour (see Carter, this volume). The Liberal Party, in opposition in 2012, contested certain parts of Bill C-38; having been elected to govern in October 2015, it may now repeal or otherwise amend the changes wrought by the Conservatives' Jobs, Growth and Long-term Prosperity Act.

Conclusion

The very fact that Alberta's oil and gas experience is nested in a global set of economic, political, and consumer relationships suggests that community experiences in Alberta are inherently connected to processes outside of communities' control. An industry that was once considered the jewel in the crown of the 'Alberta Advantage' has increasingly come to represent social relations of conflict, uncertainty, and distrust. As this study shows in relation to sour gas regulation, the construction of citizen participation in public consultations about environmental problems typically serves the functions of legitimation and disciplining. As citizens resist these constraints, government regulators and industrial actors are obliged to adapt their strategies. The key questions for democratization concern how citizens can determine the directions of the ongoing transformation of governmentality.

NOTES

1 It is worth noting, however, that in some years gambling revenue has surpassed revenue from the oil sands or natural gas royalties (see Alberta Federation of Labour 2011). This fact is explained by the minister responsible for gaming as a result of Alberta's relatively young population, high per capita income, and disposable income (see Fred Lindsay, quoted in 'Alberta Gambling Revenue Expected to Outstrip Oilsands Royalties,' *CBC News*, 27 April 2009, http://www.cbc.ca/news/canada/calgary/story/2009/04/27/cgy-alberta-gambling-revenues.html.

2 This figure is provided by the provincial government on its 'Oil Sands' website, http://www.oilsands.alberta.ca/economicinvestment.html (accessed 11 August 2011).

3 For a detailed account of these events, see Nikiforuk (2001).

4 Focus groups were conducted by the author as part of a SSHRC-funded research project. Two focus groups were held in each of Bonnyville and Pincher Creek; three focus groups were held in each of Medicine Hat and Grande Prairie.

5 To maintain confidentiality, speakers are identified only as male (M) or female (F).

6 In hearings related to the siting of new sour gas wells, strict limitations have been placed on what interveners are permitted to discuss. In the past, for example, interveners were not permitted to raise concerns related to worry and stress, as those were seen as outside the parameters of siting process, which focused solely on the physical environment and geological suitability of the chosen location.

REFERENCES

Adkin, Laurie E. 1998. *The Politics of Sustainable Development: Citizens, Unions and the Corporations*. Montreal: Black Rose.

Adkin, Laurie E. 2009. 'Democracy from the Trenches: Environmental Conflicts and Ecological Citizenship.' In *Environmental Conflict and Democracy in Canada*, ed. Laurie E. Adkin, 298–319. Vancouver: UBC Press.

Alberta Electoral Boundaries Commission. 2010. *Proposed Electoral Division*

Areas, Boundaries, and Names for Alberta. Final report to the Speaker of the Legislative Assembly of Alberta. June.

Alberta Energy. n.d.(a). 'Alberta Resource Revenues: Historical (1970 to Latest) and Budget ($ Millions).' http://www.energy.alberta.ca/About_Us/2564.asp (accessed 18 May 2015).

Alberta Energy. n.d.(b). 'Natural Gas Facts.' http://www.energy.gov.ab.ca/NaturalGas/726.asp (accessed 18 May 2015).

Alberta Energy and Utilities Board Gas Department (AEUB). 1990. *GASCON2: A Model to Estimate Ground-Level H2S and S02 Concentrations and Consequences from Uncontrolled Sour Gas Releases.* Prepared by Concord Environmental Corporation. Calgary: AEUB.

Alberta Energy and Utilities Board Gas Department (AEUB). 1995. *Sour Gas Processing Plants in Alberta.* Calgary: AE UB.

Alberta Energy Regulator. 2014. *What We Do.* http://www.aer.ca/about-aer/what-we-do (accessed 23 April 2014).

Alberta Federation of Labour (AFL). 2011. 'Hooked on Sin: Alberta Is Addicted to Revenue from Gambling and Boozing.' *Union Magazine Online,* 1 January. http://www.afl.org/index.php/January-2011/hooked-on-sin-alberta-is-addicted-to-revnue-from-gambling-and-boozing.html (accessed 11 August 2011).

Alberta Health and Wellness (AHW). 2006. *Health Effects Associated with Short-Term Exposure To Low Levels of Hydrogen Sulphide* (H_2S)*: A Technical Review.* http://www.health.alberta.ca/documents/Health-HS2-Exposure-2002.pdf (accessed 12 October 2009).

Alberta Heritage. 2010. *Alberta's Resource Inventory.* http://www.abheritage.ca/abresources/inventory/resources_hydro_gas_deposits.html (accessed 5 May 2010).

Amin, Ash. 2004. 'Regions Unbound: Towards a New Politics of Place.' *Geografiska Annaler* 86: 33–44.

Beck, Ulrich. 1992. *Risk Society: Towards a New Modernity*. London: Sage.

Bellah, Robert N., Richard Madsen, William M. Sullivan, Ann Swidler, and Steven M. Tipton. 1991. *The Good Society*. New York: Knopf.

Bickerstaff, Karen, and Gordon Walker. 2005. 'Shared Visions, Unholy Alliances: Power, Governance and Deliberative Processes in Local Transport Planning.' *Urban Studies* 42: 2133–44.

Blaikie, Piers M., and Harold C. Brookfield. 1987. *Land Degradation and Society*. London: Methuen.

Callaghan, Gill D., and Gerald Wistow. 2006. 'Publics, Patients, Citizens, Consumers? Power and Decision Making in Primary Health Care.' *Public Administration* 83: 583–601.

Canadian Association of Petroleum Producers. 2008. *Sour Gas*. http://www.capp.ca/default.asp?V_DOC_ID=767 (accessed 12 March 2008).

Canadian Association of Petroleum Producers. 2010. *Basic Statistics*. http://www.capp.ca/library/statistics/basic/Pages/default.aspx#HZV52lf86bIy (accessed 5 May 2010).

CBC News. 2013. 'Sour Gas Leak in Turner Valley, Alta., Now Contained.' http://www.cbc.ca/news/canada/calgary/sour-gas-leak-in-turner-valley-alta-now-contained-1.1304093 (accessed 23 April 2014).

Cruikshank, Barbara. 1999. *The Will to Empower: Democratic Citizens and other Subjects*. Ithaca: Cornell University Press.

Dean, Hartley. 2004. 'Popular Discourse and the Ethical Deficiency of "Third Way" Conceptions of Citizenship.' *Citizenship Studies* 8: 65–82.

Delli Carpini, Michael X., Fay Lomax Cook, and Lawrence R. Jacobs. 2007. 'Public Deliberations, Discursive Participation and Citizen Engagement: A Review of the Empirical Literature.' *Annual Review of Political Science* 7: 315–44.

Draper, Diane, and Bruce Mitchell. 2001. 'Environmental Justice Considerations in Canada.' *Canadian Geographer* 45: 93–8.

Energy and Utilities Board (EUB). 2000. *Public Safety and Sour Gas: Findings and Recommendations Final Report*. Calgary, AB: Provincial Advisory Committee on Public Safety and Sour Gas. Accessed October 12, 2009. http://www.publicsafetyandsourgas.org/mr04-FinalRec.htm.

Energy and Utilities Board (EUB). 2007. *Public Safety and Sour Gas Final Report March 2007*. Calgary: EUB. http://www.ercb.ca/docs/documents/reports/PSSG_FinalReport_2007-03.pdf (accessed 12 October 2009).

Environmental Resources Conservation Board (ERCB). 1994. *Report and Recommendations to the ERCB on Public Safety and Sour Gas*. Advisory Committee to the ERCB Reviewing Public Safety and Sour Gas. Calgary: ERCB.

Evans, Joshua, and Theresa Garvin. 2009. '"You're in Oil Country": Moral Tales of Citizen Action against Petroleum Development in Alberta, Canada.' *Ethics, Place and Environment* 12: 49–68.

Fluet, Colette, and Naomi Krogman. 2008. 'Changing Alberta's Oil and Gas Tenure System: Lessons from an Oil Dependent State.' *Projections: MIT Journal of Planning* 8: 120–39.

Freudenburg, William R. 1993. 'Risk and Recreancy: Weber, the Division of Labor and the Rationality of Risk Perceptions.' *Social Forces* 74: 909–32.

Freudenburg, William R., and Susan K. Pastor. 1992. 'Public Responses to Technological Risks: Toward a Sociological Perspective.' *Sociological Quarterly* 33: 389–412.

Garvin, Theresa. 2001. 'Analytical Paradigms: The Epistemological Distances between Scientists, Policymakers and the Public.' *Risk Analysis* 21: 443–55.
Garvin, Theresa, and Renee G. Lee. 2003. 'Reflections on the "Policy Relevant Turn" in Social Justice Research.' *Social Justice* 30: 40–53.
Giddens, Anthony. 1984. *The Constitution of Society: Outline of the Theory of Structuration*. Cambridge: Polity.
Gismondi, Michael, Mary Richardson, and Joan Sherman. 1995. 'Participation: Local Knowledge versus Expert Knowledge: at the Environmental Public Hearings for a Pulp Mill in Northern Alberta.' In *Debating Development Discourse: Institutional and Popular Perspectives*, ed. David Moore and Gerald Schmitz, 230–48. London: Macmillan.
Government of Alberta. 2008. *Natural Gas*. http://www.energy.alberta.ca/NaturalGas/Gas_Pdfs/BrochureHUB.pdf (accessed 5 May 2010).
Government of Alberta. 2009a. *Alberta's Energy History*. http://www.energy.alberta.ca/About_Us/1133.asp (accessed 12 October 2009).
Government of Alberta. 2009b. *Alberta Energy: General Natural Gas Frequently Asked Questions*. http://www.energy.alberta.ca/NaturalGas/742.asp#sour_for (accessed 10 May 2010).
Government of Alberta. 2010. *Alberta Land Use Framework FAQs*. http://www.landuse.alberta.ca/AboutLanduseFramework/LUFFAQs/Default.aspx (accessed 10 May 2010).
Government of Alberta. 2011. *Talk about Natural Gas*. http://www.energy.alberta.ca/NaturalGas/Gas_Pdfs/FactSheet_NGFacts.pdf (accessed 7 February 2012).
Guidotti, Tee L. 1996. 'Hydrogen Sulfide.' *Occupational Medicine* 46: 367–71.
Hendriks, Carolyn M. 2005. 'Participatory Storylines and Their Influence on Deliberative Forums.' *Policy Sciences* 38: 1–20.
Hierlmeier, Jodie L. 2008. '"The Public Interest:" Can It Provide Guidance for the ERCB and NRCB?' *Journal of Environmental Law and Practice* 18: 279–311.
Iannantuono, Adele, and John Eyles. 1997. 'Meanings in Policy: A Textual Analysis of Canada's "Achieving Health for All" Document.' *Social Science and Medicine* 44: 1611–21.
Inserra, Steven G., Betty L. Phifer, W. Kent Anger, Michael Lewin, Roberta Hilsdon, and Mary C. White. 2004. 'Neurobehavioural Evaluation for a Community with Chronic Exposure to Hydrogen Sulphide Gas.' *Environmental Research* 95: 53–61.
International Agency for Research on Cancer (IARC). 1992. *Monographs on the Evaluation of Carcinogens: Sulphur Dioxide*. Lyons: IARC.
Jaremko, Gordon. 2013. *Steward: 75 Years of Alberta Energy Regulation*. Calgary: Energy Resources Conservation Board. http://www.aer.ca/about-aer/who-we-are.

Keeling, Arn. 2001. 'The Rancher and the Regulator.' In *Writing Off the Rural West*, ed. Roger Epp and Dave Whitson, 279–300. Edmonton: University of Alberta Press.

Kieltyka, Matt. 2014. 'B.C. Horse Breeder Recounts Fracking Leak Scare.' *Metro* (Toronto). 7 May 2014. http://metronews.ca/features/health-and-fracking/1024905/a-close-call-b-c-horse-breeder-recounts-fracking-gas-leak-scare/ (accessed 12 May 2014).

Kilburn, Kaye H. 1997. 'Exposure to Reduced Sulfur Gases Impairs Neurobehavioral Function.' *Southern Medical Journal* 90: 997–1097.

Leiss, William. 2001. *In the Chamber of Risks: Understanding Risk Controversies*. Montreal/Kingston: McGill-Queen's University Press.

Luginaah, Isaac, N.S. Martin Taylor, Susan J. Elliott, and John D. Eyles. 2002. 'Community Responses and Coping Strategies in the Vicinity of a Petroleum Refinery.' *Health and Place* 8: 177–90.

Mang, Robert. 2006. *Energized: Canada*. Special information supplement for the Canadian Association of Petroleum Producers. Calgary: CAPP.

Marr-Laing, Tom, and Chris Severson-Baker. 1999. *Beyond Eco-Terrorism: The Deeper Issues Affecting Alberta's Oilpatch*. Drayton Valley, AB: Pembina Institute.

Massey, Doreen. 2004. 'Geographies of Responsibility.' *Geografiska Annaler B* 86: 5–18.

Masuda, Jeffrey, and Theresa Garvin. 2008. 'Whose Heartland? The Politics of Place at the Rural-Urban Interface.' *Journal of Rural Studies* 24: 118–23.

Masuda, Jeffrey, Tara McGee, and Theresa Garvin. 2008. 'Power, Knowledge, and Public Engagement: Constructing "Citizenship" in Alberta's Industrial Heartland.' *Journal of Environmental Policy and Planning* 10: 359–80.

Neumann, Roderick P. 1998. *Imposing Wilderness: Struggles over Livelihood and Nature Preservation in Africa*. Berkeley: University of California.

Nikiforuk, Andrew. 2001. *Saboteurs: Wiebo Ludwig's War against Big Oil*. Toronto: Mcfarlane, Walter and Ross.

O'Donnell, Sarah. 2010. 'Still No Charges in Pipeline Bombings.' *Vancouver Sun*, 11 January.http://www.vancouversun.com/news/Alberta+Mounties+finish+searching+Ludwig+farm/2429658/story.html (accessed 7 May 2010).

Paulson, Susan L., Lisa L. Gezon, and Michael Watts. 2005. 'Politics, Ecologies, Genealogies.' In *Political Ecology across Spaces, Scales and Social Groups*, ed. Susan Paulson and Lisa L. Gezon, 2–64. New Brunswick, NJ: Rutgers University Press.

Petroleum Communication Foundation. 2000. *Sour Gas: Questions and Answers*. Calgary: Petroleum Communication Foundation.

Pulido, Laura. 1996. *Environmentalism and Economic Justice*. Tucson: University of Arizona Press.

Robbins, Paul. 2004. *Political Ecology: A Critical Introduction*. Oxford: Blackwell.
Roth, Sheldon, and Verona Goodwin. 2003. *Health Effects of Hydrogen Sulphide: Knowledge Gaps*. Edmonton: Alberta Environment, Science and Standards Branch. http://www.environment.gov.ab.ca/info/library/6708.pdf (accessed 15 February 2011).
Rowe, Gene, and Lynn J. Frewer. 2005. 'A Typology of Participation Mechanisms.' *Science, Technology, and Human Values* 30: 251–90.
Scott, Harvey M. 1998. 'Effects of Air Emissions from Sour Gas Plants on the Health and Productivity of Beef and Dairy Herds in Alberta, Canada.' Doctoral diss., University of Guelph.
Taylor, Dorceta E. 2000. 'The Rise of the Environmental Justice Paradigm: Injustice Framing and the Social Construction of Environmental Discourses.' *American Behavioral Scientist* 43: 508–80.
Therborn, Goran. 1978. *The Ideology of Power and the Power of Ideology*. London: Verso.
Torgerson, Douglas. 1986. 'Between Knowledge and Politics: The Three Faces of Policy Analysis.' *Policy Sciences* 19: 33–59.
Watts, Michael. 2009. 'Political Ecology.' In *Dictionary of Human Geography*, 5th ed., ed. Derek Gregory, Ron Johnston, Geraldine Pratt, Michael J. Watts, and Sarah Whatmore, 545–7. Oxford: Wiley-Blackwell.
Western Interprovincial Scientific Studies Association (WISSA). 2006. *Western Canada Study of Animal Health Effects Associated with Exposure to Emissions from Oil and Natural Gas Field Facilities: Interpretive Overview by the Science Advisory Panel*. http://www1.agric.gov.ab.ca/$Department/deptdocs.nsf/all/webdoc12577/$FILE/bkgrnd-keyfindings1.pdf (accessed 5 May 2010).
Yanow, Dvora. 1996. *How Does a Policy Mean? Interpreting Policy and Organizational Actions*. Washington, DC: Georgetown University Press.
Zaup, Brent Z. 2008. 'Negotiating Through Nature: The Resistant Materiality and Materiality of Resistance in Bolivia's Natural Gas Sector.' *Geoforum* 5: 1734–42.

10 Mobilizing to Address the Impacts of Oil Sands Development: First Nations in Environmental Governance

BRENDA PARLEE

Just as political ecology studies of the global south have sometimes focused on the experiences of Indigenous peoples affected by rapid industrialization and its associated impacts, the political ecology of 'first world' contexts must consider the unique and complex histories of Canada's Indigenous peoples. We could not arrive at an adequate understanding of the political ecology and governance of Alberta without taking into account the particularities of relations between the state and the growing population of Cree, Dene, and Métis peoples in the oil sands region. The province's resource-extraction-based economic development has arguably unsettled, undermined, and transformed the circumstances of Indigenous communities more than it has those of any other peoples in the province. Although adversely impacted ecologically, economically, and socially, many Indigenous peoples in Alberta and Canada have been important actors in provincial environmental governance – and have in diverse ways been working to address these effects in ways that matter both to the sustainability of their communities and to the protection of the environment. In February 2008, chiefs representing Treaties 6, 7, and 8 of Alberta met and passed a resolution stating that no new oil sands projects should be approved until a comprehensive watershed management plan was in place and approved by their Nations (AEN 2008). This resolution demonstrated the collective interest of forty-three communities to work together to improve the environmental management of the Athabasca River Watershed.[1] Among the most active in this fight are Mikisew Cree First Nation and Athabasca Chipewyan First Nation of Fort Chipewyan.

But the issue is not just a local fight. This David-and-Goliath scenario has garnered attention worldwide, capturing the imagination and concern of celebrities, award-winning scientists, and authors; it has spurred the creation of a federal panel of inquiry into water quality monitoring and led to audiences with the United Nations Standing Committee on the Rights of Indigenous Peoples. The oil sands have become a symbolic edge or frontline for a growing number of environmental groups and allies whose names, 'Stand with Fort Chipewyan' or 'Rethink Alberta,' appear to resonate with many across Canada, the United States, and abroad. Although welcomed and sometimes invited by local communities, this support has also led to questions about whose voices are being privileged in this highly volatile debate that has become as much about the world's hunger for oil as it is about the health of the people and resources of the Athabasca River watershed.

Oil sands mining in the watershed has led to the disruption of thousands of square kilometres; in the Muskeg River sub-watershed alone, almost 900 km^2 of land may be disturbed as a result of approved and planned oil sands mining activities (Government of Alberta 2008b). Little has been done to recover mined areas; by one government estimate, less than 1 per cent of the 50,000 km^2 mined by the largest lease holders has been reclaimed (Government of Alberta 2008a). Among the most significant concerns is the health of the water itself, particularly for the communities living downstream of mining activity. As told by Dene elder Pat Marcel from Fort Chipewyan, 'Oil sands development in the Athabasca region has had devastating effects on our people. We are afraid to drink the water or eat the fish from the river as we have always done. The fish have strange tumours, and cancer rates in our community have increased dramatically in the last 10 years' (quoted in Holroyd 2008, 30).

The land, which historically formed the basis of life and livelihood in the region, is now viewed by many in the community as the source of significant environmental harm. Although vulnerable, the people of Fort Chipewyan have also been innovative in their efforts to influence decisions about where, when, how, and – most importantly – *if* further oil sands mining will be approved by provincial regulatory bodies. The term 'environmental governance' is used here to refer to the suite of institutions that have direct or indirect decision-making power over the access, use, and distribution of benefits from natural resources. Provincial and federal governments, including elected officials and bureaucracies, are seen as only some members of a more complex set of actors influencing decisions in the oil sands regions. Multinational corporations, non-governmental organizations (NGOs), voters, consumers, and investors across different regional, national, and global scales also have degrees of influence over natural resource decision making.

What follows is a discussion of five different strategies of Indigenous environmental governance at play in the oil sands region. The first deals with the 'requirement to consult,' which has created opportunities for input into planning and management and has sometimes given way to other kinds of institutional arrangements, including bilateral agreements and participation in multi-stakeholder initiatives. In addition to these formal arrangements, the chapter also discusses some of the alternative ways in which communities aim to understand and deal with environmental changes around them, including community-based monitoring initiatives and public outreach and advocacy. Reflecting on these multiple strategies of engagement with oil sands development, this chapter indicates there is no 'one size fits all' approach to Indigenous environmental governance. Through multiple rather than single strategies, Mikisew Cree and Athabasca Chipewyan First Nations have arguably had a greater degree of influence over how development occurs in their region.

Background

The Athabasca Watershed is home to over fifteen thousand Cree, Dene, and Métis peoples, six thousand of whom reside in the Wood Buffalo region. The Athabasca watershed is 1,231 km long and encompasses an area of over 95,000 km^2 in Alberta and Saskatchewan. Beginning in Jasper National Park with the Athabasca glacier, this watershed includes 94 rivers, 150 creeks, and 153 lakes that eventually drain north into the Mackenzie River (figure 10.1).

Reserves and Métis Settlement areas comprise just over 5 per cent of the land area. Traditional land-use studies, however, reveal that a much larger proportion of the land in the region was – and continues to be – used and occupied by First Nations and Métis peoples (Government of Alberta 2007). As a result of the strong social, economic, cultural, and spiritual ties to the land and resources in the region, First Nations and Métis have suffered disproportionately from loss of access to land and from degradation of the environment. Oil sands mining in the region has significantly altered the delta and watershed landscape through open-pit mining, deforestation, contamination and de-watering of rivers and lakes, degradation and fragmentation of wildlife habitat, and disruption of cultural and spiritual sites (Kelly et al. 2009; Nikiforuk

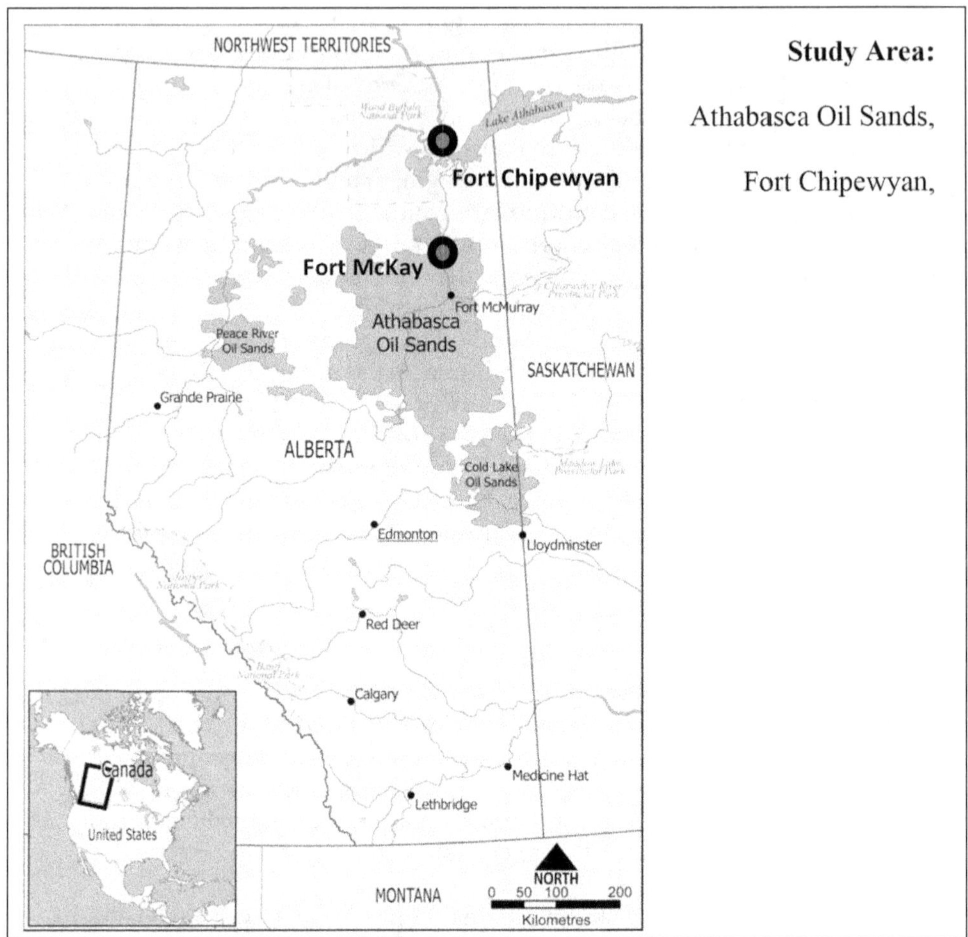

Figure 10.1. Study area (map of Alberta).

2010; Timoney and Lee 2009, 2001). According to many biophysical scientists, the environmental effects have been significant: forest and aquatic habitats have been lost and fragmented; the health of local and downstream fisheries, waterfowl, and wildlife populations has been diminished; and the cumulative effects have led to grave concerns about regional biodiversity (Dyer et al. 2001; Kelly et al. 2009; Schneider et al. 2002). In Fort Chipewyan, environmental effects of oil sands mining are blamed for a spike in cancer rates since 1999, including rare forms of lymphoma and intestinal cancers (Alberta Cancer Board April 2008; Chen 2009).

Resource development has also restricted Indigenous communities' access to land. Industrial land leases surround most Aboriginal communities in the oil sands region, making it unsafe and difficult for harvesters to use those areas. Such leases create both physical and institutional barriers to uses such as hunting, fishing, and trapping, as well as travel to areas outside lease areas still considered to be viable for traditional practices.

The socio-economic benefits to First Nations also do not seem in any way to offset these environmental losses. Available statistics show that the incomes of First Nations and Métis peoples in this region of northern Alberta are low by both national and provincial standards (Armstrong 2001; INAC 2010; McHardy and O'Sullivan 2004). Significant disparities between the well-being of non-Aboriginal and Aboriginal people exist with respect to almost every social and economic indicator (INAC 2010). Using the Community Well-Being Index, the federal government reports that Alberta First Nations score second lowest of all First Nations across Canada (INAC 2010). Some of the greatest disparities are between remote on-reserve communities in northern Alberta and urban non-Aboriginal peoples. During the boom years of the mid-2000s, the unemployment rate in the communities closest to Fort McMurray ranged from 11 to 33 per cent (Government of Canada 2006); this was significantly higher than the 3 to 4 per cent averages for the Wood Buffalo/Cold Lake regions and for the province as a whole.

Other kinds of indicators of well-being and cultural continuity confirm these circumstances of vulnerability. Epidemiological and public health studies suggest how disruptions in family, community, and culture, visible in the histories and experiences of First Nations, strongly influence health outcomes (Chandler and Lalonde 1998; RCAP 1996). The added discontinuity of identity associated with the loss and degradation of lands and resources may be understood as a third level of stress on an already vulnerable population. The inability of youth, adults, and elders to see themselves reflected in their own community and environment may explain a range of social illnesses that are very present in the oil sands region, including Fort Chipewyan. This discussion of environmental governance is thus not only about the sustainability of the environment, but also about the well-being of current and future generations of Cree, Dene, and Métis peoples in this region.

Participation in Government-Initiated Consultation Processes

Consultation in the Oil Sands

The entry point for many Aboriginal communities into decision making around resource development and its effects begins with the federal government's fiduciary obligation to consult Aboriginal peoples wherever and whenever crown decisions or actions have the potential to affect adversely treaty or Aboriginal rights. The requirement stems from treaty obligations and section 35.1 of the Canadian Constitution; the specific details on the meanings, triggers for, and practice of consultation have been spelled out in greater detail by the courts, but there are still many unknowns. This issue has been defined and discussed at length by scholars of history, political science, sociology, and other fields; this section focuses more narrowly on how the requirement to consult has played out in the oil sands region.

For some Aboriginal groups, it is an uphill battle both to prove the right to the lands and resources that may be affected by development and to be included in the scope of consultation defined by the crown. In other cases, where rights are clearly established, the crown has a duty to ensure that a development does not unnecessarily infringe upon those rights, and where such infringement does occur or is anticipated, that some efforts towards accommodation are made (Isaac and Knox 2003; Natcher 2001).

Legally, there are two levels of obligation: consultation is not only required where there is the possibility of infringing on substantive rights (e.g., hunting, fishing, and trapping) but is also triggered by procedural rights to be informed and considered in the decision-making process. This is clearly established in the case of the *Mikisew Cree First Nation v. Canada* (2005). Mikisew Cree First Nation had challenged the decision of the minister of Canadian Heritage to approve the construction of a winter road through a portion of the Mikisew Reserve, located within the Wood Buffalo National Park. The basis of the challenge was the lack of adequate consultation and efforts to minimize the impact of the road on their treaty rights to hunt, trap, fish, and carry out their traditional mode of life, pursuant to Treaty 8. As noted by the court, 'were the Crown to have barreled ahead with implementation of the winter road without adequate consultation, it would have been in violation of its procedural obligation, quite apart from whether or not the Mikisew could have established that the winter road breached the Crown's substantive treaty obligations as well' (2005, 57–8).

In principle, consultation is intended to rebalance the decision-making process on resource development such that Aboriginal rights and interests are weighed alongside other land and resource interests (Hunter 2005, 19). However, achieving 'balance' between traditional practices of hunting, trapping, and fishing and the interests of industry seems a difficult goal. More than 90 per cent of the 110,000 km^2 defined by the federal government as lands owing to the Mikisew Cree First Nation under treaty entitlement negotiations were unilaterally 'taken up' by the province for development (Mikisew Cree First Nation 2007, 11).[2] Many members view all of their traditional lands as completely unhealthy and diminished in terms of traditional use value, due to the environmental effects of oil sands mining (ACFN 2000; Fort McKay First Nation 1994).

Apart from these fundamental questions of purpose, the *how* of consultation is also not well defined. Timing and adequacy of community-level information and resources needed to meaningfully participate in consultation are key issues. There are various gaps and inconsistencies in the procedures for consultation as well as in anticipated outcomes or accommodations of Aboriginal interests. Of fundamental concern is *when* consultation should take place and how much time is taken in the process. The basic measure of consultation is often one of quantity. Regulators seem to draw positive correlations between the number of interactions and 'consultedness': the greater the amount of time taken or number of attempts made to meet and communicate with an Aboriginal community, the more likely the regulator is to determine that the community has been consulted (or that a reasonable effort was made to consult), regardless of the *nature* of the interactions and the outcomes.

There is also a problem of 'consultation fatigue,' particularly in communities and regions like the oil sands where there are multiple projects being proposed at any given time. Consultation fatigue can be aggravated by the amount of information that the community needs to consider. Following the Supreme Court decision *Halfway River v. British Columbia* (1999), it was determined that government has

> a positive obligation to reasonably ensure that [First Nations] are provided with all necessary information in a timely way so that they have an opportunity to express their interests and concerns and to ensure that their representations are seriously considered and wherever possible demonstrably integrated into the proposed plan of action. (note 12 at para. 160)

Adequacy of information, however, can be a double-edged sword. Although details about project development are necessary for understanding a potential project and its effects, too much information can be worse than not enough. As noted by Natcher (2001), the sheer number of referrals sent for community review limits the extent to which communities are informed and able to interpret and communicate their concerns about the impacts of multiple projects.

Other frustrations are with the sequencing of consultation in the development process; while best practice involves meeting with a community *prior to* applications for water licences, land use permits, or leases, this is not necessarily the norm for industry and government in Alberta or British Columbia. In some cases, efforts to consult take place only *after* companies have gained access to traditional lands for exploration or development – it seems too little, too late given that such consultation takes place well after effects have been rendered on the landscape or in the community (Fidler 2010).

Lack of community-level resources to enter into consultation arrangements can aggravate such consultation fatigue and frustration. Although there is no specific requirement to fund Aboriginal participation in the consultation process, this has been an implicit consideration as a result of *Taku River First Nation v. British Columbia* (2004) and *Dene Tha' First Nation v. Canada* (2006). This court ruling stated that adequate funding for research, preparation of briefs, and representation in hearings is essential to ensuring that communities can meaningfully engage in the consultation process, and bring forward their knowledge and perspectives on such issues as environmental impacts.

In general, the lack of clarity and the tensions that exist over the details of how, why, when, and with whom consultation should take place have limited the ability of Aboriginal communities to meaningfully influence decision making. The Government of Alberta began to address this issue in 2003, when it introduced *The Traditional Use Studies Initiative*. This initiative provided funding to First Nations for the collection of data that will assist government decision making about land management and resource development applications. Other initiatives include traditional knowledge studies funded by industry. What happens to the traditional knowledge once shared by the community, however, is a perennial question. Even in situations where consultation has occurred, there is often no evidence that traditional knowledge had an impact on management or regulatory decision making.

It is important to note that consultation does not give Aboriginal groups a veto over the proposed project, as was established in the *Haida Nation v. British Columbia* (2004) decision. Although there are some instances – as discussed in the *Delgamuukw*[3] decision – where consent is required before a project goes ahead, this is a rare decision-making opportunity afforded Aboriginal communities, particularly in the oil sands region. Even small changes to project proposals are rare. Where there is little possibility of changing the status quo, the *de facto* mechanism of accommodation has become financial compensation. As noted by Hunter (2005), this practice seems fundamentally at odds with the concept of legal intent of accommodation (see *Haida Nation v. British Columbia* 2004).

In recent years, the Government of Alberta has aimed to address confusion over consultation by developing more prescriptive guidelines for consultation. Following commitments to consultation that were made in 2003, the provincial government produced the First Nations Consultation Policy on Land Management and Resource Development, approved in 2005 (Government of Alberta 2005). Since the policy's release, First Nations have been working with the province on a framework for consultation guidelines. As of spring 2015, proposals for guidelines have been rejected by many First Nations, including the Treaty 8 First Nations of Alberta, because of inconsistencies between the provincial consultation guidelines and Supreme Court decisions. Treaty 8 First Nations of Alberta released its own consultation policy and guidelines in 2005 with the aim of clarifying their positions on traditional use and the relationship between themselves and the crown.

While the provincial government, First Nations, Métis, and industry continued to dialogue about the meanings, legal obligations, and purposes of consultation, the provincial government was drafting policy frameworks and regulatory processes to guide broader public consultation about land and resource development. These include the Alberta Land Use Framework (ALUF), the Mineable Oil Sands Strategy (MOSS), and the Fort McMurray Oil Sands Integrated Resource Management Plan (IRP). These frameworks have been characterized by a notable lack of attention to issues of consultation and accommodation of Aboriginal rights. Ross and Potes observed in 2007 that 'Alberta is failing in its duty to consult Aboriginal peoples in oil sands development' (1).

While problematic in many respects, the requirement to consult has created opportunities for communities to engage in the governance of development in ways that did not exist prior to the landmark *Del-*

gamuukw or – more locally – *Mikisew* decisions. The capacity and willingness of First Nations to engage in consultation vary considerably, however, and the situations on the ground are complex.

Outsiders' generalizations about community experiences in consultation are also problematic. Depending on the community, the company, and the development activity, there are different issues at stake, and perceived roles and terms of engagement also vary significantly. On one end of the spectrum, communities may play a passive role in defining issues and the means for their accommodation. On the other end of the spectrum, communities play a proactive role in consultation, defining not only the issue or focus but how the consultation process itself should unfold. Formal or informal 'consultation protocols,' for example, have been developed by all of the Athabasca communities in the oil sands region. These protocols detail procedures regarding the timing/sequencing of consultation relative to permitting, information-sharing methods, and resourcing. The terms and conditions of consultation protocols, as well as the degree of their 'enforcement' by communities, vary according to the policies of the corporations with which they are negotiating, and the level of interest in the project on the part of the community concerned. Recent and impending court cases in Alberta, British Columbia, and the Northwest Territories reveal that communities are becoming more sophisticated and proactive rather than reactive in respect of the 'duty to consult' requirement. For some government agencies and industry leaders, such proactivity is welcomed as it clarifies the needs and interests of communities at a very early stage in the development process. Other pro-development actors are wary of the extent to which communities use consultation protocols to prescribe the way in which development should take place in the region (Ellis 2009).

The extent to which consultation provides real opportunities for First Nations like Athabasca Chipewyan and Mikisew Cree First Nation to influence the course of development in the region can thus be questioned on many levels. Consequently, many groups have looked beyond the immediate gains that can be made in individual consultation exercises and are seeking other vehicles for defending or asserting their interests, as we see in the following sections.

Bilateral Agreements

The consultation process, insofar as it is a fiduciary obligation of the crown, is a public process in which actors are required to demonstrate some level of transparency and accountability. However, the public process has given way, in practice, to private agreement making in involving representatives of Aboriginal groups and corporations (see Zalik, this volume). These negotiated arrangements range from soft agreements (memorandums of understanding) to formal agreements, such as impact and benefit agreements. In the oil sands region, these include 'good neighbour agreements,' which focus on respect and trust, and more formal agreements involving compensation for impacts on traditional lands and resources, as well as benefits related to training, employment, and business development. While regulatory agencies are often interested in whether these agreements have been made, as some measure of accommodation, the *contents* of the agreements are not part of the public record. For example, the Joint Panel/Energy and Utilities Board, in their decision on the Imperial Oil Kearl Lake project, noted only that 'agreements' had been made with Aboriginal groups with rights and interest in the lands and resources to be impacted by the project: 'The Athabasca Chipewyan First Nation (ACFN) reached an agreement with Imperial Oil such that the ACFN did not object to the KOS Project' (Government of Canada 2007).

Conventionally, these kinds of bilateral agreements have tended to focus almost exclusively on issues of employment and business opportunities; more recently, they have evolved into more comprehensive socio-cultural, economic, and environmental contracts that may specify participation in environmental monitoring and management (Gibson 2008; O'Faircheallaigh 2007; Sosa and Keenan 2001). In other jurisdictions, those issues classified as 'environmental' in nature have fallen under the scope of 'environmental agreements.' O'Faircheallaigh (2006), in a review of five environmental agreements in the Canadian north, observed their valued role in filling holes or gaps in environmental regulation and management, as a vehicle for communication and decision making, in providing resources for Aboriginal participation in environmental management, and in recognizing Aboriginal rights and interests in lands and resources. Depending on the status of Aboriginal title to lands, agreements like impact and benefit agreements may be a means of 'accommodation' because they address what will be done to offset the infringements of Aboriginal rights.

To the extent that such agreement making is a voluntary process, it affords both communities and corporations greater flexibility and innovation that are not necessarily possible in more rigid institutional contexts. For example, provincial water-use licensing and regulation require compliance monitoring that affords little role for traditional knowledge or community participation. Alternative approaches to water-quality monitoring that are informed and guided by elders may be made possible through bilateral agreements. While not replacing or duplicating provincial regulatory requirements, such alternative monitoring initiatives may be a more culturally appropriate and relevant means of consulting local communities. One example of such traditional knowledge initiatives is the Athabasca Oil Sands Project (AOSP)[4] agreement with the Athabasca Chipewyan First Nation, made in 2005. A posting by Natural Resources Canada stated:

> The AOSP management made a set of commitments to the Athabasca Chipewyan First Nation (ACFN) that will build their environmental capacity and strengthen their self-sufficiency, as well as provide opportunities to include traditional ecological knowledge into environmental assessment. Based on the same information it suggests that ACFN will have the opportunity to review and affect the design of the environmental monitoring programs' (Natural Resources Canada 2010).

Not all research activities focused on traditional knowledge live up to their potential. The outcomes of the AOSP project are unknown, as none of the contracting parties have yet provided evidence of its success.

The growing number of companies and the consistency of concerns within and among Aboriginal communities in the region in the 1990s led to the development of more elaborate agreement making. The All Parties Core Agreement between the Athabasca Tribal Council and Athabasca Regional Developers (an association of oil and gas companies operating in the oil sands) involved multiple corporate and Aboriginal actors. The stated intent of the All Party Core Agreement was to provide communities with resources and means of responding to and addressing the increasing level of resource development in the region. Among its key goals, according to the Athabasca Tribal Council vision, was to improve community capacity to ensure 'responsible development' of the oil sands (ATC 2010). The agreed-upon means was an Industry Relations Corporation (IRC) (see Zalik, this volume). Each of the five First Nations in the Athabasca oil sands region operates an IRC. The functions of each IRC vary, but most are charged with overseeing the consultation process at the local level as well as with representing the First Nation in environmental governance initiatives at a regional level.

To what extent have the IRCs enhanced the bargaining power of First Nations communities with the oil and gas companies operating in the region? Notably, they have no real regulatory power; the IRC has only a pseudo-advisory role through processes of consultation and interventions in environmental assessments and regulatory processes. A further drawback to the agreements is their potential to limit the ways in which communities are able to assert their interests. For example, some First Nations have been prevented from challenging oil sands development because of clauses in these agreements securing their support for the initial project (Sosa and Keenan 2001). Finally, there is a very large question about the nature of consent when the resources and power of two parties to an agreement are so unequal. Although for many communities the issue is not simply an either/or set of contingencies associated with environmental sustainability, one can easily imagine how those seemingly trapped by poverty and unemployment would welcome the short-term benefits offered by bilateral agreements at the cost of longer-term environmental harms (Caine and Krogman 2010). Increasing the transparency with which these kinds of terms and conditions are negotiated and are implemented would potentially address such conflicts and increase the value of such agreements as tools of environmental governance.

Multi-stakeholder Arrangements (Co-management)

The growing number of industry players in the region and increasing public concerns about the cumulative effects of oil sands development led the Government of Alberta to adopt a more pluralistic approach to the planning, monitoring, and management of oil sands activity. As noted by Kennett (2007, 1), 'faced with a significant increase in project applications and planned development, key players quickly recognized the limitations of a project-by-project approach to environmental regulation.'

In 2000, the Alberta government created the Cumulative Environmental Management Association (CEMA), which dovetailed with its Regional Sustainable Development Strategy (RSDS) (Alberta Government 1999). Ten years later, nearly all aspects of planning and management were characterized by a kind of multi-stakeholderism in which industry, the provincial government, and representatives from each of the First Nations were involved. The case is made that multi-party initiatives are better able to address the integrated nature of development activity and the cumulative nature of environmental effects. The multi-stakeholder model is also said to be more efficient in terms of time and resources, particularly related to developing regional monitoring indicators and transparency to stakeholders about government management activities. At the same time, multi-stakeholder initiatives may ultimately serve to diffuse authority and responsibility for addressing and managing effects (Hoberg and Phillips 2010; Philip 2008). The case of CEMA is instructive here.

In 1998, Alberta government consultation related to the Regional Sustainable Development Strategy (RSDS) led to the identification of seventy-two environmental issues (falling under fourteen themes and three priority categories) needing further study. The diversity of environmental values and interests in the region spawned another multi-stakeholder forum to implement the strategy: the Cumulative Effects Management Association (CEMA). In 2000, CEMA was established with members from industry, the governments of Alberta and Canada, the Municipality of Wood Buffalo Region, non-governmental organizations, and First Nations and Métis peoples. Its primary purpose was to collect scientific information and make recommendations to government on how best to manage the cumulative environmental impacts of industrial development in the region.

Various working groups were created within CEMA to carry out technical monitoring of environmental effects. One working group that has attracted great scrutiny is the Regional Aquatic Monitoring Program (RAMP). The aim of the RAMP was to integrate aquatic monitoring activities across different components of the aquatic environment, different geographical locations, Athabasca oil sands, and other developments in the oil sands region so that long-term trends, regional issues, and potential cumulative effects related to oil sands and other development can be identified and addressed. In 2004, a peer-reviewed study concluded that RAMP had no ability to assess the impacts of oil sands development on the Athabasca River (Ayles et al. 2004). It further found that the effectiveness of the monitoring program was limited due to 'lack of details of methods, failure to describe rationales for program changes, examples of inappropriate statistical analysis, and unsupported conclusions' (iv).

Phillips (2008) argues that CEMA as well as other related initiatives have been dominated by the interests of two key groups of actors – government and industry – and that multi-stakeholder bodies have done little to alter the historic government-industry model of interest representation. In 2007 and 2008, several Aboriginal groups, including the Mikisew Cree First Nation, pulled out of CEMA due to concerns about failures of the organization to effect any meaningful change. The Mikisew spokesperson at the time, Sherman Sheh, stated: 'CEMA is a parking lot where everything, all the major issues, are placed. Meanwhile approvals [for new oil sands projects] are given' (quoted in Woodford 2007). In August 2008, the Pembina Institute, the Toxics Watch Society of Alberta, and the Fort McMurray Environmental Association (FMEA) also withdrew from CEMA, proposing a new multi-stakeholder approach to environmental management that would see further oil sands development suspended.

In an analysis of CEMA and its role in governance, Steven Kennett (2007) noted three problems with respect to its purpose and development. First, much of the mandate and constitution of CEMA facilitated and supported the interests of industry and government rather than challenging them: 'The incentives driving stakeholder initiatives like CEMA are in large measure a product of government policy, or lack thereof' (10). CEMA essentially made it easier and faster for government and industry to 'consult' with Aboriginal groups as well as to appear to be considering the issue of cumulative effects, while having little obligation to act on such concerns. In support of this conclusion, Kennett highlighted the fact that no mechanism was ever created for implementing and enforcing CEMA's decisions. Similarly, following the release of the RSDS document in 1999, no progress was ever made within government to build the legal, policy, and institutional framework for managing cumulative effects in relation to the pace of oil sands development (Kennett 2007). These criticisms explain why Aboriginal groups like the Mikisew Cree and Athabasca Chipewyan First Nations withdrew their support and have sought other means to address their concerns about environmental conditions in the Athabasca watershed.

Consultation, agreement making, and multi-stakeholder arrangements such as CEMA have afforded communities only negligible influence over the pace and scale of oil sands development. George Poitras, a former chief of the Mikisew Cree First Nation, has said:

> The Mikisew Cree have formally participated in hearings for the past six years and several multi-billion dollar projects, where we provided very sophisticated arguments and evidence both through western science and traditional Indigenous knowledge and more recently have strongly advised decisions of 'no more new approvals' until many of the concerns

> and issues raised by the Mikisew Cree and other non-Indigenous groups are addressed; every hearing that we participated in resulted in an approval with little to no consideration of the Mikisew Cree. (Poitras 2009)

For some communities, 'multi-stakeholder fatigue' has begun to accompany the phenomenon of 'consultation fatigue,' exacerbated by the experience and perception that regional bodies have not led to desired local outcomes. Consequently there has been an effort to find alternative paths to environmental governance that better honour and accommodate the priorities and traditional knowledge of local communities.

Community-Based Monitoring of the Oil Sands

Community-based monitoring is increasingly being considered by First Nations as a means of gathering information about oil sands impacts in ways that are more trusted to represent local interests. The Mikisew Cree, for example, suggest that community-based monitoring may be more meaningful to local traditional knowledge holders as well as more consistent with established systems of monitoring. For the Mikisew Cree First Nation, the issues are about trust in the systems of monitoring and management and ultimately the decisions made about development projects. Athabasca Chipewyan First Nation has conceptualized a system of monitoring that is about watching and protecting the land: *ni ho ghe di.* The focus of the process is on engaging communities in collecting, recording, and sharing information.

While scientific information is an element, the *ni ho ghe di* program is conceptually and practically grounded in local and traditional knowledge and practices of land users, including hunters, trappers, fishers, and plant harvesters. Such programs are guided by elders and carried out by community members employed as 'guardians of the land.' They are planned and implemented independent of existing monitoring efforts by organizations such as CEMA and RAMP, and the results of these programs are shared primarily at the local level to build community capacity for participation at larger scales of land-use decision making. Methods for collecting and reporting information are based on traditional and local knowledge systems but are supported by the scientific sampling protocols and techniques used by industry, government, and research organizations. Results may also be shared with other communities and organizations, thereby contributing to understanding of environmental conditions and trends at a regional level.

Ni ho ghe di and similar programs may be perceived by First Nations' members and others critical of regional monitoring efforts to date as offering a more legitimate perspective on oil sands impacts than data that are provided by government- or industry-run monitoring programs. Part of this legitimacy comes from the ability of those doing the monitoring to speak to scale and locale. The ACFN and Mikisew Cree have first-hand knowledge of the effects they experience from living downstream of the oil sands mines. The other important element of legitimacy comes from the diachronic or long-term relationship between these communities and the ecosystems in which they presently live. As noted by many ecologists such as Berkes et al. (2008) and sociologists (e.g., Bradshaw 2003), land-based peoples who have lived in a particular place for generations and have developed a 'sense of place and belonging' have both superior local ecological knowledge and a vested interest in ensuring the survival of that place. As noted by the late Dene elder Maurice Lockhart, 'Some people who don't care so much won't notice the changes' (quoted in Parlee et al. 2005).

Community-based monitoring and management approaches like those proposed by the Mikisew Cree and Athabasca Chipewyan First Nation are not new, but they are increasingly valued and recognized for the contributions they make to the science of ecological change as well as to the broader goals of sustainability (Berkes et al. 2000, 2008; Danielson et al. 2009). For scholars of common pool resources in particular, community-based resource management has in many case studies shown itself to be more effective in achieving conservation goals than private or state-led management. This is due to the benefits of inclusivity in decision making, flexibility and capacity to adapt to variable ecological conditions, and tight and efficient feedback loops between learning and decision making (Ostrom 1992; Poteete and Ostrom 2008). While there is a growing weight of evidence regarding success, community-based approaches to resource management are also under scrutiny in some regions for not living up to their promise (Blaikie 2006; Leach et al.1999). The unwillingness of government to forgo responsibility or devolve responsibility to the local level and the weak role of communities, particularly Aboriginal communities, in the socio-political landscape are among the reasons found for the poor performance of some community-based resource management programs. Due to the lack of power and capacity to play David to the Goliath of multinational corporations, communities may become unknowing Trojan Horses. As described by Blaike (2006), in the guise of decentralized management and greater autonomy in decision making, communities may be at greater risk from outside interests using symbolic agreement to capitalize on local resources. Insights into whether real power may be exercised at the community level can be garnered by studying the experiences of decentralization or devolution of environmental governance. It is important to consider what power and authority does *not* rest at the community level. Arun Agrawal wisely observes: 'It is precisely by an examination of what is not devolved that the hidden politics of decentralization becomes visible, and the influence of these hidden interests becomes amenable to analysis' (1999, 21).

While not successful on all levels, local participation or engagement in monitoring may provide community members with a means of understanding and dealing with the rapid pace and scale of environmental change in ways that would otherwise not exist. Acceptance and accommodation of community perspectives and 'science' generated from community-based monitoring at larger scales is certainly the underlying purpose of such efforts. Although government and industry are unlikely to embrace either the process or the outcomes, the knowledge generated from monitoring may aid Mikisew Cree First Nation and Athabasca Chipewyan First Nation in influencing governance in a different way – through public outreach and advocacy.

Public Outreach and Advocacy

In the last ten years, media attention to the health concerns of downstream communities and to the environmental harms associated with the oil sands has reached widening audiences. While the messages disseminated by films such as *Dirty Oil* and *H2Oil* may be ignored or denied by some actors, they provide fuel for international social movement campaigns. As noted by Hoberg and Phillips (2010) in regard to the nuclear power industry in the United States, the 'public imaging' created about a particular industry can play a fundamental role in the governance of industries. In the oil sands region, Aboriginal communities are increasingly making efforts to influence the image of the oil sands presented to public audiences beyond the province. In 2005, ACFN and Mikisew Cree representatives travelled to Geneva to make a presentation before the United Nations on Human Rights and Climate Change, making the following recommendation:

> The Permanent Forum, through ECOSOC [Economic and Social Council]

> call on the UN General Assembly to convene an emergency world session to fully explore, with all branches of the UN, and relevant treaty bodies, in particular UNCERD, the multiple impacts of climate change and its link to fossil fuel development and the human rights of Indigenous Peoples, to include the topics of, but not limited to social, economic, cultural, environmental, health, food security, land and water rights, and treaty rights.

Identification of and networking with allies in the environmental movement have amplified the message of communities like Mikisew Cree and ACFN. An organization that represents such alliance efforts is the Keepers of the Athabasca Watershed, which aims to protect water and watershed lands along with ecological, social, cultural, and community health and well-being. As noted in this address by Peter Cyprien, the organization seeks to suspend oil sands development and to address the environmental pollution that they believe is harming local communities: 'Please stand in solidarity with us all and call on your government to say "No" to new approvals for tar sands and demand that they deal with the pollution that is killing the people of Fort Chipewyan and the environmental impacts that are destroying our way of life. We need your voice and we need a time-out of the tar sands now' (quoted in Healy 2008).

More generally, linkages between the oil sands and global climate change have created grounds for alliances between local Indigenous communities and people everywhere who believe in the necessity of drastically reduced greenhouse gas emissions. Athabasca River communities have found willing partners in organizations such as Pembina Institute, Keepers of the Water, and Forest Watch. In February 2010, Mikisew Cree and Athabasca Chipewyan First Nations released a poster coinciding with US president Barack Obama's visit to Canada. 'President Obama,' the poster read, 'you'll never guess who's standing between us and our new energy economy.' It branded Canada's tar sands as 'the dirtiest oil on earth.' An advertisement paid for by the First Nations and Forest Ethics and broadcast on American television in February 2009 contained a message for the American public: 'Your voice counts. Please let President Obama know that he should ask Canada to clean up the Tar Sands' (*CBC News* 2009). As described by George Poitras, the purpose of the campaign was educational: 'This is the beginning of a process to educate the American people and the Obama administration on the issue of the tarsands and its impacts on our people' (quoted in *CBC News* 2009).

The use of media to influence the imaging of oil sands and its effects has profound transnational mobilization possibilities. Of particular value are those individuals within communities who are able to use both old and new technologies to communicate local concerns, for example, by drawing on the traditional knowledge and perspectives of those grounded in land-based practices such as hunting, fishing, and trapping as well as on their knowledge of the global economy and media.

The reimaging going on through efforts such as the 'Dear Mr. Obama' and subsequent campaigns asks us to see oil sands exploitation not simply as the benign production of a valued commodity but as the cause of environmental harms that matter to everyone. Such imaging may serve to influence governance by increasing the 'reputational risk' of corporations. Oil sands companies are highly exposed to such risk given that a significant proportion of project budgets is drawn from public shareholder investment. While Exxon and more recently British Petroleum seem to have emerged relatively unscathed from single disasters such as the Exxon Valdez oil spill in Alaska (1989) or the Deepwater Horizon oil well blowout in the Gulf of Mexico (2010), a sustained campaign that links such disasters – including the cumulative environmental harms of the oil sands – to the political economy of fossil capitalism has potential to shift the model of development in a more ecologically sustainable direction.

The Mikisew Cree First Nation and Athabasca Chipewyan First Nation have been particularly active in connecting the oil sands to larger scales. As a result of this 'policy imaging' effort, First Nations have attracted new allies in their efforts to achieve greater social and environmental sustainability in the region for current and future generations.

Conclusion

The release in February 2008 of the Treaties 6, 7, and 8 Joint Resolution to halt oil sands development, pending a more comprehensive management strategy for the Athabasca River watershed, signaled the inadequacies of the existing institutional means available to First Nations to influence the course of oil sands development. The governance structure for the oil sands has been characterized as a relatively closed, bipartite policy subsystem where industry has enjoyed a 'policy monopoly' (Baumgartner and Jones 1991, 1070). The partial recognition by the provincial government of Aboriginal rights and interests underpins the opportunities for Aboriginal engagement in environmental governance in the ways outlined above, including consultation, bilateral agreement making, and 'multi-stakeholderism.' However, many communities see themselves as having a long journey ahead to a place and time when they have meaningful roles in the decision making over when, how, and if development continues within their traditional territories.

Aboriginal communities have historically been marginalized or framed as victims of environmental change, whereas they have complex identities and exercise many forms of agency. In this chapter I have attempted to discuss and honour the diverse efforts made by the Mikisew Cree First Nation and Athabasca Chipewyan First Nation to prevent, mitigate, manage, or adapt to the changes brought about by oil sands development. While the Government of Alberta has made some efforts in the last ten years to establish better governance of the Athabasca region, including the Athabasca Watershed Planning Advisory Council and the Lower Athabasca Regional Plan, mining in the oil sands regions continues unaffected by the rhetoric of sustainable planning and management.

Given the socio-economic conditions they face, made chronic by their dispossession of resources under Canadian law, isolated First Nations communities have had limited bargaining leverage and few rights with which to defend their interests. Yet as I have described in this chapter, they continue to learn from, and innovate upon, the opportunities available to them to influence decisions that may further encroach upon their bases of survival. Through engagement with government, corporate actors, and the courts, they have tried to set limits on the scope and rate of resource extraction, and to obtain recognition of their rights. Moreover, Indigenous actors have stepped beyond these frameworks, with greater confidence, onto the global stage. In expressing their fears for their communities they rely on their knowledge of the meaning of 'development,' and offer their visions of a more truly sustainable future.

NOTES

1 The 2008 resolution is one among many kinds of collective action used by Indigenous peoples in Canada to influence the progress of resource development on traditional lands. In other regions of Canada, Aboriginal peoples have been celebrated as key players in successful fights against clear-cut

logging in British Columbia, low-level flying exercises by the Canadian military, and the banning of a variety of contaminants known to bio-accumulate in Arctic marine ecosystems. The forms of resistance have ranged from blockades, sit-ins, and legal actions to negotiation, education, and community-based efforts of reconciliation. But the successes of the Nuu-chah-nulth of Clayoquot Sound, the Innu Nation of Labrador, and the Inuit Tapiirit Kanatami are paralleled by an equal number of frustrations, such as those experienced by the Lubicon Cree in northern Alberta.

2 As signatories to Treaty 8, the Mikisew Cree and Athabasca Chipewyan were promised the 'right to pursue their usual vocations of hunting, trapping and fishing throughout the tract surrendered ... saving and excepting such tracts as may be required or taken up from time to time for settlement, mining, lumbering, trading or other purposes.'

3 *Delgamuukw v. British Columbia* describes the scope of protection afforded Aboriginal title under subsection 35(1) of the *Constitution Act, 1982*; it also defines how Aboriginal title may be proved; and outlines the justification test for infringements of Aboriginal title. The decision is referred to often as a groundbreaking ruling due to the fact it was the first definitive statement from the Supreme Court on the content of Aboriginal title in Canada. For more information see Hurley (2000).

4 The Athabasca Oil Sands Project (AOSP) is a joint venture among Shell Canada, Chevron Canada Limited, and Marathon Oil Sands L.P. It consists of the Muskeg River mine, located about 75 kilometres north of Fort McMurray, Alberta, and the Scotford Upgrader, located near Fort Saskatchewan, Alberta.

REFERENCES

Agrawal, Arun. 1999. 'Accountability in Decentralization: A Framework with South Asian and West African Cases.' *Journal of Developing Areas* 33: 473–502.

Alberta Cancer Board. 2009. *Cancer Incidence in Fort Chipewyan, Alberta (1995–2006)*. Edmonton: Government of Alberta. http://www.albertahealthservices.ca/files/rls-2009-02-06-fort-chipewyan-study.pdf (accessed 1 January 2010).

Alberta Environmental Network. 2008. 'Unanimous Passing of No New Oil Sands Resolution at the Assembly of Treaty Chiefs Meeting.' *Alberta Environment Network*, 25 February 2008. http://www.aenweb.ca/node/2131 (accessed 15 June 2010).

Armstrong, Robin P. 2001. *The Geographical Patterns of Socio-Economic Well-*

Being of First Nations Communities in Canada. Agriculture and rural working paper no. 46. Ottawa: Statistics Canada.

Athabasca Tribal Council. 2010. 'History of the All Parties Core Agreement.' http://atc97.org/history-of-apca (accessed 1 October 2011).

Ayles, Burton, Monique Dubé, and David Rosenberg. 2004. *Oil Sands Regional Monitoring Program (RAMP): Scientific Peer Review of the Five Year Report (1997–2001)*. Winnipeg. http://www.andrewnikiforuk.com/Dirty_Oil_PDFs/RAMP%20Peer%20review.pdf (accessed 15 June 2010).

Baumgartner, Frank, and Bryan Jones. 1991. 'Agenda Dynamics and Policy Subsystems.' *Journal of Politics* 53 (4): 1044–74.

Berkes, Fikret. 2008. *Sacred Ecology*. New York: Routledge.

Berkes, Fikret, Johan Colding, and Carl Folke. 2000. 'Rediscovery of Traditional Ecological Knowledge as Adaptive Management.' *Ecological Applications* 10 (5): 1251–67.

Blaikie, Piers. 2006. 'Is Small Really Beautiful? Community-Based Natural Resource Management in Malawi and Botswna.' *World Development* 34 (11):1942–57.

Bradshaw, Ben. 2003. 'Questioning the Credibility and Capacity of Community-Based Resource Management.' *Canadian Geographer* 47 (2): 137–50.

CBC News. 2009. 'Alberta First Nations Place Anti-oilsands Ad in Major U.S. Paper.' 17 February. http://www.cbc.ca/news/canada/edmonton/alberta-first-nations-place-anti-oilsands-ad-in-major-u-s-paper-1.852412 (accessed 28 March 2016).

Chandler, Michael, and Christopher Lalonde. 1998. 'Cultural Continuity as a Hedge against Suicide in Canada's First Nations.' *Transcultural Psychiatry* 35 (2): 191–219.

Chen, Y. 2009. *Cancer Incidence in Fort Chipewyan, Alberta, 1995–2006*. Edmonton: Alberta Cancer Board, Division of Population Health and Information-Surveillance.

Danielson, Finn, Neil F. Burgess, Andrew Balmford, Paul F. Donald, Mikkel Funder, Julia P.G. Jones, et al. 2009. 'Local Participation in Natural Resource Monitoring: A Characterization of Approaches.' *Conservation Biology* 23 (1): 31–42.

Delgamuukw v. British Columbia [1997] 3 S.C.R. 335.

Dene Tha' First Nation v. Canada (Minister of the Environment), 2006 FC 265, [2006] 4 F.C.R. D-28.

Dyer, Simon J., Jack P. O'Neill, Shawn M. Wasel, and Stan Boutin. 2001. 'Avoidance of Industrial Development by Woodland Caribou.' *Journal of Wildlife Management* 65 (3): 531–42.

Ellis, Steve. 2009. 'NWT First Nations Launch Court Proceeding against

Canada, Water Board & North Arrow Minerals.' *Mostly Water.* http://mostlywater.org (accessed 15 October 2010).

Fidler, Courtney. 2010. 'Increasing the Sustainability of a Resource Development: Aboriginal Engagement and Negotiated Agreements.' *Environment, Development and Sustainability* 12 (2): 233–44.

Fort McKay First Nation. 1994. *There Is Still Survival Out There.* Fort McKay: Arctic Institute of North America.

Frideres, J.S. 1996. 'The Royal Commission on Aboriginal Peoples: The Route to Self-Government?' *Canadian Journal of Native Studies* 16: 247–66.

Gibson, Ginger. 2008. *Negotiated Spaces: Work, Home and Relationships in the Dene Diamond Economy.* Doctoral diss., University of British Columbia.

Government of Alberta. 1999. *Regional Sustainable Development Strategy.* Edmonton: Department of Sustainable Resource Development.

Government of Alberta. 2005. 'The Government of Alberta's First Nations Consultation Policy on Land Management and Resource Development.' Edmonton: Government of Alberta. http://www.aboriginal.alberta.ca/images/Policy_APROVED_-_May_16.pdf (accessed 15 June 2010).

Government of Alberta. 2007. 'Applying Traditional Use Data: A Case Study.' *Fast Facts on First Nations Consultation* 4 (1): 1. http://www.assembly.ab.ca/lao/library/egovdocs/2007/alii/164713_jan07.pdf (accessed 10 October 2008).

Government of Alberta. 2008a. 'Alberta Issues First-Ever Oil Sands Land Reclamation Certificate.' Press release, 19 March. Edmonton: Alberta Environment.

Government of Alberta. 2008b. 'Muskeg River Interim Management Framework for Water Quantity and Quality.' Edmonton: Alberta Environment.

Government of Canada. 2006. 'Aboriginal Community Profiles.' Ottawa: Statistics Canada.

Government of Canada. 2007. 'Joint Review Report/EUB Decision 2007-013 Imperial Oil Resources Ventures Limited, Application for an Oil Sands Mine and Bitumen Processing Facility (Kearl Oil Sands Project) in the Fort McMurray Area (February 27, 2007).' http://www.ceaa.gc.ca/050/documents/19660/19660E.pdf (accessed 15 June 2010).

Haida Nation v. British Columbia (Minister of Forests), [2004] CarswellBC 2656, 2004 SCC 73, 245 D.L.R. (4th) 33.

Halfway River First Nation v B.C., [1999] BCCA 470 .

Healy, Teresa. 2008. 'The Harper Record.' Ottawa: Canada Centre for Policy Alternatives. http://www.policyalternatives.ca/sites/default/files/uploads/publications/National_Office_Pubs/2008/HarperRecord/Downstream_From_the_Tar_Sands_People_are_Dying.pdf (accessed 15 June 2010).

Hoberg, George, and Jeffrey Phillips. 2010. 'Playing Defense: Early Responses to Conflict Expansion in the Oil Sands Policy Subsystem.' Energy series working paper no. 10-002. Evanston, IL: Northwestern University, Buffett Centre for International and Comparative Studies.

Holroyd, Peggy. 2008. 'Northern Leaders Tour Oil Sands, Downstream Environmental Risks Studied.' Media release 1647. Pembina Institute. http://www.pembina.org/media-release/1647 (accessed 10 January 2010).

Hunter, John L. 2005. 'Impact of the Haida and Taku River Decisions.' Presentation at Pacific Business and Law Institute Conference, 26–7 January 2005, Vancouver. http://www.litigationchambers.com/pdf/JJLH%20Paper%20Compensation.pdf (accessed 10 January 2010).

Hurley, Mary C. 2000. 'Aboriginal Title: The Supreme Court of Canada Decision in *Delgamuukw v. British Columbia (BP-459E).*' Ottawa: Library of Parliament. http://www2.parl.gc.ca/content/lop/researchpublications/bp459-e.htm (accessed 13 October 2010).

INAC (Indian and Northern Affairs Canada). 2010. 'Regional Variation and Disparities of First Nations and Inuit Well-Being.' *First Nation and Inuit Community Well-Being: Describing Historical Trends (1981–2006).* http://www.ainc-inac.gc.ca/ai/rs/pubs/cwb/cwbdck-eng.asp (accessed 13 October 2010).

Isaac, Thomas, and Anthony Knox. 2003. 'The Crown's Duty to Consult Aboriginal People.' *Alberta Law Review* 49 (1): 49–77.

Kelly, Erin, Jeff Short, David W. Schindler, Peter Hodson, Mingsheng Ma, Alvin Kwan, and Barbara Fortin. 2009. 'Oil Sands Development Contributes Polycyclic Aromatic Compounds to the Athabasca River and Its Tributaries.' *Proceedings of the National Academy of Sciences of the United States* 106 (52): 22346–51.

Kennett, Steven. 2007. *Closing the Performance Gap: The Challenge for Cumulative Effects Management in Alberta's Athabasca Oil Sands Region*. Calgary: Canadian Institute of Resources Law, University of Calgary.

Leach, Melissa, Robin Mearnes, and Ian Scoones. 1999. 'Environmental Entitlements: Dynamics and Institutions in Community-Based Resource Management.' *World Development* 27 (2): 225–47.

Marcel, Pat. 2008. *Northern Leaders Tour Oil Sands.* Media release. Edmonton: Pembina Institute. http://www.pembina.org/media-release/1647 (accessed 8 October 2008).

McHardy, Mindy, and Erin O'Sullivan. 2004. *First Nations Community Well-Being in Canada: The Community Well-Being Index (CWB).* http://dsp-psd.pwgsc.gc.ca/Collection/R2-344-2001E.pdf (accessed 13 October 2010).

Mikisew Cree First Nation. 2007. 'Response to the Multi-Stakeholder Committee Phase II Proposed Options for Strategies and Actions – Submission to the Government of Alberta for the Oil Sands Strategy.' Fort McMurray: Mikisew Cree First Nation.

Mikisew Cree First Nation v. Canada (Minister of Canadian Heritage), [2005] SCC 69 [*Mikisew*], paras. 57–8.

Natcher, D.C. 2001. 'Land Use Research and the Duty to Consult: A Misrepresentation of the Aboriginal Landscape.' *Land Use Policy* 18: 113–22.

Natural Resources Canada. 2010. 'Aboriginal Engagement in the Athabasca Oil Sands Project.' Ottawa: Natural Resources Canada. http://www.nrcan.gc.ca/mms-smm/abor-auto/eng-eng/tal-hyd-eng.htm (accessed 15 June 2010).

Nikiforuk, Andrew. 2010. *Tar Sands: Dirty Oil and the Future of a Continent.* Vancouver: Greystone.

O'Faircheallaigh, C. 2006. 'Mining Agreements and Aboriginal Economic Development in Australia and Canada.' *Journal of Aboriginal Economic Development* 5 (1): 74–91.

O'Faircheallaigh, C. 2007. 'Environmental Agreements, EIA Follow-up and Aboriginal Participation in Environmental Management: The Canadian Experience.' *Environmental Impact Assessment Review* 27 (4): 319–42.

Ostrom, Elinor. 1992. 'Community and the Endogenous Solution of Commons Problems.' *Journal of Theoretical Politics* 4 (3): 1.

Parlee, Brenda, Micheline Manseau, and Lutsel K'e Dene First Nation. 2005. 'Understanding and Communicating about Ecological Change: Denesoline Indicators of Ecosystem Health.' In *Breaking Ice: Integrated Ocean Management in the Canadian North*, ed. Fikret Berkes, Rob Huebert, Helen Fast, Micheline Manseau, and Alan Diduck, 162–82. Calgary: University of Calgary Press.

Phillips, Jeffery. 2008. 'Collecting Rent: Political Culture and Oil & Gas Fiscal Policy in Alberta, Canada and Norway.' Master's thesis, University of British Columbia.

Poitras, George. 2009. Speaking pointss for a conference call with the Natural Resource Defence Council and American and Canadian journalists prior to President Obama's visit to Canada. 17 February. http://www.deadducklake.com/?p=60 (accessed 28 March 2016).

Poteete, Amy, and Elinor Ostrom. 2008. 'Fifteen Years of Empirical Research on Collective Action in Natural Resource Management.' *World Development* 36 (1): 176–95.

RCAP (Royal Commission on Aboriginal Peoples). 1996. *Gathering Strength.* Vol. 3. Ottawa: Ministry of Supply and Services.

Passelac-Ross, Monique, and Veronica Potes. 2007. *Crown Consultation with*

Aboriginal Peoples in Oil Sands Development: Is it Adequate, Is it Legal? CIRL occasional paper no. 19. Calgary: Canadian Institute of Resources Law.

Schneider, Richard R., J. Brad Stelfox, Stan Boutin, and Shawn Wasel. 2003. 'Managing the Cumulative Impacts of Land Uses in the Western Canadian Sedimentary Basin: A Modeling Approach.' *Conservation Ecology* 7 (1): 8.

Sosa, Irene, and Karyn Keenan. 2001. *Impact Benefit Agreements between Aboriginal Communities and Mining Companies: Their Use in Canada*. Unpublished research paper.

Taku River Tlingit First Nation v. British Columbia (Project Assessment Director), [2004] 3 S.C.R. 550, 2004 SCC 74.

Timoney, Kevin, and Peter Lee. 2001. 'Environmental Management in Resource-Rich Alberta, Canada: First World Jurisdiction, Third World Analogue.' *Journal of Environmental Management* 63 (4): 387–405.

Timoney, Kevin, and Peter Lee. 2009. 'Does the Alberta Tar Sands Industry Pollute? The Scientific Evidence.' *Open Conservation Biology Journal* 3: 65–81.

Woodford, Peter. 2007. 'Health Canada Muzzles Oilsands Whistleblower.' *National Review of Medicine* 4 (6): 1.

11 Duty to Consult or Licence to Operate? Corporate Social Practice and Industrial Conflict in the Alberta Tar Sands and the Nigerian Niger Delta

ANNA ZALIK

This chapter considers the relational implications of particular community negotiating institutions supported by private firms in two significant and controversial oil-exporting regions: Alberta's tar sands and Nigeria's Niger Delta. These institutions serve as expressions of a broader context of attenuated regulatory structures intended to facilitate the social 'licence to operate' in state jurisdictions that have a significant presence of oil and gas capital (see also Zalik 2015). The regulation of multinational corporations, as shaped by firms, states, and industrial lobbyists, is historically and contemporarily a global endeavour. Studying corporate regulation in only one context may therefore lead the analyst to overlook explanations for the directions observed that arise from global corporate strategy. The two industry-community governance mechanisms examined in this chapter – the Global Memorandum of Understanding (GMOU) employed across clan-affiliated communities in the Nigerian Delta, and the Industrial Relations Corporations (IRCs) entered into by the Aboriginal communities of the Athabasca Tribal Council in northern Alberta – exemplify the globalization of neoliberal, quasi-privatized policy production in the oil and gas industry, and the political conditions that shape its character at a variety of scales (Peck 2003, 229). But in contrast to some readings, which view the potential outcomes of these neoliberal regulatory practices somewhat optimistically (Idemudia 2009; Slowey 2008), this chapter argues that these practices lessen the likelihood of substantive regulation. Instead, my analysis supports calls for stringent, legally enforceable state monitor-

I am grateful to Janet Fishlock for her assistance with the first stage of this research as well as to Dave Vasey for subsequent interviews.

ing policy. This requires the creation of institutions that are rigorously autonomous from industry so as to operate as overseers of social and environmental initiatives in extractive sites. Following the insights of critical scholars Laureen Snider and Frank Pearce (1997), among others, the chapter supports the criminalization of corporate malfeasance, in lieu of the criminalization of protest actions against extractive capital.

This chapter drew initially from fieldwork conducted in 2008 concerning the Industrial Relations Corporation, which was established to facilitate industrial negotiations with First Nations of the Athabasca Tribal Council and is still in partial effect. The chapter is informed, more broadly, by field research conducted in 2006, 2007, and 2013 (in Nigeria) and 2008, 2010, and 2012 (in Alberta), as well as by interviews in various locations conducted by colleagues and through correspondence, and document analysis in the intermittent periods. The chapter's analysis thus largely predates the 2012 overhaul of Canadian Environmental Assessment Act (CEAA) that further skewed the environmental regulatory terrain in favour of industrial actors. CEAA 2012 and a subsequent federal bill, C-38, were heavily critiqued by First Nations and environmental NGOs for violating Aboriginal territorial rights and for reducing the public's opportunities to challenge the activities of the Canadian oil and gas industry. As one expert put it, within the international context 'the Canadian government's environmental assessment legislation (under CEAA 2012) stands as a particularly extreme example of regressive changes' (Gibson 2012). Subsequently, Alberta's Responsible Energy Development Act, passed in November 2012, created the Alberta Energy Regulator and further reduced the provincial regulator's responsibility to ensure consultation with Aboriginal peoples and the public's ability to intervene in application processes (see chapter 17, this volume). More recent provincial and federal elections leading to majority governments for the NDP in Alberta and the Liberals in Ottawa, a number of governmental changes in Nigeria, and the 2014 drop in oil prices that endures into 2016 have once again shifted the regulatory terrain on which the actors and institutions described below operate.

The Drivers of Corporate Social Responsibility in Oil-Producing Jurisdictions

Despite the evident socio-historical and cultural differences between Nigeria and Canada, the populations of oil-producing zones face conditions common to regions of petroleum extraction with long colonial histories. Global advocacy efforts have recently underlined these similarities. In Canada, as in Nigeria, regional conflicts over the allocation of oil industry revenues and royalties are increasingly manifest (Miller 2007), and local communities' opposition to the deterioration of human and environmental health mounts. Both countries experience the boom-bust cycles of a resource-extraction-dependent economy, as well as the regional effects (labour migration, inflation, etc.). Similarly, the populations that have suffered the socio-ecological consequences of this extraction reside in frontier locations, spatially removed from major national economic centres, and are subject to the ongoing, divisive consequences of imperialism and colonialism.

As other chapters in this book have also described, climate justice movements – impelled by Canadian Aboriginal populations and a range of environmental justice and protection organizations – have drawn global attention in recent years to the Canadian tar sands as a globally destructive endeavour. International advocacy against tar sands expansion has built upon local and transnational activism focusing on the oil and gas industry's violations of human rights in the global south and against class-or-race-marginalized, 'fence-line' communities in the global north. Boycott campaigns have met with private industry response in the form of both harsh repression and new public affairs strategies.

In Nigeria, following the Ogoni uprising of the 1990s and the executions of Ken Saro-Wiwa and eight others (with Shell's complicity) in 1995, the Niger Delta became a key target for global critics of corporate practices and a central pivot in the adoption of the stakeholder accountability agenda (Osaghae 1996). Consequently, 'corporate social responsibility' in the late 1990s was heavily influenced by Shell's experiences in Nigeria and the North Sea. Shell's 1997 report in the aftermath of these events, 'Profits or People: Does There Have to Be a Choice?' was a major document in the evolution of 'corporate social responsibility' (CSR) (Wheeler, Fabig, and Boehle 2002) as voluntary self-regulation in the context of neoliberal globalization.

In this same period, Aboriginal communities in Canada increasingly challenged the state for territorial sovereignty, while criticizing provincial and federal government regulators and private operators for their complicity in environmental deterioration (Carter, this volume; Fluet and Krogman 2009). The increasing significance of 'unconventional' oil, and the role of the re-dubbed 'oil sands' (as opposed to the dirtier-sounding 'tar') for North American 'energy security,' have triggered broad-based movements opposing various pipeline projects – in particular, Enbridge's Northern Gateway and Trans-Canada's Keystone XL.

Corporate public affairs programs aim to achieve what oil industry parlance refers to as the 'social licence to operate.' These programs are also shaped by emerging legal norms. Examples of the latter include the directive of 'free prior and informed consent of indigenous people,' a recommendation of the World Bank Extractive Industries Review (Szablowski 2010), and stated in articles of the UN Declaration on the Rights of Indigenous Peoples (see also Stendie and Adkin, this volume).

In Canada, the juridically established 'duty to consult' arising from section 35 of the 1982 Constitution Act mandates consultation with First Nations communities on energy projects on Aboriginal lands (Sossin 2009).[1] Informed by a growing number of Canadian legal precedents, this duty is now understood as the state's (or crown's) obligation – one often devolved to the corporation wishing to gain regulatory approval for a development – to confer with Aboriginal groups concerning activities that will affect their constitutionally protected rights. Legal arguments in favour of the duty to consult (as a possible contributor to administrative fairness) hold that the duty amounts to a formal recognition of possible future land claims, and may shape consultation processes in ways that result in better outcomes for Aboriginal communities. Procedural justice via consultation may correspond in form with a right to collective bargaining (Sossin 2009).

That said, since the implementation of the 2012 CEAA, a number of judicial decisions in Canada's western provinces appear to contradict the duty to consult and thus fail to uphold First Nations territorial rights. In a number of commentaries, legal expert Nigel Bankes has questioned post-CEAA 2012 federal and provincial judicial decisions.[2] But the historic June 2014 Canadian Supreme Court Tsilqot'in decision may change these dynamics in favour of First Nations' claims.

This chapter focuses on two formal public consultation and agreement mechanisms funded by industry: the Global Memorandum of Understanding (GMOU), employed across clan-affiliated communities in the Nigerian Delta, and the Industrial Relations Corporations (IRCs)[3] created by the Aboriginal communities of the Athabasca Tribal Council in northern Alberta. Both structures reflect their regional colonial histories, expressed in contemporary forms of indirect rule and (neo) colonial government. As one interviewee observed, the colonial dimensions of corporate practice are explicitly manifest in the names of the firms – for instance, 'Imperial Oil' or 'Royal Dutch Shell' – just as they are in the practices through which the state facilitates private industrial extraction. By examining these cases concurrently, the chapter shows how the 're-regulation' of market actors (after neoliberal deregulation) occurs locally, but is shaped by global actors – in this case, within the global regime of oil extraction (McMichael 1990). Equally, both Athabasca Chipewyan First Nation (ACFN) and the Mikisew Cree First Nation (MCFN) have launched legal challenges or challenged institutional elements via their IRCs (or Government and Industrial Relations (GIR) as renamed by MCFN) that have also reshaped these dynamics.

In the following section I provide a brief overview of the political economy of the regulatory structure surrounding multinational operations in southern Nigeria and northern Alberta. Subsequently, I argue that the GMOU (Nigeria) and IRC (Alberta) are regionally specific, yet globally informed, private-industry-funded forms of governance that seek formal, 'legal' consent for extraction. The variations between the two models, I argue, were clearly shaped by specific territorialized identities and social resistance (including juridical claims) against the state and industry, and by demands for increasing corporate accountability in each site. Moreover, the GMOU and the IRC models provide evidence of how corporate-funded institutions become contested. The state-industry pact favouring formal rather than substantive consultation runs up against sovereignty movements' demands for ecological protection. The result is double edged: while the GMOUs and IRCs can provide new openings for community voice and protest against industry, their presence as part of a broader set of bilateral agreements between firms and bands also offers an additional channel through which the state may marginalize and criminalize the blockading strategies central to successful claims against industry. The mediating role that the states attribute to such institutions and agreements allows state and industry to claim, as liberalism would have it, that adequate spaces for negotiation are in existence, and should be used in lieu of protest (Nader and Grande 2002).[4]

The Political Economy of Petro-Regulation

Debates over the regulation of the Canadian tar sands have captured both global and national attention in recent years, with the development of the 'most destructive project on the planet' shadowing Canada's long-offered self and external image as a socially and ecologically exemplary country (Berland 2009). In the period preceding the 2008 financial crash, the low royalty rates paid by private operators in the tar sands garnered attention from both the left and the right of the political spectrum (Acuña 2007; Boychuk 2010). Indeed, private firm interests have dominated the fiscal regime. A provincial increase to royalties in 2008 was partially reversed in the spring of 2010 following corporate protest and threats to move oil and gas investment to British Columbia and Saskatchewan. The business press and all levels of government also argue that Alberta oil revenues have been diminished due to insufficient access to US markets, claiming that Alberta bitumen is sold at discounted prices at Midwestern regional rail transportation hubs due to inadequate pipeline infrastructure. The increased supply of oil and gas arising from the US shale boom deepens the pricing problem.

Moreover, mounting concern with the lack of substantive environmental regulation – including the striking fact that not a single proposed tar sands project in Alberta has ever been denied – has buttressed a broad coalition of 'no-new approvals' advocates. Overlapping jurisdiction between federal and provincial regulators shapes a context of regulatory buck-passing on environmental problems and land governance (Harrison 2005; author interviews[5] with community representatives/NGO staff in Alberta, July 2008, and British Columbia, October 2009). Given this context, and the costliness and technical demands associated with participating in public review panels on oil-and-gas-related projects, environmental organizations and activists have pursued public protest that includes civil disobedience. Among the objectives of these protests is to heighten public concern about the rapid expansion of tar sands development in the region.

Actions in the courts, in which Aboriginal land claims and the enforcement of Canada's GHG emissions reduction/Kyoto commitments figure prominently (i.e., Beaver Lake v. the Crown, Athabasca Chipewyan First Nation v. the Province of Alberta, Mikisew Cree First Nation and Ecojustice v. Kearl Tar Sands), are indicative of the failures of the regulatory process to mitigate and redress social and environmental impacts on nearby communities and the global environment. The time delays arising from judicial cases make avoidance of the courts central to current oil industry social strategy. While this is clearly now the case in Alberta, this strategy of avoidance is also closely tied to the British Columbian context, where the absence of treaties has led to significant, successful land claims, as in the Tsilhqot'in decision referred to above. Thus, novel approaches to gaining a 'social licence' are especially important as the British Columbia gas and pipeline industry expands, and where Aboriginal sovereignty claims and environmental advocates have a historical record of strength.[6] The IRC, a corporate-community negotiating model employed to address this context, is discussed in the next section.

In the Nigerian context, the nature of state-industry negotiations is epitomized by the 'joint-venture' contract, in which transnational corporations and the Nigerian National Petroleum Corporation participate (Osaghae 1998). While these contracts – existing in some form in various OPEC member states – give a degree of resource sovereignty to the landlord states, leases on extraction are monopolized by the transnationals (Labban 2008; Mommer 2002). In the Nigerian context, Royal Dutch Shell was the first entrant (Frynas 2000) to the operating process, and continues to dominate land-based operations. On the basis of the risks associated with onshore extraction, various new agreements between transnationals and the Nigerian government employ what is known as a 'production sharing contract' in which the oil firm pockets the great majority of the profits in high-risk ventures. Such contracts are also being employed in the deep offshore, however, which is relatively insulated from the social volatility onshore.[7]

From 2006, and declining somewhat as of 2013, a Niger Deltan insurgency under the umbrella of MEND (Movement for the Emancipation of the Niger Delta), has pursued a contraband oil trade (employing a notion of 'direct' resource control) in which youth of the region – marginalized from oil industry employment – have participated in a sort of protection racket with military and oil industry middle men. The arming of the insurgent movement in Nigeria emerged from Nigeria's long history of repression of minority groups seeking redress for social and ecological damages (Obi 2001; Okonta 2007; Omeje 1996). These groups became globally known through the Movement of the Survival of the Ogoni People (MOSOP) in the 1990s.

As in Canada, Nigerian minority communities have also sought compensation for ecological damages through the national courts.[8] A 2005 Nigerian Supreme Court decision declared the practice of gas flaring by the multinationals in the Niger Delta illegal and a gross violation of the environmental rights of the resident population. Accounting for 70 per cent of sub-Saharan Africa's CO_2 emissions in 2002 – larger than all other sub-Saharan African sources combined (World Bank 2002) – gas flaring remains ubiquitous in the Delta. Its phase-out, argue joint venture partners,[9] was delayed due to the work stoppages arising from facility bombings by MEND and to armed conflict between rebels and

Nigerian federal security forces. Notwithstanding these obstacles, the oil industry has made record profits, in part due to the previous upswing in prices resulting from media attention to violence in the Nigerian oilfields, Iraq, and other strategic sites of extraction (Zalik 2004, 2011).

Following a major Nigerian military offensive against Western Delta militia groups in 2009, an offer of amnesty to rebel leaders in exchange for the surrender of weapons may have brought a fragile peace to the region. The rise of Goodluck Jonathan to the Nigerian presidency resulted in oil revenue shifting to the south of the country, thus helping to quell the youth-led insurgency. However, the armed Boko Haram movement is now destabilizing the northern and central region of the country. Meanwhile, gas flaring continues; communities seeking benefits from oil industry engagement remain a potential barrier to access to the oil installations. It was to reduce the risk of interruptions and to gain social compliance that corporations operating in Nigeria initiated the Global Memorandum of Understanding (GMOU), discussed in detail below.

The Model(s)

The formalization of the GMOU and IRC models emerged in roughly the same period, following the 11 September 2001 attacks on the World Trade Center and the US Pentagon, and in the context of rising oil prices.

In Alberta, the Industrial Relations Corporations of the Athabasca Tribal Council (ATC) were institutionalized under the All Parties Core Agreement (APCA), negotiated in 2002 (All Parties Core Agreement 2003; Erskine 2003). First Nation members of the ATC, in conjunction with various oil and mining corporations, federal and provincial government ministries, and some forestry operators in the region, formally signed the agreement in early 2003. The APCA emerged in part from mounting pressures to meet the legally enforceable 'duty to consult.' However, it was also modelled on earlier initiatives in which the oil companies were involved, including the Fort McKay [First Nation] Industrial Relations Corporation (IRC), established in the early 1990s, and the piloting of 'whole community'–style strategies in sites like the Soku Gas Plant on the Niger Delta (discussed in Zalik 2011).

The APCA offers formal recognition of the significance of 'traditional ecological knowledge' (referred to as TEK) as well as an opportunity for Aboriginal communities to more actively participate in processes of industrial development. The APCA offered contracts and employment to the members of the Athabasca Tribal Council as a condition of access to Aboriginal territories. Critics view the APCA as fulfilling the legal requirement of government (delegated to the corporations) to 'consult' with First Nations, but providing, in the end, no meaningful community consent or lasting benefit (Natcher 2001; O'Faircheallaigh 2007). Arguably, the Industrial Relations Corporation model constituted a consultative process in which significant alteration of industrial plans was unlikely, except in regard to a First Nation's buy-in to the region's oil economy. That is, the 'right to say no' to mining (Adkin 2009; Kuyek 2008; Laforce, Lapointe and Lebuis 2009; Podur 2008) has been simply unattainable through any state-supported mechanism, including via such bodies as the IRC/GIR (in 2010 the Mikisew Cree First Nation changed the name of their IRC to Government and Industry Relations [GIR[, to more accurately reflect its mandate), although the IRC's co-council management structure may enable residents to make more radical demands. As a community member put it to a key tar sands operator at a 2008 IRC-facilitated meeting:

> We know what has happened to indigenous peoples due to the oil industry worldwide ... I won't stand for that, I won't allow that to happen to me, to my kids, to my grand-kids. I have two hands: one hand can sign a Memorandum of Understanding on a piece of paper; the second can motion to another thousand people behind me saying: We won't allow that to happen here. We're not going to give you that land.

Tar Sands Operations, the All Parties Core Agreement, and the Athabasca Tribal Council

The Athabasca Tribal Council was formed in the Municipality of Wood Buffalo in 1988. Made up of the Chipewyan Prairie Dene First Nation, Fort McMurray 468, Mikisew Cree First Nation, Athabasca Chipewyan First Nation, and Fort McKay First Nation, the ATC's mission is 'to provide advisory/program services and specified strategies to assist affiliate aboriginal bands and their membership in building capacity and the protection of Aboriginal & Treaty rights and traditional territories.'[10] The Fort McKay First Nation is the largest of the ATC First Nations and the closest in physical proximity to long-standing Syncrude and Suncor tar sand mining sites. It is often viewed as the most 'empowered' of the ATC members. Fort McKay's IRC, established long before the APCA, and its anomalous position in comparison to the other four bands of the ATC, are frequently referenced in discussions. Partnership contracting arrangements entered into by the Fort McKay First Nation underpin its stronger capacity in industry-state negotiations. Yet the community's decades-long embeddedness in the socio-ecological impacts of tar sands mining could also be read as having contributed to a weak *historical* bargaining position. That is, Fort McKay's formal relations with the oil industry were established just prior to the growing prominence of global and national indigenous rights claims and sovereignty movements. The band's involvement in tar sands development today closely ties Fort McKay's revenue to the rise and fall of industry profits, even as this involvement has strengthened the band's influence in regional planning and decision making.

On the industry side, initial APCA signatories included Alberta-Pacific Forest Industries Inc., Albian Sands Energy Inc. (since merged with Shell), Shell, the ATCO Group of Companies, Canadian Natural Resources Limited, Conoco Phillips Resources Corp., Deer Creek Energy Ltd., Enbridge, EnCana, ExxonMobil Canada, Japan-Canada Oil Sands, Nexen Petroleum, OPTI, and Petro-Canada, as well as long-time tar sands operators Suncor and Syncrude. Government signatories were the Municipality of Wood Buffalo, the Province of Alberta, and the Government of Canada under Indian and Northern Affairs Canada (INAC).

A key provision of the agreement was the creation of Industrial Relations Corporations under each band council, with base funding of $230,000 per IRC in 2003 (APCA 2003). Concurrently, a key priority area for the APCA was the pursuit of 'sustainable employment' for ATC members. Implicitly and explicitly, priority was placed on formal employment in the region's industrial development through contract provision to, and skills training by, the oil industry or its partners.[11]

A corporate representative spelled out the importance of industry-community relations to the business strategy of his firm:

> Our engagement is more time intensive. In order to increase the number of Aboriginal employees that we have, we've had to intervene during that whole fifteen-year period to help mentor school children about the importance of staying in school and to give them some vision of what they need to have to be able to access employment opportunities. It's not about intervening in the college level with training programs; it's about going right in at the early-stage education, so they have role models and assistance (like

> tutoring around reading). They need support all the way through the education process ... 'cause you don't even have kids staying in school.' (summarized from phone interview conducted by Janet Fishlock, Fall 2008)

The IRCs serve the function of corporate knowledge production, shaping a model of industrial social practice. IRCs provide companies and governments with important information for successfully completing environmental impact assessment (EIA) processes and for incorporating TEK – which is a requirement of the EIA process on Indigenous territories.

According to key informants, significant architects of the IRC model included a Calgary-based oil industry consultant who had experience with EIA processes. This same consultant has also overseen TEK components of oil industry EIAs, as well as proposing major industrial infrastructural investment for the region, involving construction of new towns and river route alteration in the Athabasca Delta (interviews by author, Alberta, 2008). The involvement of such networks of corporate knowledge production has shaped the industry's social practice regionally and globally.

Legally, the IRCs are non-profit corporations with boards of directors that include the chief of council and at least one band councillor. As initially constituted, these IRCs were to report to a management committee made up of the directors of the IRCs, the CEO of the Tribal Council, corporate community relations staff, and government representatives. In its initial structure, this management committee was overseen by an executive committee that met quarterly to approve its business plans. Established on a chief-to-chief basis, the APCA executive committee representatives included the chiefs of the five ATC bands, corporate vice-presidents, and high level directors of government ministries as well as the IRC director. Ultimately, this process broke down; negotiations now occur between the corporate and government representatives and participating First Nation ATC members (i.e., with subsets of the ATC).[12]

Under the initial model, budgets were approved at the level of the Athabasca Tribal Council for the ATC as a whole, ideally avoiding the fragmentary, competitive dynamics characteristic of oil industry negotiations with different groups (particularly in post/neo-colonial contexts). In principle (as seen in the account of the GMOU, below), 'cross-community' negotiation aims to ensure greater industry transparency and to avoid the classic divide-and-rule approach associated with colonial government wherein communities are not aware of the terms agreed to between industry and their neighbours. But budgets related to each IRC and for particular consultation processes have been negotiated individually and remain confidential. Like the GMOU in Nigeria, specific payment information between firms and communities remains protected, confidential, or unattainable in the IRC – even to community members, who are often unaware of openings to access this information. The divisions characteristic of welfare negotiations therefore persist (see also Caine and Krogman 2010; Westman 2013).

Although funded by the corporations, the IRCs (one for each of the signatory First Nations) are supposed to act as advisors for the bands. They review industry proposals and respond to industry's calls for 'consultation.' Based on these processes of consultation, partnerships between the bands and the industrial operators are then to be established with corporations on a case-by-case basis. For instance, arrangements could involve the provision of some community infrastructure or perhaps employment. These sorts of programs create the 'stakeholder' status among community members central to what oil industry strategy refers to as a 'leveraged buy-in' (Lapin 2000).

While the bands may negotiate separately with industry and government partners, the companies collectively strategize their 'bottom-line' with regard to community demands. For instance, both the APCA and a number of key industry networks in the tar sands (notably the RIWG, or Regional Infrastructure Working Group, also known as the Oil Sands Development Group at the time of the APCA's establishment) formed a corporate 'common front' with regard to First Nations negotiations, so as to avoid 'setting precedents' with Aboriginal communities that other contractors would then be pressed to meet (author's interview with an industry representative, 2008). Through the APCA the industry sought greater developmental 'certainty' (as described to the author in 2008 by one interviewee), via a 'collaborative' approach among companies 'to avoid [any one] company setting an unrealistic precedent in the community.'[13] The notion of developmental 'certainty' has become increasing prevalent in industry and government discourse concerning relations with Aboriginal communities in Western Canada (Hayes 2010; Woolford 2005, 2009).

According to various informants interviewed in 2008 and 2011, a key challenge facing the IRCs is the rapid pace of development in the region (although projects have stalled from time to time, as they did following the 2008 financial crisis).[14] Rapid rural development requires IRCs to carry out assessments of numerous plans, stretching their staff beyond capacity (an outcome that was considered by some informants to be part of industry strategy). Indeed, the 'inability' of the IRCs to respond to industry timelines is referred to in industry reports and by the APCA. Significantly, as explained by one interviewee to the author (July 2008), an unreturned phone call or materials slipped under an office door may be listed by companies as evidence (in EIA submissions) of their efforts to 'consult' with First Nations. The timelines, according to interviewees, are set by the corporations, not by the IRCs. Various interviewees (summer and fall 2008, 2012) indicate that it would require far more staff to deal with the number of requests at the pace sought by industry.[15]

With regard to funding of the IRCs, the APCA offers the following 'note' under the section entitled 'appropriate capital and operating investment': 'The financial arrangement should not be considered a corporate contribution or "funding" as it is an arrangement between equal parties who are both investing in the consultation process for joint benefits' (All Parties Core Agreement, 3). Notable in the language used here is how the Aboriginal demand for equal status in negotiations, as a means of asserting sovereignty and territorial control, is employed in a form that denies the huge differential in access to capital and financial/legislative power between industry and the members of the Athabasca Tribal Council.

The reference to 'joint benefits' suggests that the development will proceed (or else there is no benefit for the corporate partner in the APCA). It thus begs the question of the affected residents' or Aboriginal communities' capacities or rights to *reject* projects (Adkin 2009; Kuyek 2008). This is a clear area of dispute surrounding conflicts between First Nations and the mining industry across the country. In Alberta, this dispute is embodied in the 2008-formed coalition of chiefs (Treaty 7 and 8 and other Alberta stakeholders) opposing any further government approvals of tar sands development.

To avoid accusations of non-inclusive consultation, companies often depict band councils as potentially hostile and/or 'insufficiently representative.' Companies negotiate with various subsectors of communities that are clearly non-monolithic, but whose 'unity' or 'disunity' has been exploited and reshaped via colonial rule and via the results of legally mandated consultation (see Daly and Napoleon 2003). This form of subgroup negotiation is a strategy common to industrial divide-and-rule tactics globally. As put by an industry representative in a 2008 interview: 'The IRC is supposed to be representing everyone [in the community], not just the band council.'[16] Thus, corporate public relations staff will hold separate meetings with different groups in the community – for instance, with youth or elders. In one case observed by the author, such meetings were intended to break the council's consensus position on a particular industry project. This tactic is central to a model agreement proposed by industry to replace the APCA after 2006, called the 'Consultation and Regional Benefits Agreement,' or CRBA. The CRBA proposal, ultimately cancelled, held certain traits in common with the GMOU in the Niger Delta.

The IRC structure also shapes a sort of 'mutual responsibility' between the First Nations[17] and the firms, creating pressure for regional industrial negotiations and compliance among Aboriginal 'stakeholders.' As we will see, industry's strategy in the Alberta tar sands resembles that advanced through the GMOU in the Nigerian context, to which we now turn. In both contexts, participating communities have been pressured to comply with the terms accepted by others or are required to negotiate among themselves over budgetary allocations; consequently, conflict over restricted funds is devolved to the affected populations.

The GMOU Model, Multinationals, and Oilfield-Wide Development Committees

In the GMOU (Global Memorandum of Understanding) model, Chevron and Shell subsidiaries (Shell Petroleum Development Corporation of Nigeria, or SPDC), alongside other major transnational operators, launched an initiative to establish oil-field-wide corporate negotiation and governance structures. The 'Global' MOU (GMOU), now instituted in many Nigerian oil-producing areas, seeks to displace previous community-by-community agreements that were criticized for fostering intra-communal resentment and territorial violence.

As discussed below, the GMOU process partially overlaps with traditional local governance structures – whether chieftaincies or community development councils – and displaces them with oil industry initiated and sponsored governance structures funded by the industry sponsor of the particular GMOU.[18] As opposed to previous community-by-community agreements, the Global MOU offers to pay out (unequal) benefits to communities according to whether these communities are identified as 'landlords' of particular wellheads and installations or not. The GMOU also creates penalties for disruption or protest by any one community by making development benefits (e.g., schools and hospital buildings, microcredit projects) legally dependent on the maintenance of a safe operating environment for the companies *across all of the communities covered by the GMOU*. That is, grievances and social mobilization in one area could jeopardize industry-offered benefits in another area governed by the same GMOU. Earlier, oil-field-wide negotiating committees served as a base for collective action on the part of communities united under one production project or oil field, for example, the Soku gas plant on the border of Rivers and Bayelsa states.[19] It is such collective action that the new GMOU process explicitly forbids, as evidenced in the agreements described below.

The emergence of the GMOU model followed upon the various transitions from community assistance to community development to sustainable community development that have been promoted by Shell and other Niger Delta operating companies since 2001. The GMOU also reflects more explicit attention to the relationship between industrial 'security' and possible disputes over compensation payments. In the period when the GMOU model was under development, Shell began to implement a policy of non-compensation for spill damage caused by sabotage. This is problematic on a number of levels. Various NGOs (e.g., the Centre for Social Corporate Responsibility, the organization representing Catholic Church shareholders in Shell) have documented cases in which sabotage was falsely claimed by corporations. More importantly, this policy acts as a form of 'collective punishment' in which all residents are punished for the transgressions of a few. It may be viewed as extending the practices of colonial indirect rule, under which local chiefs were responsible for maintaining 'order' in their enclaves. In the current GMOU model – as with the sabotage policy – the '*global*' treatment of 'collective stakeholders' is formalized by the creation of legally binding requirements that community signatories permit and facilitate ongoing operations, thereby outlawing protest should local residents be dissatisfied. If one signatory from the 'recipient' side (affected communities) breaks the GMOU's terms requiring that a safe and unobstructed operating environment be provided to the industrial project, the corporation is then not legally bound to fulfil its obligations to any of the other 'non-offending' signatories.[20]

The continuity between current corporate governance methods and colonial governance practices is evident in both the GMOU and northern Alberta's IRC models. As with the Albertan Consultation and Regional Benefits Agreement process promoted by industry, the GMOU aims to achieve 'assurance' that communities will negotiate disputes rather than stall project development. In both sites there are reports that GMOU and IRC staff members perceived as being too critical of industry were removed following industry pressure.

In 2006, the GMOU implemented by Chevron and Shell consisted of a somewhat opaque set of committees constituted by representatives of various levels of government, state bureaucrats, community representatives, and oil industry operations and public affairs staff. The complex negotiation and implementation process was mediated by NGOs, but determined by the corporations. In the case of both Chevron and SPDC the GMOU is ultimately projected to encompass their entire areas of operations, replacing previous community-by-community agreements.

In its structure, the GMOU process employs an incentive system similar to the 'milestone' system utilized in Shell's earlier community development and sustainable community development models (see Zalik 2004). Under this system, financing for future stages of implementation is not provided until 'community' representatives demonstrate that earlier stages have been completed. Similar to the CRBA model proposed by industry in Alberta, the GMOU's project advisory committees (PACs), whose members are paid a sitting allowance by the company, are to eventually receive their own operating budgets. But for this to occur, and analogous to the IRCs, members must arrive at an intra- and inter-community agreement with regard to the expected 'benefits' (infrastructure, jobs) associated with a given longer-term project. Based on this process, binding agreements of approximately five years are to be made with the company. These agreements establish a set of promised developmental benefits to communities, the complete provision of which will depend upon these communities ensuring a 'non-conflictual' operating environment for the company. In some cases, the GMOUs contemplate the completion of the Niger Delta's so-called legacy projects (otherwise known as abandoned projects, as one colleague put it), although certain informants indicate that responsibility for implementing these will be assumed by state governments.

A review of another 2006 GMOU[21] demonstrates how binding terms are quite differently applied to corporations and communities. A selection from the company's obligations includes the following:

> The company agrees to use its *best endeavours*[22] to ensure that its Contractors award appropriate subcontracts to community contractors commensurate with their capabilities and in line with [the Company's] contracting procedures and Nigerian/Local Content Development goals.

The term 'best endeavours' was similarly applied, in this agreement, to clauses regarding community participation.

In contrast, a selection of the affected communities/clans' obligations reads:

> The [affected regions/clans] and their representatives *will ensure* that the activities including those executed by [the company's] contractors are free from any form of harassment and disruptions by [the affected clans] throughout the duration of activities.
>
> The [affected clans] agree to *actively assist* the police and other law enforcement agencies in handling and prosecuting matters that may cause public disturbance or impede [the company or the company's contractors'] ability to operate.
>
> *No other benefits will be requested from the company or its contractors.* (italics added)

Other examples of the differential language employed in the agreement are offered in a table summarizing the company's versus the affected communities' commitments. Here we see that the language imposed upon the company with regard to its commitments is less stringent than that applied to the communities:

> [The company] shall put a process in place to *ensure compliance* with environmental standards in collaboration with the communities and regulatory agencies.
>
> The affected regions: provide [the company] and her contractors *at all times* with a conducive atmosphere for its operations and to take responsible steps to forestall and avoid disruptions to [the company's] activities.

The agreement spells out specific infrastructural commitments, although without a timeline, that would be provided by the company, as well as a timeline of four annual payments to the implementing body (or PAC) for the GMOU of the region, and specific amounts to be offered in homage and courtesy payments, at times of official company visits to the communities. It is worth noting that key informants interviewed by the author in 2007 indicated that although this particular GMOU had been signed in 2006, funds had yet to be disbursed by the corporation,[23] replicating the problems that plagued earlier rounds of corporate-community negotiations in the Delta (Aaron and George 2010; Zalik 2004).

And from the final section on penalty for breach of GMOU:

> All the benefits including SCD projects and programmes are conditioned to uninterrupted operations that may accrue to beneficiary communities and all parties agree that funding for these projects and programmes shall be suspended if there is any disruption to [the company] and/or its contractors' operations during the period of the G-MOU. Any breach of this G-MOU on the part of [the company] especially from delay in disbursement of funds (taking into account Joint Venture Partner funding by more than 6 months), will attract a penalty, which is equivalent to an income-generating project to the amount of N5.[24]
>
> *[The affected regions] expressly agree that breach or perceived breach of the GMOU by [the company] or its Contractor shall not be a ground for disruption of [the company] or the Contractor operations but will use the grievance handling procedure to address their concern herein attached.* (italics added)

Thus, communities are not even permitted to engage in non-violent protest in the face of the company's breach of contract. If the grievance process takes too long, or yields unsatisfactory results, communities have no recourse to action that does not incur penalties imposed by the corporation.

The messy and highly charged business of distributing compensatory funds from industry to a range of stakeholders is downloaded to the affected residents themselves. In the process, community representatives serving on various implementing and advisory committees become accountable to the corporation in the case of disputes. Thus, in the Shell Nigeria GMOU, problematic distribution questions are transferred to cluster boards that are intended to 'represent communities': 'The Cluster Development Board representing the communities determines the sharing formulae based on the existing local understanding, with support from government and SPDC' (SPDC Nigeria GMOU, as described above; see also Zalik 2011).

Ongoing global and local social protest and legal claims have shaped the form and content of the GMOU model in Nigeria. Yet to date, the model reflects a top-down approach to corporate-community negotiation in which the payment of sitting allowances to multi-stakeholder participants is made via industrial coffers, and the language employed in the memoranda is weighted legally on the side of the corporation.

For these reasons, one might well conclude that privately endorsed structures like the GMOU or the IRCs merely provide a paper 'social licence to operate' for industry. As described by one interviewee:

> In the case of the IRCs, Industry uses/presents in the EA/project approvals process, the fact of their participation in the ATC and the funding of the IRCs as a 'mitigation' or means of addressing social impacts. Alberta seems to accept this as sufficient, although how one can say that funding a consultation process mitigates social impacts (which should, in theory be identified through the EA and consultation processes) is questionable. This 'paper' license to operate works in Alberta because the Alberta government itself does not require anything greater (written comment to author, May 2013).

While social negotiations or substantive consultation with affected populations may be absent, the institutional processes and bureaucratic structures that they mandate and support have consequences for the empowerment, development, and income of affected communities.[25]

Conclusion

In a review of the duty to consult and accommodate in the Canadian context, Laurence Sossin argues that 'while procedural justice holds considerable promise, its limitations increase as time passes without significant procedural enhancements such as the Crown's "duty to consult and accommodate" Aboriginal communities leading to more just outcomes' (Sossin 2009, 3). This chapter suggests that the reshaping of corporate-social relations via oil industry negotiation models reveals such limitations – one local and one global.

The local limitation to procedural justice involves shifting territorial identification so that it corresponds to that of a particular industrial *oil field* or *corporate project*, thus reshaping the interests of the local 'public,' or community residents, to reflect the naming employed by the private firm. Indeed, the clusters and the project advisory committee created within the Nigerian GMOU process change the way the population has laid claim to its natural resource wealth. Similarly, the IRC stakeholder model, operating through the Athabasca Tribal Council, seeks to formalize – although not always with success – company-to-band agreements on *specific projects*. The proposed Consultation and Regional Benefits Agreement (CRBA) sought to advance government and industry's interest in a 'non-obstructionist' operating environment. In Canada, 'obstructionism' is more likely to take the form of litigation than of direct action, although the rise of the Idle No More movement in 2012 expressed First Nations resistance to the anti-democratic nature of both legislative and judicial processes. While the Nigerian GMOU discourse claims that GMOU 'communities are united either on historical (clan) or local government basis as approved by the relevant state government,' key civil society informants indicate that in various cases only a handful of 'host communities' – out of fifteen or more pertaining to a particular clan – have been included in the committee structure for a particular GMOU.[26]

Thus, in both the GMOU and IRC cases, the negotiating process – despite claims to inclusiveness and equal treatment – creates further fragmentation both between and within industry-affected communities. Importantly, the processes in both Canada and Nigeria also aim to keep corporate-community disputes out of the courts, aiming to prevent the time delays that negatively affect the corporate bottom line. Communities are offered a quid pro quo for foregoing protest or litigation that takes the form of business contracts for their members, employment in the oil and gas project, and construction of needed infrastructure.

A 'global implication' of the IRC and GMOU models involves the 'institutional' merging of public and private interests, facilitating the portrayal of the multinational oil company as 'socially responsible' in its public, internationally available discourse. This is exemplified in both sites through strategic partnerships involving industry, state institutions, and non-governmental organizations. Indeed, Shell Nigeria indicates that the GMOU is part of its 'sustainable community development' effort. Emerging from the oil industry's globalized endorsement of corporate-driven sustainable development, the good corporation is portrayed in industrial public affairs pamphlets as a central actor reshaping or responding to socio-environmental norms in each region. Corporate actors are discursively constructed as promoting 'sustainable development' in both Alberta and the Niger Delta, in a form that ties the health of local economies to oil industry contracts and the maintenance of a non-'obstructionist' operating environment in which physical protest is restricted or forbidden.

NOTES

1 Section 35 provides for the recognition and affirmation of the existing treaty and Aboriginal rights of the Aboriginal peoples of Canada.
2 Pieces by Professor Nigel Bankes and his colleagues may be consulted at ablawg.ca. These include critical commentaries on judicial decisions vis-à-vis the Athabasca Chipewyan First Nation's challenge to Shell Oil; see acfnchallenge.wordpress.com (Zalik 2015).
3 The Mikisew Cree First Nation employs the name GIR (Government and Industry Relations) to better depict the work of the institution.
4 The contours of such relations were similarly described in a July 2008 interview by the author with Fred Lennarson, who has extensively researched the Lubicon Cree case and has acted as an advisor to the Lubicon Cree.
5 The anonymity of these interviewees is protected, as per the project's ethics requirements.
6 Recent contentious, proposed projects include Shell's coal-bed methane projects, the Enbridge Gateway pipeline project, and, to a lesser extent, LNG (of which one major project has been approved near Kitimat, BC). A representative of an environmental NGO interviewed by the author in northern British Columbia in October 2009, when discussing the proposed Enbridge Gateway pipeline from the tar sands through the BC interior, suggested that the companies did not realize what they were up against in BC: 'this isn't Alberta, this is "NO" country.'
7 That said, the International Maritime Organization, a major industry-dominated association, depicts the Nigerian offshore as a highly risky location – although production in the offshore has often proceeded when onshore production has been shut-in (see Zalik 2009).
8 A notable case was the successful law suit against Shell (Onoh 2006).
9 There are a number of joint venture operations throughout Nigeria. The main operators onshore or in the shallow offshore are Shell, ChevronTexaco (previously two separate firms), Agip, and TotalFinaElf. Exxon Mobil is an important operator in the deep offshore. In all oil and gas projects the Nigerian National Petroleum Corporation holds a majority stake but the operations are carried out largely by the private transnationals.
10 Membership in the region is estimated at approximately five thousand. The full ATC mission statement as spelled out in the APCA is as follows: 'Enhance and promote the general well-being of our people by providing programs, services and opportunities; Foster the growth, prosperity and development of First Nations communities through capacity building;

Maintain and protect our Treaty rights and freedoms; Promote, maintain and protect the integrity of our relationship with Mother Earth, the land, water, ice, air, and resources; Promote and protect our origins, territories, environment, culture, customs, history and languages as First Nations peoples; Work together in harmony and unity, supporting each other politically, socially, economically and culturally; Develop meaningful and productive relationships with our stakeholders' (APCA 2003, schedule A).

11 For instance, the Northern Alberta Institute of Technology (NAIT) has offered circulating training programs to First Nations.

12 Key interviewee communication with author, 2013.

13 Literature on risk is most salient here. Lack of 'certainty' arising from (approximately five hundred) unresolved Aboriginal land claims in Canada is cited in the business literature as a factor discouraging industrial development and investment. In Alberta, however, industrial representatives indicate that land claims uncertainty does not pose the challenges it does in BC (author interview with key informants) because of the existence of the historical number of treaties that provided, at least on their face, for the surrender of all land rights, in return for treaty rights and Indian reserves. Rather, the question on the table and 'always contested' concerns the distribution of resources from development (industry representative, interview by J. Fishlock, July 2008).

14 Resulting industry cutbacks are likely to only deepen the general lack of ecological remediation in the region. Critics point out that in the forty-one-year period of oil sands development, only 0.2 per cent of damaged land has been reclaimed.

15 Syncrude's sustainability report cites having held 263 'consultative activities' with members of Aboriginal communities.

16 An industry representative stated that when it came to neighbouring Métis communities, they wished to deal with everyone together, so as not prompt divisiveness. Of course, unity and disunity among colonized groups is employed as a tool by different sectors seeking support – both industry and sectors of the environmental movement.

17 It is important to note that the term 'First Nations' may be preferred by industry/government because it is seen to exclude Métis populations and therefore minimizes their obligations to a wider Aboriginal population.

18 Shell and Chevron community briefing documents, 2006, provided to author by informants in the Niger Delta.

19 A region in which the author has conducted research saw an ultimatum issued collectively by the youth associations of three communities, related

to the fact that basic infrastructure – promised via the inter-community mediation – had yet to be completed. 'In the six-point petition, the groups stated that a lump sum of N15 million (US $115,000) be paid to each of the three youth associations to prevent an impending feud between the oil company and the host communities due to the oil giant's insensitivity and neglect' (Mike Oduniyi, quoted in Lagos 2005).

20 In the case of the Chevron model in Delta state, a new set of community 'foundations ' are to be established through the GMOU, which are also to seek funds from non-governmental and multilateral sources, including the World Bank and private donors. This, explained a corporate representative, aims to address the fact that industry cannot pay for all the region's development needs. These new foundation-like bodies will operate alongside government-affiliated institutions, but in their initial establishment are directly tied to particular oil field projects, with participant-representatives paid for their time via direct sitting allowances from the corporation.

21 The parties to this agreement cannot be identified, as the sources requested confidentiality.

22 Italics added. It should be noted that 'best endeavours' may be considered a fairly high level of obligation in legal discourse; nevertheless, the varying level of stringency in requirements facing the communities versus the firm is notable.

23 SPDC states, for instance, that the GMOU's key aspects include 'involvement of all production, assets and pipeline communities (Whole Community Concept); encouragement of people-based socio-economic development plans at communities and cluster levels; encouraged focus on Community Development Plan and Local Economic Empowerment and Development Strategy (LEEDS) as a basis for projects.'

24 'N5' is presumably a typographic error that should read N5000 or 50,000.

25 In the context of volatility in Nigeria and strategic mobilization of activists and industry in Alberta, these dynamics require ongoing investigation. One direction, were researchers able to access a forthright account, would be via industry dissemination (at corporate conferences) of the results of these community engagement activities; another would be detailed community and organizational ethnographies and observation to assess their implications to date.

26 For instance, the 2006 draft GMOU in the Gbarain Ekpetiama area, extending over the Local Government Area (LGA) of the same name as well as parts of Yenagoa LGA, states that it 'covers the range of SPDC's activities within Gbarain and Ekpetiama under the area impacted by the Gbarain Integrated Oil and Gas Project.'

REFERENCES

Acuña, Ricardo. 2007. 'Read between Encana's Lines.' *Calgary Herald*, 7 October.

Adkin, Laurie E. 1998. *Politics of Sustainable Development: Citizens, Unions and the Corporations*. Montreal: Black Rose.

Adkin, Laurie E. 2009. 'Democracy from the Trenches: Environmental Conflicts and Ecological Citizenship.' In *Environmental Conflict and Democracy in Canada*, ed. Laurie E. Adkin, 298–319. Vancouver: UBC Press.

All Parties Core Agreement. 2003. Athabasca Tribal Council.

Allub, Leopoldo. 1983. 'Heterogeneidad Estructural, Desigualdad Social y Privación Relativa En Regiones Petroleras.' *Revista Mexicana de Sociología* 65 (1): 169–90.

Amin, Ash. 2004. 'Regulating Economic Globalization.' *Transactions of the Institute of British Geographers* (29): 217–33.

Auty, Richard. 1993. *Sustaining Development in Mineral Economies: The Resource Curse Thesis*. London: Routledge.

Barker, Andrew. 2008. 'UK Offers Nigeria Help to Train Security Forces.' *Financial Times*, 16 July.

Bauman, Zygmunt. 1998. *Globalization: The Human Consequences*. New York: Columbia University Press.

Bernstein, Steven. 2001. *The Compromise of Liberal Environmentalism*. New York: Columbia University Press.

Bond, Patrick. 2006. 'Resource Extraction and African Underdevelopment.' *Capitalism Nature Socialism* 17 (2): 5–25.

Boychuck, Regan. 2010. 'Searching for a Reason for the Tories' Royalty Cuts.' *Edmonton Journal*, 3 April.

Bridge, Gavin.2008. 'Global Production Networks and the Extractive Sector: Governing Resource-Based Development.' *Journal of Economic Geography* 8: 389–419.

Bridge, Gavin, and Andrew Wood. 2005. 'Geographies of Knowledge, Practices of Globalization: Learning from the Oil Exploration and Production Industry.' *Area* 32 (2): 199–208.

Castree, Noel. 2004. 'Differential Geographies: Place, Indigenous Rights and 'Local' Resources.' *Political Geography* 23 (2): 133–67.

Chastko, P.A. 2004. *Developing Alberta's Oil Sands: From Karl Clark to Kyoto*. Calgary: University of Calgary Press.

Davidson, Debra J., and Norah A. MacKendrick. 2004. 'All Dressed Up with Nowhere to Go: The Discourse of Ecological Modernization in Alberta, Canada.' *Canadian Review of Sociology & Anthropology* 41 (1): 47–65.

Dyer, Simon, Jeremy Moorhouse, Katie Laufenberg, and Rob Powell. 2008. *Under-Mining the Environment: The Oil Sands Report Card*. Drayton Valley, AB: Pembina Institute/WWF.Fluet, Colette, and Naomi Krogman. 2009. 'The Limits of Integrated Resource Management in Alberta for Aboriginal and Environmental Groups: The Northeast Slopes Sustainable Resource and Environmental Management Strategy.' In *Environmental Conflict and Democracy in Canada*, ed. Laurie E. Adkin, 123–39. Vancouver: UBC Press.

Frynas, George, Matthias P. Beck, and Kamel Mellahi. 2000. 'Maintaining Corporate Dominance after Decolonization: The "First Mover" Advantage of Shell in Nigeria.' *Review of African Political Economy* 27 (85): 407–25.

Gibson, Diana. 2007. *The Spoils of the Boom: Incomes, Profits and Poverty in Alberta*. Edmonton: Parkland Institute.

Gibson, R.B. 2012. 'In Full Retreat: The Canadian Government's New Environmental Assessment Law Undoes Decades of Progress.' *Impact Assessment and Project Appraisal* 30 (3): 179–88.

Harrison, Trevor, ed. 2005. *The Return of the Trojan Horse: Alberta and the New World (Dis)order*. Montreal: Black Rose.

Hayes, Ali. 2010. "No Olympics on Stolen Land": Neoliberal Reterritorialization, Colonial Dispossession and the Vancouver 2010 Winter Games. *MES Major Research Paper.* Toronto: Faculty of Environmental Studies, York University.

Idemudia, U. 2009. 'Oil Extraction and Poverty Reduction in the Niger Delta: A Critical Examination of Partnership Initiatives.' *Journal of Business Ethics* 90 (1): 91–116.

Labban, Mazen. 2008. *Space, Oil and Capital*. New York: Routledge.

Laforce, Myriam, Ugo Le Pointe, and H. Lebuis. 2009. 'Mining Sector Regulation in Quebec and Canada.' *Studies in Political Economy* 84: 47–78.

Lapin, Deirdre. 2000. *The Leveraged Buy-in: Creating an Enabling Environment for Business through Strategic Social Investments*. Richardson, TX: Society for Petroleum Engineers.

Lipschutz, Ronnie. 2005. 'Power, Politics and Global Civil Society.' *Millennium* 33 (3): 747–69.

MacKendrick, Norah, and Deborah Davidson. 2007. 'State-Capital Relations in Voluntary Environmental Improvement.' *Current Sociology* 55 (5): 674–95.

McCarthy, James, and Scott Prudham. 2004. 'Neoliberal Nature and the Nature of Neoliberalism.' *Geoforum* 35: 275–83.

McCullum, Hugh. 2005. *Fuelling Fortress America: A Report on the Athabasca Tar Sands and U.S. Demands for Canada's Energy*. Ottawa: Canadian Centre for Policy Alternatives/Parkland Institute/Polaris Institute.

McKillop, Jennifer. 2002. 'Toward Culturally Appropriate Consultation: An Approach for Fort McKay First Nation.' Master's thesis, University of Calgary.

McMichael, P. 1990. 'Incorporated Comparison in a World Historical Perspective: An Alternative Comparative Method.' *American Sociological Review* 55 (3): 385–97.

Miller, Byron. 2007. 'Modes of Governance, Modes of Resistance: Contesting Neo-Liberalism in Calgary.' In *Contesting Neo-Liberalism: Urban Frontiers*, ed. J. Helga Leither, James Peck, and Eric Sheppard, 223–49. New York: Guildford.

Mommer, Bernardo. 2002. *Global Oil and the Nation State*. Oxford: Oxford University Press.

Nader, L. 1972. 'Up the Anthropologist: Perspectives Gained from Studying Up.' In *Reinventing Anthropology*, ed. Dell Hymes, 284–311. New York: Pantheon.

Nader, L., and E. Grande. 2002. 'From the Trenches and the Towers: Current Illusions and Delusions about Conflict Management – In Africa and Elsewhere.' *Law and Social Inquiry* 27: 573–94.

Natcher, David. 2001. 'Land Use Research and the Duty to Consult: A Misrepresentation of the Aboriginal Landscape.' *Land Use Policy* 18: 113–22.

Obi, Cyril. 2008. 'Enter the Dragon? Chinese Oil Companies and Resistance in the Niger Delta.' *Review of African Political Economy* 117: 414–34.

O'Faircheallaigh, Ciaran. 2007. 'Environmental Agreements, EIA Follow-up and Aboriginal Participation in Environmental Management: The Canadian Experience.' *Environmental Impact Assessment Review* 27: 319–42.

Okonta, Ike. 2008. *When Citizens Revolt: Nigerian Elites, Big Oil and the Ogoni Struggle for Self Determination*. Trenton: Africa World Press.

Okonta, Ike, and Oronto Douglas. 2001. *Where Vultures Feast: Shell, Human Rights and Oil in the Niger Delta*. San Francisco: Sierra Club.

Omeje, Kenneth. 2006. *High Stakes and Stakeholders: Oil Conflict and Security in Nigeria*. Burlington: Ashgate.

O'Rourke, Dara, and Stephen Connolly. 2003. 'Just Oil? The Distribution of Environmental and Social Impacts of Oil Production and Consumption.' *Annual Review of Environmental Resources* 28: 587–616.

Osaghae, E. 1995. 'The Ogoni Uprising: Oil Politics, Minority Agitation and the Future of the Nigerian State.' *African Affairs* 94 (375): 325–44.

Pearce, F., & Snider, L. 1997. 'Corporate Crime, Contemporary Debates.' *Capital & Class* 21 (2): 213–14.

Podur, Justin. 2008. 'Canada's Latest Political Prisoners.' *The Bullet*, no. 95, 31 March. http://podur.org/node/613.

Rose, Nicholas. 1996. 'The Death of the Social: Refiguring the Territory of Government.' *Economy and Society* 25 (3): 327–56.
Rose, Nicholas. 1999. *Powers of Freedom*. Cambridge: Cambridge University Press.
Schnarch, Brian. 2004. '"Ownership, Control, Access, Possession" or Self Determination Applied to Research.' *Journal of Aboriginal Health* 1 (1): 33–45.
Seidman, Gaye. 2003. 'Monitoring Multinationals: Lessons from the Anti-Apartheid Period.' *Politics and Society* 31 (3): 381–406.
Seymour, Frances. 1998. 'Go with the Flows? What the Changing Landscape of Development Finance Means for the Public Interest Community.' International Financial Flows and Environment Project. Washington, DC: World Resources Institute.
Shell. 2007. *Annual Report for the Year Ending December 31, 2007*. http://www.shell.com/home/content/investor/financial_information/annual_reports_and_publications/archive/2007/ (accessed 26 March 2012).
Slowey, G. 2008. *Navigating Neoliberalism: Self-Determination and the Mikisew Cree First Nation*. Vancouver: UBC Press.
Soederberg, Susanne. 2003. 'The Promotion of "Anglo-American" Corporate Governance in the South: Who Benefits from the New International Standard?' *Third World Quarterly* 24 (1): 7–27.
Sossin, Lorne. 2009. 'Duty to Consult and Accommodate: Procedural Justice as Aboriginal Rights.' Paper presented at Osgoode Hall Law School 5th Annual National Forum on Administrative Law and Practice, Toronto, 16 October.
Spivak, Gayatri Chakravorty. 2006 [1988]. 'Subaltern Studies: Deconstructing Historiography.' Chap. 12 in *In Other Worlds: Essays in Cultural Politics*, 270–305. New York: Routledge.
Szablowski, David. 2007. *Transnational Law and Local Struggles: Mining, Communities and the World Bank*. Oxford: Hart.
Szablowski, David. 2010. 'Operationalizing Free, Prior, and Informed Consent in the Extractive Industry Sector? Examining the Challenges of a Negotiated Model of Justice.' *Canadian Journal of Development Studies* 30 (1–2): 111–30.
Tickell, Andrew, and Jamie Peck. 1995. 'Social Regulation after Fordism: Regulation Theory, Neoliberalism and the Global-Local Nexus.' *Economy and Society* 24 (3): 357–86.
Treacy, Heather, Tara Campbell, and Jamie Dickson. 2007. 'The Current State of Law in Canada on Crown Obligations to Consult and Accommodate Aboriginal Interests in Resource Development.' *Alberta Law Review* 44 (3): 571–618.
Turner, Terisa E. 1997. 'Oil Workers and Oil Communities in Africa: Nigerian

Women and Grassroots Environmentalism.' *Labour, Capital and Society* 30 (1): 66–89.

Ukiwo, Ukoha. 2007. 'From "Pirates" to "Militants": A Historical Perspective on Anti-State and Anti-Oil Company Mobilization among the Ijaw of Warri, Western Niger Delta.' *African Affairs* 106 (425): 587–610.

Vitalis, Robert. 2002. 'Black Gold, White Crude: An Essay in American Exceptionalism, Hierarchy and Hegemony in the Gulf.' *Diplomatic History:* 185–213.

Vlavianos, Nickie. 2004. 'Human Rights, Health and Resource Development in Alberta: A Summary of Current and Emerging Law.' *Resources,* no. 85 (Winter). Calgary: Canadian Institute of Resources Law.

Vlavianos, Nickie. 2006. 'Albertans' Concerns About Health Impacts and Oil and Gas Development: A Summary.' Human rights paper no. 3 (February). Calgary: Canadian Institute of Resources Law.

Vlavianos, Nickie. 2007. 'Public Participation and the Disposition of Oil and Gas Rights in Alberta.' *Journal of Environmental Law and Practice* 17(3): 205.

Watts, Michael.2004. 'Resource Curse? Governmentality, Oil and Power in the Niger Delta, Nigeria.' *Geopolitics* 9 (1): 50–80.

Westman, C.N. 2013. 'Cautionary Tales: Making and Breaking Community in the Oil Sands Region.' *Canadian Journal of Sociology* 38 (2): 211–32.

Wheeler, David, Heike Fabig, and Richard Boehle. 2002. 'Paradoxes and Dilemmas for Stakeholder Responsive Firms in the Extractive Sector: Lessons from the Case of Shell and the Ogoni.' *Journal of Business Ethics* 39: 297–318.

Woolford, Andrew J. 2005. *Between Justice and Certainty: The British Columbia Treaty Process*. Vancouver: UBC Press.

Woolford, Andrew J. 2009. 'Ontological Destruction: Genocide and Canadian Aboriginal Peoples.' *Genocide Studies and Prevention* 4 (1): 81–97.

Woynillowicz, Dan, and Marlo Raynolds. 2005. *Oil Sands Fever: The Environmental Implications of Canada's Oil Sands Rush*. Drayton Valley, AB: Pembina Institute.

Woynillowicz, Dan, and Chris Severson-Baker. 2006. 'Down to the Last Drop: The Athabasca River and Oil Sands.' Oil sands issue paper no. 1. Calgary: Pembina Institute.

Yergin, Daniel. 1992. *The Prize: The Epic Quest for Oil, Money and Power.* New York: Free Press.

Zalik, Anna. 2004. 'The Niger Delta: Petro-Violence and Partnership Development.' *Review of African Political Economy* 31(101): 410–24.

Zalik, Anna. 2011. 'Labelling Oil, Contesting Governance: Legaloil.com, the GMOU and Profiteering in the Niger Delta.' In *Oil and Insurgency in the*

Niger Delta, ed. C. Obi and S.A. Rustaad, 184–99. New York: Zed/Nordic Africa Institute.

Zalik, Anna. 2015. Resource sterilization: reserve replacement, financial risk, and environmental review in Canada's tar sands. *Environment and Planning A*, 47 (12): 2446-2464.

12 'All Against the Haul': The Long Road to the Athabasca Tar Sands

LAURIE E. ADKIN AND BENJAMIN COURTEAU

Imperial Oil was not prepared, in 2010, for the widespread and determined opposition that greeted its plan to transport Korean-made components for the construction of its new Kearl oil sands mine in the Athabasca region[1] through Idaho and Montana. An energy and resources fund manager for Sentry Select Capital Corporation, based in Toronto, commented that oil company executives were more accustomed to public opposition being stirred up by 'Greenpeace, carbon dioxide and dead birds' (Laura Lau, quoted in Carrie Tait 2012).[2] Yet this sparsely populated region, comprised of high mountains and deep river valleys, became an important front in the environmental campaign against the Alberta tar sands. This chapter traces the development of this campaign, focusing on the discursive struggles to frame what was at stake, and offering an explanation for the outcomes that takes into account historical and institutional factors. We ask what this movement means for the international political-ecological movement that has mobilized around reliance on non-conventional fossil fuels.

The 'All Against the Haul' movement was rooted in Missoula's local tradition of grassroots environmental activism. Yet it is also part of a far-reaching web of sites of opposition to 'fossil capitalism.' Challenges to fossil capitalism have mushroomed across North America, encompassing opposition to proposed oil pipeline routes, the transport of megaloads of tar sands equipment on state highways, the expansion of refineries in the Great Lakes states, subsidies to the coal industry, and many other campaigns. An industrial development as massive as the Alberta tar sands has a footprint that far exceeds the physical limits of the mining areas. As other chapters in this book describe, major battles are underway to secure transportation routes for oil and gas from their points of extraction to refineries and markets as well as to get solvents to the sites of bitumen extraction. This chapter focuses attention on a segment of the arterial web in which fossil fuel extraction in Alberta's tar sands is situated. We report here on opposition in the United States to the local and global costs of maintaining and expanding tar sands exploitation in Canada (which in turn feeds fossil fuel consumption in the United States). The opposition to the 'high and wide' shipment route for oil sands equipment through Idaho and Montana exemplifies the ways in which the Albertan/Canadian 'energy superpower' project is vulnerable to multiple disruptions and challenges. This movement constitutes part of a broader movement that calls upon citizens and policy makers to rethink North America's energy future.

The Conflict and the Stakes

Imperial Oil, a subsidiary of Exxon-Mobil, has planned for many years to mine bitumen from the Kearl tar sands, located on the northeast quadrant of the Athabasca tar sands. The company obtained final environmental approvals for this new mine in 2008 (following a legal challenge of its initial permits by Canadian ENGOs), and aimed to get construction underway soon thereafter. The conflict with civil society organizations in the USA began with Imperial Oil's proposal to transport over two hundred giant pieces of equipment for bitumen processing (machinery, specialized pressure vessels, and heat exchangers – referred to in reports as 'modules' over nine metres in height – from South Korea across the Pacific Ocean to the Port of Vancouver, Washington, and then on to its Kearl mine site (Gilbert 2011a) (see figure 12.1).

Greenhouse gas emissions linked to the extraction of bitumen from the oil sands and the burning of fossil fuels were among the concerns motivating ENGO opposition. In addition, the Northern Rockies contain some of the largest intact ecosystems in the western United States, and there were concerns about the local impacts on waterways, wildlife habitat, and Aboriginal and recreational uses of the area that would result from road-widening and heavy traffic. Imperial Oil's proposal was followed by one from ConocoPhillips, seeking approval for the transportation of oversized loads (pieces of coke drums) from the Port of Lewiston in Idaho to its refinery in Billings, Montana.

Imperial Oil's Kearl Module Transportation Project (KMTP) required two hundred truckloads weighing up to 580,000 pounds each, and

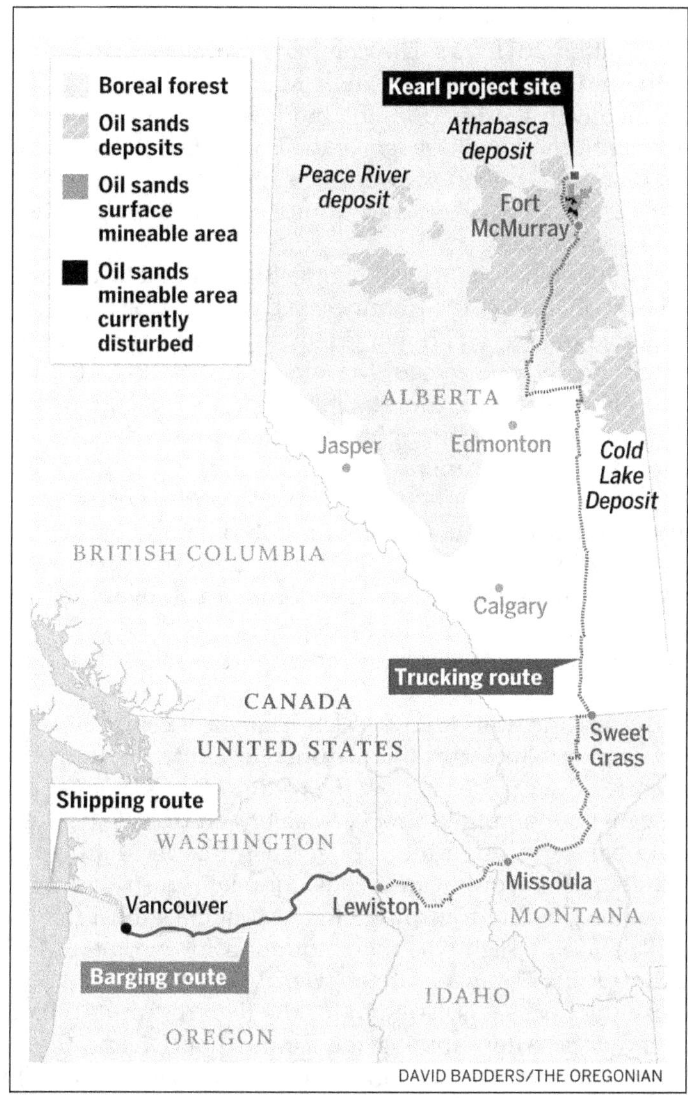

Figure 12.1. Map of the transportation route for the Kearl oil sands mine modules. (*Image*: David Badders, *The Oregonian*, 20 March 2010. http://www.oregonlive.com/portland/index.ssf/2011/03/native_american_tribes_gather.html. Accessed 28 March 2011. © Oregonian Publishing Co. Reprinted with permission.)

measuring up to 210 feet long and 24 feet wide (Montana Department of Transportation 2011, 25). The company's preferred route for transporting the modules passed through Idaho and Montana.[3] From the port of Vancouver, Washington, the modules were to be sent by barge up the Columbia and Snake rivers to the Port of Lewiston, Idaho. Imperial Oil's plan was to then transport the modules by tractor trailer up Highway 12, over Lolo Pass into Montana, and through Missoula to Montana Highway 200.

Highway 200 snakes its way along the Big Blackfoot River to Roger's Pass and from there over the Continental Divide to US Highway 287. US 287 cuts across the region known locally as the Rocky Mountain Front, to US Highway 89 at Choteau, MT, then past the Blackfoot Reservation, until finally entering Alberta at the Port of Sweetgrass, Montana. This winding route runs through some of the most wild and culturally significant parts of the American West. From the Nez Perce Nation in Idaho to Missoula, the route hugs the banks of the Lochsa, a National Wild and Scenic River and a centre for both river and backcountry recreation. In addition, this is the historic trade route between the Nez Perce and the Salish, the path of the Corps of Discovery led by Lewis and Clark, and the route of Chief Joseph and his band of Nez Perce as they fled the US army in a doomed attempt to reach British territory in 1877.

All of the historic trails found on US Highway12 skirt the edge of the largest contiguous forest in the continental United States. Atop Lolo Pass, a centre of winter recreation, the route enters Montana. Down the winding canyon along Lolo Creek, a celebrated trout fishery, the route eventually passes through the City of Missoula. In Missoula the high and wide modules would cause considerable infrastructure issues for that community before transiting along Montana's most famous river, the Big Blackfoot, which was the inspiration for Norman MacClean's book *A River Runs Through It*. It, too, is a National Wild and Scenic River and, like the Lochsa, is essential for the region's multi-billion-dollar tourism industry. After crossing the Continental Divide, the modules would enter Montana's ecological heartland, the Rocky Mountain Front. Home to North America's last remaining wild prairie grizzlies and intact prairie/mountain ecosystem, 'the front' (as it's known to Montanans) symbolizes the state's natural heritage and culture.

Substantial modifications to infrastructure would be required to allow the passage of the shipments. In order to adhere to state traffic laws that forbid shipment-related traffic delays of more than fifteen minutes, the slow-moving vehicles required turnouts along the roadways. In Montana this meant the construction or expansion of seventy-five turnouts (Montana Department of Transport 2010, 21). In Idaho, Imperial Oil had already, by 2010, improved at least nine turnouts on US HWY 12 and had identified at least fifty more to ensure limited delays (Idaho Transportation Department 2010).

In Missoula and other communities along the route, any object (utilities, traffic signals, street lighting, signs, and trees) obstructing the movement of the large shipments would need to be moved or altered. No assessment was made in Montana or Idaho of wear and tear to road surfaces or bridges; these would likely be ongoing costs to the state because the oil sands shipments would effectively create a long-term industrial shipment corridor between the Port of Lewiston and the Alberta tar sands.

Opponents were also concerned about the implications for human rights and political transparency of local dependence upon short-term Imperial Oil investment for economic vitality – particularly in light of Montana's historical experience of resource-based boom and bust. Lastly, the consequences of the bitumen mining operations for both human and planetary health made these shipments targets of environmental opposition.

In June 2012 a significant victory was won by local communities and environmental NGOs opposing the KMTP. Following a series of legal challenges and delays in obtaining permits in Montana and Idaho, Imperial Oil decided to withdraw its applications for permits to use its preferred route. Instead, the corporation would break down into smaller parts ('disassemble') the Korean-made modules and transport these loads via other highways. In the face of delays, in November 2011 Imperial Oil had already applied for and received three hundred permits from the Montana Department of Transportation to ship these smaller loads using an alternative route (interstate routes 90 and 15) (Gilbert 2011). Thus, the diverse coalition of businesses, citizens' groups, and ENGOs succeeded in thwarting Imperial Oil's access to a particularly environmentally sensitive passage to Alberta's tar sands, and in increasing the costs of the Kearl mine project for Imperial Oil.[4] However, there is concern that Imperial Oil (or perhaps other companies) may try again to secure access to this route.[5] Citizens involved in the 2010–12 battle with the oil giant continue to watch developments closely.[6] Moreover, some activist groups have continued to try to obstruct Imperial Oil's heavy load shipments as they make their way from Port Lewiston to the Canadian border.

The following sections are intended to shed light on the roots of environmental activism in Montana and its specific characteristics. We see, first, that Montana has a long history of destructive resource extraction by corporate monopolies. Like Alberta, it can in many ways be characterized as a rentier state. But despite the significant dependence of the population on employment and revenue generated by the copper mining industry, Montana has produced a vibrant environmental movement. What factors explain this development?

Montana, the Recovering Resource Colony

From the asbestos contamination at W.R. Grace's Libby mine to the Anaconda Mining Company's operations in Butte, Montanans are still cleaning up after the developments that made its elite extremely wealthy. Like the western territories of California, Nevada, Colorado, and Idaho, originally settled by pioneering placer miners,[7] Montana's first boom came from the mining of gold, silver, and copper. Butte, Montana, in particular, was widely known as 'the richest hill on earth' for its massive quartz veins containing high quantities of copper. Butte's hill was the source of such massive wealth that the hill's owners dominated regional politics for decades. Since the 1870s Montana has been a 'rentier state' (see Carter and Zalik, this volume), only shedding the title beginning in the 1970s. It may now be entering another era of rentierism in the form of coal, oil, and gas development.

The Montana Territory in the 1860s was much like Canada's Northwest Territories and the Yukon today. It was isolated from industrial centres and sparsely populated. The economy was based on fur trading, scattered gold mining camps, logging, and limited farming and ranching. Almost thirty years remained before the first railroads would chug their way over Montana's passes. At the time, the territory did not seem to require such infrastructure, until prospectors discovered copper deposits in the high, subalpine valley now known as Butte-Silverbow. Butte's enormous metal wealth would drive Montana's political destiny through World War II, with disastrous consequences for democracy, the environment, and social justice.

Montana's 'robber barons' – a label given to the lords of capitalist monopolies across the United States at the turn of the twentieth century – were in the copper business. William Andrews Clark and Marcus Daly of the Anaconda Mining Company, through their mining wealth in Butte and Anaconda, led Montana's fledgling political culture into the downward spiral of rentierism and corruption (see Swibold 2006). What Andrew Nikiforuk calls 'the resource curse' (2008, 153) and Terry Lynn Karl (1997) 'the paradox of plenty,' in reference to oil economies, was also true of the copper economy. Historian Michael Malone describes the consequences of a long-lasting political feud between Montana's 'copper kings' Clark and Daly in these words: 'Remote and little known Montana gained in all of this an unenviable reputation as a political and economic pawn – bought, sold and sullied by mining wealth' (1981, 80). From the beginning, Montana's entire political infrastructure was devised and controlled by the copper kings, and in 1889 when Montana graduated from a territory to statehood, a special election was held to approve a state constitution and to elect representatives to a state legislature. 'Rumor had it that the "Big Four" were pouring $300,000 into the contest in an effort to gain control of the state government and of the two [federal] Senate seats' (Malone 1981, 90). Battles over control of the state government, as well as over the prestige associated with winning the seat of the new state capital, raged over the next decade. Daly advocated the move of the capital from Helena to his company town of Anaconda, while Clark allied himself with the powerful business interests centred in Helena.

Malone's description of the extent of the graft during this contest is quite startling. Leading up to the 1894 elections,

> both sides reached out to all corners of the state for support. And the contaminating flow of mining money, which had hitherto funneled mainly into the legislature, now spread out into the corpus of the body politic itself. Nobody will ever really know how much money the copper kings spent in this contest; but if journalist Christopher Connolly's guess of $2,500,000 by Daly and $400,000 by Clark is correct, then they spent over $50 for each vote cast! ... In any case, the money flowed freely, and its effects proved to be both intoxicating and debilitating. (99)

As the Anaconda mine's copper ran low and prices slumped, Butte and Anaconda's heady days of domination faded. This decline was hastened by the nationalization of Anaconda's holdings in Chile following the 1973 election of a socialist government. In 1976 Atlantic Richfield (ARCO) bought the company and its assets in Butte-Anaconda.

Michael Malone's final thought on Butte's lasting political legacy in Montana is poignant:

> Beyond a doubt, Anaconda held to its iron-handed, secretive methods for so many years in large part because of the desperate, traumatic battles through which it passed in its formative years. Just as certainly, the company's antediluvian behavior fed the festering Montana mood of anger, radicalism, and despairing resignation which has persisted in weakening form even into the second half of this century... It corrupted the political culture of an American state and shocked the sensibilities of a nation that was not easily shocked. (217)

The extraction industry in Montana left behind not only a wake of political and economic damage; there are also environmental scars, affecting thousands of square kilometres around the state. The worst of the damage and contamination was left by the mining and smelting activities around Butte and Anaconda. A large quantity of mining waste was washed down the Clark Fork River by a flood in 1908. This flood deposited about 6.6 million cubic yards of sediment containing heavy metals and arsenic behind the Milltown Dam, which created an arsenic plume in the local groundwater and contaminated drinking water (US Environmental Protection Agency 2003, 3–4). For Missoula, just downstream of the dam site, the contaminants were a constant reminder of the long-term consequences of unbridled resource extraction.

Thanks to the American Superfund law, ARCO was required to assume Anaconda's environmental liability – from over one hundred years of mining activities – when it purchased that company. When ARCO filed for bankruptcy in the 1990s, the Clark Fork Coalition, a water quality watchdog group, made sure that the cost of cleaning up the mess was included in the Chapter 11 bankruptcy proceedings. This kept ARCO on the hook under the Environmental Protection Agency's (EPA) Superfund law. This success, among others, emboldened Missoula's lively environmental community.

The birth of the national environmental movement in the 1970s coincided with the weakening hold of the Anaconda Mining Company on state politics. In 1972 voters approved the holding of a constitutional convention to rewrite the state's constitution – a document that had been written under the supervision of the copper kings in 1889. The drafters of the 1972 document, knowing the heavy toll paid by Montanans for the benefit of profit from resource extraction, codified the right to a clean and healthy environment. In article 2, section 3, 'Inalienable Rights,' the current Montana constitution states:

> All persons are born free and have certain inalienable rights. They include the right to a clean and healthful environment and the rights of pursuing life's basic necessities, enjoying and defending their lives and liberties, acquiring, possessing and protecting property, and seeking their safety, health and happiness in all lawful ways. In enjoying these rights, all persons recognize corresponding responsibilities.[8]

The institutionalization of the right to a 'clean and healthful environment' has been an important resource for environmentalists.[9] Montana Supreme Court cases established in 1999 and 2001 that the environmental rights guaranteed in the Montana Constitution were enforceable against public and private actors respectively.[10]

Indeed, the incorporation of environmental rights in Montana's 1972 constitution has been viewed as an obstacle to economic development by business interests in the state, backed by Republican politicians. They have lobbied hard for amendment of the constitution to substantially weaken the environmental rights clause. One such proposal was defeated in a state referendum in November 2010. However, Republicans then sponsored an amendment to the constitution in the form of a house bill (292) tabled in January 2011. This bill proposed to change the phrase 'the right to a clean and healthful environment' to 'the right to a clean, healthful, and economically productive environment.' According to Montana journalist Mike Porco (2011), 'HB 292 has strong support from Republican legislators who believe in abandoning environmental protections to bolster the economy and create jobs. Under the current phrasing, some energy development projects have been denied because they do not adhere to Montana's inalienable rights. HB 292 advocates hope that the amendment will help sway decisions in the favor of energy development.'[11]

However, the legal battles concerning Imperial Oil's KMPT have been fought mainly by appeal to the state of Montana's Environmental Protection Act (MEPA). The Montana Environmental Information Center took the lead in challenging the Montana Department of Transportation's authorization of Imperial Oil's megaloads, arguing that the department had performed an inadequate environmental assessment and was in violation of the MEPA. This case is discussed in more detail below.

An additional important resource for the environmental movement in Montana is the University of Montana, located in Missoula. Much of the intellectual leadership and activist energy in the struggle against the Imperial Oil Kearl Mine megaload shipments has come from the university.

Because of its unique history, its resource-based political economy, the legal resources provided by the state constitution, and the intellectual leadership provided by the University of Montana, the state of Montana has given rise to a substantial environmental movement. These factors also help us to understand the specific character of the environmental opposition to Imperial Oil's transportation of oil-sands-bound megaloads, to which we now turn.

Montana's Environmental Movement

Members of the grassroots alliance fighting Imperial Oil can be placed within three categories: the local and national environmental groups centred in Missoula, Montana (listed in table 12.1); the local residents living along the proposed route, some with environmental group affiliations but most simply locals concerned about the impact the shipments will have on their daily lives and businesses; and the indigenous nations (as they are called in the USA) along the route supported by a continent-wide coalition of indigenous peoples and social justice groups. The environmental groups and the indigenous nations have focused on the Kearl mine module shipments as a symbol of the larger tar sands development and what it means for planetary and local ecosystems as well as for Aboriginal communities in Alberta. Dozens of local organizations have supported the goals of All Against the Haul, but only a few have directly organized grassroots actions. The Northern Rockies Rising Tide, an environmental organization endorsing direct action to stop climate change, took an early lead in organizing community meetings on the issue (Sakariassen 2011). Students at the University of Montana (UM), long known for their support of environmental activism, have been central to the movement. A student organization, UM Climate Action Now (UMCAN), targeted the shipment issue due to the contributions of oil sands mining to climate change.

The Clark Fork Coalition has a long history of success in cleaning up rivers and streams in the Clark Fork Basin. Their involvement in the shipments stems from the potential degradation of water quality from highway modifications. The Clark Fork Coalition is also concerned

Table 12.1 Member Organizations of All Against the Haul

Adventure Cycling Association (MT)
Advocates for the West (ID)
Alliance for the Wild Rockies (MT)
Alternative Energy Resources Organization (AERO) (MT)
American Rivers (MT)
American Whitewater (MT)
Association of NW Steelheaders (OR)
Big Blackfoot Riverkeeper (MT)
Buffalo Field Campaign (MT)
Clark Fork Coalition (MT)
Climate Solutions (MT)
Columbia Riverkeeper (OR)
Community Action for Justice in the Americas (MT)
Conservation Northwest (WA)
Deer Creek Valley Natural Resources Conservation Association (OR)
Dunrovin Ranch (MT)
Earthjustice (MT)
Environment Montana (MT)
Federation of Western Outdoor Clubs (MT)
Fighting Goliath – The Rural People of Highway 12 (ID)
Friends of Lolo Creek and the Blackfoot River (MT)
Friends of the Clearwater (ID)
Friends of the Earth (DC)
Friends of Two Rivers (MT)
Hellgate Hunters and Anglers (MT)
Hells Canyon Preservation Council (OR)
Honor the Earth (MN)
Ice Age Floods Institute (MT)
Idaho Conservation League (ID)
Idaho Rivers United (ID)
Indigenous Environmental Network (MN)
Lewis and Clark Trail Adventures (MT)
Missoula Valley Outfitters (MT)
Montana Audubon (MT)
Montana Chapter of the Sierra Club (MT)
Montana Environmental Information Center (MT)
Montana Environmental Student Alliance (MT)
Montana Trout Unlimited (MT)
National Wildlife Federation (MT)
Native Fish Society (OR)
Natural Resources Defense Council (MT)
Nez Perce Tribe (ID)
No Shipments Network Missoula (MT)
Northern Plains Resource Council (MT)
Northern Rockies Rising Tide (MT)
Northwest Energy Coalition (WA)
Northwest Guides and Anglers Association (OR)
Northwest Steelhead and Salmon Conservation Society (WA)
Orca Network (WA)
Other Nations (MT)
Pacific Coast Federation of Fishermen's Associations (OR)
People for Puget Sound (WA)
Polebridge Mercantile (MT)
Rainforest Action Network (CA)
Robert Gentry Law (MT)
Row Adventures (ID)
Save Our Wild Salmon Coalition (ID)
Sierra Club (MT)
Six Pony Hitch (MT)
Smith's Backhoe Service, Inc. (MT)
Square Dance Center and Campground (MT)
Summer Nelson Law (MT)
The Lands Council (WA)
University of Montana Climate Action Now (MT)
Wapiti River Guides (ID)
Washington Environmental Council (WA)
Western Environmental Law Center (MT)
WestRidge Creative (MT)
Wild Steelhead Coalition (WA)
Women Donors Network (CA)
350.org

Source: All Against the Hall, http://allagainstthehaul.org/the-all

about recreation access along rivers that are of central importance to the region's tourism industry.

National groups fighting the shipments include the National Wildlife Federation, Sierra Club, Friends of the Earth, and the Indigenous Environmental Network. State and national organizations have Missoula offices with paid staff and volunteer resources dedicated to the campaign.

This alliance of environmental groups, local non-natives, and indigenous nations (as they are known in the USA) – first known as the 'No Shipments Network' – evolved from humble beginnings in the back room of the Jeannette Rankin Peace Center in Missoula to the well-organized grassroots entity now known as All Against the Haul.

In the spring of 2010, the plans for the Kearl mine shipments through western Montana became public knowledge. At that time, even with the many active environmental grassroots groups in Missoula, most people in the wider community were not aware of the Alberta tar sands deposits. For this reason, the initial focus of the No Shipments Network was to educate Missoula and the other communities along the route about the environmental and social risks posed by these shipments.

It took several months for the extensive list of groups (approximately eighty) to form a coalition in support of the grassroots movement against the shipments, and to link those shipments to opposition to the tar sands. In the early days, activists met in the Jeannette Rankin Peace Center (founded in memory of the first female member of the US House of Representatives from Missoula).[12] The small meeting room was framed by shelves with books on activism, feminism, and equality. About twenty individuals (veteran activists, young students from the university, young professionals, and a smattering of middle-aged, concerned citizens diving into their first political battle) talked about how hard it was going to be to fight the behemoth Exxon-Mobil. The discussion focused on actions that might bring attention to the issue. Northern Rockies Rising Tide had hosted a rally at the Montana Department of Transportation (MDT) offices in Missoula on 3 June 2010 (Chaney 2010), but the group was thinking of a bigger and more visible demonstration. Holding a protest march on Missoula's busiest thoroughfare, Reserve Street, at rush hour, was suggested. A similar demonstration held years earlier on a different issue had proven highly successful, and so this idea was adopted. It was the first of many high-profile public demonstrations meant to draw attention to the issue, and along with

Figure 12.2. Protest in Missoula against the Imperial Oil megaloads, June 2010. (*Photo*: Tom Bauer, *The Missoulian*, 30 June 2010. http://missoulian.com/news/state-and-regional/article_ce7dfb08-84c2-11df-8a71-001cc4c03286.html.)

the MDT office protest, earned the movement its first front-page and prime-time news coverage (see figure 12.2).

It is fair to say that most people in western Montana and central Idaho now know of the tar sands and the shipments. The level of news coverage from local papers, especially *The Missoulian,* has been substantial. Not since the Milltown Dam removal debate has there been this level of controversy over a local environmental issue. In this regard, the grassroots movement has enjoyed substantial public relations success against the oil giant Exxon-Mobil. Even so, the tar sands shipments issue has become a polarizing topic in some communities hit hard by the recession, as the oil industry continues its steady drum beat of 'jobs, jobs, jobs.'

Since ExxonMobil/Imperial Oil's retreat to an alternative route for its heavy-load shipments in 2011, small groups of activists who view themselves as part of a global movement for climate justice have continued to try to obstruct the shipments moving from the Port of Lewiston, Idaho, to the Canadian border. The activists use social media and alternative media (like Radio Free Moscow, or movement websites) to share information about the shipments bound for the Kearl oil sands mine in Alberta. These activists are associated with Wild Idaho Rising Tide, the Indigenous Environmental Movement, Northern Rockies Rising Tide, Peaceful Uprising, Cascadia Rising Tide, and the youth climate movement, among others.[13] There have been sit-down blockades in the path of the trucks,[14] arrests, and detentions. One woman who was detained during a March 2012 blockade in Moscow, Idaho, said: 'We're not going to be accessories to genocide and climate change and increased cancer rates and all the other ecological damages that the tar sands intends to cause.'[15]

Pro-KMPT Discourse

When Imperial Oil made known its plan to use the Lewiston-to-Sweetgrass route for transportation of its oil sands modules, it stated that the project would generate $67.8 million in economic activity in Montana, including $11.4 million for road modifications and the construction of fifty-three new turnouts, and $21.6 million for utility relocations. The company also claimed that the KMPT would create eighty-two full-time jobs (Sakariassen 2011). The governors of Montana and Idaho welcomed this promise of job creation and economic growth.

Likewise, some owners of businesses in small towns along the route thought they might benefit from the spending of the rig drivers and shipping personnel in hotels and restaurants (Sakariassen 2011). Imperial Oil suggested, as well, that it would be employing highway patrol officers or local sheriffs to escort the loads. Managers and directors of the Ports of Vancouver and Lewiston actively supported the KMPT because of the revenue the project would generate for their enterprises (Sakariassen 2011). A coalition of Idaho- and Montana-based business associations was formed in fall 2010 to support the KMPT, calling itself the Drive Our Economy Task Force (http://www.driveoureconomy.org/). This coalition grew from an initial membership of thirteen to a membership of almost fifty organizations by fall 2011 (see table 12.2).[16]

Drive Our Economy, co-chaired by the Idaho Association of Commerce and Industry and the Motor Carriers of Montana, was formed during the time that the Montana Department of Transportation was considering ExxonMobil/Imperial Oil's application for the megaload permits. At that time, Alex Sakariassen, a Missoula-based journalist who followed the 'heavy haul' conflict from beginning to end, asked what impact 'such a weighty conglomerate of advocates' might have

Table 12.2 Drive Our Economy Task Force Membership

American Waterway Operators	Montana Association of Livestock Auction Markets
Associated General Contractors of Washington	Montana Building Industry Association
Associated Logging Contractors of Idaho	Montana Business Leadership Council
Association of Washington Businesses	Montana Chamber of Commerce
Beartooth Stock Association	Montana Coal Council
Bill Rodgers Lowbed Services	Montana Contractors Association, Co-Chair
Central Washington Home Builders Association	Montana Farm Bureau Federation
Food Producers of Idaho	Montana Farmers Union
Home Builders Association of Tri-Cities	Montana Grain Growers Association
Idaho Association of Commerce & Industry	Montana Logging Association
Idaho Association of General Contractors	Montana Mining Association
Idaho Chamber Alliance	Montana Petroleum Marketers & Convenience Store Association
Idaho Consumer-Owned Utilities Association	Montana Stockgrowers Association
Idaho Cooperative Council	Montana Women Involved in Farm Economics
Idaho Farm Bureau Federation	Montana Wood Products Association
Idaho Grain Producers Association	Motor Carriers of Montana, Co-Chair
Idaho Hay and Forage Association	Northwest Mining Association
Idaho Housing and Finance Association	Port of Lewiston
Idaho Mining Association	Sweetgrass Development
Idaho Petroleum Marketers & Convenience Store Association	Tri-Cities Regional Chamber of Commerce
Idaho Recreation Council	United Property Owners of Montana
Idaho Trucking Association	Valley Vision
Inland Northwest AGC	Washington Trucking Association
Milk Producers of Idaho	Wenatchee Valley Traffic Association
Montana Agri-Women	Western Environmental Trade Association
	Western L0egacy Alliance

Source: http://www.driveoureconomy.org/member-list/ (accessed 21 March 2016)

on the MDT's decision (2010). The 'conglomerate' claimed that the ExxonMobil/Imperial Oil project would result in new investment of about $80 million in the two states (Sakariassen 2010). On its website, the coalition broke down the economic 'benefits' of the project in a way that suggested a substantially larger investment, including:

- $13 million in economic activity for Idaho, the result of $9 million in direct spending;
- $67 million in economic activity for Montana, the result of $40 million in direct spending;

- $35 million in building new turnouts for rural highways in the region;
- $30 million on contracting, trucks, drivers, and state police escorts; and
- $80,000 per month in handling and receiving cargo for the Port of Lewiston.

Sakariassen was struck by the apparent incongruity of the membership in the coalition of the Western Legacy Alliance (WLA) – the sole member with a 'conservation' mandate (Sakariassen 2010). Although the WLA states that its mission is to support and promote 'sustainable land-use solutions to ensure social and economic benefits for local communities and the nation,' an examination of the organization's website (http://westernlegacyalliance.org/) suggests that its core interest is the defence of economic uses of public lands and of private property rights when these come into conflict with state or ENGO-led conservation initiatives.[17]

The Drive Our Economy Task Force adopted a rhetorical strategy very similar to that of the Western Legacy Alliance, characterizing environmental opponents of the KMTP as 'outsiders' with little knowledge of 'what's best for Idaho' and equally little respect for local democracy (Sakariassen 2010). Interestingly, another key strategy has been to frame the conflict as one between the livelihood interest of local citizens (dependent upon free access to the highways) – defended by Drive our Economy – and the 'political' interest of national environmental organizations (purportedly in excluding economic activities from rural areas and transportation routes). At the same time, the opponents of the KMPT were arguing (as we will see below) that the megaloads – and the possible creation of a permanent route for heavy industrial shipments along the states' scenic highways – would *impede* local citizens' use of their roads, as well as the movement of emergency vehicles. Yet according to Drive Our Economy:

> This project is about protecting open access to our roadways. The Drive Our Economy taskforce is principally dedicated to maintaining commercial usage of highways in Montana and Idaho for all industries. By turning the shipping permitting process into a political game and allowing outside pressure to influence these shipments, we're creating an anti-business framework in our state. What's worse, we risk letting national interests highjack local decisions about our economic future. Our highways have important functions for commerce, and we need to maintain those uses. If the Kearl shipments are shut down for political purposes, outside groups opposed to any and all trucking activity could use this precedent

to shut down farming activity, timber production, construction growth, and the movement of other goods critical for our industries. (http://www.driveoureconomy.org/facts/raw-fact/#more-81)

Montana's Governor Schweitzer, a Democrat, similarly took up the job creation and economic growth themes in support of the KMPT, but added a 'national security' rhetoric to the discourse of the project's supporters. In May 2010, as the controversy over the Kearl shipments gathered steam throughout Montana, Schweitzer told a *Missoulian* reporter: 'It's jobs, jobs, jobs. It's $68 million. If somebody else finds $68 million that can be invested in Montana in 2010–2011, I'll give them my home phone number' (quoted in Briggeman 2010c).

Although Montana's economy is underpinned by agriculture (especially wheat), tourism and service industries, and three major cities,[18] fossil fuel extraction and transport continues to be a driving force in state politics. Coal mining and, recently, hydraulic fracturing have been promoted by recent administrations in the name of economic growth and job creation. As a member of the Montana Land Board in 2010, Schweitzer approved the leasing of the state's portion of 1.3 billion tons of coal in southeast Montana's Otter Creek drainage area to the mining company Arch Coal. The executive director of the Montana Environmental Information Center (MEIC), Jim Jenson, claims that burning this coal would 'double Montana's present total carbon dioxide emissions for the next half century' (Jenson 2012). Governor Schweitzer's response to this criticism was that if Montana didn't mine and export the coal, another producer would, echoing the arguments that have been made by Alberta politicians with regard to bitumen exports. The MEIC and Sierra Club have filed an appeal in the Supreme Court of the State of Montana, claiming that the Land Board's 2010 decision 'violates Montana's environmental constitutional provisions, as applied, by exempting the Otter Creek coal leases from environmental review at the stage of the development process that opens the door to mining and burning 1.3 billion tons of coal and resulting environmental harm.'[19]

Exploitation of the portion of the Bakken rock formation that underlies the northeast part of the state has produced a new oil boom (Vock 2011). Governor Schweitzer argued that 'Montana is in the middle of one of the most important energy corridors on the planet, not just because we have wind energy and we have oil and gas and coal, but because our neighbors have all of those as well … In order to produce energy, you need to transfer energy' (quoted in Briggeman 2010c).

The governor's political support for making Montana a key transportation corridor for the oil industry also made use of a 'national security' rationale. Schweitzer, having lived for years in Saudi Arabia as a young man, argued for the need for Americans to wean themselves off foreign oil. In an interview with Canadian Press on 30 September 2010, Schweitzer dismissed the concerns raised by director James Cameron about the effects of the Alberta tar sands on indigenous peoples:

> 'I would say this is conflict-free oil and I don't want to send one more son or daughter from Montana to defend an oil supply from one of these dictators and become dependent on that energy supply,' he said in an interview with The Canadian Press from his office in Helena ... 'For the most part Alberta is the exception. Most places around the world that are oil exporters have some of the most brutal regimes, their people have the fewest civil liberties and the individual citizens gain no particular wealth from the oil. Alberta oil is conflict-free energy' ...
>
> Schweitzer said Cameron needs to put things into perspective. 'You think this [Alberta's tar sand development] is bad? Have you been to Nigeria? Do you see how they produce their crude oil there? They've poisoned every town along every pipeline route. The big rivers and lakes that used to produce large quantities of fish – the fish are all dead ... Their pipelines leak. They've got rebels that are blowing up the pipelines all the time and running [oil] into the rivers. Really? You're going to compare the oil sands to that? Don't joke with me.'
>
> The Montana governor said he is tired of the United States having to prop up dictatorial regimes to guarantee his nation's energy needs. 'Any of these people who say they don't like the oilsands, you ought to ask them if they'll invite you to their house, and unless they're living naked in a cave and eating nuts, they are totally dependent on petrol.
>
> 'That petrol is either going to be produced in places like Montana and North Dakota and Alberta and Saskatchewan or it's going to be produced in places like Venezuela and Saudi Arabia and Nigeria. I would prefer it to come from friends than enemies.'[20]

Similar arguments about US energy security and the relative 'cleanness' of Alberta's tar sands crude have been used by proponents of the Keystone XL pipeline extension through the United States, and by the Government of Alberta (see Adkin and Stares, and Stendie and Adkin, this volume). What such arguments ignore, of course, are the possibilities for alternative solutions to energy demand that do not require a drastic worsening of global warming trends.

Governor Schweitzer's successor, Steve Bullock, also a Democrat, affirmed his support for the oil and coal industries in the state, while making the case for a diversified economy that has long-term resilience. It is not yet clear how the new governor will respond to future demands from corporations for a megaloads corridor through the state, should these arise. However, Bullock supported the Keystone XL pipeline project and did not express public disagreement with Governor Schweitzer's support for the KMPT.

Public Consultation in Montana's Environmental Assessment of the Kearl Oil Sands Shipments Proposal

Montana's Department of Transportation (MDT) chose to prepare a full environmental assessment (EA) of Imperial Oil's proposal, under the Montana Environmental Protection Act (MEPA). Idaho, however, opted not to do so. During the thirty-day comment period, from 13 April 2010 to 14 May 2010, the MDT received comments from approximately 7,200 individuals and organizations via public hearings, letters, email, and telephone (MDT 2011, 6). In Idaho, the Idaho Transportation Department (ITD) received, between 14 June and 14 July 2010, 216 comments on the shipments proposal (ITD 2010, 1). These comments covered a wide range of points and concerns, but for the purposes of this chapter we will focus on the three most common themes.

A reading of the submissions to Idaho's Department of Transport reveals that the foremost concern of persons living along the corridor has been access to their highways. This includes normal access to the corridor by residents and tourists, as well as emergency vehicle use. One resident of Kooskia, Idaho stated:

> This section of US Hwy. 12 is north-central Idaho's only route east. Loss of this route for practically any length of time could be devastating to this area's primary industry – recreation/ travel/tourism ... The route in question consists of almost 100 miles of narrow, winding, mountain roadway that was built in the 1960s. Gargantuan loads such as are planned today were in no way envisioned at that time and the roadway was in no way constructed to handle such loads. The sighting [*sic*] of a suitable crane would be impossible along perhaps 80% of this route – the roadbed and/ or right of way is simply too narrow or too subject to slippage or collapse to handle the weights involved ... Regardless of the wishes of politicians or whoever else helped birth this idea, the people of north central Idaho trust that Idaho's Department of Transportation will demand that viable

> and adequate emergency preparedness and response plans are in existence before any transport permitting takes place. (ITD 2010, 8–9)

Local environmental and safety impacts were also frequently cited as a major concern of citizens. One central Idaho resident wrote:

> Most of the campgrounds are right next to the highway and the trucks roar by at anytime day or night … What kind of future hazards would this precedent produce to the people who live, work, and play along this corridor? As you and other officials make your decision about this license, please consider the effect to the humans, the environment, and to the long-standing economy of this area, not to mention the sacred land of the Nez-perce [*sic*] Indians. (ITD 2010, 13)

In Montana, officials claimed that due to the high number of responses to a public consultation process overseen by the Department of Transportation, individual comments could not be made available to the public. However, the MDT did publish a summary of the feedback, along with its *rebuttals* to the objections raised by citizens. From this source we learn that citizens' concerns included: 'The analysis should have included environmental impacts of the Alberta Oil Sands development'; 'Some commenters expressed concern that the Kearl Module Transportation Project (KMTP) would have negative visual/scenic impacts on Highway 12 and 200'; and 'Some commenters expressed concern that the proposed project would adversely impact Lolo Creek and/or the Blackfoot River' (MDT 2011, 10, 14, 16).

A further issue raised by opponents of the Kearl Module Transportation Project was the refusal of the Government of Montana to produce an environmental impact statement concerning the project. The National Environmental Policy Act and Montana's Environmental Protection Act require the respective federal and state agencies to undertake one of the following actions in response to project applications: a categorical exclusion (CE), an environmental assessment, or an environmental impact statement (EIS). In the first case, no significant environmental impacts are foreseen and no investigation is required. This is the decision that was made by the State of Idaho regarding the KMTP. An EA 'is prepared when it is unclear whether a proposed action may generate impacts that are significant. If the EA determines that the proposed action will have significant impacts, an EIS must be conducted prior to initiation of the proposed action' (MDT 2011, 8). If the EA process finds no evidence of impacts significant enough to require an EIS, the agency releases a 'finding of no significant impact' (FONSI). This is exactly the finding that MDT released in February 2011, paving the way for the issuance of permits for the shipments to pass through Montana. The FONSI usually updates the EA with comments gathered during the public comment period, along with corresponding rebuttals. The relevant agency (in this case, MDT) must cover all potential courses of legal action with sufficient arguments to back up their FONSI. Many of the public comments submitted to the Montana process argued that the MDT should have produced an environmental impact statement (MDT 2011, 8).

The environmental review process is tiring and generally weighted in favour of an agency or contractor. Indeed, the environmental review process specified under the Montana Environmental Protection Act often serves to fortify a decision made before the review process was initiated. The difference between Idaho and Montana is that Montana has a stronger environmental movement; therefore, the MDT felt obliged to offer some opportunity for public input on the KMTP. While this review procedure may be burdensome to intervenors, and its neutrality is in question, it also gives opposition groups one of their only legal routes to victory.

In the spring of 2011, the Montana Environmental Information Centre, the Montana Chapter of the Sierra Club, Missoula County, and the National Wildlife Federation filed a lawsuit with the Missoula County District Court challenging the MDT's permitting decision, arguing that the permits had been approved 'without adequately identifying and analyzing all relevant impacts of the project and without analyzing a reasonable range of alternatives as required by the Montana Environmental Policy Act' (Briggeman 2011). In February 2012, Judge John Larson ruled in favour of MEIC and the other plaintiffs, concluding that MDT had failed to properly consider the environmental impacts of the seventy-five road turnouts Imperial would have to build in order to haul the oversized loads over two-lane mountain highways. MDT did not contemplate whether those turnouts will be temporary or permanent – a consideration important in determining whether the route would become (as its opponents feared) a 'full-fledged industrial corridor for years to come' (Vanderklippe 2011). Moreover, the MDT had failed to consider alternative routes.

Another victory for the shipment opposition was a ruling by an Idaho judge in August 2010 that there was insufficient public input into the shipment permit approval process. A temporary restraining order was issued by Second District judge John Bradbury against the ConocoPhillips heavy loads, based on the possibility that the 'Idaho Department of Transportation may be violating its own regulations.' Laird Lucas, attorney for the plaintiffs (the Fighting Goliath citizens group), added: 'The deeper issue here is, is this roadway the right path? What other alternatives are there? Nobody seems to know, and that's the real problem. We don't think it's reasonable that Highway 12 is the only place you can haul this equipment' (quoted in Dvorak 2010).

A key strategy of the KMPT opponents has been to try to oblige the Montana and Idaho Transportation Departments to produce full environmental impact statements. The opponents were aware that, for Imperial Oil, approval delays are costly and could force the company to reroute the shipments.

Indigenous Peoples' Discourses and Strategies regarding the KMPT

The Nez Perce[21] is one of the indigenous nations along the route, and the only nation whose territory is physically crossed by the trucks carrying the megaloads. In 2010, the Nez Perce General Council passed a resolution on the shipments that was submitted to the Idaho Transportation Department's public consultation process. They objected that 'the project would establish a dangerous and unacceptable precedent in one of the most beautiful and pristine federally protected corridors in the U.S.' and that oil sands mining constitutes 'an environmentally destructive method that will have negative impacts on the First Nations of Alberta' (LaDuke 2011).

As in Canada, indigenous nations also have legal courses of action available to them. They have a special legal status in relation to both state and federal governments, due to treaty obligations. At the federal level, Executive Order 13175, issued 6 November 2000, requires federal agencies to 'establish regular and meaningful consultation and collaboration with tribal officials in development of Federal policies that have tribal implications.' In their consultations, agencies are required to 'respect Indian tribal self-government and sovereignty, honor tribal treaty and other rights, and strive to meet the responsibilities that arise from the unique legal relationships between the Federal Government and Indian tribal governments.' This order refers specifically to federal rather than state-level agencies; however, similar rules apply at the state level, and indigenous nations do have relations with state governments and agencies.

Executive Order 13175 was used in 2010 by the Nez Perce First Nation to overturn a decision by a National Forest Service[22] supervisor to allow utility relocation (contracted by Imperial Oil) within their territory. According to the Nez Perce attorney, Mike Lopez, 'the United States has a responsibility by executive order through its unique relationship with tribes that each federal agency consult with tribes on action that may affect tribal interests, on or off reservations' (quoted in Briggeman 2010). In this case, 'the tribe was not given an opportunity to consult on the project, either under the National Historic Preservation Act or under the government-to-government consultation policy the Forest Service has with tribes' (quoted in Briggeman 2010). The approval was therefore rescinded, and Imperial Oil was obliged to resubmit its request for a permit.

All Against the Haul organizers observed that 'the Kearl Module Transport Project (KMTP) or Heavy Haul proposes to travel through 70 miles of [the Nez Perce] reservation in Idaho. The route also passes through the ancestral territory of the Confederated Salish-Kootenai Tribes and, in northern Montana, the Blackfeet Reservation.'[23] In June 2010, Tribal Council chair E.T. Bud Moran, of the Confederated Salish and Kootenai Tribes (CSKT), sent a letter to the Montana Department of Transportation calling upon the department to reject Imperial Oil's requests for permits,[24] in which he wrote:

> The planned route for these shipments passes directly through the aboriginal territories of the Salish, Pend d'Oreille, and Kootenai people. These areas remain of great importance to our people. Our use of them is guaranteed by treaty ...
>
> From the Lochsa River to the Rocky Mountain Front, literally dozens of our traditional place names line the planned route. Many of these names are rooted in our creation stories, reflecting the spiritual importance of these places. Vital resources for fishing, hunting, and the gathering of food and medicinal plants lie along the route. Two of the most abundant places for one of our most important traditional foods, camas, would be directly traversed by these shipments, at Lolo Pass and in the Potomac Valley ...
>
> The tar sands development, and any facilitation of it, is contrary to the cultural values of the Confederated Salish and Kootenai Tribes. Our elders have always stressed that we must respect the land and waters and ensure that they will be here in good condition for future generations.

All Against the Haul's website drew attention to the effects of tar sands development on Aboriginal communities in Alberta as a further reason to oppose the Kearl Mine shipments.

Conclusions

The opponents of ConocoPhillips' and ExxonMobil/Imperial Oil's megaload transport plans have relied on a combination of local protests along the transportation routes and legal actions to delay shipments and their authorizations. The protests succeeded in generating substantial public awareness of the proposed heavy loads, of the local costs and risks they entailed, and of their connection to the larger issues of oil sands exploitation and global warming. The multiple challenges to the oil companies' permit approvals served to delay the shipments, and in the case of the KMPT, to cause Imperial Oil to abandon its preferred route through Idaho and Montana. Ultimately, Imperial Oil resorted to a plan B, which involved a costly disassembling of the modules shipped from Korea, the transportation of smaller-size loads on interstate highways, and the reassembling of components at their destination. The development of the Kearl oil sands mine was not stopped by the determined resistance of citizens groups and ENGOs in the American northwest, but the multiplication of such points of resistance is increasing the political and economic costs of fossil fuel extraction. For some of these groups, this was not the primary concern; that is, the victory they sought was to prevent US Highway 12 from being turned into a one-time – or worse, permanent – heavy industrial transportation route. In this, they have (as of the time of writing) been successful, using every legal and procedural means available to them. However, the connection between oil sands exploitation in another country and threats to local culture and environments have become apparent to all members of the networks formed during this struggle.

Moreover, even a cursory survey of the websites of opponents of the KMPT reveals that linkages are being made among organizations fighting shipments of oil sands mine equipment, those fighting the Keystone XL pipeline, and those defending Aboriginal rights and territories. To a significant extent, common interests are being identified in finding alternatives to the fossil fuels whose extraction is imposing intolerable local and global costs. The Rising Tide, indigenous, and youth climate justice organizations constitute a kind of intellectual and activist vanguard in this broad-based movement, as they consistently link the local, North American, and global scales of the political ecology of fossil capitalism. The involvement of indigenous actors has been critical to identifying the dependence of corporate-driven resource exploitation upon the dispossession of indigenous peoples, and hence the importance of alliances among non-native environmentalists/communities and indigenous peoples.

The KMPT conflict reveals, once again, the permanent contestation between capitalist and 'alternative development' actors for any foothold within state institutions that may win them an advantage. Environmental organizations must repeatedly defend the legacy of the American environmental movement of the 1960s and early 1970s – its success in shifting public opinion and securing environmental legislation – as we see in the case of Montana's 1972 constitution. Corporations and 'pro-development' governments or parties persistently attempt to weaken environmental legislation or bypass regulatory processes. These are indeed 'obstacles to development,' as the Republicans and business associations contend. The constitutional rights of 'tribal' peoples in the United States provide another defence against environmental degradation. However, as in Canada, we see that in the US the constitutional duty of state agencies to 'consult' First Nations regarding development projects that might affect their lands, livelihoods, or culture may not be respected. The Nez Perce and the Confederated Salish and Kootenai Tribes have had to be vigilant in defending their rights, despite the demands on their scarce resources.

The stance of Montana's Democratic administrations towards fossil fuel exploitation as a driver of economic growth differs little from that of the Conservative governments in Alberta. The combination of expanding coal production and hydraulic fracturing (fracking) in Montana may lead a growing number of communities to question the extractive model of development. In Canada, the recent assertion of leadership of the environmental movement by First Nations, articulated to demands for the respect of constitutional rights and the treaties made with the crown (the Idle No More movement), holds the potential to fuel a broad-based movement for ecological democracy. The Indigenous Environmental Network – which spans the US-Canada border – is playing a significant role in the Idle No More movement. In Alberta, this movement challenges the expansion of oil sands mining, and is connected to many other sites of struggle against fossil capitalism – including communities located along the proposed routes of pipelines, near existing or planned refineries, next to fracking wells, or along US Highway 12. Thus, ExxonMobil/Imperial Oil may have succeeded in constructing and expanding its Kearl mine, but the Stop the Haul movement – like the mobilizations against the Keystone XL and Northern Gateway pipelines – has raised the costs of production while picking at the seams of hegemonic discourse. In the last decade the energy industry has become considerably more vulnerable to disruptions and obstructions of its supply routes and market access. The extraction and transportation of oil is revealed to be anything but 'conflict free.'

NOTES

1 The Kearl oil sands project is jointly owned by Imperial Oil (operator) and ExxonMobil Canada. In 2012 the project had regulatory approval for up to 345,000 barrels a day of production (of diluted bitumen) and is one of Canada's largest open-pit mining operations. The first phase of production at the site was scheduled to begin at the end of 2012, but in the end the mine did not begin producing bitumen until April 2013. See 'Kearl Overview' and related pages, http://www.imperialoil.ca/Canada-English/operations_sands_kearl_overview.aspx (accessed 15 January 2013), and 'Operations,' http://www.imperialoil.ca/canada-english/operations_sands_kearl_overview.aspx (accessed 21 March 2016). In lieu of building an upgrader at Kearl, Imperial Oil 'is using a patented paraffinic froth treatment system to produce a solids-free bitumen that will be blended with diluent and shipped by pipeline to North American refineries' (Dave Cooper, 'Imperial Gets Last of Kearl on Road,' *Edmonton Journal*, 13 March 2012, http://www.standingstonedevelopments.com/news/imperialgets.html).

2 'Dead birds' is a reference to the approximately sixteen hundred ducks that drowned in Syncrude 'tailings ponds' in 2008, an event that – because recorded on film – fixed attention around the world on the ongoing deaths of wildlife caused by these lakes of toxic sludge.

3 Imperial Oil preferred the route because of adequate clearance and cost savings (Vanderklippe 2012b). As it turned out, Imperial was obliged to disassemble and reassemble the modules, which required an estimated labour cost of $500,000 per module (Gilbert 2011b).

4 This claim of increased costs has been made a number of times by Impe-

rial Oil and reported in the business press. A February 2013 report quoted Imperial Oil as saying that the cost of the first phase of the Kearl Mine had risen by two billion dollars due to 'delays in transporting equipment for the project, and the harsh winter in northern Alberta.' See Carrie Tait, 'Imperial Bumps Up Cost of Kearl Oil Sands Project,' *Globe and Mail* (*Report on Business*), 2 February 2013, B1, B7. A 16 June 2015 report by Dan Healing in the *Calgary Herald* repeated the two-billion-dollar figure, while explicitly labelling "trouble getting Korean-built modules through Idaho and Montana" as the cause of the cost overrun (http://calgaryherald.com/business/energy/kearl-oilsands-mine-completion-continues-bitumen-boom).

5 Imperial Oil planned a second-phase expansion of the Kearl mine which was completed in mid-2015.

6 The founders of Idaho's Fighting Goliath citizens' group argued that 'Imperial/Exxon's decision to use interstates constitutes a victory for Idaho and Montana opponents of the transformation of the pristine Wild & Scenic Lochsa-Clearwater US12 corridor and scenic rural 2-lane roads in Montana into an industrial megaload truck route. However, Imperial/Exxon still has its eye on the US12 route for its permanent "high and wide" corridor, as do other corporations. Thus, opponents remain vigilant' (2 November 2011 statement, home page, http://www.fightinggoliath.org/).

7 'Placer miners' is a term used for miners who excavate or pan for minerals, precious metals (especially gold), or gemstones from alluvial deposits (sand and gravel deposited by streams, rivers, or glaciers).

8 Legal scholars have remarked upon the placement of environmental rights in the same section as rights to liberty, property, and happiness and before the rights to non-discrimination, religion, and assembly. One commentator claimed that because of this placement, environmental rights have 'priority' over these other rights (Cross 1990). John L. Horwich (2001, 269) observed that 'while the [Montana Supreme] Court ... [has] concluded that the environmental rights in Montana's 1972 Constitution are fundamental and that they create enforceable limits at least on legislative action, the Court has also held that these rights are subject to a balancing against other, important public values such as economic and social development. As with other fundamental rights, they are subject to infringement in appropriate circumstances.' The potential conflict between the obligation of land trust managers to maximize revenue to fund public goods, on one hand, and the obligation to protect environmental rights, on the other hand, is examined by Sienkiewicz (2006).

9 It is worth noting that no such environmental rights are entrenched in the Canadian constitution, although Yukon, NWT, and Ontario have environmental bills of rights, and similar rights are embodied in Quebec's Environmental Quality Act. In 2010 the NDP and ENGOs tried, unsuccessfully, to get an environmental bill of rights passed in the Canadian Parliament.

10 See *Montana Envtl. Info. Center v. Dep't. of Envtl. Quality*, 988 P.2d 1236 (Mont. 1999); *Cape-France Enterprises v. In re Estate of Peed*, 29 P.3d 1011 (Mont. 2001).

11 See also Dennison (2011). HB 292 failed on third reading on 28 March 2011, as it did not receive the votes of two-thirds of the members of the legislature required for amendments to the constitution.

12 Benjamin Courteau was a Missoula-based participant in All Against the Haul, and attended the meetings described here.

13 For reports of local actions, see 'It's Getting Hot in Here: Dispatches from the Youth Climate Movement,' http://itsgettinghotinhere.org/tag/heavy-haul/.

14 The protests generally took place at night because the trucks transporting the heavy loads travelled at night, when traffic was reduced.

15 Quoted in 'Breaking: Two Arrested for Blocking Tar Sands "Megaloads" in Idaho,' https://risingtidenorthamerica.org/2012/03/breaking-four-arrested-for-blocking-tar-sands-megaloads-in-idaho/.

16 Although Drive Our Economy's website was still active as of 21 March 2016, the organization's last updates and blogs were posted in September 2011.

17 In other words, the WLA appears to be an offshoot of the American Wise Use movement. See McCarthy (2002) for a description and analysis of this movement.

18 Billings is a trade centre and transport hub for Montana's agriculture and fossil fuel industry; Great Falls is supported by agriculture and a strong military and aerospace presence; Missoula is a centre of education, trade, health care, technology, and tourism.

19 *Montana Environmental Information Center and Sierra Club, Plaintiffs and Appellants, vs. Montana Board of Land Commissioners, Ark Land Company, Inc., and Arch Coal, Inc.*, on Appeal from the Montana Sixteenth District Court, Powder River County, Cause Number DV-38-2010-2481, The Honorable Joe L. Hegel, presiding, 12 July 2012, http://meic.org/wp-content/uploads/2012/06/Opening-Brief-of-MEIC-and-Sierra-Club.pdf.

20 'Schweitzer Chastizes Avatar Director over Oilsands Critique,' *Billings Gazette*, 30 September 2010, http://billingsgazette.com/news/state-and-

regional/montana/schweitzer-chastizes-avatar-director-over-oilsands-critique/article_76b0e728-cced-11df-a7be-001cc4c03286.html.

21 It should be noted that 'Nez Perce' is an Anglicization of the name given to this people by French explorers. Their name in their own language is Ni Míi Puu (http://www.nezperce.org/history/mainhistory.html).

22 The National Forest Service is a subagency within the US Department of the Interior (the same department in which the US Bureau of Indian Affairs resides).

23 All Against the Haul, http://allagainstthehaul.org/the-haul/the-heavy-haul/tribal.

24 Moran, Tribal Council chair, Confederated Salish and Kootenai Tribes, 'Memo to Jim Lynch, Director, Montana Department of Transportation,' 21 June 2010. The excerpts quoted here are published on the website of All Against the Haul, under the heading 'Impacts of an Industrial Corridor and Tar Sands Development on Indigenous Communities,' http://allagainstthehaul.org/the-haul/the-heavy-haul/tribal (accessed 15 January 2013).

REFERENCES

Balzel, Bruce, Susan Casey-Lefkowitz, Kate Colarulli, Bruno Kenny, and Elizabeth Shope. 2010. 'Tar Sands Invasion: How Dirty and Expensive Oil from Canada Threatens America's New Energy Economy.' Corporate Ethics International, Natural Resources Defense Council, and Sierra Club issue brief, May.

Briggeman, Kim. 2010a. 'Lolo National Forest Rescinds Power Line Burial Decision for Big Rig Route.' *Missoulian*, 29 July.http://missoulian.com/news/local/article_2de47b3c-9ac9-11df-91da-001cc4c002e0.html (accessed 28 March 2011).

Briggeman, Kim. 2010b. 'Protest Shows How Big Oversized Loads Will Be.' *The Missoulian*, http://missoulian.com/news/state-and-regional/article_ce7dfb08-84c2-11df-8a71-001cc4c03286.html (last modified 30 June).

Briggeman, Kim. 2010c. 'Schweitzer: Oilfield Transportation Project Will Help Montana Economy.' *Missoulian*. http://missoulian.com/news/state-and-regional/article_b3b8d954-563d-11df-879d-001cc4c03286.html (last modified 2 May).

Briggeman, Kim. 2011. 'Missoula County, Environmental Groups Sue to Stop Kearl Megaloads.' *The Missoulian*, 2 April. http://missoulian.com/news/local/missoula-county-environmental-groups-sue-to-stop-kearl-megaloads/article_6357b1ae-5c8d-11e0-8466-001cc4c002e0.html.

Briggeman, Kim. 2012a. 'Imperial Oil/Exxon Mobil Withdraws Megaload Permit Application.' *The Missoulian*, 20 June.

Briggeman, Kim. 2012b. 'Imperial Oil to Ramp Up Kearl Tar Sands Operation.' *The Missoulian*, 15 July 2012.

Carter, Cameron, and Kyle Karinen. 2001. 'NOTE: A Question of Intent: The Montana Constitution, Environmental Rights, and the MEIC Decision.' *Public Land & Resources Law Review* 22: 97–132.

Chaney, Rob. 2010. 'Protesters Rally against Oil Equipment Trucking at Missoula MDT Office.' *The Missoulian*, 3 June.http://missoulian.com/news/local/article_b60c49f0-6f80-11df-a29e-001cc4c03286.html (accessed 28 March 2011).

Cross, C. Louise. 1990. 'The Battle for the Environmental Provisions in Montana's 1972 Constitution.' *Montana Law Review* 51 (2): 449–57.

Dennison, Mike. 2011. 'Montana Legislature: Montana Legislature Back in Session.' *Billings Gazette*, 3 January.

Dvorak, Todd. 2010. 'Idaho Judge Halts Big Rig Shipments on Highway 12.' *Missoulian*, 18 August. http://missoulian.com/news/local/article_e0ad9326-aa36-11df-91b4-001cc4c002e0.html (accessed 28 March 2011).

Gilbert, Richard. 2011a. 'Oilsands Components Find Alternate Route to Alberta.' *Journal of Commerce*, 9 November, http://www.joconl.com/article/id47472/--oilsands-components-find-alternate-route-to-alberta (accessed 15 January 2013).

Gilbert, Richard. 2011b. 'Oilsands Modules May Finally Arrive in Alberta.' *Journal of Commerce*, 15 August. http://journalofcommerce.com/Home/News/2011/8/Oilsands-modules-may-finally-arrive-in-Alberta-JOC046202W/.

Horwich, John L. 2001. 'MEIC v. DEQ: An Inadequate Effort to Address the Meaning of Montana's Constitutional Environmental Provisions.' *Montana Law Review* 62: 269–99.

Idaho Transportation Department (IDT). 2010. 'Public Meetings Generate Comments on U.S. 12 Transport Proposal.' http://itd.idaho.gov/transporter/2010/070210_Trans/070210_US12meetings.html (accessed 26 March 2016).

Jenson, Jim. 2012. 'Governor's Stance on Otter Creek Irresponsible.' *Independent Record*, 29 August. http://helenair.com/news/opinion/governor-s-stance-on-otter-creek-irresponsible/article_3ae27f3c-f1a2-11e1-b1a4-001a4bcf887a.html.

LaDuke, Winona. 2011. 'Big Oil Wants to Truck through Nez Perce Land.' *Indian Country Today*, 24 March. http://indiancountrytodaymedianetwork.com/2011/03/big-oil-wants-to-truck-through-nez-perce-land/ (accessed 28 March 2011).

Malone, Michael P. 1981. *The Battle for Butte: Mining and Politics on the Northern Frontier, 1864–1906*. Seattle: University of Washington Press.

McCarthy, James. 2002. 'First World Political Ecology: Lessons from the Wise Use Movement.' *Environment and Planning A* 34 (7): 1281–1302.

Montana Department of Transportation (MDT). 2010. 'Kearl Module Transportation Project: Environmental Assessment.' April. Helena: MDT.

Montana Department of Transportation (MDT). 2011. 'Kearl Module Transportation Project: Finding of No Significant Impact (FONSI).' February. Helena: MDT.

Nikiforuk, Andrew. 2008. *Tar Sands: Dirty Oil and the Future of a Continent*. Vancouver: Douglas & McIntyre.

Porco, Mike. 2011. 'A Montanan's Right to a Clean and Healthful Environment.' *The Bozeman Magpie* 30 January. http://www.bozeman-magpie.com/perspective-full-article.php?article_id=194.

Sakariassen, Alex. 2010. 'New Regional Task Force Backs Big Rigs.' *Missoula Independent*, blog, 11 November. http://missoulanews.bigskypress.com/IndyBlog/archives/2010/11/11/new-regional-task-force-backs-big-rigs.

Sakariassen, Alex. 2011. 'From Economic Optimism to Outright Opposition, Big Oil's Proposed "Heavy Haul" Has Divided Cities and Towns along the Route.' *Missoula Independent*, 20 January. http://missoulanews.bigskypress.com/missoula/crossroads/Content?oid=1371483 (accessed January 2013).

Sakariassen, Alex. 2012. 'Montana Gets Into the Megaload Business.' Missoula Independent, 7 June. http://missoulanews.bigskypress.com/missoula/montana-gets-into-the-megaload-business/Content?oid=1659049 (accessed January 2013).

Sienkiewicz, Alex. 2006. 'A Battle of Public Goods: Montana's Clean and Healthful Environment Provision and the School Trust Land Question.' *Montana Law Review* 67 (1): 65–87.

Swibold, Dennis L. 2006. *Copper Chorus: Mining, Politics, and the Montana Press, 1889–1959*. Helena: Montana Historical Society.

Tait, Carrie. 2012. 'Imperial Oil Plan Meets Heavy Traffic.' *Globe and Mail*, 23 August.

US Department of Energy. 2000. 'Executive Order 13175 – Consultation and Coordination with Indian Tribal Governments.' 6 November. http://ceq.hss.doe.gov/nepa/regs/eos/eo13175.html (accessed 28 March 2011).

US Environmental Protection Agency. 2010. 'National Environmental Policy Act (NEPA).' 21 October. http://www.epa.gov/oecaerth/basics/nepa.html (last accessed 28 March 2011).

US Environmental Protection Agency and Montana Department of Environmental Quality. 2003. 'Superfund Program Clean-up Proposal: Milltown

Reservoir Sediments Operable Unit of the Milltown Reservoir/Clark Fork River Superfund Site.' April.

Vanderklippe, Nathan. 2011. 'U.S. Judge Puts Brakes on Imperial Oil Sands Gear.' *Globe and Mail,* 20 July.

Vanderklippe, Nathan. 2012a. 'Arctic Seen as Possible Shipping Route for Massive Industrial Components.' *Globe and Mail,* 23 August.

Vanderklippe, Nathan. 2012b. 'Transportation Woes Threaten to Delay Imperial Oil Sands Project.' *Globe and Mail,* 24 August.

Vock, Daniel C. 2012. 'Montana's Oil Boom Reshapes Race for Governor.' *Stateline,* Daily News Service of the Pew Centre on the States, 17 October 17. http://www.pewstates.org/projects/stateline/headlines/montanas-oil-boom-reshapes-race-for-governor-85899423756.

Wilson, Bryan P. 2004. 'Comment: State Constitutional Environmental Rights and Judicial Activism: Is the Big Sky Falling?' *Emory Law Journal* 53 (2): 627.

13 In the Path of the Pipeline: Environmental Citizenship, Aboriginal Rights, and the Northern Gateway Pipeline Review

LARISSA STENDIE AND LAURIE E. ADKIN

Canada's emergence as a global energy powerhouse – *the* emerging 'energy superpower' our government intends to build ... is an enterprise of epic proportions, akin to the building of the pyramids or China's Great Wall. Only bigger.

Stephen Harper, 14 July 2006[1]

We have inhabited and governed our territories within the Fraser watershed, according to our laws and traditions, since time immemorial. Our relationship with the watershed is ancient and profound, and our inherent Title and Rights and legal authority over these lands and waters have never been relinquished through treaty or war ...We will not allow the proposed Enbridge Northern Gateway Pipelines, or similar Tar Sands projects, to cross our lands, territories and watersheds, or the ocean migration routes of Fraser River salmon. We are adamant and resolved in this declaration, made according to our Indigenous laws and authority. We call on all who would place our lands and waters at risk – we have suffered enough, we will protect our watersheds, and we will not tolerate this great threat to us all and to all future generations.

Save the Fraser Declaration, 2 December 2010[2]

Just as it would link Alberta's oil sands to Asian markets for bitumen and synthetic crude oil, the Northern Gateway pipeline has already generated many linkages of a geo-political-ecological nature. As the statements quoted above demonstrate, the pipeline is – for its proponents – critical to Alberta's prosperity and Canada's future as an 'energy superpower.' For its opponents, it is a path to ecological destruction at multiple scales as well as a threat to sustainable livelihoods and the survival of Aboriginal cultures.

Many questions have been raised about the necessity and risks of this project, as well as about the legitimacy of the decision-making process that has been used to determine whether or not the pipeline's construction may go ahead. Our focus in this chapter is the structure and functioning of the Joint Review Panel (JRP) appointed by the Canadian Environmental Assessment Agency (CEAA) and the National Energy Board (NEB) to assess the environmental impacts and economic benefits of the proposed project. Specifically, we ask what the decision-making process implemented by the federal government tells us about the nature of environmental citizenship in Canada today, and how this model of citizenship accommodates, excludes, or otherwise relates to the sovereignty claims of Aboriginal peoples.[3] In addition to assessing how well the JPR process conforms to the criteria of deliberative democracy, we ask how well the panel's procedures fulfil the criteria of free, prior, informed consent-consultation with local communities and indigenous peoples that are found in a number of international agreements as well as in Canadian laws.

The Western Canadian 'energy corridor' consists of numerous proposed natural gas and coal mine projects as well as four pipeline projects currently slated for construction or expansion across Alberta and British Colombia. Multiple pipeline projects (west, east, north, south) have been proposed to alleviate the otherwise landlocked 'bottleneck' said to be constraining Alberta's capacity to exponentially increase oil and gas exports. These come not only from the oil sands, but also from fracking, and the Western energy corridor encompasses liquid natural gas projects as well as the movement of other products. The pipeline projects include:

- Kinder Morgan's Trans Mountain pipeline, connecting Alberta to Burnaby, BC
- TransCanada pipeline's Keystone extension, connecting Alberta to Gulf Coast refineries
- TransCanada pipeline's Energy East pipeline, connecting Alberta and Saskatchewan to Quebec and Ontario
- Chevron's Pacific Trail pipeline to deliver gas from Summit Lake, BC, to the Kitimat LNG facility
- Pembina Pipelines Corporation's multiple expansion plans
- Enbridge's Eastern Canadian Refinery Access Initiative ('Line 9' reversal)
- Imperial Oil Resources' Mackenzie Gas pipeline

While this chapter focuses on the Enbridge Northern Gateway pipeline (NGP), the counter-movement that we describe opposes each of these proposed corridor projects.

The NGP would link the pumping station in Bruderheim, Alberta, to the port-terminal town of Kitimat, British Columbia, where the bitumen or synthetic crude will be loaded into 'Very Large Crude Carriers' (VLCCs) (with capacities of at least two million barrels), bound for Asian refineries and markets. At an estimated construction cost of almost eight billion dollars, the twinned Enbridge pipelines would traverse over 1,177 kilometres of mountainous terrain. Assuming a thirty-six-inch-diameter pipe, the westward line would provide a capacity of 525,000 barrels of diluted crude bitumen oil (dilbit) per day (CERI 2011, 6). The second pipeline would carry approximately 200,000 barrels of imported chemical condensate mix – required to make the diluted bitumen flow – back to the upgraders located in the tar sands.

The NGP project demands unique feats of engineering. It will cross more than one thousand salmon-bearing waterways, tunnel through two mountains in remote and seismically unstable zones, traverse areas known for record snowfalls, and face other technical challenges even before reaching the marine terminal (Allan 2012). The route proposed for the tankers that will carry the Alberta bitumen would put at risk 900 kilometres of temperate rainforest coastline. To reach open water, the tankers would have to navigate the Douglas Channel, known as the 'fourth most dangerous waterway in the world,' where small passenger ferries have foundered and sunk in violent storms and record high waves (Environment Canada 2012, 109). The tanker's route requires ninety-degree turns through shallow, rocky shoals where, at multiple points, there would be only a three-metre clearance from the ocean floor. Approval of the NGP would effectively end a thirty-year voluntary Tanker Exclusion Zone agreed to by US and Canadian authorities that was created to steer oil tankers around the west side of Haida Gwaii as they plied back and forth from Valdez, Alaska. If Kitimat becomes a major oil terminal, at least 6,600 oil tankers (over 200 a year) of unprecedented size will cross between Kitimat and the outer edge of Hecate Strait over the next thirty years. Critics of this proposal, such as the Raincoast Conservation Foundation and the Heiltsuk Tribal Council, call attention to the fact that this route provides only one emergency anchorage large enough to harbour an oil tanker in the 'certain event of a surprise storm' (Kopecky 2013, 61). The environmental concerns that have been raised encompass not only the possibilities of pipeline ruptures or leaks along the route, or oil spills affecting waterways and coastal areas, but also the greenhouse gas emissions that will be produced if oil contained in Alberta's bituminous sands is burned.[4]

The Joint Review Panel

The quasi-judicial JRP was composed of three National Energy Board members (two permanent and one temporary). The JRP coordinated the governmental authorities responsible for this project. Its broad mandate was to consider whether the project is in the public interest and the likelihood of detrimental environmental impacts (NEB news release, 4 December 2009). The delegation of the socio-environmental approval process to a Joint Review Panel is in itself significant. Prior to 2006, a development proposal of this nature would have gone first to the CEAA for approval, then to the NEB. As Enbridge acknowledged in its application to the JPR:

> Under the CEA Act, an environmental assessment must be carried out before a federal authority such as the NEB can issue a permit or license, or grant an approval to a project. Because more than 75 km of new RoW [right of way] will be constructed, the Project is subject to a comprehensive study level of environmental assessment pursuant to the Comprehensive Study List Regulations under the CEA Act. (Enbridge 2010, 6-1)

However, Enbridge asked that the minister of environment 'enter into an agreement with the NEB to establish a JRP pursuant to subsection 40 (2) of the CEA Act' (which permits the minister of environment to delegate responsibilities under the Canadian Environmental Assessment Agency to other government agencies) (6-1). A joint process would mean a 'single set of environmental filing requirements and public hearing proceedings.' The new Conservative minister of environment agreed to this in September 2006. Terms of reference for the joint review panel were drafted. The process was put on hold, however, because the Northern Gateway consortium was not yet ready to proceed with the project. In February 2009, the draft JPR agreement (terms of reference) was presented to the public for a sixty-day comment period. Public comments on the draft agreement were considered between 9 February and 14 April 2009, to inform aspects of the Terms of Reference and JRP agreement. On 4 December 2009 the CEA Agency and the NEB issued the JRP agreement, including the terms of reference and scope of factors for the environmental and regulatory review of the proposed project.[5]

Pressure from business associations in the resource sector to 'streamline' the socio-environmental approval processes for development projects has been a constant theme since the CEA Act was implemented in 1995. During the review of the CEA Act in 2003, the Canadian Energy Pipeline Association was a key player in what seasoned environmentalist observers of that process described as the 'clamour from much of the industrial sector around "duplication and overlap"' (Schneider, Sinclair, and Mitchell 2007, 3).[6] This push gained momentum following the election of a Conservative minority government in 2006. Upon taking office, the federal Conservatives undertook a series of steps to accommodate the resource industries' wishes for quicker turnarounds on applications for permits from regulatory authorities. The federal government budget of 2007 announced a Major Project Management Office that was subsequently established by the minister of natural resources, Gary Lunn, in October 2007. The minister of Indian and northern affairs created the Northern Regulatory Improvement Initiative a month later, appointing the former chair of Alberta's Energy and Utilities Board, Neil McCrank, to lead this initiative. NEB officials appointed by the government called for more to be done to make Canada 'competitive in the world,' including legislative changes to coordinate environmental assessments with regulatory reviews, improve consultation with Aboriginal peoples, and reduce the time taken to make regulatory decisions.[7]

There has been pressure, particularly, to have environmental assessments delegated to agencies other than the CEA Agency, or for what is called 'substitution' of the CEA Agency by another agency, such as the Canadian Nuclear Safety Commission (CNSC), the Canada-Newfoundland Offshore Petroleum Board, or, since 2006, the National Energy Board (Schneider, Sinclair, and Mitchell 2007, 2–3). This was the preference of companies in the resource sector because the criteria of environmental sustainability were hardest to meet under the CEA Act, and because several review processes might be reduced, thereby, to one. Under the CEA Act (section 4 in both pre-2006 and post-2012 versions), federal authorities have a mandate to 'exercise their powers in a manner that protects the environment and human health and applies the precautionary principle.' Under the NEB Act (section 52[1]), on the other hand, criteria for certification include that the project 'will be required by the *present and future public convenience and necessity*'; its report must 'set out' the environmental assessment that has been carried out (if required) under the CEA Act (see also Benevides, Clenaghan, and Lindgren 2011, 5–6).

In recent years there have been further moves by the federal Conservative government to 'streamline' (some would say rubber-stamp) the processes through which private investors must pass in order to obtain environmental and 'public interest' approvals (or 'social licence to operate') for their projects. The implications of these changes for the Northern Gateway pipeline process are discussed in subsequent sections.

The Process

The JRP hearing order released on 5 May 2011 (see table 13.1) detailed the four ways in which the public and Aboriginal peoples could participate. Anyone could submit a Letter of Comment by 31 August 2012, and approximately 9,500 letters were received.[8] Individuals or groups could register by 6 October 2011 to make an oral statement of ten minutes (though possibly longer) at open public hearings to share knowledge, views, or concerns about the project. During the oral statement participation from January to April 2012, interventions were made by thirty-nine indigenous groups (represented by 291 individuals); six community-based or environmental-NGO groups (33 speakers); one labour union; four governmental agencies (represented by 11 persons); one industry association (with 2 spokespersons); and two other groups.[9] By the end of the community hearings in February 2013, 1,179 participants had spoken. (Notably, only two persons – a former mayor and a former MLA – spoke in favour of the NGP [Gilchrist 2013].) Intervenors could be either groups or individuals; those registered as of 14 July 2011 included forty-eight indigenous groups; eighty-five individuals; three unions; four land-owner groups; eleven environmental, conservation, or community organizations; thirty-six energy industry or business associations; and three miscellaneous parties. Finally, there were thirteen government participants.

Letters of comment or oral statements did not qualify participants to make arguments at the final hearings from September 2012 to May 2013, which are reserved for parties – identified as intervenors, government participants, the panel, and the proponent (Enbridge). According to the JRP process advisor, 389 witnesses gave oral evidence, with 60 representing parties.

Figure 13.1. Protestors from the Carrier Sekani Tribal Council outside the Joint Review Panel hearing in Fort St. James, BC, 2 February 2012. (*Photo*: Larissa Stendie.)

JRP community sessions were open to the public except in Vancouver and Victoria, BC, where the panel feared disruption due to multiple earlier protest actions, and allowed only one guest per participant, in addition to media reporters, into the hearings. Instead, live-feed public viewing rooms were set up in nearby venues. Table 13.1 outlines the JRP's major dates (which had to be expanded beyond the original schedule to accommodate all those who registered to participate) along with participation details.

Table 13.2 introduces two sets of criteria by which the democratic nature of the JPR process may be evaluated. This comparison of the precepts of international human rights standards on meaningful "free, prior, informed consultations" with standards of deliberative democracy was drawn by German scholar Almut Schilling-Vacaflor (2012).

Table 13.1. Enbridge Northern Gateway Proposal: Joint Review Panel Schedule

Dates*	Event/Stage of Process	Details of Participation
(Pre-JRP) 2002–05	Aboriginal introduction to and engagment in the project by Enbridge	Details are sparse about what this actually entailed, but it was intended to identify potentially affected communities and within 80-km radius of the project. These 'relationship-building' activities also included information sharing and the undertaking of traditional knowledge studies.[i]
4 Dec 2009	Northern Gateway pipeline project Joint Review Panel Agreement issued	Public comments on the draft agreement were considered between 9 February and 14 April 2009 to inform aspects of the Terms of Reference (ToR) and JRP agreement
27 May 2010	Northern Gateway Pipelines Inc. (proponent) applied to National Energy Board	Sought authorization to construct and operate (1) an oil export pipeline and associated facilities; (2) a condensate import pipeline and associated facilities; (3) a tank terminal and marine terminal to be located near Kitimat, BC
5 July 2010	Joint Review Panel issues procedural direction for comments	• 27 communities had public access to hard copies of the complete proponent proposal • 25 notifications published in local, regional, and national papers • 200+ written comments between June and Sept 2010 • comments accepted on the draft List of Issues; additional information that proponent should file, and location(s) for the oral hearings
10, 31 Aug and 8 Sept 2010	Panel sessions held in Whitecourt, AB; Kitimat, BC; Prince George, BC	• 70 participants
19 Jan 2011	JRP decisions on procedural directions	• Required proponent to file additional information re: engineering challenges and risks, impacts on subsistence communities, high volumes being transported • Issued the revised List of Issues • determined locations of oral hearings

Table 13.1. (*Continued*)

Dates*	Event/Stage of Process	Details of Participation
5 May 2011	Hearing order released	Detailed four options for participating in the joint review process: (1) submit a letter of comment; (2) make an oral statement; (3) become a party-intervenor; or (4) qualify as a government participant (federal, provincial, territorial, or municipal government bodies)
Nov 2010– Sept 2012	'Letters of comment' period	5,680 letters received at this time (Feb 2013 total 9,500)
10 Jan–17 April 2012	Community hearings (oral evidence from parties and oral statements from registered participants; oral traditional knowledge; personal knowledge or experience about potential effects to the individual and his/her community)	• Held in 17 communities along the proposed route • Oral evidence (3 hr max time limit per submission): 380 people representing 54 groups; 50 oral statements by independents • Interventions focus on what cannot be submitted in written evidence
26 March– 10 Aug 2012	Community hearings: oral statements from registered participants – knowledge, views, concerns, position as to whether in the public interest	• 14 communities (closest along the proposed route) • Oral statements: 681 individuals • Oral evidence from parties: 5 (10-min time limit for oral statements) • Stick to the issues list, not beyond the panel's scope. Participants not cross-examined
30 May 2012	Conference on process for the parties involved in the final hearings, Calgary, AB	• 34 participants • Allowed responses on range of questions related to procedural issues, process design particulars

Table 13.1. (*Concluded*)

Dates*	Event/Stage of Process	Details of Participation
4 Sept–15 Dec, 2012	Final technical hearings: questioning phase for parties and expert or lay witnesses	• Edmonton, AB, Prince George, BC, and Prince Rupert, BC • 28 parties, JRP legal counsel, and witnesses • Parties ask any outstanding questions to test credibility of the evidence filed and witnesses on record and about the issues on the List of Issues
4 Jan –1 Feb 2013	Final community hearings: oral statements from registered participants to hear their knowledge, views, concerns, positions as to whether the project is in the public interest; recommendations	• Victoria, Vancouver, and Kelowna, BC (communities further from the proposed route) • Oral statements: 482 • 10 min time limit. Stick to the issues list, not beyond the panel's scope. • Participants not cross-examined
17 June 2013 (+ 2 weeks)	Final arguments for parties. Written arguments from all parties by 31 May 2013. Oral argument responses to other parties' written arguments begin 17 June 2013	• Session held in Terrace, BC • Parties express views, summarize evidence, cite law to persuade the panel whether proposal is in the public interest, recommend approval or denial, or suggest conditions for mitigation in written argument. Intervenors and government participants have 1 hour for oral argument, proponent has 2 hours (given burden of proving case and responding to all other parties) • Northern Gateway and 56 other parties submitted final arguments
Dec 2013	JRP report to federal cabinet	Report available online at http://gatewaypanel.review-examen.gc.ca/clf-nsi/dcmnt/rcmndtnsrprt/rcmndtnsrprt-eng.html

*These were the dates originally mapped out for the JRP process. However, there were extensions along the way. In February 2013 the 'questioning phase for parties and expert or lay witnesses' was still in progress in Prince Rupert, BC.

[i] Enbridge Preliminary Information Package for the NGP, section 4. Consultation. October 2005, p. 4-4. https://www.neb-one.gc.ca/ll-eng/livelink.exe/fetch/2000/90464/90552/384192/384008/384170/A0S2C2_-_Consultation.pdf?nodeid=384180&vernum=0 (accessed 10 February 2013).

Sources: Enbridge Northern Gateway Joint Review Panel, Public Registry, https://www.neb-one.gc.ca/ll-eng/livelink.exe/fetch/2000/90464/90552/384192/620327/customview.html?func=ll&objId=620327&objAction=browse&sort=-name; Colette Spagnuolo, process advisor, Northern Gateway Project JRP Secretariat, email to L.Stendie, 11 February 2013.

Table 13.2 Comparison of International Human Rights Standards on Meaningful Prior Consultations and Standards of Deliberative Democracy

	International Human Rights Standards (Free Prior Informed Consultation and Consent)	*Standards of Deliberative Democracy*
General characteristics of consultation/ deliberation	Genuine and constant dialogue between representatives of state institutions and indigenous communities, carried out previously to a planned legislative or administrative measure, in a climate of confidence, mutual respect, and good will	Climate of the deliberations should be respectful, without discrimination; mediators in the negotiation should be accepted by all parties involved
Participants	Representative institutions of the indigenous peoples and legitimate representatives of all affected communities participate in the process; consultations must take into account affected groups' own norms and procedures	Everybody affected or their legitimate representatives should have the right to participate; deliberators should have equal opportunity to present interests and preferences; all significant interests should be represented
Information/ transparency	Complete information about the planned project must be submitted to the local communities, including information about the expected environmental and socio-cultural risks and impacts	Access to broad information contributes to fair deliberations; the deliberative process should be transparent and scrutinized by media and other citizens; informed public debates also in informal arenas
Cultural, social, and linguistic adequacy	Consultation should be adapted to the social and cultural models of the indigenous peoples (e.g., values, conceptions, handling of time, reference systems, and modes to conceive the consultation)	Arguments should be presented and decisions justified so that they are understood by all participants
Establishment and binding character of agreements	Consultations must have the sincere objective of achieving a common agreement; consultations should help shape the planned measure; agreements obtained as a result of the consultations are binding	Deliberative processes should reach shared understandings of common good; all participants must accept final decisions as binding; principle of revisibility of decisions

Source: Schilling-Vacaflor 2012

Creating a fair participation process for the JRP, like many such consultations concerning resource exploitation in Canada, entails the accommodation of more than the diverse values or interests of a single public. In a settler-colonial context, Aboriginal groups constitute distinct communities with unique rights and status vis-à-vis the state. There are, moreover, significant differences among Aboriginal communities with regard to their interests and their resources for participation. In general, the consultation process should accommodate differences in cultural norms, as sketched briefly in table 13.2. Our account of the inclusiveness of participation, the panel's effectiveness in dealing with power dynamics and differences in the resources available to participants, and the ultimate meaning, or influence, of participation with regard to the outcomes, begins with the engagement of First Nations communities that will be affected by the NGP project, should it be successfully completed.

Consultation with First Nations

Elsewhere in this volume, Brenda Parlee describes the multiple strategies of engagement for First Nations involved in contesting extractive developments, including precedence-setting case law based on the principle of mandating free prior informed consultation-consent (FPIC). While these rights are enshrined in many widely adopted international agreements, not all of these have been ratified by Canada, for example the ILO Convention 169. The United Nations Declaration on the Rights of Indigenous Peoples (UNDRIP) was adopted in 2007 after twenty years of negotiations, but the former 'frontier-colonies' of Canada, the US, New Zealand, and Australia did not sign until 2011. As a signatory of the UNDRIP, the Canadian government has a duty to adequately consult and engage in meaningful dialogue with First Nations under FPIC. Moreover, Canada has obligations under its own constitutional provisions.

Aboriginal rights are enshrined in section 35(1) of the 1984 Canadian Constitution, and have been affirmed and further defined in a series of high-level court decisions. Legal rulings on Aboriginal land rights are coming into direct conflict with the 'streamlining' of environmental assessment/approval processes embodied in recent federal government legislative changes. The 26 June 2014 ruling of the Supreme Court of Canada in *Tsilqot'in Nation v. British Columbia*, 2014 SCC 44, means that provincial governments will face stiffer tests to obtain legal approval for industrial developments on lands to which an Aboriginal people holds title. In the absence of the consent of the Aboriginal group to the 'development,' the government will have to establish to the court's satisfaction that a project is required to fulfil a 'compelling and substantial governmental objective and that the government action is consistent with the fiduciary duty owed by the Crown to the Aboriginal group.'[10]

The JRP issued its final report in December 2013, and in June 2014 the federal cabinet as most observers anticipated – approved the NGP's construction (Government of Canada 2014). The approval was made contingent upon Enbridge's fulfilment of the 209 conditions set out in the panel's report. These include requirements that Enbridge report on employment and other economic benefits for Aboriginal groups, that it notify Aboriginal groups of any anticipated effects of pipeline construction on traditional land use, and that it report to the NEB on mitigation measures or agreements made with the affected communities (see 'ranges' 39–46 of the JRP's recommended conditions).[11] At the time of writing, it is unclear how the June 2014 Supreme Court decision will shape the outcome of the NGP conflict, although it does appear that the decision has strengthened the judicial resources of First Nations in their dealings with governments and corporations.

While the crown has a constitutional duty to consult with Aboriginal peoples about development proposals that may affect existing or potential land claims or traditional Aboriginal rights, governments have routinely delegated this responsibility to the project proponents. In the case of the NGP, the duty to consult was effectively delegated to Enbridge (notwithstanding the federal government's statement to the JRP noted below), and those consultations have been conducted privately. The section of the JPR report (volume 2, section 4) entitled 'Aboriginal interests and consultation with Aboriginal groups,' however, records numerous complaints from First Nations about the quality or absence of consultation with them on the part of Enbridge.[12] Among the First Nations that reported there had been inadequate consultation were the Driftpile First Nation, Coastal First Nations, Michel First Nation, Swan River First Nation, and the Wet'suwet'en Council of Hereditary Chiefs (which complained to the CEAA Panel Manager in June 2011 that Enbridge had falsely claimed to have consulted them about the pipeline[13]).

The new Canadian Environmental Assessment Act of 2012 does not clearly specify which bodies are responsible for consulting with Aboriginal groups or for determining the adequacy of consultation – the CEA Act, the review panels, the minister, or the governor in council (cabinet)? Project development, Aboriginal, and regulatory lawyer Sandy Carpenter, in his review of Bill C-38 (2012), observes that nothing in CEA Act 2012 specifies who 'is responsible for what in carrying out or assessing the duty to consult in relation to federal environmental assessments' (253). Carpenter comments further that, notwithstanding the lessons of the Berger Inquiry of 1977, 'until recently, there does not appear to have been any real recognition of the role of Aboriginal interests in [energy project proceedings], let alone attempts to consider how to address them. As a result, the role of Aboriginal interests in energy project approval processes seems to be developing on an ad hoc basis, mostly as a result of obligations being judicially superimposed on existing structures' (252–3).

Thus, the criteria of proper consultation are still ambiguous in Canada: Does consultation mean merely being invited to attend an information session? Provided with written information about a project? Or – as in the JRP case – being offered an opportunity to submit evidence and give oral testimony in a public hearing, or question evidence during the final hearings? The CEAA did provide some funds to help First Nations participate in the review process. Coastal First Nations (representing nine bands from the BC coast and Haida Gwaii) requested $520,000 for expenses, including studies, lawyers, and attendance at the hearings (Moore 2013). In 2009 they were allotted $286,000, including $25,000 for legal costs (Moore 2013). Enbridge also provided some funding to First Nations intervenors, to the tune of $13 million (Moore 2013). Yet, in February 2013 the Coastal First Nations withdrew from the review process, saying that they had already spent more than three times the amount of funding they had been allocated, and that this funding did not allow them to participate on the same terms as Enbridge, said to be spending $250 million on the regulatory review process. The executive director of Coastal First Nations, Art Sterritt, said:

> We simply have not been provided with the funding necessary to engage in this process meaningfully or effectively ... This is extremely distressing and disappointing to us, as we have a great deal at stake in these proceedings and in particular this panel ... It seems the only party that can afford this long and extended hearing process is Enbridge itself, and perhaps the Crown. The average citizen can't afford to be here, and certainly the Coastal First Nations can't afford it. (quoted in Moore 2013)

As a result of their withdrawal, Coastal First Nations did not participate in the final technical hearing in Prince Rupert, BC, which examined the possibility of a marine spill and Enbridge's readiness to respond. Sterritt said that Coastal First Nations were very concerned that the process was going ahead without the 'necessary scientific studies' (Moore 2013). It is likely that bands along the pipeline route have had experiences much like that of the Coast First Nations. Given, as Sterritt puts it, that the Northern Gateway consortium has had 'a battery of lawyers' to make its case, it would be difficult to conclude that the playing field for participants has been level.

Despite Canadian Supreme Court decisions (*Haida* and *Taku River Tlingit*), the definition of meaningful consultation has remained vague, and certainly permits no First Nations veto. Supreme Court Justice Antonio Latimer commented in 1997 that 'the development of agriculture, forestry, mining and hydroelectric power, the general economic development ..., the building of infrastructure and the settlement of foreign populations to support those aims, are the kind of objectives that ... can justify the infringement of aboriginal title' (*Delgamuuk v. R,* para 165 of the Chief Justice's opinion, 11 December 1997). Nor does the more recent *Tsilqot'in Nation v. British Columbia* (June 2014) decision interpret Aboriginal title to a land area (where treaties do not exist) to confer an absolute right to allow or deny provincially – or federally – approved industrial development; the decision refers instead to the criteria for meeting reasonable consultation and accommodation set out in earlier rulings. In *Grassy Narrows v. Ontario* (July 2014), the Supreme Court reaffirmed the supremacy of provincial land ownership for forestry or other uses within the numbered treaties, basing accommodation on addressing the fiduciary responsibility of consulting on 'the burden' of minimal harm obligations, but with the capacity to override First Nations concerns in the interests of the 'public good.'[14]

The pressure on First Nations to engage in consultations with private interests is relentless. In his testimony to the Joint Review Panel in Smithers, BC, 16 January 2011, Chief Adam Gagnon said: 'We have representatives for Enbridge sitting here. They've asked for the last eight years about this pipeline with the Wet'suwet'en. I sat in many meetings. The answer was unilaterally "no" right from the beginning. What part of "no" do they not understand?' (JRP Hearing Order OH-4-2011, para. 6178). A similar view of the 'consultation' was expressed by Wet'suwet'en member Richard Sam of the Laksilyu Clan, in his testimony to the panel at the hearing in Smithers, BC, on 16 January 2012:

> Our Chiefs have decided, after considering all the information available to them, that this pipeline will not go through Wet'suwet'en land, and it's up to people like myself to try and see that that wish is done, that wish is carried out, that this pipeline will not go through … [O]ur Prime Minister of Canada goes on the television and says, 'Oh, I support this and I'm going to do everything I can to push Enbridge through, our pipeline through.' How is it that he can pre-decide what his decision is going to be when he hasn't even spoke with us? How is that consulting in good faith with Aboriginal people? ... The Supreme Court has set out that the Canadian government must consult in good faith but yet our Prime Minister goes ahead and says he supports it and that's all there is to it. So there's no good faith. (JRP Hearing Order OH-4-2011, paras. 5659–77)

Many were frustrated during the JPR hearings by the federal government's treatment of Aboriginal peoples as a subset of the general public, rather than as a group holding unique land title and rights. West Coast Environmental Law (WCEL) concluded in January 2011 that

> the JRP does not respect the decision-making authority of Indigenous peoples. The JRP was unilaterally imposed on First Nations by the federal government. The terms of reference and list of issues to be considered were developed without meaningful consultation. The JRP lacks the authority to fully assess potential impacts on Aboriginal Title and Rights, and there is still no established process outside of the JRP to assess these impacts. As a result of these flaws in consultation, there continues to be significant, on-going legal risk to the Enbridge project. (WCEL 2011, 1)

Some First Nations, like the Carrier-Sekani, refused to participate in the JRP process at all, on the grounds that the review panel was not a substitute for direct consultation with the federal government on resource and land issues (Moore 2013). Alexander First Nation was among numerous First Nations reporting that they had made repeated requests for government involvement in consultation, but that provincial and federal levels of government had essentially delegated consultation to the project proponent.[15] They had in fact been told that no government consultation would take place until after the JRP had rendered its decision.[16] In its final written submission to the JRP, the Haisla Nation argued that 'Canada has unilaterally designed the process so as to negate the possibility of Crown consultation with the Haisla Nation to occur at the earliest time possible, to prevent open, transparent and timely discussion and to frustrate Canada's obligation to provide full information ... The Panel is not an emanation of the Crown, or its agent, and as such, does not have an independent duty to consult and accommodate the Haisla Nation' (Haisla Nation 2013, 59).

In response to these concerns, the Government of Canada said that it had not delegated 'aspects of Canada's consultation or accommodation obligations' and that it was 'relying on the Panel's process and Northern Gateway's broader consultation efforts, to the extent possible, to assist the Crown in fulfilling its legal duty to consult.' Moreover, 'early consultation should have resulted in Aboriginal groups understanding their opportunity to be meaningfully consulted on the project.' Asked by First Nations what type of government-to-government consultation would occur following the release of the JRP's report, the government replied that participant funding would be available to Aboriginal groups through the CEEA for 'Phase IV' consultation, and that it would 'take into consideration ... outstanding concerns before making any final decisions on the project.'[17]

The Nuxalk Nation withdrew from the JRP process following statements by the prime minister and the minister of natural resources that led them to have 'a reasonable apprehension that approval of this project has been "predetermined" by federal government [*sic*], and that the Enbridge regulatory process is not part of a good faith effort to consult First Nations.'[18] The Nuxalk elders stated that 'the process chosen by the Crown does not allow any scope for Nuxalk to be recognized as a decision-maker in a true government-to government decision-making process.'[19] This, indeed, is the message of the Idle No More movement, which is rejecting band-by-band 'consultations' with corporations over the terms of resource exploitation within their territories. Idle No More demands direct negotiation *with the crown* that would be more akin to the nation-to-nation spirit of the original treaty negotiations (Palmater 2013).

Under circumstances like these, it cannot be said that the FPIC criterion of 'genuine and constant dialogue between representatives of state institutions and indigenous communities, carried out previously to a planned legislative or administrative measure, in a climate of confidence, mutual respect and good will' has been met by the JPR process.

Containing Opposition by Limiting Participation

For meaningful citizen participation in public hearings, ample opportunities must be provided, with participation actively encouraged and supported. Citizens must be supplied the relevant information about the proposal, as well as comprehensive details about how the process operates. The process for gaining standing in the JRP hearings was rather legalistic. To facilitate public involvement, civil society organizations and environmental law groups provided simplified, online forms for letters of comment (Forest Ethics) and helped participants prepare for giving oral testimony (WCEL). The 'Mob the Mic' campaign, organized by a coalition led by the ENGO Dogwood Initiative, helped approximately four thousand citizens and groups to obtain speaker standing, forcing the extension of community hearings and prolonging the JRP process by about a year.

However, frustrated by these 'delays' in approval of the Northern Gateway pipeline, the federal Conservative government included amended environmental legislation in the omnibus bills C-38 (the Jobs, Growth and Long-term Prosperity Act) and C-45 (489 and 443 pages in length, respectively) that it pushed through Parliament in 2012.[20] These bills contained amendments to the Canadian Environmental Assessment Act (CEAA), Fisheries Act, Navigable Waters Protection Act, National Energy Board Act, Species At Risk Act, Parks Canada Agency Act, Canadian Oil And Gas Operations Act, Nuclear Safety Control Act, Canada Seeds Act, and the Canadian Environmental Protection Act, and repealed the Kyoto Protocol Implementation Act and the National Roundtable on the Environment and the Economy Act, as well as the Fair Wages and Hours Labour Act. Five of the amended acts had been identified explicitly by resource industry lobbyists as legislation needing to be reformed so as to create a 'more modern, integrated, efficient framework of environmental regulation.'[21] The claims and concerns of these industry actors were consistent with those they had been expressing since the 1990s, but the new focus was clearly on the pipeline conflicts. While we cannot, within the limits of this chapter, review all of the changes introduced within these omnibus bills, the discussion of changes to the CEA and NEB acts below provides a general picture of the direction of these changes. In brief, they have essentially narrowed the criteria of participation (standing) as well as the scope of the assessments in the name of greater 'efficiency.'

The National Energy Board Act was amended in 2012 to limit future hearings to an eighteen-to-twenty-four-month period, although the new federal government elected in 2015 has initiated a review of these timeline and other procedural changes.[22] The minister for Natural Resources at the time, Joe Oliver, insisted: 'We want to allow everyone who has a *direct* interest in a particular project to have the time. What we don't need, frankly, is thousands of people belonging to the same organization coming and repeating the same packaged presentation.'[23] The new CEA Act incorporated in Bill C-38 (section 52) defined an interested party as 'a person who is directly affected by the carrying out of the designated project or a person with relevant information or expertise' (Vittal 2012, 3).[24]

Freedom to Determine the Scope of the Inquiry, or 'Scoping'[25]

James Bohman argues that legitimacy is not conferred merely by being granted access to the system, but also by the degree to which participants may initiate, introduce, and influence discussion points, problems, or even alternative proposals (Bohman, in Young 2001, 686). In the case of the JRP, this freedom was substantially circumscribed. On 19 January 2011, the JRP released the List of Issues for the environmental assessment that it would *not* consider.[26] West Coast Environmental Law (2011) summarized these as including:

- the broad climate change and greenhouse gas implications of the project and the related increase in tar sands production, or the impact of the Enbridge project on Canada's international commitments to reduce greenhouse gas emissions;
- the land, water, air, health, and social impacts of the increased tar sands developments facilitated by this pipeline;
- the environmental and climate change impacts of burning the oil and fuel that travels through Enbridge pipelines and tankers; and
- the question of whether this tar sands pipeline scheme should be a part of Canada's energy future, given the need to transition away from fossil fuels.

Though alternatives to the pipeline were part of the JRP's considerations, discussions about a national energy strategy to supply eastern Canada, the minimal bitumen upgrading being done domestically (which results in losses of value-added processing and employment opportunities), and the option of transporting bitumen via rail were not aired anywhere. The panel accepted Enbridge's argument that the transportation of the bitumen to the west coast was 'necessary' to 'give Canadian oil producers full value for their oil by diversifying market access, and preventing condensate shortages.' Thus, in its scoping decision, the JRP stated that it would not consider alternatives to the pipeline that were 'not consistent with the project's need and purpose' (JRP, Panel Session Results and Decision, 19 January 2011, 10). In the view of WCEL, this narrow interpretation of the review's scope precludes the alternative of denying the project altogether, which ultimately undermines a genuine evaluation of the real worth of the project to the public interest, which is supposed to be foremost in the panel's decision (WCEL 2011, 3). Indeed, WCEL describes how, over several years of federal consultations to draw up the Northern Gateway JRP's terms of reference, concerned citizens and First Nations repeatedly explained the importance of including up- and down-stream considerations, but the JRP rejected these requests, claiming that the issues were beyond the scope of the review (WCEL 2011, 2). As a result, the chair of the JRP repeatedly cut off or redirected participants who wished to raise such issues.

Civil society actors who persisted in trying to use the JRP hearings as forums to educate the public or voice opposition to the pipeline on grounds 'outside of' the panel's narrow terms of reference were characterized as irrational extremists by federal politicians. On 9 January 2012, federal Minister of Natural Resources Joe Oliver published an open letter that stated:

> Unfortunately, there are environmental and other radical groups that would seek to block this opportunity to diversify our trade. Their goal is to stop any major project no matter what the cost to Canadian families in lost jobs and economic growth. No forestry. No mining. No oil. No gas. No more hydro-electric dams.
>
> These groups threaten to hijack our regulatory system to achieve their radical ideological agenda. They seek to exploit any loophole they can find, stacking public hearings with bodies to ensure that delays kill good projects. They use funding from foreign special interest groups to undermine Canada's national economic interest. They attract jet-setting celebrities with some of the largest personal carbon footprints in the world to lecture Canadians not to develop our natural resources. Finally, if all other avenues have failed, they will take a quintessential American approach: sue everyone and anyone to delay the project even further. They do this because they know it can work. It works because it helps them to achieve their ultimate objective: delay a project to the point it becomes economically unviable.[27]

Iris Young observed that it is a 'common rhetorical move of official powers to paint all protest action with the tar of "extremism"' (2001, 675–6). This rhetoric may be understood, she argued, as a 'power ploy whose function is to rule out of bounds all claims that question something basic about existing institutions and the terms in which they put political alternatives,' and one that ought to be 'resisted by anyone committed to social justice and reasonable communication' (675–6). Characterizing critics of government projects in this way, while limiting their opportunities to participate in the 'authorised' process, are tactics used to delegitimize and marginalize opposition (Schilling-Vacaflor 2012, 16). Ironically, such strategies often have the effect of further radicalizing one's critics. In regard to the Northern Gateway project, now associated with a systemic institutional and political campaign on the part of the Canadian Conservative government to accelerate resource exploitation and exports in every part of the country, such attempts to marginalize opponents are fuelling new alliances and tactics among subordinated groups. 'I am not a radical,' Lucy Gagnon, Chief Dunehn and manager of the Moricetown Band, told the JRP public hearing in Smithers, BC, on 16 January 2012: 'I am a protector of the land. However, it is my responsibility to stand with my Nation to join other opponents to this devastating project to protect our territory from this pipeline.'[28]

Other First Nations responses to the pipeline project have included the Coastal First Nations Declaration that bans tar sands tankers on the north coast (March 2010), the Save the Fraser Declaration[29] banning tar sands pipelines and tankers in the Fraser River watershed and on the north and south coasts (signings in November–December 2010 and December 2011), and the St'át'imc Chiefs council resolution (October 2010). Green Party leader Elizabeth May's response to Joe Oliver's statement was that 'by characterizing this issue as environmental radicals versus Canada's future prosperity you have done a grave disservice to the development of sensible public policy' (May 2012).

Often contrasted to the JPR is the Berger Inquiry of the 1970s, which reviewed the proposal for a natural gas pipeline that would have established an energy corridor from the Beaufort Sea across the Northwest and Yukon Territories to Albertan refineries and Alaskan tanker ports (Berger 1988). In fewer than three years, the Berger Commission drew together a broad segment of concerned northern residents, instigating practices that provided the time and opportunity for all to speak. These included funding intervenors to travel, interpretation of native language testimony, holding hearings in thirty-five remote affected communities, encouraging media coverage, no strict time limitations, and making proponent information widely available. According to Canadian scholar Frances Abele, this commission initiated 'an important moment in this broad political transformation, which was both institutional and attitudinal' with regard to the processes of creating policy for developments and expanding participation (2014, 1). Justice Thomas Berger's approach moved towards Habermas' 'ideal speech situation' referring to 'free and equal discussion, unlimited in duration, constrained only by consensus which would be arrived at by the force of the better argument' (Fishkin 1991, 36).

Due to the widespread opposition that Berger heard from Aboriginal communities, he recommended that there should never be a pipeline along the Yukon coast, and that a ten-year moratorium on pipelines would facilitate time to settle indigenous land claims and determine a responsive, appropriate development program for residents. Although decades later, projects have gone through the northern corridor, this exercise in inclusive, participatory decision making not only allowed local people to make their case – giving voice to the indigenous peoples of the north for the first time – but also gave the entire Canadian population the opportunity to better understand and to relate to the peoples of the north, and revealed the issues to be far more complex than the opposition of pro- and anti-development interests. As Thomas Berger stated recently in an interview regarding the NGP: 'Democracy consists of more than just voting the government in or out every four years. Inquiries can be a critical part of the democratic process. It allows people to have a say about their future. If you consult local people, the people affected, you get better projects; that's a lesson in democracy. That's why it is so important that we provide for the fullest possible consultation in any major project.'[30]

The Limitations of the JRP as a Vehicle of Environmental Citizenship

Understanding the value pluralism that exists in Canada, one must ask whether the JRP was an adequate venue for satisfying the criteria of either deliberative democracy or of free, prior, informed consent-consultation (FPIC) with affected indigenous peoples. First, like many such processes, the panel was not charged with identifying a consensus, but with making recommendations after listening to testimony, reviewing evidence, and considering its mandate under relevant legislation. Second, while the hearings were ongoing, the federal government in June 2012 passed Bill C-38 (mentioned above), which amended the NEB Act so as to allow the federal cabinet to overrule an environmental assessment decision taken by the NEB (Olthafer and Slipp 2012). Given the federal government's strident support for the NGP, this immediately called into question the integrity of the JRP process. In either case, as in other cases observed by Archon Fung, 'officials commit to no more than receiving the testimony of participants and considering their views in their own subsequent deliberations' (2012, 615).

Another major limitation on citizenship in this process was the exclusion of certain values through the 'scoping' of the inquiry. The exclusion of the climate change implications of the pipeline from the assessment prevented citizens from speaking to their ethical obligations to people in other parts of the world, or to future generations. Nor could they speak to the implications of the pipeline in terms of environmental rights, as these are not entrenched constitutionally in Canada.

Aboriginal entities may assert collective, treaty-based or constitutional rights that are separate from 'Canadian' individual citizenship rights. Yet – as so many groups observed – the JRP was not really the appropriate venue for the contestation of the pipeline on these juridical grounds, and, indeed, the conflict will now move to the courts. At the time of writing, eight First Nations and nine other legal challenges have been initiated against Enbridge and the governments over Northern Gateway pipelines and tankers (WCEL 2014).

The values considered relevant to the review, under the terms of the JRP agreement, included those set out in the CEEA and the NEB Act. That is, the panel should determine whether significant adverse environmental effects might be caused and, if so, whether the project might nevertheless be justified by reasons of 'public convenience and necessity.' But deeply rooted, significant value differences concerning what is in the public interest, or 'necessary' – stemming from the divergent interests and beliefs among Canadians as well as the existence of cultural pluralism – could not be equally represented or influential within the JRP process. In addition to the questions of 'resources' and 'scoping' that have already been discussed, the reality is that indigenous cultural or eco-centric values are subordinated to other values in the government's definition of 'public interest.' Records of the hearings and speeches made at protest rallies are replete with statements by indigenous community members and leaders concerning their land-based cultures, the importance of place, species interdependence, and the non-negotiability of ecological integrity. Many of these values were echoed by environmental organization representatives. Yet the government's assertions that the future development of Canada as an 'energy superpower' was at stake made the hegemonic position clear. The outcome of the review process was really never in question, although the conditions imposed upon the pipeline consortium might be made more or less stringent. Thus the NGP review process confirms Cesar Rodríguez-Garavito's observation that 'neoliberal multiculturalism, for its part, recognizes cultural differences and collective rights, as long as they do not give rise to this type of entitlement and do not question, as indigenous claims do, the conventional conceptions of economic development' (2010, 280).

In his study of the potentially ambiguous impacts of FPIC in Colombia, Rodriguez-Garavito elucidates how indigenous consultations tend to substitute procedural rules (regarding timelines, affidavits, and attendees' legal standing) for meaningful discussion of the difficult issues – such as ethnic rights, land rights, and the terms of natural resource exploitation (2010, 273). In other words, the process appears to be fair because the same legal framework applies to everyone, but the origins and functions of the rules themselves are not up for discussion, and these may effectively privilege some interests (e.g., those with better resources to work within this system) over others. Rodriguez-Garavito concluded that constant, iterative, and meaningful involvement of all parties would remedy the 'displacing effect' legalism can have on consultations, that 'transform[s] substance into form' while it could still retain 'its capacity to offer a point of contact among actors defending extremely different, even antagonistic, positions'(292).

In other words, to be capable of incorporating diverse interests and values in a meaningful way, consultation processes must permit the serious consideration of differing interpretations of common good or public interest and take account of multiple forms of knowledge. But a 'consultation process' is not equivalent to a deliberating body charged with producing decisions; it does not necessarily entail the organized representation of interests, or a process for dealing with conflicting interests and values, or for making consensus or majority-based decisions. There is no commitment on the part of government decision makers to accept the recommendations produced by a representative body of 'stakeholders.' Instead, the review panel and its public hearings serve the purposes of generating knowledge (research, reports, testimonies) for public record, as well as providing a platform for different voices to be heard by the general public (mostly via media reporting). Government decisions may be more or less informed by, and accountable to, such knowledge and voices or shifts in public opinion.

It is clear that many of those who participated in the JRP process (as well as those who boycotted) did not consider it to be a truly democratic exercise in the sense that there was a possibility of substantially changing the outcome of the approval decision. Hence the continued, widespread opposition to the Northern Gateway pipeline project 'by other means,' including municipal council motions, demonstrations, and declarations by Aboriginal organizations that they will use court actions and blockades to obstruct pipeline construction.[31] Almut Schilling-Vacaflor has observed in other contexts that the 'acceleration of consultation procedures frequently proves counterproductive, as in the long term their deficiencies are likely to come to light and to instigate anger, protest activities, social conflicts or even the revoking of obtained agreements' (2012,17). Because of these issues, there is also no guarantee of shorter reviews due to the time it takes to assess who is qualified to participate, and due to the potential risk of numerous appeals of board rulings – both points made by Carpenter (2012) in his review of Bill C-38.

Science-Based Policy and Environmental Citizenship

Prime Minister Stephen Harper and Minister of Natural Resources Joe Oliver have frequently stated that the NGP decision will not be based on 'politics' but on 'science.'[32] The Joint Review Panel issued a procedural direction that it 'makes its decisions on the facts or evidence in front of it' (JRP Procedural Direction no. 3, Community Hearings, p. 2). Like judges hearing evidence in a courtroom, the panel members are represented as being impartial, that is, they will not be swayed by emotion or rhetoric. Many scientists also enthusiastically support the call for 'science-based' public policy, seeing this as the opposite of policy decisions based on 'political interests,' that is, impartial and free of corruption. However, reviewing the use of science in other environmental conflicts in Canada, Adkin concludes that the term 'science-based policy' 'conveys the ideas of impartiality and superior rationality, and may serve to circumscribe debate' (Adkin 2009, 305–7).

First, only certain kinds of scientific questions are asked, or included, in the 'scoping' of environmental and social impact assessments or in the mandates of various policy consultation processes. The Joint Review

Panel was interested to know what might happen to salmon stocks if oil spills of various sizes occur in the Douglas Channel or Hecate Strait; it was not interested to know how much CO_2 equivalent will enter the atmosphere within the next thirty years as a result of transportation of crude oil from the Alberta oil sands to markets via the NGP.

A second problem with the 'science-based' approach is that much of the research presented typically concerns risk and probability estimates. The tolerance of affected communities for risk depends not only on these numbers, but on what communities value – how they perceive, subjectively, the value of what is threatened or the alternatives available. There is no 'objective' position from which scientists or political decision makers can decide what constitutes an 'acceptable' risk for a given community. In other words, policy choices are culturally, socially, and ethically based decisions that make use of scientific and other forms of knowledge.

Moreover, many so-called scientific or technical claims are really little more than appeals for faith – in technologies yet to be invented, perfected, or tested. The question about the likelihood of oil spills and their possible effects on marine life offers an instructive example. When confronted with arguments that spills would be nearly inevitable, Enbridge responded that it would contain, clean up, and compensate for any future harms to salmon stocks (and those Aboriginal communities and other fishers who depend upon them). In other words, the public should rely upon the company's technical knowledge and abilities to 'clean up' any messes created by leaks from pipelines or damaged tankers.[33] Significantly, these assurances also assume that such harms are recompensable in monetary form, bringing us back to the point made above about the role of values in individuals' and communities' responses to risks and uncertainties. The insistence of many Aboriginal communities that monetary and cultural-ecological values are incommensurable has been the hallmark of their resistance to the Northern Gateway project.

Another example of the kind of scientific risk analysis that played a significant role in the hearings is the debate about the population density threshold for caribou. A professor representing BC Nature and Nature Canada challenged Enbridge's data on 'scientific' grounds, charging that it relied on a single source that was unpublished and not peer reviewed.[34] Science can provide more or less thorough, comprehensive, and accurate estimates of various phenomena, and these kinds of criticisms help us (scientists and non-scientists) to evaluate the claims presented to us. Indeed, in the JPR process we witnessed some highly questionable omissions of data and other techniques to influence public opinion. Enbridge produced, in December 2011, a video of the proposed tanker path into Kitimat port that omitted a thousand kilometres of islands from the graphic map image, depicting the channel as wide open. In reaction, a scientist launched a formal complaint with the federal Competition Bureau, and a forty-thousand-person petition was presented to Enbridge by a Prince Rupert city councillor, requesting the company to stop its misleading advertising.[35] In May 2014, more than three hundred scholars from multiple disciplines wrote to the prime minister, urging him to reject the JPR's report on the grounds that it 'cannot be considered scientific, empirical, quantitative, or objective as it relies on the subjective assessments of interested parties ... and the undocumented manipulation of data, without any described methodology, by a paid consultant of the proponent.'[36] The authors claimed that the panel had dismissed the concerns raised by many expert intervenors, indicating 'that they, in fact, are using a separate process for evaluating the risks of the project – one that is not scientific or based on a technical analysis of the evidence.'

Such challenges are not unimportant, but they may become the strategic equivalent of 'using the master's tools to dismantle the master's house'[37] because they deflect attention from more fundamental challenges to corporate sustainable development discourse. For example, the panel was interested to know why the pipeline is 'necessary' to transport the huge reserve of bitumen from the oil sands and how much revenue this will generate for the provincial and federal governments in royalties (although these are, arguably, economic questions rather than 'scientific' ones). It was not interested to know what the potential is for substituting renewable sources of energy for oil, to meet energy demands in Canada (a scientific-technical question). Surely a 'science-based' energy policy should examine all options and their environmental implications. The point here is that whatever 'science' is presented is necessarily selective – it answers certain questions but not others. Citizens need to know who decides *the questions*, and to be *as* aware of the *absent* questions as they are of those that have been prioritized and included in predetermined consultation processes.

None of the urgent socio-environmental problems facing us can be answered by scientific knowledge alone. At the heart of conflicts like those surrounding the Northern Gateway pipeline, or the exploitation of the oil sands, are values, and visions of what conditions make possible or constitute a good life. Yes, the pipeline will allow oil companies to ship their products long distances. But is this 'necessary'? Why, and to whom? Will this make our lives better? How? Are there other ways to make our lives better that might entail less harm or risk to others (human and non-human) and to future generations? Such questions of values and ethics are, indeed, the starting points of indigenous peoples' interventions in these conflicts: If we are of 'good mind,' which path do we take? Which choices are in harmony with the well-being of the natural world? Which choices will preserve this world for the generations to come after us? How much can we take without causing harm – without disturbing the balance between our needs and those of other beings? Scientific knowledge can be helpful in identifying what is possible, but it cannot tell us what is the right ethical choice, or what is the best way to live. The privileging, in processes like the JRP, of the competence of experts, professionals, and scientists really puts the cart before the horse. The prior questions are ethical and societal, not scientific. Deliberation about these is both necessary and difficult because there is more than one answer to questions about societal priorities and directions.

Moreover, there may be multiple *communities* with asymmetrical rights and status vis-à-vis the state, as is the case in settler states like Canada. The JRP – like many other government-initiated consultation processes in this country – created 'special conditions' for First Nations participation. These included the panel's acceptance of 'oral traditional knowledge' in the form of oral narratives from Aboriginal elders (as affirmed in *Delgamuukw v. British Columbia* [1997] 3 S.C.R. 1010). In the world of Western scientific rationalism, such knowledge occupies a kind of no-man's land between the empirical and the cultural. The panel and the federal cabinet may choose to 'accept' or 'hear' or act upon traditional ecological knowledge at their pleasure. Enbridge, as Tyler McCreary and Richard Milligan (2013) point out, included traditional knowledge in its application. However, in the view of these authors, the selection and organization of such knowledge actually functioned to reproduce colonial beliefs about indigenous culture and interests, and thereby to minimize indigenous concerns and rights claims. For example,

> efforts to account for sites of Indigenous cultural importance in Enbridge's ATK report effectively compartmentalize Indigenous geographies to a set of discrete legible places worthy of protection. Incorporating an Indigene-

> ity presumed to lack sovereign authority in order to sanction development on unceded territories, this move to recognition works quietly to reestablish a *terra nullius* open again to development but mildly constrained by discrete, localized patches of Indigeneity.' (122)

Recognition of indigeneity is ultimately merely token if the dominant model of development cannot meaningfully accommodate it – or worse, steadily erodes the conditions for indigenous ways of being to flourish in this world. Nothing in the JRP process altered structural power asymmetries and biases, although the possibility that indigenous voices will be heard by the larger public may promote 'dialogical understanding' (Young, in Smith 2003, 57).

Concluding Thoughts

Deliberative democracy actively encourages contributions, discussion, and public defence of rationally derived perspectives and opinions from as wide a swath of society as possible, and requires that those most affected by a decision have a substantial say in the outcome. Participants are free of coercion, and open to changing their minds if so persuaded by the force of good arguments about what is in the public interest. In light of the analysis we have made, how does the JRP measure up against the criteria set out in table 13.2?

Impartiality

The panel could be impartial only in the narrow procedural sense, that is, within the parameters that were determined for the inquiry's scope prior to public hearings. Political biases do not have to be found within the biographical profiles of the panelists themselves (although this would not be impossible and might be relevant) because 'political biases are built into the process itself' (Amy 1987, 196). The existence of transparent, formalized procedures for the JRP did not guarantee that the substantive outcomes of the decision-making process would be free of political bias or privileged influence on the part of particular 'stakeholders.' Indeed, the reasons behind the decision taken by the federal cabinet have not been transparent, due to cabinet secrecy and the difficulty of obtaining full knowledge of which particular interests had access to decision makers, or the forms of leverage exerted by these actors.

Inclusiveness

Scope: A utilitarian, economic definition of 'public interest' predominated in the JPR's terms of reference from the outset; the full range of values and visions of what might best serve the public interest was never in play. The operational definition was, moreover, not a consensus arrived at by citizens engaged in a fully inclusive (in a representational sense) deliberation, but rather by policy makers with a particularly close relationship to the proponent (and by means that were not fully transparent).

Participation: Participation in the public hearings was reasonably open, although obstacles were created by the legalistic application procedures. Differential access was given, however, at different stages of the process, with the proponent and its experts having twice as many opportunities to present their cases and respond to evidence, such as within the cross-examinations for the scientific and final reviews portions. Also, after the JRP's technical hearings, Enbridge submitted documents proposing to double the marine terminal capacity – while no other intervenors were permitted to include late evidence.[38] Resources to influence the panel's report were substantially unequal, if one compares the legal, technical-engineering, and scientific expertise that the proponent could mobilize within a relatively short time frame to the resources available to citizens' groups, individuals, and First Nations intervenors (notwithstanding the funding provided to some First Nations to enable their participation). Due to the participation of environmentalists and First Nations, voices were raised for the non-human, although the interests of non-human nature had no standing in the process comparable to that of 'the economy' (standing in for human interests within Canadian borders). Not represented were non-Canadian citizens whom we can expect to be affected by the export of synthetic crude or dilbit to other parts of the world, and by the consequent rise of global greenhouse gas emissions. The attempts of citizens in the USA to draw attention to these issues and to support environmental intervenors in the JRP hearings were condemned by federal government officials as 'foreign interference.'

Free Prior Informed Consultation and Consent for Indigenous Peoples

The criteria from table 13.2 reveal that deficiencies related to FPIC emerged in the NGP consultation. The neutrality of the JRP has been questioned, given the perceived intervention by the federal government – which for many reflects a lack of 'mutual respect or good will.' The inabilities of First Nations to contribute meaningfully to agenda formation or to veto the project were fundamental flaws in the way consultations were established. The high costs in terms of both time and money were prohibitive to active, equal participation, and few opportunities existed to fund studies other than those produced by the proponent. The closed-door nature of many Enbridge–First Nations negotiations has fomented divisions within communities regarding which voices legitimately represent them (Holland 2012; Unistoten Camp 2012).

From 'Done Deal' to 'No Deal'?

At the end of this expensive, three-year inquiry – entailing formal procedures, intervenor funding, lengthy environmental assessment studies, and extensive oral testimony – we are struck by the outcomes that it has already produced (or at least failed to prevent): deepened public mistrust in the impartiality and democratic accountability of the federal government, a radicalized Aboriginal rights movement, and a new convergence of social movements.

In addition to the NGP, governments and industry continue to pursue alternative pipeline routes to move Alberta oil to the west coast as well as to refineries in Eastern Canada and in the USA (Cousineau 2013). In the meantime, more and more oil is being shipped by rail, and there is growing concern in communities across the country about the risks of train derailments, as well as pressure on the federal government to improve and enforce its rail transportation standards. The greatest obstacle to the expansion of production in the oil sands, and hence to new investment in the mining and pipeline sectors is now the concerted effort of indigenous and environmental movements to block every new route to markets.

The mobilization around the Northern Gateway pipeline project suggests that there exists a growing public consensus about the need to broaden the discussion of Canada's developmental path beyond the needs of extractive capitalism. Particularly in British Columbia, there has been a loud and sustained call for transition towards a greener economy and more meaningful, participatory citizenship. A green transition will mean embracing ways of living within natural limits, and subordinating markets to the needs of resilient, sustainable communities. The transformation of our democracy, including the institutionalization of deliberative practices, will be critical to achieving these ends. But in the current movements, we also see the potential role of civil disobedience and the creation of 'really existing' local alternatives as forms of ecological citizenship that will carry this process forward. The Idle No More movement is articulating many of these threads, combining ecological, indigenous, and democratic discourses. Given the crises we face, we may expect that challenges to the status quo, as Jorge Valadez says, will no longer be 'sporadic, but systemic and continual '(2001, 66).

Lastly, it must be said that the account we offer here of the Northern Gateway pipeline conflict is necessarily incomplete, and only touches upon the peaks of other mountainous questions. There are dissertations to be written, for example, about the Enbridge-led consortium's negotiations with Aboriginal communities in Alberta and British Columbia – research that would contribute greatly to the kind of work being undertaken by Anna Zalik on corporate strategies with regard to the social licence to operate. Nor do we yet have adequate knowledge of how high-level strategic and policy decisions are being negotiated among corporate and political elites. In years to come, the Northern Gateway pipeline review process is likely to generate a body of scholarship comparable to that produced by the Mackenzie Valley pipeline inquiry some forty years ago. Indeed, while their contexts have changed significantly, they are the same conflict, and the stakes for the planet, and for indigenous peoples, have never been higher.

NOTES

1 Address to the Canada-UK Chamber of Commerce, London, England, posted on http://www.pm.gc.ca/eng/media.asp?id=1247 (accessed 8 December 2012).
2 http://savethefraser.ca/ (accessed 23 April 2012).
3 'Aboriginal peoples' and 'Indigenous Peoples' are terms used interchangeably, although the latter is more commonly used in an international context. In the Canadian context, 'Aboriginal peoples' refers to both status and non-status Indians, Inuit, and Métis nations, whereas 'First Nations' refers more narrowly to status and non-status Indians. The term 'First Nation' began to be used in the 1970s to replace 'band' or 'Indian.'
4 See, for example, the calculations made by Barry Saxifrage in 'Alberta's Oilsands Pipelines Promise Massive Non-stop Brand-Canada Carbon Spills,' *Vancouver Observer*, 21 June 2012, http://www.vancouverobserver

.com/blogs/climatesnapshot/albertas-oilsands-pipelines-promise-massive-non-stop-brand-canada-carbon?page=0,1).

5 The JRP agreement may be found on the website of the CEAA: http://www.ceaa-acee.gc.ca/050/document-eng.cfm?document=39960.

6 Gary Schneider, John Sinclair, and Lisa Mitchell were members of the Environmental Planning and Assessment Caucus of the Canadian Environmental Network, representing the Environmental Coalition of Prince Edward Island, the Green Action Centre of Manitoba, and East Coast Environmental Law, based in Nova Scotia, respectively. They noted that there was not much evidence to support the claims of the resource industry. On the contrary, a 1999 study for the CEA Agency had concluded that 'EA duplication is rare, sometimes positive and avoidable where there is a will on the part of the parties involved' (2007, 5).

7 Gaetan Caron, chair and CEO of the National Energy Board, quoting a previous board member, Jean-Paul Théorêt, in 'North American Energy and Responsible Development: The Role of the National Energy Board as a Partner,' presentation to the Legal and Regulatory Affairs Committee of the Interstate Oil & Gas Compact Commission, Calgary, Alberta, 6 May 2008, http://www.neb-one.gc.ca/clf-nsi/archives/rpblctn/spchsndprsnttn/2008/nrthmrcnnrgrspnsbldvlpmnt/nrthmrcnnrgrspnsbldvlpmnt-eng.html.

8 JRP process advisor, personal communication to Larissa Stendie, 11 February 2013. This account of the JRP process draws upon research for Stendie's master's thesis (2013).

9 Compiled by Larissa Stendie from official documentation and transcripts archived on the website of the National Energy Board, https://www.neb-one.gc.ca/ll-eng/livelink.exe/fetch/2000/90464/90552/384192/620327/customview.html?func=ll&objId=620327&objAction=browse&sort=-name&redirect=3.

10 The Supreme Court decision elaborates: 'This means the government must act in a way that respects the fact that Aboriginal title is a group interest that inheres in present and future generations, and the duty infuses an obligation of proportionality into the justification process: the incursion must be necessary to achieve the government's goal (rational connection); the government must go no further than necessary to achieve it (minimal impairment); and the benefits that may be expected to flow from that goal must not be outweighed by adverse effects on the Aboriginal interest (proportionality of impact)' (http://scc-csc.lexum.com/scc-csc/scc-csc/en/item/14246/index.do).

11 The conditions may be found in Attachment B of the JRP's report: https://docs.neb-one.gc.ca/ll-eng/llisapi.dll/fetch/2000/90464/90552/384192/620327/624909/942629/A346-5_-_Panel-Commission_-_Attachment_B_-_Collection_of_potential_conditions_-_A3G7X1.pdf?nodeid=942306&vernum=-2.
12 See sections 4.3.1 and 4.3.2 of the JPR report's 'considerations,' at http://gatewaypanel.review-examen.gc.ca/clf-nsi/dcmnt/rcmndtnsrprt/rcmndtnsrprtvlm2chp4-eng.html.
13 David de Wit, natural resource manager, Office of the Wet'suwet'en, letter to panel manager, Canadian Environmental Assessment Agency, Review Panels Division, 20 June 2011. Letter on file with the NGP JRP Public Registry, https://www.neb-one.gc.ca/ll-eng/livelink.exe/fetch/2000/90464/90552/384192/620327/624910/699647/697977/D157-1-1_-_Office_of_the_Wet'suwet'en_-_Letter_-_A1Z9S8.pdf?nodeid=697978&vernum=0.
14 *Grassy Narrows First Nation v. Ontario (Natural Resources)*, 2014 SCC 48. For commentaries on this decision, see Hayden King, 'Land Ruling's Message to First Nations: You Have No Place in Confederation,' *Globe and Mail*, 14 July 2014; Roy Millen, Laura Cundari, and Bryan Hicks, 'Canada: Ontario Can "Take Up" Lands under Treaty 3 without Seeking Federal Approval,' 14 July 2014, http://www.mondaq.com/canada/x/327230/indigenous+peoples/Ontario+Can+Take+Up+Lands+Under+Treaty+3+Without+Seeking+Federal+Approval.
15 See note 12.
16 Caroline O'Driscoll, 'Final Written Argument of the Alexander First Nation in the Matter of the Enbridge Northern Gateway Project Joint Review Panel,' 31 May 2013, https://www.ceaa-acee.gc.ca/050/documents/p21799/89915E.pdf; Testimony of Chief Herb Arcand at the JRP Hearing in Terrace, BC, 17 June 2013, Hearing Order OH-4-2011, 605-685, http://www.ceaa-acee.gc.ca/050/documents/p21799/90736E.pdf.
17 These responses are made in the JRP's report, volume 2, section 4. See note 12.
18 Andrew Andy, chief elect, Nuxalk Nation, letter to Secretariat of the Joint Review Panel, 4 April 2012, filed with the JRP's Public Registry.
19 Andy, letter to Secretariat of the Joint Review Panel, 4 April 2012.
20 Natural Resources Minister Joe Oliver accused the pipeline's opponents of 'stacking public hearings with bodies to ensure that delays kill good projects' (Natural Resources Canada 2012).
21 Greenpeace Canada obtained a letter written by the Canadian Energy Pipelines Association, Canadian Gas Association, Canadian Petroleum

Products Institute, and Canadian Association of Petroleum Producers to the Canadian ministers of environment and natural resources, dated 12 December 2011, http://www.greenpeace.org/canada/Global/canada/pr/2013/01/ATIP_Industry_letter_on_enviro_regs_to_Oliver_and_Kent.pdf (accessed 12 January 2013).

22 See http://www.neb-one.gc.ca/clf-nsi/rpblctn/ctsndrgltn/ct/jbsgrwthprsprty/jbgrwthprsprtyfq-eng.html#fq01.

23 'Minister on Pipeline Review,' *Power and Politics*, CBC, 30 March 2012, http://www.cbc.ca/player/Shows/ID/2217656262/.

24 The National Energy Board review process for the Kinder Morgan pipeline proposal, issued in 2014 and operating under the changes effected by the Harper government's omnibus Bill C-38, has been met with concerted opposition from environmentalists, First Nations, and other communities along the route. See Sierra Club BC, *Credibility Crisis*, 29 June 2015, http://sierraclub.bc.ca/wp-content/uploads/2015/08/NEB-Flaws-Report-June-27.pdf.

25 'Scoping is the term used to describe the process of a federal authority deciding what is included in a project or what will be considered in the environmental assessment of a project' (Doelle and Kwasniak 2007, 1).

26 JRP Sessions Decision, 19 January 2011, p. 10, https://www.neb-one.gc.ca/ll-eng/livelink.exe/fetch/2000/90464/90552/384192/620327/624910/695919/727809/D122-3-06_-_Living_Oceans_Society,_Raincoast_Conservation_Foundation_and_ForestEthics_-_Tab_C__Panel_Session_Results_and_Decision,_19_January_2011_A2F2U5_.pdf?nodeid=728030&vernum=0.

27 Honourable Joe Oliver, 'An Open Letter from the Minister of Natural Resources,' Natural Resources Canada, 9 January 2012, http://www.nrcan.gc.ca/media-room/news-release/2012/1/3520 (accessed 10 March 2012).

28 JRP Hearing Order OH-4-2011, Smithers, BC, 16 January 2012, para 5577.

29 West Coast Environmental Law, http://wcel.org/sites/default/files/file-downloads/Coastal%20First%20Nations%20Tanker%20Ban%20Declaration.pdf.

30 'Living Democracy from the Ground Up; Part 3 – Honorable Thomas Berger,' *YouTube* video, posted by West Coast Environmental Law, 27 November 2012, http://www.youtube.com/watch?feature=player_embedded&v=n0bsqUlPEkM (accessed 7 December 2012).

31 Some municipalities along the pipeline's path have voted to reject the project – most notably, the port terminal town of Kitimat.

32 See, for example, 'Harper Defends Independence of Pipeline Approval Process,' *CBC News*, 7 August 2012, http://www.cbc.ca/news/politics/

story/2012/08/07/pol-gateway-tuesday-harper-bc.html (accessed 15 September 2012).

33 Questions put to Enbridge by the JRP resulted in the company committing to spending five billion dollars more for safety measures (to prevent and respond to spills) than had been in its initial proposal. See WCEL 2011; JRP Panel Session Results and Decisions, 19 January 2011, 18.

34 Chris Tollefson, quoted in Mark Hume, 'Enbridge's "Errata" on Caribou Could Prove a Costly Error,' *Globe and Mail*, 11 November 2012, http://www.theglobeandmail.com/news/british-columbia/enbridges-errata-on-caribou-could-prove-a-costly-error/article5188396/ (accessed 13 November 2012).

35 Lori Waters, letter of complaint to the Competition Bureau of Canada against Enbridge for false or misleading advertising and deceptive marketing practices, 16 September 2012, https://www.box.com/s/16509628de91608d12e1#/s/16509628de91608d12e1/1/355824541/2876446089/1 (accessed 17 September 2012).

36 A link to this letter may be found at http://www.cbc.ca/news/canada/british-columbia/northern-gateway-pipeline-report-flawed-300-scholars-tell-pm-1.2663230 (accessed 18 May 18).

37 This phrase was used by American feminist, poet, and essayist Audre Lorde, in an essay entitled 'The Master's Tools Will Never Dismantle the Master's House,' in *Sister Outsider*, 110–13 (Berkeley: Crossing Press, 1984). Lorde used the phrase in her argument that white, middle-class feminists would not be able to achieve the emancipation of women from patriarchy unless they also recognized, and confronted, the relations of domination operating along racial, class, sexuality, and other axes. Here, we use the phrase to suggest that fighting science with science – without examining the ways in which modern science is socially determined and cannot substitute for other forms of wisdom – is an inadequate strategy to challenge and transform the hegemonic model of development.

38 Robin Rowland, 'Expansion of Proposed Kitimat Bitumen Terminal Urgent to Get Offshore Markets, Enbridge Tells JRP,' *Northwest Coast Energy News and Issues*, 4, February 2013. http://nwcoastenergynews.com/2013/02/04/4184/expansion-proposed-kitimat-bitumen-terminal-urgent-offshore-markets-enbridge-tells-jrp/ (accessed 22 February 2013).

REFERENCES

Abele, Frances. 2014. 'The Immediate and Lasting Impact of the Inquiry into the Construction of a Pipeline in the Mackenzie Valley, 1974–77.' In *Commis-*

sions of Inquiry and Policy Change: A Comparative Analysis, ed. Greg Inwood and Carolyn Johns, 88–112. Toronto: University of Toronto Press.

Adkin, Laurie, ed. 2009. *Environmental Conflict and Democracy in Canada*. Vancouver: UBC Press.

Amy, Douglas J. 1987. *The Politics of Environmental Mediation*. New York: Colombia University Press.

Benevides, Hugh, Theresa Clenaghan, and Richard Lindgren. 2011. 'Submission on the National Energy Board's November 2010 Draft Update of the Environmental and Socioeconomic Assessment Section of Filing Manual.' Calgary: Canadian Environmental Law Association (CELA). http://www.neb-one.gc.ca/clf-nsi/rpblctn/ctsndrgltn/flngmnl/flngmnlpdtcmmnt/cndnnvrnmntllwssctn.pdf (accessed 20 March 2013).

Berger, Thomas. 1988 [1977]. *The Report of the Mackenzie Valley Pipeline Inquiry: Northern Frontier, Northern Homeland*. Rev. ed. Vancouver: Douglas & McIntyre.

Bohman, James. 1996. *Public Deliberation*. Cambridge: MIT Press.

Canada, Parliament of. 2000. 'Aboriginal Title: The Supreme Court of Canada Decision in Delgamuukw v. British Columbia.' Parliamentary Research Branch, 37. Ottawa: Library of Parliament.

Carpenter, Sandy. 2012. 'Fixing the Energy Project Approval Process in Canada: An Early Assessment of Bill C-38 and Other Thoughts.' *Alberta Law Review* 50 (2): 229–70.

Concerned Professional Engineers. n.d. 'Flawed Risk Analysis: Northern Gateway's Risk Calculations Are Flawed and Unscientific.' http://www.concernedengineers.org/flawed-risk-analysis/ (accessed 30 August 2014).

Davis, Megan. 2012. 'Identity, Power, and Rights: The State, International Institutions, and Indigenous Peoples in Canada.' In *The Politics of Resource Extraction*, ed. Suzana Sawyer and Edmund Terence Gomez, 230–52. London: Palgrave Macmillan.

Doelle, Meinhard, and Arlene Kwasniak. 2007. *Backgrounder on Scoping under CEAA*. Ottawa: Canadian Environmental Network. http://rcen.ca/sites/default/files/uploads/epa_scoping_backgrounder07.pdf (accessed 15 March 2012).

Enbridge. 2010. *Volume 1: Overview and General Information*. Enbridge Northern Gateway Project. Sec. 52 Application. http://www.northerngateway.ca/assets/pdf/application/Master_Vol%201_Final_11May10.pdf (accessed 20 March 2012).

Enbridge. 2011a. 'Commitment to Public Engagement Facts Sheet.' Document No. NGP-FS-04-001, 17 March. http://www.northerngateway.ca/assets/

pdf/Public%20Consultation/NGP-FS-04-001_Commitment%20to%20Public%20Engagement.pdf (accessed 13 November 2012).
Enbridge. 2011b. 'Aboriginal Benefits Facts sheet.' Document no. NGP-FS-05-002, 9 February. http://www.northerngateway.ca/aboriginal-engagement/benefits-for-aboriginals/ (accessed 15 January 2012).
Fishkin, James. 1991. *Democracy and Deliberation: New Directions for Democratic Reform*. Yale: University of Yale Press.
Fung, Archon. 2012. 'Continuous Institutional Innovation and the Pragmatic Conception of Democracy.' *Polity* 44 (4): 609–24.
Haisla Nation. 2013. 'Final Argument: Northern Gateway Pipeline Project.' 31 May. https://docs.neb-one.gc.ca/ll-eng/llisapi.dll/feth/2000/90464/90552/384192/620327/624910/693017/960020/D80-104-2__Haisla_Nation_Final_Written_Argument_-_31_May_2013_-_A3I0V0.pdf?_gc_lang=en&nodeid=959797&vernum=0.
Joint Review Panel (Enbridge Northern Gateway Pipeline). 2012. 'Role and Mandate of the JRP.' http://gatewaypanel.review-examen.gc.ca/clf-nsi/prtcptngprcss/nfrmtnsssn/rlndmndt-eng.html (accessed 30 March 2012).
Kopecky, Arno. 2013. 'The $273 Billion Question.' *Reader's Digest Canada*, February, 52–71.
McCreary, Tyler, and Richard A. Milligan. 2013. 'Pipelines, Permits, and Protests: Carrier Sekani Encounters with the Enbridge Northern Gateway Project.' *Cultural Geographies* 21 (1): 115–29.
Moore, Dene. 2013. 'Coastal First Nations Quit Northern Gateway Pipeline Review.' Canadian Press, 4 February. http://www.huffingtonpost.ca/2013/02/04/coastal-first-nations-northern-gateway-pipeline_n_2616287.html?view=print&comm_ref=false (accessed 10 February 2013).
Natural Resources Canada. 2012. 'An Open Letter from Minister Oliver on Our Energy Markets and the Regulatory Process.' http://www.nrcan.gc.ca/media-room/news-release/2012/1/3525.
Rodriguez-Garavito, Cesar. 2010. 'Ethnicity.Gov: Global Governance, Indigenous People and the Right to Prior Consultation in Social Minefields.' *Indiana Journal of Global Legal Studies* 18 (1): 263–305.
Schilling-Vacaflor, Almut. 2012. *Democratizing Resource Governance through Prior Consultations? Lessons from Bolivia's Hydrocarbon Sector*. Hamburg: German Institute of Global and Area Studies.
Schneider, Gary, John Sinclair, and Lisa Mitchell. 2007. 'Final Substitution Paper – Environmental Assessment Process Substitution: A Participant's View.' Toronto: Environmental Planning and Assessment Caucus, Canadian Environmental Network. http://rcen.ca/sites/default/files/uploads/epa_final_substitution.pdf (accessed 15 November 2012).

Sinclair, John A., Alan Diduck, and Patricia Fitzpatrick. 2008. 'Conceptualizing Learning for Sustainability through Environmental Assessment: Critical Reflections on 15 Years of Research.' *Environmental Impact Assessment Review* 28: 415–28.

Smith, Graham. 2003. *Deliberative Democracy and the Environment*. New York: Routledge.

Stendie, Larissa K. 2013. 'Public Participation, Petro-Politics and Indigenous Peoples: The Contentious Northern Gateway Pipeline and Joint Review Panel Process.' Master's thesis, University of Oslo. https://www.duo.uio.no/handle/10852/35958.

Valadez, Jorge. 2001. *Deliberative Democracy: Political Legitimacy and Self-Determination in Multicultural Societies.* Boulder: Westview.

Vittal, Priyanka. 2012. 'The Canadian Environmental Assessment Act (CEAA) 2012: A Plain Language Summary.' Toronto: Environmental Planning and Assessment Caucus, Canadian Environmental Network. http://rcen.ca/sites/default/files/vittal_ceaa_2012_plain_language.pdf (accessed 15 August 2012).

West Coast Environmental Law. 2009. *Legal Backgrounder: The Crown's Approach to First Nations Consultation on the Enbridge Gateway Pipeline*. Vancouver: WCEL.

West Coast Environmental Law. 2011. 'Backgrounder – Joint Review Panel's Decision on the Scope of the Environmental Assessment for Enbridge Northern Gateway Pipelines.' http://wcel.org/sites/default/files/publications/West%20Coast%20-%20JRP%20Decision%20on%20Scope%20-%20Backgrounder_0.pdf (accessed 15 February 2012).

West Coast Environmental Law. 2014. 'First Nations Legal Challenges to Enbridge Pipelines and Tankers Project: Legal Backgrounder.' 16 July. http://wcel.org/sites/default/files/publications/Legal%20backgrounder_First%20Nations%20launch%20challenges%20against%20Enbridge_0.pdf.

Young, Iris Marion. 2001. 'Activist Challenges to Deliberative Democracy.' *Political Theory* 29 (5): 670–90.

14 Social Movements Scaling Up: Strategies and Opportunities in Opposing the Oil Sands Status Quo

RANDOLPH HALUZA-DELAY AND ANGELA V. CARTER

Alberta's energy production trajectory has left civil society within and beyond the borders of the province struggling to reconcile wealth accumulation with ecological and social concerns. In this chapter we focus on social movements resisting the status quo of oil sands production and the 'petro-capitalism' that dominates the political culture of the province. The focus of analysis in this chapter – civil society organizations – complements the analyses of actors in petro-politics found in other chapters of this volume, including Adkin and Stares' focus on conservative governments' hegemonic politics, Carter's and Zalik's summaries of corporate strategies, and Davidsen's focus on media roles. While social movements are often viewed primarily as political entities, here we focus on social movements as discursive and cultural agents that challenge the legitimacy of the oil sands status quo and that therefore have real, material impacts on power relations.[1]

Structures, institutions, resources, and fortuitous events constitute an 'opportunity structure' within which social movement action is strategized (McCammonet al. 2007; Meyer and Minkoff 2004; Poloni-Staudinger 2008). Social movement actors' responses to these opportunities (or closure of opportunities) and to the actions of other actors have created a complex trajectory for oil sands activism in Alberta and beyond. The struggle against the unbridled expansion of oil production from the Alberta oil sands has taken on a transnational character and has brought a diverse set of social movement actors to the campaign. Until recently the engagements of labour, indigenous movements, church groups, and environmental non-governmental organizations (ENGOs) have had limited traction in the province itself.

In this chapter, we describe the movement actions and the transnationalization of these local struggles against the status quo of Alberta's intensive reliance on hydrocarbon extraction, particularly struggles against the exploitation of non-conventional oil reserves and its weak environmental monitoring. We identify successes and weaknesses in the discursive strategies, programs, and organizational practices of these movements in the context of Alberta's unique political culture. Ultimately, we argue that the cultural politics of Alberta are as contested as the policies regarding the status quo development and regulation of the oil sands. Effectively countering Alberta's petro-capitalism will require rethinking the material processes of production and social reproduction *as well as* reimagining the cultural foundations and collective identities of Albertans beyond being producers of energy. Opposition and reconstruction are required on the terrain of both the economic-political and the cultural-ideational.

Studying Social Movements

Scholarship on social movements has been dominated by research focusing on political outcomes (Giugni, McAdam, and Tilly 1999; McAdam, Tarrow, and Tilly 2001). However, social movements may pursue socio-cultural change – that is, seek to change the values, norms, behaviours, symbols, and self-identifications of movement members, supporters, or the society at large (Earl 2004; Hart 2001). Contrary to the relatively limited attention they receive in the literature, such cultural politics are crucial elements of the political-economic changes sought by social movements. Recent research into the utility of the concept of collective identity in understanding public responses to social movement mobilization (Polletta and Jasper 2001) reinforces the importance of attending to the cultural politics of Alberta in association with more conventional political-economic approaches.

Donald Moore, Jake Kosek, and Anand Pandian define 'cultural politics' as 'an approach that treats culture itself as a site of political struggle, an analytic emphasizing power, process and practice' (2003, 2). The effort to constitute an environmental citizenship in Canadian jurisdictions such as Alberta implies such cultural action by diverse actors, including social movement organizations (Adkin 2009a; Haluza-DeLay 2009). The social movements contesting the oil sands status quo may challenge accepted social norms and other cultural features in the effort to communicate their alternatives, but for the most part they have tried merely to situate themselves within the political culture of Alberta rather than actively to shape it.

Corporate and government actors have worked in concert to present the narrative that 'Alberta is energy' – which is the title of a two-year marketing campaign by the Canadian Association of Petroleum Producers (CAPP) and a clear attempt to produce a collective provincial identity. This is the cultural context in which oil sands opposition movements operate. Industry's efforts are, of course, reinforced by a provincial government that is heavily invested in the oil sands, that devotes extensive money, bureaucracy, and other resources to attract corporate investment, and that staunchly defends and promotes the industry. The oil industry has manipulated and heightened these cultural tendencies such that the hegemonic 'common sense' notion of Albertan identity has become intimately connected with energy (Haluza-DeLay 2014, forthcoming). That is, one could say that Albertans are interpellated by industry and governmental discourse as *Homo energeticus* (Kowalsky and Haluza-DeLay 2013), having deeply rooted identifications with the interests and values of the resource extraction industries.

Over the last half-century of Alberta's history, a capital-state partnership has developed (Davidson and Gismondi 2011; Gailus 2012; Richards and Pratt 1979) redolent of what Michael Watts (2005) describes as an 'oil complex' wherein the intimate association of supermajor energy companies and governments structures the political economy of the jurisdiction. This partnership appears to have deepened in the past decade (Carter, this volume), clearly shaping a particular political culture for the province. This political culture affects corporate involvements and institutional arrangements with other social actors, such as Aboriginal communities (Zalik, this volume). It is on this landscape of petro-capitalism that social movement actors must oppose the reigning hegemony and present alternatives that are sufficient to evoke acceptance by the populace (Crehan 2002).

In this way, we understand social movements as culture workers and not merely political actors. Such recognition is especially relevant for thinking about counter-hegemonic strategy in Alberta. And while the influence of discursive practices is sometimes exaggerated (Carroll 2006), discursive campaigning is core to the 'war of position' described by Antonio Gramsci (1971) in his reflections on how to bring about profound and far-reaching social change in the context of industrial capitalism. The war of position was Gramsci's way of describing the slow, patient effort of transforming civil society, pulling citizens towards an alternative hegemonic worldview (Carroll 2006; Crehan 2002; Ekers, Loftus, and Mann 2009). 'Cultural politics,' in the context of hegemonic struggle, aims to change public understanding, values, and practices regarding an issue – along with government policies and associated power relations (Golding 1992).

In our analysis of the diverse resistance to Alberta's 'petro-hegemony,' we draw upon our research on social movement actions and counter-responses.[2] Our approach is similar to what Paul Gellert and Jon Shefner (2009) call 'structural fieldwork' wherein 'the deep familiarity with people and locales' offers analytic traction on the political-economic world-system. Between 2007 and 2012 we each conducted participant-observation and interview research with key figures related to the oil sands, and with citizens involved in the oil sands opposition movements. Our research draws upon participatory action research, content analysis of media responses to social movements' actions, discourse analysis of these actors' public statements, and focus groups with Albertan citizens (Carter 2011; Haluza-DeLay and Berezan 2013; Haluza-DeLay and Carter 2014; Haluza-DeLay, Ferber, and Wiebe-Neufeld 2013; Le Billon and Carter 2012). Our research programs inform the analysis presented here to elucidate the complicated features of social movement action and its effects upon Alberta's social landscape.

Social Movement Activism in Alberta and Canada[3]

The landscape of oil sands opposition is constantly shifting; any mapping of it is quickly outdated. It is complicated by a multiplicity of actors, strategies, and the scales on which they operate. Therefore, while giving sufficient detail to illuminate the case, we focus on explaining the outcomes of social movement activism, particularly as it relates to the production of environmental citizenship and greater social and environmental justice (D'Arcy et al. 2014). As monolithic and as undemocratic as Albertan politics may appear to casual observers, resistance to status quo environmental regulation in the oil sands does exist, has expanded to include new movement allies, has changed its strategies over time, and has sought to develop more encompassing demands for a social order that is less carbon and petroleum intensive.

Social movement actors involved in opposition to the oil sands include environmental, labour, Aboriginal, and religious organizations. We briefly describe each of these and their strategies in the following sections, along with the coalition building and transnational organizing that has occurred, and the tentative steps to develop an alternative hegemonic project resonant with the public. The description enables the analysis of contestation in the realm of cultural politics that concludes the chapter.

Aboriginal Activism

Aboriginal communities have been longstanding sources of resistance, since many of them are surrounded by, or downstream of, oil sands developments or in the path of pipelines carrying gas to the projects or bitumen from them. While seeking to benefit economically and socially from oil sands extraction, First Nations have protested the cumulative environmental health impacts of the projects, the degradation of water and air quality, the increased toxicity of subsistence food such as fish and game, and the limitation of access to their traditional lands. First Nations argued that the effects of the oil sands projects contravene the rights associated with treaties signed with the Canadian crown. Some bands, such as the Fort McKay First Nation and the Athabasca Chipewyan First Nation, derive substantial revenue from contracts with the resource industries. However, disappointment with government and industry inaction on their concerns has increased over time. A case in point is the withdrawal of the Mikisew Cree First Nation and Athabasca Chipewyan from the Cumulative Effects Management Association (CEMA) in 2007 and 2006, respectively, in protest against the association's lack of meaningful progress (see Parlee, this volume). While they have since rejoined CEMA, the Aboriginal groups point to the association's failure to bring about effective environmental regulation of the resource sector, and the way First Nation participation has been used to justify claims that the constitutional duty to consult Aboriginal peoples has been fulfilled (Mikisew Cree First Nation 2007).

Public attention has focused on the possible health impacts of the oil sands on Aboriginal communities downstream, although First Nations and Métis organizations have tried to raise other issues of justice as well. Widespread attention to health problems in Fort Chipewyan, one of the downstream communities, was raised by the community's physician, Dr John O'Connor, in 2006. O'Connor raised an alarm about higher-than-expected rates of rare cancers in Fort Chipewyan. He was subsequently threatened with disciplinary action by the Alberta College of Physicians and Surgeons (Loyie 2009). A series of reports by fed-

Table 14.1 Topics 'Justice for Fort Chipewyan' Rally Speakers Referred to as Oil-Sand-Related Injustices Occurring in Their Communities, 1 March 2008

- Health
- Community development (Are local people benefitting?)
- Changes to lifestyle (Local foods contaminated, requires more money to live)
- Changes to culture
- Treaty rights (They are still valid, just like the 1800s laws governing the mining)
- History-colonialism (Here we go again ...)
- Environment
- Destruction of lands
- Damage to other creatures
- Massive carbon emissions
- Democratic deficit (Is government listening to the people?)
- Corporate power (Who's in charge?)
- Tar sands development has outpaced social development (Fort McMurray begging for slowdown)

eral and provincial ministries, health agencies, and independent and university researchers have not settled questions about the risks generated by oil sands emissions, especially for those who rely on traditional foods. Since governmental reports have appeared politicized, and because of the pressure exerted on Dr O'Connor, distrust of government and industry science has grown within the communities (Brooymans 2009; Kopecky 2014).

Despite the ongoing contestation over the scientific evidence for health effects and increased pollution in the region, the provincial government has downplayed or denied health and environmental risks and taken little action on air or water contamination. Furthermore, the ongoing media attention to health has de-emphasized the social justice and environmental racism dimensions of the conflict. A case in point is the March 2008 'Justice for Fort Chipewyan' rally held on the steps of the Alberta legislature. Speakers referred to an extensive range of issues (listed in table 14.1). However, in news reports from CBC TV, CTV, and the *Edmonton Journal*, only 'health' was mentioned as an issue that had been raised at the rally. According to research by Leith Deacon and Jamie Baxter (2009), mainstream media routinely ignore equity and justice framing in favour of health themes. This is a pattern that reduces the opportunity for the general public to be challenged by the environmental racism, class, or justice aspects of sustainability as they are experienced by marginalized populations in Canada.

Industry officials and apologists promote the social and economic benefits of oil sands development for Aboriginal peoples in the region in terms of jobs, community facilities, and other forms of economic development (for example, see Flanagan and Jose 2014). Indeed, First Nations have both sought firmer economic returns from the extractive industries operating in their vicinities *and* increased their general opposition to oil sands development (see Parlee, Zalik, this volume). First Nations' criticisms of the oil sands have, therefore, been framed by some as being hypocritical – as biting the hand that feeds them. This kind of discourse depicts First Nations as mere recipients of benefits from corporate operations. Yet First Nations are seeking more than this. Beyond profit sharing, job guarantees, or participation in consultative bodies, they want power sharing and joint decision making about the direction and types of development (Slowey 2008). They also insist upon more environmental criteria for project development.

That adequate consultation is not occurring has led to an increased use of litigation against provincial or federal governments. Mikisew Cree First Nation's landmark Supreme Court case in 2005 established an important precedent requiring a 'duty to consult' about land management on traditional territories or where treaty rights might be infringed. Yet Aboriginal peoples in the oil sands region feel they have been shut out of planning, such as in the process associated with the Lower Athabasca Regional Plan (Weber 2014). Moreover, both the provincial and federal governments say that First Nation concerns need to be addressed by the other level of government. In 2008 the Beaver Lake Cree Nation filed a law suit against the Alberta and Canadian governments, claiming that oil sands expansion – combined with multiple other projects operating within their traditional territory – is infringing on their constitutional and treaty rights. According to a reporter (Kopecky 2014, 37), both governments sought to have the case dismissed: "To appreciate the spirit of their approach, consider that one argument Canada advanced was that it was Alberta that issued industrial permits in the oil sands and therefore the federal government should not be held accountable; meanwhile, Alberta argued that treaty rights were a federal matter and therefore not the province's responsibility." The Court of Appeal of Alberta in April 2013 rejected the numerous objections of the Alberta and Canadian governments and permitted the case to go to trial.

Aboriginal efforts to benefit more from oil sands development and to achieve better protection of their rights and environments in the oil sands status quo have frequently been blocked by the neoliberal socio-political landscape that would see the land liberated from most forms of regulation – including those associated with treaty rights. In light of the limitations of localized actions, as well as limited financial resources, northern Alberta Aboriginal groups have sought allies who have partnered in taking the fight to the courts, to Ottawa, and to corporate offices, governments, and publics in the United States, United Kingdom, and Europe. The Indigenous Environmental Network has been particularly effective at partnering with other movement actors to draw attention to human rights abuses, treaty rights, corruption, and violence often associated with oil extraction around the world (Watts 2005; Zalik 2009). Inside and outside of Canada, they have modified strategies, incorporating more symbolic actions to raise attention. Examples of both coalition building and strategic shifts include the totem pole journey from the Lummi Tribe of Washington State to Beaver Lake First Nation in September 2014. Speakers at various events along the way called on indigenous peoples, faith groups, and environmentalists to unite for justice and sustainability. Another example is the annual Tar Sands Healing Walk that began in 2010 and has drawn supporters in increasing numbers and profile. The walk in 2014 involved around five hundred people circling the sixteen kilometres around a former tailings pond near Fort McMurray. According to organizers, this was the final walk, not because the issue has been resolved or the land healed, but 'because the story of the Athabasca region is only one small piece of the immense scope of this issue, these practices, of the land that's being abused, and of the people who are being so brutally mistreated' (http://www.healingwalk.org/). Outside of Canada, the environmental justice frame seems to have played a greater role in mobilizing public opposition to the oil sands.

Environmental Organizations

Numerous ENGOs have mobilized against the oil sands since the mid-2000s. Before this time, the Pembina Institute – a unique hybrid of energy consulting firm and environmental non-profit organization – was nearly the only group actively working on the oil sands and has continued to be central in the debate by providing extensive critical research and policy analysis. Opposition to the oil sands by ENGOs escalated mid-decade and then expanded in geographic scale as well. Active ENGOs now range from local, regional, and provincial groups – like the Keepers of the Athabasca or Toxics Watch Society of Alberta – to nationally organized groups – like Greenpeace or the Sierra Club of Canada and its local Prairie Chapter – to international organizations like Forest Ethics or the Rainforest Action Network. Along with the geographical expansion of opposition to the oil sands, ENGO strategies have changed.

According to environmental activists, their initial strategies sought to be collaborative with corporations and government, often consisting of recommendations to 'tweak' industrial practices, specific projects, or policy regimes. ENGOs also participated in government-initiated consultation processes. However, by 2008 ENGOs had experienced numerous flawed provincial government consultation processes, leading them to believe that such efforts were wasteful of their time and limited resources. These consultation processes included the Special Places 2000 and Boreal Forest Conservation Strategy in the 1990s (Schneider 2000), the North East Slopes Land Use Strategy process (Fluet and Krogman 2009), citizen advisory groups for forest management (Parkins 2002), and sour gas consultations (Masuda, McGee, and Garvin 2008). ENGOs felt that CEMA followed the same pattern, and by 2008 most had withdrawn from the association. According to political scientists George Hoberg and Jeffrey Phillips, following data collected in a study of the oil sands consultation processes, stakeholder consultation in Alberta is designed to provide only a 'selective opening' – primarily to 'bolster the legitimacy of the policy process while maintaining control over decision rules and venues,' rather than permitting genuine participation in environmental management (2011, 507). Since ENGO experiences have indicated that meaningful opportunities for political engagement on environmental policy are limited, ENGOs have changed strategies. They began to employ more direct action and more legal action, creating active coalitions with other citizen groups, targeting specific components of the oil sands extraction, distribution, and production system (a strategy called 'segmented localism' by political scientist Stephen Hoffman [2012]), and conducting activism external to the province as a way of putting pressure on internal provincial regulation or oil sands development and profit making in general.

The first significant strategy shift is more frequent use of direct action. Greenpeace made headlines with creative 'stunts' that drew national and international attention to the environmental consequences of the oil sands operations. Greenpeace quickly became a central organization in the oil sands opposition after it opened an Alberta office in

2007 (staffed by an Albertan). Greenpeace has blocked production sites, unfurled a provocative banner at one of Premier Stelmach's fundraising dinners, held activist training (that by 2014 had become three-day tar sands/climate action camps), and organized rallies on the steps of the legislature or at corporate offices in Calgary and elsewhere. Premier Stelmach implied that Greenpeace is an 'ecoterrorist' organization, and he and Solicitor General Fred Lindsay promised to ensure that Greenpeace activists were prosecuted to the full extent of the law (Berry 2010). Suggestions by government figures that Greenpeace's actions constituted 'security threats' that might call for use of terrorism legislation raised concerns among some citizens that legitimate civil disobedience was being criminalized (Adkin et al. 2009b; Le Billon and Carter 2012). Other activities by ENGOs that have caught media attention include an online video game shooting clumps of oil at Prime Minister Stephen Harper and then Opposition leader Michael Ignatieff, billboards after the BP oil spill in the Gulf of Mexico that hailed Alberta as 'the OTHER oil disaster' (Turner 2012), and an 'AvaTarSands' campaign that compared Alberta to the storyline in the popular *Avatar* movie (Haluza-DeLay, Ferber, and Wiebe-Neufeld 2013). Such campaigns have engendered increasing recognition among Albertans that, beyond the province's borders, the oil sands status quo is viewed as being problematic. That these actions have been controversial in the cultural politics of the province is evidenced by the denial of billboard space for some ENGO campaigns – even a relatively tame one promoting solar energy over fossil fuels (Krashinsky 2012).

A second strategic shift is the increased use of legal action, which Aboriginal groups have also employed. An aggressive litigation strategy is a variation of the culture of environmental direct action (Bevington 2009). For example, when it became known than hundreds of ducks (originally reported as 500 but later determined to be closer to 1,600) had died on a Syncrude tailings pond in April 2008, Ecojustice launched a private prosecution against the company, forcing provincial and federal environment ministries to lay charges against the company. By summer 2012, environmental organizations had filed more than a dozen statements of concern on oil sands development projects; the effectiveness of the strategy was such that the Alberta Ministry of Environment and Sustainable Resource Development (ESRD) denied standing to a coalition of ENGOs in a hearing on an oil sands project on the grounds that they were not 'directly affected' by such projects (Pembina Institute 2013; Weber 2013). Court of Queen's Bench justice

Richard Marceau ruled later that year that the Ministry's decision was 'in violation of the rules of natural justice' by selectively disallowing groups known to be critical of oil sands expansion (*Pembina v. Alberta* 2013, 19). Despite this ruling, ERSD applied the same reasoning on the same case the following spring, contravening the judge's recommendation to 'err on the side of openness' (Weber 2014). The judicial strategy is expensive and has had mixed success, as governments have not always acceded to judicial decisions, further demonstrating the closed political opportunity structure.

A third shift in ENGO strategy is building both long- and short-term coalitions with other organizations. ENGOs frequently draw on scientific research and expertise from university researchers and policy institutes. The Pembina Institute generates well-researched reports that have been a central resource for other ENGOs and have often received media attention. The Parkland Institute, a political-economy research institute housed at the University of Alberta, has also produced reports focused on energy security and royalty regimes (Boychuk 2010). ENGOs have actively partnered with Aboriginal organizations and with organizations not primarily environmentally oriented; Public Interest Alberta and the Canadian Centre for Policy Alternatives are two such examples, showing how opposition to the oil sands grew from narrow environmental concerns to become a broader, national movement. The scale of oil sands opposition movements extends from local to provincial, national, continental, and transnational (Hoffman et al, 2014).

One form of coalitional politics employs the environmental justice frame. Aboriginal groups have used 'environmental justice' as a way of describing the inequitable impacts of the oil sands; as ENGOs have chosen to partner with Aboriginal groups, they also have used the term. However, a search of a database of press releases produced by the national offices of the Sierra Club and Greenpeace showed that the concept of environmental justice is not commonly deployed by these organizations.[4] Greenpeace-National used the word 'justice' only three times between 2006 and 2008 (all in 2008), and Sierra-National did so six times, mostly in 2008.[5] The term was still not prominent on either organization's website in 2011. The Sierra Club Prairie Chapter website once prominently featured a description of 'environmental justice,' but the term was not prominent in 2014. Of course, the absence of the term does not mean that social justice considerations are absent from ENGO analyses of the oil sands, but it does continue a pattern wherein 'envi-

Table 14.2 Comparison of National ENGO Press Releases That Refer to Oil Sands or Tar Sands

	Press Releases			
	Sierra Club of Canada		Greenpeace Canada	
Year	Tar Sands	Oil Sands	Tar Sands	Oil Sands
1996	0	0	n/a	n/a
1997	0	0	n/a	n/a
1998	0	0	n/a	n/a
1999	0	0	n/a	n/a
2000	0	0	n/a	n/a
2001	0	0	n/a	n/a
2002	0	0	n/a	n/a
2003	0	1	n/a	n/a
2004	4	0	n/a	n/a
2005	11	4	n/a	n/a
2006	7	3	1	0
2007	10	7	7	1
2008	16	6	13	4
Total	48	21	21	5

ronmental justice' has not had much discursive presence in Canada despite the existence of many environment-related injustices (Agyeman et al. 2009; Haluza-DeLay and Fernhout 2011).

Another form of coalitional politics was the call for a moratorium on new oil sands projects in 2007–8. As expansion of oil sands operations accelerated and the cumulative effects of resource development became clearer, over seventy environmental, labour, church, social development, and Aboriginal groups signed the call for a temporary moratorium on new project approvals (see appendix 14.1). On the organizational website, the list of signatories is classified into two categories: Alberta-based organizations and national organizations. Discursively, this division has two significant implications. First, it signalled that scale-crossing had occurred, that is, that the oil sands were seen as an issue of national as well as provincial importance. National ENGOs began pronouncing on the oil sands only in the mid-2000s, long after oil sands development had ramped up in the late 1990s (for trends of investment in the oil sands, see table 6.1 in Adkin and Stares, this volume). Table 14.2 tabulates mention of the terms 'oil sands' and 'tar sands' in the press releases of the national offices of two of the major players in the environmental movement between 1996 and 2008.

As these data show, the Sierra Club national office did not mention the oil sands issue until 2003. However, according to an interview with Sierra Prairie Chapter director Lyndsay Telfer, the Prairie Chapter office had actively campaigned on this issue for a longer period of time.[6] Greenpeace Canada press releases were only available from 2006 to 2008 but show increasing attention to the oil sands, beginning with a single press release in 2006. These data show that national-level opposition to the oil sands has grown and become more diverse.

Second, the delineation of Alberta organizations in the list of moratorium signatories shows a homegrown environmental concern that belies government and industry representations of environmentalists as 'outsiders.' The diversity of the list shows that oil sands exploitation deeply concerned a variety of Albertans and their organizations. Environmental engagement has always been challenging in Alberta, where provincial elites have not only argued that citizens' prosperity is dependent upon the continuing exploitation of natural resources but that it must be defended against 'revenue-greedy' federal governments (see Adkin and Stares' concept of 'neoliberal nativism,' this volume). Albertan collective identity includes aspects of this provincial nationalism, famously represented in the view – expressed during the 1980s conflict with the federal government over the latter's National Energy Program – that Alberta should 'let the Eastern bastards freeze in the dark.'[7] In addition, a cowboy 'frontier' identity also meshed readily with the neoliberal conception of citizenship promoted by the Klein Conservatives. Its emphasis on individualism and independence (from the state) and its 'leave it to the market' mantra effectively removed collective action and environmental concerns from the sphere of citizenship. The widely accepted official discourse is that intensive hydrocarbon extraction is the only thinkable model of development for the province and that Alberta's prosperity on this basis is crucial to the entirety of Canada's prosperity (Gibbins and Roach 2010). It is this discourse (also articulated by the federal Conservatives), allied to tenets of Albertan identity and culture as well as the closed political opportunity structures of the province, that confronts ENGOs. Under these conditions, many ENGOs have turned away from participation in government-initiated and organized consultation processes, investing instead in campaigns to shape public opinion. In addition, rather than seeking to influence the Alberta or federal Conservative governments directly, environmentalists have mobilized opposition to the oil sands at multiple points in their supply and distribution system. This strategy goes hand in hand with the expansion of organizational networks at national and transnational scales.

The successes of ENGO strategies have provoked counterattacks. These have included verbal declarations by federal ministers that foreign-funded 'environmental and other radical groups … threaten to hijack our regulatory system' and to 'undermine Canada's national economic interest' (Oliver 2012). There has been a public alignment of federal trade and environment ministries with 'allies' in industry and against 'adversaries' in the environmental and Aboriginal organizations (Vanderklippe 2012). Federal budgets from 2012 onward have increased audit resources for the Canada Revenue Agency to investigate environmental and other registered charities – with the threat of loss of charitable status and other punitive actions (*Globe and Mail* 2012).

Religiously Based Activism

Religious organizations have traditions that may be mobilized in support of social justice. Although Christian churches have been implicated in colonial practices such as the residential schools, some Christian organizations have supported Aboriginal groups and other justice concerns in recent decades (Heinrichs 2013; Lind and Mihevc 1994). In general, religious groups have increasingly attended to environmental issues, developing theologies of creation-care, stewardship, and eco-justice (Gottlieb 2006). This work has extended to encompass a critique of the oil sands. However, religious groups approach the issue in very different ways than environmental and Aboriginal groups.

The immediate constituencies of religious activism might be seen to be congregants; however, the impact of religious groups is broader than this relatively small proportion of the population, as these organizations have publicly addressed the oil sands issue. Two notable incidents in 2009 (discussed in greater detail below) drew considerable media attention: the January statement by the Roman Catholic bishop of the northeastern Alberta oil sands region about the oil sands, and a week-long tour of the oil sands in May by church leaders, conducted by the ecumenical justice and development organization KAIROS.[8] In addition, religious organizations have their own media, study groups, workshops, and other events that are not often noticed by outside, non-religious observers. For example, the Edmonton-based blogger David

Climenhaga pointed out that there was no mainstream media coverage of the World Council of Churches' decision to divest from fossil fuels.[9] The World Council of Churches (WCC), which represents denominations numbering about a half billion of the 2.1 billion Christians in the world, has an active climate justice program and was following the urging of United Nations climate chief Christina Figueres.

While several religious groups are represented among those calling for a moratorium on new oil sands development (appendix 14.1), the most visible interventions by religious actors began with a high-profile and contentious statement by Roman Catholic bishop Luc Bouchard, whose Diocese of St Paul includes the Fort McMurray area. Declaring a responsibility to help Catholics in his parishes to think about the moral issues associated with the oil sands, Bishop Bouchard released an extensively researched pastoral letter titled 'The Integrity of Creation and the Athabasca Oil Sands' (Bouchard 2009). The document presented scriptural and theological reasons for viewing the safeguarding of the natural environment as a religious obligation, and summarized the environmental effects of the oil sands. Bouchard concluded that the extent and type of oil sands development 'cannot be morally justified.' In a speech made a year later, Bishop Bouchard said he was amazed at the quantity and tone of the responses to his text (Warnica 2010; see also Muthui 2013). According to him, among the most frequent responses were declamations that the church should 'stick to morality' and 'stay out of the oil sands business.' Bouchard's speech was delivered at a conference in Edmonton sponsored by a Christian organization called the Social Justice Institute. The title of the conference – 'Living Faithfully in Oil Country' – illustrates the way some religious organizations are trying to approach the issue. Bouchard pointed out that such responses assume that issues of economics are above moral comment and that the oil sands should only be dealt with by technical experts (Kowalsky and Haluza-DeLay 2013).

KAIROS is an ecumenical coalition of ten of Canada's Christian denominations, and has a national office and local chapters in many Canadian cities (http://www.kairoscanada.org/). The organization conducts international development programs with partners in over twenty countries and Canadian-based educational campaigns on numerous social issues related to 'human rights, justice and peace, human development and ecological justice' (KAIROS 2011). Its climate change campaign continues to actively link the issue with global justice.[10] In

May 2009, KAIROS conducted a tour of the oil sands in which Canadian representatives of Christian denominations, two indigenous representatives, and one member each of Oil Watch Nigeria and Acción Ecológica-Ecuador participated. A public event attended by more than 150 persons launched the tour, with a speech by The King's University political scientist John Hiemstra. The tour was organized primarily so that church leaders could have direct experience with which to engage their own denominational constituencies on the issue. In Fort McMurray and Fort Chipewyan the participants met with industry, government, and Aboriginal leaders. A video and teaching material were produced and used for several years in churches across the country (KAIROS 2011). A concluding statement was also produced,[11] but little further public action seems to have occurred. One participant confided in an interview that while he tried to get invitations from churches in his denomination to speak about what he had witnessed, he received none. The KAIROS events did, however, attract media attention; the media analysis conducted by the Canada West Foundation (CWF) during 2009–2010 showed an upsurge of negative reporting on the oil sands during the time of the church leaders' tour.

Another religious intervention in the oil sands conflict that captured international attention took place in June 2014, when Nobel Peace Prize recipient and South African Anglican archbishop Desmond Tutu visited Fort McMurray. He participated in a conference about oil sands development and treaty rights organized by the Athabasca Chipewyan First Nation and a Toronto law firm specializing in Aboriginal law. In his keynote speech to the delegates, the archbishop said that 'climate change is the moral struggle that will define this century' (quoted in Cryderman 2014). Drawing a connection between two issues, he stated: 'The fact that this [oil sands] filth is being created now, when the link between carbon emissions and global warming is so obvious, reflects negligence and greed.' Tutu has actively campaigned for an anti-apartheid-style divestment from fossil fuels.

Mennonite involvement in the oil sands conflict sheds light on the complex terrain that religious activists navigate when they engage publicly in political or policy matters. Two Mennonite leaders were among those on the KAIROS tour. Indicating even further concern at leadership levels, a third church leader attended the 2011 meetings of the United Nations Framework Convention on Climate Change and contributed a reflection on the oil sands to both the denominational magazine and a book on ecological justice and Christian faith (Lysack and Munn-Venn 2013). The biweekly denominational magazine *Canadian Mennonite* has published seven articles between 2007 and 2012 focusing on the oil sands (Hall 2013). All acknowledge the problematic nature of the oil sands and a Christian duty to take care of the environment. But whereas several articles are extremely critical of the 'creation'-damaging nature of oil sands development and fossil fuel reliance, a recent article focuses on how divisive the issue is for Mennonite churches, especially in Alberta (Wiebe-Neufeld 2013). Following the Fort McMurray conference at which Bishop Tutu spoke, *Canadian Mennonite* published a reflection by yet another denominational leader who pointed out that the issue is 'loaded' for Albertans (Shantz 2014,18). Nevertheless, he argued, Mennonites in Alberta must re-examine assumptions and practices about resource development, lifestyles, and the duty of stewardship in the face of the moral urgency of climate change and the planet's 'exploitation as a dollar-value commodity' (19).

Clearly, members of faith communities have varied relationships to the oil economy and find the issue challenging to their interests and values – both as citizens and as religious believers (McKeon 2010). Participants in any faith group are socio-economically diverse – likely more so than members of other social movement organizations, especially environmental groups. Thus, among the consequences of the engagement of religious actors has been a widening of the scope of attention to the oil sands among the general public. Members of faith communities strive to frame the problematic aspects of the issues in ways consistent with their own interests and beliefs (Haluza-DeLay, Ferber, and Wiebe-Neufeld 2013). Clearly, religious leaders will have to continue to develop understandings consistent with their traditions, as, for example, Norwegian Christians are striving to develop an ecological theology in the context of being an oil-producing country (Anker 2013; Bjelland Grønvold 2013). Nevertheless, the identification of *Homo alberticus* with *Homo energeticus* is troubled by a theological counter-narrative rooted in an identity based elsewhere than the prevailing socio-political culture and economistic valuation.

Labour Organizations

Labour movements have also pointed to problematic aspects of oil sands development. The labour movement is in a conflicted position, supporting, on the one hand, a greener model of development while, on the other hand, attempting to generate employment from higher-value-added downstream refining of oil sands bitumen and synthetic crude. Moreover, its membership has no unified position with regard to the province's dependence on fossil fuel extraction. All workers do not benefit from oil sands expansion, and the extractive economy drives labour market polarization (Major and Winters 2013; Adkin and Miller, O'Shaughnessy and Dogu, this volume).

The Alberta Federation of Labour has taken the position that too much of the value of the crude oil from the oil sands is lost because it is shipped to refineries in other provinces or in the United States, rather than being upgraded within the province (AFL 2009, 2013, 2014). In this context, it signed the call for a moratorium on oil sands expansion in 2008 (appendix 14.1). At the same time, labour organizations have also argued that 'just sustainability' or 'just transition' to a greener economy must build in measures to minimize job and income losses.[12] The conclusions of a number of Parkland Institute reports, calling for restructuring of the province's fiscal regime, have been supported by the AFL.[13] The AFL partnered directly with Greenpeace and Sierra Club in April 2009 to produce the report *Green Jobs: It's Time to Build Alberta's Future* (Thompson 2009). Concurrently, the AFL produced the report *Lost Down the Pipeline* (AFL 2009) as part of the campaign 'Will the Oil Sands Be Used to Build a Brighter Future?' AFL's opposition to the oil sands status quo has centred on ensuring that value-added jobs from refining the oil stay in the province.

While the AFL has partnered with environmental organizations and supported more ecologically sustainable economic models, the level of support that such initiatives garner from workers is not clear. For example, when the provincial government was considering an increase in royalty rates for oil and gas producers, employers were able to mobilize workers to oppose such increases (Byfield 2007),[14] although the AFL as an organization supported the royalties increase.

As a result of internal divisions in the workforce (e.g., the polarization between highly paid, mostly male workers in resource extraction and poorly paid female, immigrant, and temporary foreign workers in the service sector), union opposition to the oil sands has primarily been about the manner in which costs and benefits are distributed. For example, the AFL has produced a fact sheet about excessive oil company profits[15] and royalty give-aways (Boychuk 2012).

Labour unions have opposed pipelines like the Enbridge Northern Gateway that would transport bitumen from Alberta to the Port of Kitimat in British Columbia. While AFL president Gil McGowan vows 'that the Alberta Federation of Labour will support First Nations and environmental groups that will almost certainly continue to fight against the pipeline,'[16] this opposition is motivated by the AFL's view that the promised economic benefits of the pipeline are unlikely to materialize for Alberta workers. Contrary to environmental groups, unionized workers may declare that 'there's a place for the oil sands' – as does an organization called 'Iron and Earth,' which bills itself as 'oil sands workers supporting renewable energy' (www.ironandearth.org).

Lastly, the labour movement is divided. The largest labour organization in the oil sands is the Christian Labour Association of Canada (CLAC), which is at odds with the rest of the labour movement and is often accused of having too cozy a relationship with corporate management to achieve the best bargaining agreements for its members. CLAC strenuously disagrees with this assessment, attributing its achievements to a less confrontational orientation than the rest of the labour movement. Interviewees associated with CLAC indicate that it has given limited consideration to the negative impacts of the oil sands. CLAC's website includes a single page about the oil sands, which it refers to as 'a source of pride.' The site narrates the benefits of the oil sands with little attention to problematic aspects.[17]

Overall, it is unclear to what extent organized labour's activism or policy reports will lead towards a greener economy for Alberta or influence individual workers' perspectives on the oil sands. However, in a speech at the founding of Canada's largest private-sector union (from a merger of the Communications, Energy and Paperworkers Union of Canada and the Canadian Auto Workers in September 2013), activist and writer Naomi Klein (2013) pointed out that both predecessors had policies on climate change, greening the economy, and the needs of workers who might be displaced by green transition. Labour has focused on improving workers' and the public's benefits from oil sands extraction, alongside advocacy that oil sands development be carried out more responsibly in terms of environmental, social, and economic impacts. Klein insisted that the labour movement can serve as a 'fixed address' providing stability to more transient and more poorly resourced and staffed movement organizations, and this would lead to broader support for labour's politics. Nevertheless, it remains to be seen how labour activism in the oil sands can become a more effective counter-hegemony to the petro-based political economy of Alberta and can contribute to social and environmental justice.

National and Transnational Activism

Opposition to the oil sands has mobilized in sites across Canada, North America, and beyond. The extension of opposition beyond provincial borders is a manifestation of globalization, as is the linking of Alberta's oil sands to the global issue of climate change. At the same time, the growing coalitions have engaged in what political scientist Stephen Hoffman (2012) termed 'segmented localism,' targeting specific elements of the oil sands system such as particular pipeline proposals, or investors, or extraction projects. As Hoffman's research team illustrates, the organizations that coalesce around particular targets may be only tenuously linked to other organizations but in the larger scheme operate as a transnational advocacy network (Keck and Sikkink 1999).

Oil sands opposition gained a broader constituency in the latter part of the 2000s; Canadian national NGOs that are not exclusively environmental became active on the oil sands issue at this time. Examples include KAIROS, the Council of Canadians, and the Polaris Institute. The latter is particularly interested in the political influence of the corporate sector on the direction of the country. It has produced reports concluding that the North American Free Trade Agreement may be an impediment to environmental protection, as oil from Alberta has become a transborder commodity (Clarke 2008; Laxer 2008; McCallum 2006).

At about the same time, American opposition to the oil sands began to build. An early success occurred when the US Conference of Mayors passed a motion in 2008 to reduce the use of fuel from the oil sands due to its high greenhouse gas intensity (De Souza 2008). This was also one of the clearest moments in which the oil sands were framed in association with climate change. Transnational groups like Oilwatch or Oil Change International specialize in this issue and comprise coalitions of indigenous solidarity, environmental, religious, and other social movement actors. However, numerous small networks also operate (Hoffman et al. 2014).

The big American ENGOs came on board in 2009 and 'the oil-sands battle went global,' according to journalist Shawn McCarthy (2012). Pipelines and climate change became two major frames for the transborder activism that developed. McCarthy notes that maps depicting the dense grid of pipelines criss-crossing the continent were effective in communicating to the public the proximity of risks. As federal and provincial governments resisted calls for oil sands regulation, activists also took their message to Europe. Aboriginal and environmental organizations have been particularly active in Norway, France, Germany, and the UK. Canada's 'dirty oil,' environmental injustice against Aboriginal peoples, and oil sands' contributions to greenhouse gas emissions are significant frames for European audiences. Canada's government has actively fought implementation of the European Union's Fuel Quality Directive that discourages importation of high-carbon-intensity fuels.

The 'tar sands tentacle' is one way of thinking of the far-reaching effects of the system of oil sands development and the pieces in the system required to extract, refine, transport, fund, and use the products of the oil sands. This is a phrase used by two Montana writers seeking to mobilize fellow citizens (Bass and Duncan 2010). They targeted one piece of the system – the transportation along rivers and highways of gigantic pieces of equipment. Their book's subtitle is indicative of the narrow focus: *Why the Pacific Northwest and Northern Rockies Must Not Become an ExxonMobil Conduit to the Alberta Tar Sands*. It is by such actions, spread across numerous pieces of the oil sands system, that communities of resistance are linked on both sides of the Canada–US border and beyond. The transportation of 'megaloads' headed for northwestern Alberta (see Adkin and Courteau, this volume) is only one example of the continental integration of the oil sands system and the vulnerability of specific components to public opposition. Hoffman et al. (2014) describe twenty-six discrete action targets (examples include a rally against the Keystone XL pipeline that will run from Alberta to Texas, multi-signatory letters to the Alberta Energy Regulator and the American Secretary of State, and comments on proposed pipeline approval and spill regulations) that draw on 243 separate organizations. Seventy percent of the organizations joined only one action; only eight organizations involved themselves in more than eleven of the twenty-six actions. Drawing so many organizations together indicates considerable investment in networking. Hoffman's analysis of the movement discourse around each issue shows them to be typically framed as local threats, albeit within a larger system. For example, a communication by the organization Environment Maine about the reversal of Line 9 (required to ship bitumen from Alberta to a tanker terminal in Maine) stated:

> ExxonMobil and Canadian oil giant Enbridge want to use an antiquated oil pipeline that passes right next to Sebago Lake to transport highly corrosive tar sands oil from Canada to Casco Bay for export. A tar sands spill in the Sebago watershed, or near any of Maine's waterways, would be utterly disastrous ... Sebago Lake is a Maine treasure – we escape there on

hot summer days, we drink its water, and we watch our kids grow up on its shores. (quoted in Hoffman et al. 2014)

The cumulative logic of these coalitions appears to be that if oil sands expansion cannot be halted at the level of the province (as failed policy consultations described above have shown), then attacking particular points in the interlinked system in which the oil sands are imbedded (pipelines, transportation of oil sands equipment, refinery expansions, financial investments) may impede the system's expansion and slow the expansion of the oil sands.

Pipelines to carry Alberta oil sands product to refineries elsewhere face particular opposition across Canada and the United States (as discussed elsewhere in this volume). A consistent concern is that pipelines merely continue production of and reliance on fossil fuels and impede the development of an alternative energy economy. This networking has provoked counter-organizing by corporations and the Alberta and Canadian governments (see Adkin and Stares, this volume, for details of government responses). One of Enbridge's responses to its opposition was to create and fund the Northern Gateway Alliance, made up of supporters of the pipeline (Hoekstra 2009). As closure of political opportunities at the provincial and then federal government levels have become obvious, social movements in Canada have increasingly turned to transnational actions, chosen new frames, and targeted different parts of the oil sands system.

Corporations and their profits have become more common targets of action. Environmentalists and Aboriginal activists have lobbied the Royal Bank of Canada (RBC), for example, to stop financing oil sands development projects (Barclay 2009). Shareholder resolutions can also have strategic efficacy, especially in improving corporate disclosure of risks and improved government monitoring (Cook 2012). That is, such activism has effects beyond the specific corporation targeted. Rainforest Action Network's (RAN) strategy is positioned within a larger inter-organizational struggle to weaken the legal, fiscal, and political 'pillars' of oil sands extraction. Ethical investors have concerns about infringement on indigenous rights associated with the oil sands. Europe is a major source of investment in the oil sands, so Canadian (and American) activists have travelled to Norway, France, Germany, and the UK to pressure investors. They have sought to position oil sands projects as risky investments for financial actors (Church Investors Group 2008; Crooks 2008). The 2012 World Outlook by the International Energy Association stated that nearly 80 per cent of known fossil fuel reserves must stay in the ground to avoid runaway climate change (IEA 2012, 3).[18] The organization Carbon Tracker (2013) has tried to reframe fuel reserves as 'stranded assets' – a risk to investors – since the equivalent of more than a trillion dollars of fossil fuel reserves must stay in the ground in order for uncontrolled climate change to be averted.

European publics have become increasingly active in opposing oil sands projects. Since fall 2008, numerous actions have occurred in which environmental, human rights, and native activists, financial institutions, and oil sands companies interacted in an increasingly public manner. Scientists of international repute, like James Hansen of NASA and former University of Alberta water scientist David Schindler, have used their reputations to help draw attention to the links between oil sands development and climate change, and to arrange high-level meetings with government and corporate officials. Norway's Statoil and France's Total have been pressed to divest their Alberta oil sands investments and projects.

Increasingly, the oil sands have been linked in movement framing with climate change because the commodity chain from extraction to end product is said to generate more greenhouse gas emissions (GHGs) than conventional petro-production.[19] Also, the oil and gas industry produces about 23 per cent of Canada's total GHGs according to CAPP (see also Huot, Fischer, and Lemphers 2011).[20] Movement organizations of all types and even the United Nations have increasingly called climate change a form of injustice. In Bishop Desmond Tutu's words, climate change is 'the most significant human rights issue of our day' (quoted in Cryderman 2014). Climate justice has been a major educational approach used by KAIROS, and is far more visible on its website than on Climate Action Network Canada's website. *USA Today*'s front-page photo on the day following the People's Climate March in New York City (more than three thousand people took part on 22 September 2014) prominently displayed a banner held by marchers declaring, 'Shut Down the Tar Sands.' The Rockefeller Foundation and over thirty other foundations announced the next day that they would begin a process of divestment of fossil fuels. While such actions have garnered considerable media coverage, they have not yet moved the Canadian government to address GHG emissions from the oil sands.

Instead, the Canadian federal government has become increasingly identified by many actors as among the most recalcitrant governments in international climate negotiations. In response to such attention,

Alberta government officials and Canadian federal representatives, specific corporations, and CAPP have responded with extensive public relations campaigns to defend the oil sands and Canada's environmental performance. These campaigns generally downplay the relative impact of the oil sands on climate change or GHGs. Well-placed scholarship argues that the oil sands are having, or will have, more impact than these campaigns acknowledge, and that their global impacts ought to be part of policy considerations, rather than excluded as they were in the Northern Gateway hearings (Erickson and Lazarus 2014; Palen et al. 2014).

Lastly, although climate change is usually framed as a *global* phenomenon, and the segmented localism of social movement initiatives might be seen as somehow detached from this larger issue, local and global scales are in fact being bridged in social movement campaigns. As Christopher Rootes says, 'Climate change provides local environmental campaigners with a new frame that provides effective bridging not merely between the local and the national but between the local and the global' (2013, 110). Framing the oil sands as a climate change and climate justice issue has led to new partnerships among oil sand opponents, and new pressures on proponents.

An Alternative Development Path

Much of the movement action described above has been oppositional, but there have also been efforts to develop alternatives to the existing fossil-fuel-extractive model of development and its 'petro-state' regime of regulation. Especially in the wake of the 2008 financial crisis, Albertan and Canadian social movement actors have begun to think through how the broad outlines of an ecologically sustainable post-carbon economic model could be implemented in the Canadian context. The difficulty has been in communicating the conclusions of these often quite concrete and well-researched reports, leaving ordinary citizens with little awareness of alternatives.

The shift in movement strategy towards a leadership role in advancing an alternative hegemony began with calls for a moratorium on oil sands project approvals in late 2007. These calls had to take into account the reality that hydrocarbon extraction is the current pillar of the provincial economy. Social movement organizations began producing recommendations about 'greening' the Alberta economy. The aforementioned *Green Jobs: It's Time to Build Alberta's Future* is just one example. Other reports are described elsewhere (Haluza-DeLay and Carter 2014). Germany, Norway, and even Ontario with its investments in renewable energy technology are used as examples of the feasibility of reduced reliance on the oil and gas industry or replacing the coal-fired electrical energy plants that give Alberta cities the highest ecological footprints in all of Canada (Anielski and Wilson 2004; Weis, Thibault, and Miller, this volume).

At one point, the Sierra Club and Greenpeace initiated a campaign called 'RePower Alberta' that began touring major cities in May 2010. While labelled 'Building the Green Economy for Alberta,' all of the material on its website was about renewable energy technologies and a 'green energy strategy' for the province. The campaign had limited impact in the province, and no labour or business organizations became engaged. The idea of 'RePowering' the province might indicate something about revitalizing democratic participation, but organizations with that expertise (e.g., Public Interest Alberta) were also unengaged. So was the provincial government. Similarly, the *Green Jobs* report received a flurry of media attention but no government attention, and the expansion of the oil sands proceeded with renewed promotional campaigns. Polls showed that ordinary Albertans were increasingly aware of the criticisms of the oil sands by 2010, but the status quo prevailed in government policy. Despite these efforts on the part of environmental and labour organizations as well as policy think tanks to develop and publicize an agenda of reforms, Alberta's citizens remain largely unaware of them. Focus group participants in one study insisted that the environmental movement has been critical of oil sands development but has not offered convincing alternatives (Haluza-DeLay et al. 2013). Perhaps the most significant challenge for oil sands opposition movements is to gain support for a post-carbon model of development and broad-based environmental citizenship.

Strategic Strengths and Future Opportunities

From the above description of the strategies and complexities of social movement action, we can draw out several major trends in the movement's strengths and weaknesses. First, opponents of the existing level of oil sands development and its regulation chose 'external' venues for mobilizing pressure on the Alberta and Canadian governments. Other researchers have found that when the domestic opportunities are relatively closed, a 'boomerang' into transnational advocacy may happen, creating external pressure (Keck and Sikkink 1999; Poloni-Staudinger 2008; Van Der Heijden 2006). Coalition building among a broader set of opposition groups and across national boundaries has put the provincial and federal governments on the defensive. Second, and to state it in Gramscian terms, industry and government public relations campaigns are a sign of social movements' success in the 'war of position' to delegitimize the hegemony of the oil complex. The likelihood of this success being parleyed into increased support for alternatives is unpredictable, but provincial culture remains a site of political struggle.

Opposition by social movements to the oil sands is a material challenge as well as a cultural one. Social movement organizations of all the types identified above are far more poorly resourced than their industry or government counterparts. The difficulty in communicating alternatives may also be a sign that media outlets are not receptive to strong critiques of the oil sands. Connected to this problem are democratic deficits in terms of the first-past-the-post electoral system, an aggressively pro-development government, and heavily funded public relations campaigns to greenwash the oil sands. In these circumstances, the potential public support for reforms of the province's economic and energy foundations is unknown. The labour movement's ambivalence towards the oil economy may be an indication that Albertans are divided over what importance to attach to what are too often presented as oppositional goals – desires for environmental protection and a strong economy. To summarize, the strategic challenges facing the opposition to the oil sands include the adequacy of green economy alternatives and the means to communicate them to the public, insufficient attention to identity narratives that might counter the *Homo energeticus* formulation, and issue framing. Each of these challenges relates to the cultural politics of oil sands activism and has implications for environmental citizenship in Alberta.

A further challenge for movements opposing the oil sands status quo is that existing proposals towards a less carbon-intensive form of capitalist economy may not be sufficient to resolve the contradictions of capitalism's drive for incessant growth and the mounting evidence of an ecological crisis (Gould, Pellow, and Schnaiberg 2008). Moreover, 'greening the economy' proposals may amount to a form of ecological modernization that does little to contest the 'accumulation by dispossession' of indigenous and other local communities. This process occurs through resource giveaways (by governments) to private economic actors (Harvey 2003). As economic presumptions are cultural artifacts and result in specific institutions, the effective presentation of an alternative hegemony is a crucial site of political struggle.

Overall, in our view, social movement actors have paid insufficient attention to the cultural politics of Alberta – especially to the close association of energy and provincial identity that has been constructed by conservative political parties and industry proponents. To be effective in the Albertan context, social movements need to engage directly with and contest this identity politics and broader cultural politics as they forge global linkages, undertake transnational activism, and present green economy frameworks. This 'war of position' is a critical task for social movement and political actors. For example, the struggles of actors situated at other sites of petro-capital accumulation (such as Nigeria and Ecuador) are in some ways distant from Alberta's political context. However, they may have some resonance for Alberta farmers, who also feel that their livelihoods and even their families' safety are being threatened by the hegemony of oil and gas interests, and that their government is unwilling to act on their concerns (Woolley 2008). Another possibility for building new identities may be found in local sustainability initiatives that, while not directly referring to the oil sands, express a generalized concern about a petroleum-fuelled lifestyle and about peak oil (Haluza-DeLay and Berezan 2013). Such associations make visible the common interests shared by different groups in social and ecological protection rather than in perpetuating the oil sands status quo.

Third and finally, the movements have an uncertain strategy in terms of framing the oil sands as a case of environmental justice (or climate justice). As described in the previous section, the 'justice' frame has been used repeatedly; Aboriginal groups are clearly positioning their activism as countering injustices to the lands and themselves, while the Sierra Club, Greenpeace, and other environmental groups have increasingly used the 'environmental justice' frame. But the frame is not used consistently at the local or national level; it appears to be used more often in supranational mobilization by Canadian ENGOs and in reference to climate change. KAIROS remains as one of the few groups repeatedly using the related term 'eco-justice' in its website, educational material, press releases, and issue analyses.

The success of social movement frames depends on their cultural resonance. Snow and Corrigall-Brown (2005) demonstrate that, to achieve this resonance and create a favourable discursive opportunity structure, the 'environmental justice' frame must include a culturally appreciated 'disempowered people' and meet other conditions of social recognition (Snow 2008). Further research is needed into Albertan and Canadian conceptions of social inequality as these relate to environmental matters (Haluza-DeLay and Fernhout 2011). We need to better understand why Canadian environmental groups have not mobilized effectively around 'environmental justice.' ENGOs also need to reflect on the causes of past failures with regard to environmentalist-Aboriginal coalitions (Ballamingie 2009; Davis 2010).

Conclusion

This chapter first described social movement actors involved in opposition to the oil sands, including Aboriginal, environmental, religious, and labour organizations. These four movements have shifted strategies over time so that they have increasingly created broader coalitions, operated as transnational advocacy networks, and reframed the oil sands as both a global issue and a system with parts that could be targeted in segments. They have been able to link Alberta's oil sands to international concerns about climate change and indigenous rights and justice and thus to put external pressure on the governments of Alberta and Canada. Three trends in the opposition to the oil sands status quo are notable: a shift in scale from local to international levels of action, a shift in strategy from 'normal' politics to more disruptive challenges to institutions, and a shift from oppositional politics to hegemonic leadership, represented by more fully developed proposals for a post-carbon model of development.

Due to frustration with dealing with an unresponsive provincial government, the opposition shifted from working within institutions to using a variety of more disruptive tactics and targeting a wider range of actors (moving from government actors to banks, investors, and consumers) at broader scales (from the provincial to national, transborder, and international levels). Arguably, the government and industry have been disquieted by the movement's scale-shifting (such as the 'dirty oil' campaigns in the United States and Europe). In response, counter-campaigns by government and industry have attempted to re-legitimize oil sands development. Industry has drawn a 'line of sight' between their investments and the purported public benefits of bitumen extraction, and this has helped to secure the petro-hegemony within the province.

Key in this re-legitimation of the oil sands is the use of Albertan culture. While petro-capitalism interpellates most of us as consumers, the energy industries hail Albertans specifically as beneficiaries of resource exploitation. The industry and the government work actively to construct a provincial identity that is inseparable from the short-term profit interests of the extractive industry, as seen in slogans like 'Alberta *is* Energy.' To counter this, the opposition to the oil sands needs a strong, coherent formulation of a new Albertan identity or culture associated with a political ecological order that is just and sustainable. Given the immense resources of the existing system and the capital-state partnership, this will be no easy task, but there is already evidence of success in problematizing the oil sands status quo, particularly via international coalitions.

APPENDIX 14.1. Signatories to a Moratorium on New Oil Sands Development (*No New Approvals* 2008)[1]

Alberta-Based Groups or Organizations

- Albertans Demand Affordable Housing
- Alberta Federation of Labour
- Alberta Wilderness Association
- Arusha Centre
- Bow River Keepers
- Calgary Presbytery, United Church of Canada
- Canadian Association of Physicians for the Environment
- Canadian Parks and Wilderness Society, Calgary/Banff
- Canadian Parks and Wilderness Society, Northern Alberta Chapter
- Citizens Advocating Use of Sustainable Energy (CAUSE)
- Citizens for Responsible Development (Industrial Heartland)
- Citizens for a Sustainable Okotoks
- Council of Canadians (Prairie Chapter)
- Edmonton Small Press Association
- Faith and the Common Good (Calgary)
- Federation of Alberta Naturalists
- Fort Saskatchewan Naturalist Society (FSNS)
- Green Communities Edmonton Association

1 The website originally accessed in November 2008 is no longer active, but similar information can be found at http://nonewoilsands.wordpress.com.

- Greenpeace Canada (Alberta Office)
- Jasper Environmental Association
- KAIROS (Calgary Chapter)
- KAIROS (Edmonton-Committee)
- KAIROS (Grande Prairie)
- Keepers of the Athabasca
- Lavesta Area Group
- Northeast Sturgeon County Industrial Landowners
- Oil Sands Truth
- Parkland Institute
- Peace River Environmental Society
- Peace Valley Environment Association
- Pembina Institute
- Polaris Institute
- S.A.L.T. (Seniors Action Liason Team Edmonton)
- Sierra Club of Canada, Prairie Chapter
- Sierra Youth Coalition, Prairie Chapter
- Solar Energy Society of Canada
- Stop the Tar Sands (Calgary)
- Stop Tar sands Operations Permanently (Edmonton)
- Tipping Point Project
- Toxics Watch Society
- The Urban Farmer
- Unitarian Church Social Justice Committee
- West Athabasca Watershed Bioregional Society

National/International Groups or Organizations

- Alternatives North
- The Atlantic Regional Solidarity Network (ARSN)
- BC Sustainable Energy Association
- Beyond Nuclear
- Canadian Youth Climate Coalition
- Conservation Society of Pohnpei (Micronesia, Pacific Ocean)
- Corporate Ethics International
- Council of Canadians
- Council of Canadians (BC/Yukon Chapter)
- Council of Canadians (London Chapter)
- David Suzuki Foundation
- Eastern Co-operative Health Organization (ECHO)

- Earthworks
- Ecology North (NWT)
- Environmental Coalition of Prince Edward Island
- Environmental Justice and Climate Coalition
- Faith and the Common Good (Calgary)
- Forest Ethics
- Friends of the Earth (Canada)
- Friends of the Earth (United States)
- Global Exchange
- Greenpeace Canada
- Greenspiration
- Indigenous Environmental Network
- International Institute of Concern for Public Health
- Natural Resources Defense Council
- Nature Saskatchewan
- Physicians for Global Survival
- Polaris Institute
- Public Citizen
- Rainforest Action Network
- Rising Tide (Australia)
- Rising Tide (North America)
- Ruckus Society
- Sierra Club of Canada
- Sierra Youth Coalition
- U.S. Climate Emergency Council

NOTES

1 Social movement responses to oil sands debates are very dynamic, and so research into these activities is a moving analytical target. In this chapter we focus primarily on the first decade of the 2000s. The socio-political terrain has shifted quickly in the last several years, as have strategies and coalitions, many of which emphasize the oil sands' contributions to carbon emissions associated with climate change and resistance to proposed pipelines (Hoffman et al. 2014).

2 Carter's research has been supported by the Social Sciences and Humanities Research Council, Cornell University's Centre for the Environment, Cornell University's Graduate School, Memorial University's Institute of Social and Economic Research, and the Institute for Biodiversity, Ecosystem Science and Sustainability. Haluza-DeLay's research has been supported by The King's University College Internal Research Committee.

3 Some of the details in this section are also to be found in Haluza-DeLay and Carter (2014). We have extended the time frame of that analysis with more recent material.

4 These data and the data in figure 14.2 are drawn from a database created by Dr Howard Ramos of Dalhousie University under the SSHRC-funded project 'Mapping the Canadian Marketplace of Ideas: Media and Political Opportunities of Aboriginal, Environmental, Gay and Lesbian, and Women's Advocacy.' Sierra Club and Greenpeace press releases on their national websites or provided to the researchers were used to make the searchable database. Research assistance was provided by Andreas Hoffbauer.

5 From a database created by Howard Ramos, a sociologist at Dalhousie University, Halifax, Nova Scotia. The 'Discursive Opportunities Project' was a SSHRC-funded research project that collected data on selected Canadian social movements.

6 Telfer began as chapter director in 2005. Organizational memory and records are inconsistent as to the amount of attention paid to the oil sands prior to her time.

7 The Trudeau-era National Energy Program, which increased transfers from Alberta to the national government, is still well remembered in provincial discourse. Then mayor of Calgary Ralph Klein coined this phrase, which is still reflective of the mindset of Alberta 'nationalism' – especially regarding Alberta's energy resources.

8 Such actions have been primarily from Christian groups so far. There has been little interfaith or non-Christian activism on the oil sands, although some other issues – climate change being one example (Lysack 2013) – have been occasions for interfaith engagement. However, broaching an issue in one faith context can influence other faith groups to broach the issue from within their own traditions (Gottlieb 2006; Veldman, Szasz, and Haluza-DeLay 2013).

9 Climenhaga's post is at http://albertadiary.ca/2014/07/climate-change-divestment-movement-gains-ground-but-not-in-canadian-media-or-political-circles.html.

10 Kairos is one of many Canadian charitable organizations that has been targeted by the Harper government for reduced funding and tax auditing for possibly engaging in levels of 'political advocacy' not allowed for registered charities. Thirty-five years of funding by the Canadian International Development Agency dissipated in November 2009. Some observers suspect that the 2009 oil sands tour and KAIROS' prominent criticism of Canada's position on international climate negotiations contributed to this decision (http://www.cbc.ca/news/politics/timeline-oda-and-the-kairos-funding-1.1027221).

11 Available at http://www.kairoscanada.org/sustainability/concluding-statement-kairos-ecumenical-delegation-to-the-athabasca-tar-sands, the 'Concluding Statement: KAIROS Ecumenical Delegation to the Athabasca Tar Sands' was signed by all members of the tour, who are listed at the above URL.
12 See, for example, the 'Just Transition' strategies of the Canadian Labour Congress to address climate change (http://www.canadianlabour.ca/news-room/publications/just-transition-workers-during-environmental-change), and the Communications, Energy and Paperworkers' Policy 915 (CEP 2000).
13 While the AFL is a funder of the Parkland Institute, the institute's research is independently directed, usually authored or co-authored by academics, and undergoes peer review by academics.
14 A fuller description of the historical formation and characteristics of the provincial labour force and its unions is not possible in this chapter, but would be critical to explain the lack of resources and other weaknesses of Alberta's unions, compared, for example, to the union movement in Quebec.
15 http://www.afl.org/index.php/Download-document/793-2013-Fact-Sheet_Oil-Companies-Profits.html.
16 'AFL Blasts Northern Gateway Decision,' news release, 13 December 2013, http://www.afl.org/index.php/Press-Release/afl-blasts-northern-gateway-decision.html.
17 'A Source of Pride,' www.clac.ca/news_posts/source-of-pride, 18 January 2011.
18 The precise statement, found in the executive summary of the 2012 World Outlook report, is: 'No more than one-third of proven reserves of fossil fuels can be consumed prior to 2050 if the world is to achieve the 2°C goal, unless carbon capture and storage (CCS) technology is widely deployed.'
19 The amount of global greenhouse gas emissions that the oil sands produce is a major point of contention. Assessments often differ in assumptions, methodologies, and figures used to make the calculations. Space precludes discussion of these conflicting assessments, but estimates vary as widely as from 10 to 80 per cent more than conventional oil production.
20 http://www.capp.ca/environmentCommunity/airClimateChange/Pages/GreenhouseGasEmissions.aspx.

REFERENCES

Adkin, Laurie, ed. 2009a. *Environmental Conflict and Democracy in Canada*. Vancouver: UBC Press.

Adkin, Laurie, et al. 2009b. 'Non-violent Civil Disobedience Is a Far Cry from Terrorism' (letter with 89 signatories). *Edmonton Journal*, October 9. http://www.edmontonjournal.com/news/violent+civil+disobedience+from+terrorism/2084494/story.html.

Agyeman, Julian, Peter Cole, Randolph Haluza-DeLay, and Pat O'Riley, eds. 2009. *Speaking for Ourselves: Environmental Justice in Canada.* Vancouver: UBC Press.

Alberta Federation of Labour (AFL). 2014. *In-Province Upgrading: Economics of a Green Field Oil Sands Refinery*. 6 October. Edmonton: AFL.

Alberta Federation of Labour. 2013. *A Public Interest Approach to Alberta's Oil Sands: The Case for Public Investment in Oil Sands Upgrading and an Economic Blueprint for Jobs, Environment and Revenue.* Edmonton: AFL.

Alberta Federation of Labour. 2009. *Lost Down the Pipeline*. Edmonton: AFL.

Alberta Federation of Labour. 2008. *Black Gold Clear Vision: A Proposed Policy Framework for the Alberta Oil Sands*. Edmonton: AFL.

Angus, Ian, ed., 2010. *The Global Fight for Climate Justice.* Halifax: Fernwood.

Anielski, Mark, and Jeff Wilson. 2004. *The Ecological Footprint of Canadian Cities.* Ottawa: Federation of Canadian Municipalities.

Anker, Peder. 2013. 'The Call for a New Ecotheology in Norway.' *Journal for the Study of Religion, Nature and Culture* 7: 187–207.

Arrowsmith, Lisa. 2008. 'Alberta's Oilsands like Mordor: Maude Barlow.' *Toronto Star*, 31 October.

Ballamingie, Patricia. 2009. 'First Nations, ENGOs, and Ontario's Lands for Life Consultation.' In *Environmental Conflict and Democracy in Canada.*, ed. Laurie E. Adkin, 84–102. Vancouver: UBC Press.

Barclay, Bill. 2009. *Financing Global Warming: Canadian Banks and Fossil Fuels.* San Francisco: Rainforest Action Network.

Bennett, Dean. 2010. 'Alberta Hopes James Cameron's Oilsands Comments Won't Open a Pandora's Box.' Canadian Press, 21 April.

Bennett, Dean. 2009. 'Most Canadians OK with Oilsands: Poll.' Canadian Press, 3 March.

Berry, Dave. 2010. 'The Disobedient Albertans.' *Alberta Views*, 134, no. 5, 32–37.

Bevington, Douglas. 2009. *The Rebirth of Environmentalism: Grassroots Activism from the Spotted Owl to the Polar Bear*. Washington, DC: Island.

Billings, Dwight B. 1990. 'Religion as Opposition: A Gramscian Analysis.' *American Journal of Sociology* 96 (1): 1–31.

Bjelland Grønvold, Jens. 2013. 'Theology and Sustainability in Oil-Producing Norway.' *Nature and Culture* 8: 265–81.

Bouchard, Bishop L. 2009. 'A Pastoral Letter on the Integrity of Creation and the Athabasca Oil Sands.' 25 January. http://www.crc-canada.org/sites/

default/files/files/BISHOP%20BOUCHARDS%20PASTORAL%20LETTER%20ON%20TAR%20SANDS.pdf (accessed 30 March 2016).

Boychuk, Regan. 2010. 'Misplaced Generosity: Extraordinary Profits in Alberta's Oil And Gas Industry.' Edmonton: Parkland Institute.

Brooymans, Hanneke. 2009. 'Cancer Downstream from Oilsands Merits Further Study: Alta. Gov't.' *Canwest News Service*, 26 February.

Byfield, Mike. 2007. 'Gil McGowan: A Union Vision for the Future Oilpatch.' *Oil and Gas Enquirer*, October.

Canada West Foundation. 2009. *Media Monitoring Report: July*. Calgary: Canada West Foundation.

Carbon Tracker. 2013. 'Unburnable Carbon 2013: Wasted Capital and Stranded Assets.' Grantham Research Institute on Climate Change and the Environment, London School of Economics and Political Science, London.

Carroll, William K. 2006. 'Hegemony, Counter-Hegemony, Anti-Hegemony.' *Journal of Socialist Studies* 2 (2), 9–43.

Carter, Angela. 2011. *Environmental Policy in a Petro-State: The Resource Curse and Political Ecology in Canada's Oil Frontier*. Doctoral diss., Cornell University.

Cattaneo, Claudia. 2008. 'Oil Sands Producers Start Web Site to Encourage "Dialogue."' *Financial Post*, 23 June.

Chapman, Ken. 2010. 'Where the Oil Sand Owners Stand.' *The Mark*. http://www.themarknews.com/articles/1932-where-the-oil-sand-owners-stand (accessed 2 January 2011).

Chetkovich, Carol, and Frances Kunreuther. 2006. *From the Ground Up: Grassroots Organizations Making Social Change*. Ithaca: Cornell University Press.

Church Investors Group. 2008. *Unconventional Oil: Tar and Shale Sands: A Briefing for Church Investors*. London: Church Investors Group.

Clarke, Tony. 2008. *Tar Sands Showdown: Canada and the New Politics of Oil in an Age of Climate Change*. Ottawa: Canadian Centre for Policy Alternatives.

Communications, Energy and Paperworkers Union of Canada. 2000. 'Policy 915: Just Transition to a Sustainable Economy in Energy.' Montreal. http://cep.unifor.org/docs/en/policy-915-e.pdf (accessed 4 April 2016).

Conference Board of Canada. 2010. *The Economic and Employment Impacts of Climate-Related Technology Investments*. http://www.conferenceboard.ca/documents.aspx?did=3586 (accessed 13 December 2010).

Cook, Jackie. 2012. 'Political Action through Environmental Shareholder Resolution Filing: Applicability to Canadian Oil Sands?' *Journal of Sustainable Finance & Investment* 2: 26–43.

Crehan, Kate A.F. 2002. *Gramsci, Culture and Anthropology*. Berkeley: University of California Press.

Crooks, Ed. 2008. 'Investors Warned of Risk to Oil Sands Plans.' *Financial Times*, 15 September.

Cryderman, Kelly. 2014. 'Tutu's Harsh Words Prompt New Focus on Oil-Sands Fight.' *Globe and Mail*, 1 June. http://www.theglobeandmail.com/news/national/tutus-harsh-words-prompt-new-focus-on-oil-sands-fight/article18942264/ (accessed 27 March 2016).

Cryderman, Kelly. 2013. 'Judge Quashes Alberta's Decision to Bar Environmentalists from Oil Sands Hearing.' *Globe and Mail*, 2 October.

Cryderman, Kelly, and Florence Loyie. 2009. 'Bishop Spurns Oilsands Development.' *Edmonton Journal*, 29 January, A3.

D'Arcy, Stephen, Toban Black, Tony Weis, and Joshua Kahn Russell. 2014. 'A Line in the Tar Sands: Struggles for Environmental Justice.' Toronto: Between the Lines.

Davidson, Debra J., and Mike Gismondi. 2011. *Challenging Legitimacy at the Precipice of Energy Calamity*. Berlin: Springer.

Davis, Lynne, ed. 2010. *Alliances: Re/Envisioning Indigenous-NonIndigenous Relationships*. Toronto: University of Toronto Press.

Deacon, Leith, and Jamie Baxter. 2009. 'Framing Environmental Inequity in Canada: A Content Analysis of Daily Print News Media.' In *Speaking for Ourselves: Environmental Justice in Canada*, ed. Julian Agyeman, Peter Cole, Randolph Haluza-DeLay, and Patricia O'Riley, 163–80. Vancouver: UBC Press.

De Souza, Mike. 2008. 'U.S. Mayors Point Green Finger at Alberta Oil Sands.' *Canwest News Service*, 23 June.

Downey, Dennis J., and Deanna Rohlinger. 2008. 'Linking Strategic Choice with Macro-organizational Dynamics: Strategy and Social Movement Articulation.' *Research on Social Movements, Conflict and Change* 28: 3–35.

Downie, Christian. 2014. 'Transnational Actors in Environmental Politics: Strategies and Influence in Long Negotiations.' *Environmental Politics* 23: 376–94.

Earl, Jennifer. 2004. 'The Cultural Consequences of Social Movements.' In *The Blackwell Companion to Social Movements*, ed. David A. Snow, Sarah A. Soule, and Hanspeter Kriesi, 508–30. London: Blackwell.

Ecojustice. 2009. *Syncrude Cannot Duck from Charges in Death of 500 Waterfowl.* http://www.ecojustice.ca/media-centre/press-releases/syncrude-cannot-duck-from-charges-in-death-of-500-waterfowl/ (accessed 6 February 2009].

Ekers, Michael, Alex Loftus, and Geoff Mann. 2009. 'Gramsci Lives!: Themed Issue: Gramscian Political Ecologies.' *Geoforum* 40 (3), 287–91.

Erickson, Peter, and Michael Lazarus. 2014. 'Impact of the Keystone XL

Pipeline on Global Oil Markets and Greenhouse Gas Emissions.' *Nature Climate Change* 4: 778–80.

Escobar, Arturo. 2001. 'Culture Sits in Places: Reflections on Globalism and Subaltern Strategies of Localization.' *Political Geography* 20: 139–74.

Flanagan, Tom, and Sue Jose. 2014. 'Success Story Powered by Self-Help.' *Edmonton Journal*, 21 April, A13.

Fluet, Colette, and Naomi Krogman. 2009. 'The Limits of Integrated Resource Management in Alberta for Aboriginal and Environmental Groups: The Northern East Slopes Sustainable Resource and Environmental Management Strategy.' In *Environmental Conflict and Democracy in Canada*, ed. Laurie Adkin, 123–39. Vancouver: UBC Press.

Friedel, Tracy. 2008. '(Not So) Crude Text and Images: Staging Native in "Big Oil" Advertising.' *Visual Studies* 23: 238–54.

Gailus, Jeff. 2012. *Little Black Lies: Corporate & Political Spin in the Global War for Oil*. Calgary: Rocky Mountain Books.

Gamson, William A. 2004. 'Bystanders, Public Opinion and the Media.' In *The Blackwell Companion to Social Movements*, ed. David A. Snow, Sarah A. Soule, and Hanspeter Kriesi. London: Blackwell.

Gellert, Paul K., and Jon Shefner. 2009. 'People, Place, and Time: How Structural Fieldwork Helps World-Systems Analysis.' *Journal of World-Systems Research* 15 (2): 193–218.

Gibbins, Dan. 2010. *Blackened Reputation: A Year of Coverage of Alberta's Oil Sands*. Calgary: Canada West Foundation.

Gibbins, Roger, and Robert Roach. 2010. *Look Before You Leap: Oil and Gas, the Western Canadian Economy and National Prosperity*. Calgary: Canada West Foundation.

Giugni, Marco, Doug McAdam, and Charles Tilly. 1999. *How Social Movements Matter*. Minneapolis: University of Minnesota Press.

Globe and Mail. 'Wildly Uncharitable Allegations.' 2012. Editorial, 7 May, A18.

Golding, Sue. 1992. *Gramsci's Democratic Theory: Contributions to a Post-Liberal Democracy*. Toronto: University of Toronto Press.

Gottlieb, Roger S. 2006. *A Greener Faith: Religious Environmentalism and Our Planet's Future*. Oxford: Oxford University Press.

Gould, Kenneth A., David N. Pellow, and Allan Schnaiberg. 2008. *The Treadmill of Production: Injustice and Unsustainability in the Global Economy*. London: Paradigm.

Gramsci, Antonio. 1971. *Selections from the Prison Notebooks of Antonio Gramsci*. Trans. Q. Hoare and G.N. Smith. London: Lawrence and Wishart.

Hajer, Maarten A. 2009. 'Ecological Modernisation as Cultural Politics.' In *The Ecological Modernisation Reader: Environmental Reform in Theory and Practice*,

ed. Arthur P.J. Mol, David A. Sonnenfeld, and Gert Spaargaren. London: Routledge.
Hall, Carrie. 2013. 'The Environmental Discussion within Canadian Mennonite: Content Analysis of Canadian Religious Periodical.' Paper presented at Canadian Conference on Scholarship and the Christian Faith, 3 May 2013, Concordia University College of Alberta, Edmonton.
Haluza-DeLay, Randolph. 2015. 'Alberta Internalizing Oilsands Opposition: A Test of the Social Movement Society Thesis.' In *Protest and Politics: The Promise of Social Movement Societies*, ed. Howard Ramos and Kathleen Rogers, 274–96. Vancouver: UBC Press.
Haluza-DeLay, Randolph. 2014. 'Assembling Consent in Alberta: Hegemony and the Tar Sands.' In *A Line in the Tar Sands: Struggles for Environmental Justice*, ed. Stephen D'Arcy, Toban Black, Tony Weis, and Joshua Kahn Russell, 36–44. Toronto: Between the Lines.
Haluza-DeLay, Randolph. 2009. 'Uncovering Canada's Environmental Cultural Politics.' *International Journal of Canadian Studies* 39 (1): 131–6.
Haluza-DeLay, Randolph, and Ron Berezan. 2013. 'Permaculture in the City: Ecological *habitus* and the Distributed Ecovillage.' In *Localizing Environmental Anthropology: Bioregionalism, Permaculture, and Ecovillage Design for a Sustainable Future*, ed. Joshua Lockyear and Jim Veteto, 130–45. Oxford: Berghan.
Haluza-DeLay, Randolph, and Angela V. Carter. 2014. 'Joining Up and Scaling Up: Analyzing Resistance to Canada's "Dirty Oil."' In *Activist Science & Technology Education*, ed. Steve Alsop and Larry Bencze, 343–62. New York: Springer.
Haluza-DeLay, Randolph B., Michael Ferber, and Tim Wiebe-Neufeld. 2013. 'Watching *Avatar* from 'AvaTarSands' Land.' In *Avatar and Nature Spirituality*, ed. B. Taylor, 123–40. Waterloo: Wilfred Laurier University Press.
Haluza-DeLay, Randolph, and Heather Fernhout. 2011. 'Sustainability and Social Inclusion? Examining the Frames of Canadian English-Speaking Environmental Movement Organizations. *Local Environment: A Journal of Justice and Sustainability* 16: 727–45.
Hart, Stephen. 2001. *Cultural Dilemmas of Progressive Politics: Styles of Engagement among Grassroots Activists*. Chicago: University of Chicago Press.
Harvey, David. 2003. *The New Imperialism*. New York: Oxford University Press.
Heinrichs, Steve, ed. 2013. *Buffalo Shout, Salmon Cry*. Winnipeg: Herald Press.
Hoberg, George, and Jeffrey Phillips. 2011. 'Playing Defence: Early Responses to Conflict Expansion in the Oil Sands Policy Subsystem.' *Canadian Journal of Political Science/Revue canadienne de science politique* 44: 507–27.

Hoekstra, Gordon. 2009. 'Propaganda Pipeline.' *Prince George Citizen*, 29 May.

Hoffman, Steven M. 2012. 'Hijacking Canada: Tar Sands and the Problem of an Effective Opposition.' Paper presented at Petrocultures Conference, Edmonton, 6 September.

Hoffman, Steven M., Paul Lorah, Joseph Janochoski, and Randolph Haluza-DeLay. 2014. 'Tar Sands, Oppositional Activity, and Transborder Networks.' Paper presented at the conference of the Canadian Sociology Association, St. Catherine's, 26 May.

Huot, Marc, Lindsay Fischer, and Nathan Lemphers. 2011. 'Oilsands and Climate Change: How Canada's Oilsands Are Standing in the Way of Effective Climate Action.' Edmonton: Pembina Institute.

Hussey, Amy. 2010. 'Seen as Canadian Oilsands Allegory, Environmentalists Want Oscar for *Avatar*.' *Canwest News Service*, 4 March.

International Energy Agency. 2012. *World Energy Outlook 2012*. Paris: IEA.

Jasper, James. 2006. *Getting Your Way: Strategic Dilemmas in the Real World*. Chicago: University of Chicago Press.

Jasper, James. 2004. 'A Strategic Approach to Collective Action: Looking for Agency in Social Movement Choices.' *Mobilization* 9: 1–16.

Johnston, William A., Harvey Krahn, and Trevor Harrison. 2006. 'Democracy, Political Institutions, and Trust: The Limits of Current Electoral Reform Proposals.' *Canadian Journal of Sociology* 31 (2): 165–82.

KAIROS: Faithful Action for Justice. 2011. http://www.KAIROScanada.org (accessed 2 January 2011).

Keck, Margaret E., and Kathryn Sikkink. 1999. *Activists beyond Borders: Advocacy Networks in International Politics*. Ithaca: Cornell University Press.

Klandermans, Bert, and Suzanne Staggenborg, eds. 2002. *Methods of Social Movement Research*. Minneapolis: University of Minnesota Press.

Klein, Naomi. 2013. 'Why Unions Need to Join the Climate Fight.' 4 September. http://rabble.ca/columnists/2013/09/why-unions-need-to-join-climate-fight.

Kopecky, Arno. 2014. 'Game Changer.' *Alberta Views*, June. https://albertaviews.ab.ca/2014/05/27/game-changer/.

Kowalsky, Nathan, and Randolph Haluza-DeLay. 2013. '*Homo Energeticus*: Technological Rationality in the Alberta Tar Sands.' In *Jacques Ellul and the Technological Society in the 21st Century*, ed. Helena Mateus Jeronimo, Jose Luis Garcia, and Carl Mitcham, 159–75. Berlin: Springer.

Krashinsky, Susan. 2012. 'Greenpeace Denied Edmonton Billboard Space for Oil Spill Ad.' *Globe and Mail*, 19 June.

Laxer, Gordon. 2008. *Freezing in the Dark: Why Canada Needs Strategic Petroleum Reserves*. Edmonton: Parkland Institute/Polaris Institute.

Le Billon, Philippe, and Angela Carter. 2012. 'Securing Alberta's Tar Sands: Resistance and Criminalization on a New Energy Frontier.' In *Natural Resources and Social Conflict: Towards Critical Environmental Security*, ed. Matthew Schnurr and Larry Swatuk, 170–92. London: Palgrave Macmillan.

Levant, Ezra. 2010. *Ethical Oil: The Case for Canada's Oil Sands*. Toronto: McClelland and Stewart.

Lind, Christopher, and Joe Mihevc, eds. 1994. *Coalitions for Justice*. Ottawa: Novalis.

Lofland, John. 1996. *Social Movement Organizations: Guide to Research on Insurgent Realities*. Hawthorne: Aldine de Gruyter.

Loyie, Florence. 2009. 'Doctor Cleared Over Suggested Link between Cancer & Oilsands.' *Edmonton Journal*, 9 November.

Lysack, Mishka. 2013. 'Stepping Up to the Plate: Climate Change, Faith Communities and Effective Environmental Advocacy in Canada.' In *How the World's Religions Are Responding to Climate Change: Social Scientific Investigations*, ed. R.G. Veldman, A. Szasz, and R. Haluza-DeLay, 157–73. New York: Routledge.

Lysack, Mishka, and Karri Munn-Venn. 2013. *Living Ecological Justice: A Biblical Response to the Environmental Crisis*. Ottawa: Citizens for Public Justice.

Major, Claire, and Tracy Winters. 2013. 'Community by Necessity: Security, Insecurity, and the Flattening of Class in Fort McMurray.' *Canadian Journal of Sociology* 38: 141–65.

Masuda, Jeffrey R., Tara K. McGee, and Theresa D. Garvin. 2008. 'Power, Knowledge, and Public Engagement: Constructing "Citizenship" in Alberta's Industrial Heartland.' *Journal of Environmental Policy and Planning* 10 (4): 359–80.

McAdam, Doug, Sidney Tarrow, and Charles Tilly. 2001. *Dynamics of Contention*. Cambridge: Cambridge University Press.

McCallum, Hugh. 2006. *Fuelling Fortress America: A Report on the Athabasca Tar Sands and U.S. Demands for Canada's Energy*. Ottawa: Canadian Centre for Policy Alternatives.

McCammon, Holly J., Harmony D. Newman, Courtney Sanders Muse, and Teresa M. Terrell. 2007. 'Movement Framing and Discursive Opportunity Structures: The Political Successes of the U.S. Women's Jury Movements.' *American Sociological Review* 72 (5): 725–49.

McCarthy, Shawn. 2012. 'The Day the Oil-Sands Battle Went Global.' *Globe and Mail*, 21 January.

McKeon, Bob. 2010. 'Base Energy Policy on Christian Values.' *Western Catholic Reporter*, 24 May, 15.

Meyer, David S., and Debra C. Minkoff. 2004. 'Conceptualizing Political Opportunity.' *Social Forces* 82 (4): 1457–92.

Mikisew Cree First Nation. 2007. *Response to the Multi-Stakeholder Committee Phase II Proposed Options for Strategies and Actions AND Submission to the Government of Alberta for the Oil Sands Strategy*. Fort Chipewyan: Mikisew Cree First Nation.

Moore, Donald S., Jake Kosek, and Anand Pandian, eds. 2003. *Race, Nature, and the Politics of Difference*. Durham: Duke University Press.

Muthui, Daniel. 2013. 'Religious Mediation: Media Framing and the Role of Christian Social Actors in Public Discourses.' Paper presented at the Canadian Conference on Scholarship and the Christian Faith, 3 May, Concordia University College of Alberta, Edmonton.

Nikiforuk, Andrew. 2008. *Tar Sands: Dirty Oil and the Future of a Continent*. Vancouver: Greystone.

No New Approvals for Tar Sands Developments. 2008. www.nonewapprovals.ca (accessed 3 November 2008).

Oliver, Joe. 2012. 'An Open Letter on Canada's Commitment to Diversify Our Energy Markets and the Need to Further Streamline the Regulatory Process in Order to Advance Canada's National Economic Interest.' Press release, Natural Resources Canada, 9 January. www.nrcan.gc.ca/media-room/news-release/2012/1/3520.

Ominayak, Bernard, and Kevin Thomas. 2009. 'These Are Lubicon Lands: A First Nation Forced to Step into the Regulatory Gap.' In *Speaking for Ourselves: Environmental Justice in Canada*, ed. Julian Agyeman, Peter Cole, Randolph Haluza-DeLay, and Patricia O'Riley, 111–22 . Vancouver: UBC Press.

Palen, Wendy J., Thomas D. Sisk, Maureen E. Ryan, Joseph L. Árvai, Mark Jaccard, Anne K. Salomon, Thomas Homer-Dixon, and Ken P. Lertzman. 2014. 'Energy: Consider the Global Impacts of Oil Pipelines.' *Nature* 510: 465–7.

Parkins, John. 2002. 'Forest Management and Advisory Groups in Alberta: An Empirical Critique of an Emergent Public Sphere.' *Canadian Journal of Sociology* 27: 163–84.

Pembina Institute. 2013. 'Court Rules Alberta Improperly Excluded Pembina Institute from Oilsands Regulatory Process.' Media release, Pembina Institute, 2 October. http://www.pembina.org/media-release/2484.

Pembina Institute. 2008. *Poll: Canadians Want Action on Global Warming Despite Economic Downturn*. http://www.pembina.org/media-release/1736 (accessed 2 January).

Pembina Institute. 2007. *Backgrounder: Albertans' Perceptions of Oil Sands Development: Poll Part 1: Pace and Scale of Oil Sands Development*. http://pubs.pembina.org/reports/GHGPoll_Env_2007_MediaBG.pdf (accessed 2 January).

Pembina v. Alberta [Environment and Sustainable Resource Development], [2013] ABQB 567.

Phillips, Shannon. 2010. 'Women's Equality a Long Way Off in Alberta: Alberta Most Unequal Province in Canada.' Fact sheet. Parkland Institute, Edmonton. http://parklandinstitute.ca/research/summary/womens_equality_a_long_way_off_in_alberta/.

Polletta, Francesca, and James M. Jasper. 2001. 'Collective Identity and Social Movements.' *Annual Review of Sociology* 27: 283–305.

Poloni-Staudinger, Lori M. 2008. 'The Domestic Opportunity Structure and Supranational Activity: An Explanation of Environmental Group Activity at the European Union Level.' *European Union Politics* 9 (4): 531–58.

Richards, John, and Larry Pratt. 1979. *Prairie Capitalism: Power and Influence in the New West*. Toronto: McClelland and Stewart.

Rootes, Christopher. 2013. 'From Local Conflict to National Issue: When and How Environmental Campaigns Succeed in Transcending the Local.' *Environmental Politics* 22: 95–114.

Schneider, Richard R. 2002. *Alternative Futures: Alberta's Boreal Forest at the Crossroads*. Edmonton: Federation of Alberta Naturalists.

Shantz, Jim. 2014. 'As Long as the Rivers Flow.' *Canadian Mennonite* 18 (16): 18–20.

Slowey, Gabrielle. 2008. *Navigating Neoliberalism: Self-Determination and the Mikisew Cree First Nation*. Vancouver: UBC Press.

Snow, David A. 2008. 'Elaborating the Discursive Contexts of Framing: Discursive Fields and Spaces.' *Studies in Symbolic Interaction* 30: 3–28.

Snow, David A., and Catherine Corrigall-Brown. 2005. 'Falling on Deaf Ears: Confronting the Prospect of Nonresonant Frames.' In *Rhyming Hope and History: Activists, Academics, and Social Movement Scholarship*, ed. David Croteau, William Hoynes, and Charlotte Ryan, 222–38. Minneapolis: University of Minneapolis Press.

Spencer, Bruce. 1995. 'Old and New Social Movements As Learning Sites: Greening Labor Unions and Unionizing the Greens.' *Adult Education Quarterly* 46 (1): 31–42.

Thompson, David. 2009. *Green Jobs: It's Time to Build Alberta's Future*. Edmonton: Alberta Federation of Labour/Greenpeace/Sierra Club.

Thomson, Graham. 2010. 'Greenpeace Antics Give Pause for Thought: Public Increasingly in Favour of Greening of the Oilsands. So Why Not Do It?' *Edmonton Journal*, 4 August.

Timoney, Kevin. 2007. *A Study of Water and Sediment Quality as Related to Public Health Issues, Fort Chipewyan, Alberta*. Fort Chipewyan: Nunee Health Board Society.

Timoney, Kevin, and Peter Lee. 2009. 'Does the Alberta Tar Sands Industry

Pollute? The Scientific Evidence.' *The Open Conservation Biology Journal* 3: 65–81.

Turner, Chris. 2012. 'The Oil Sands PR War.' *Marketing*, 13 August 13, 28–32.

Van Der Heijden, Hein-Anton. 2006. 'Globalization, Environmental Movements, and International Political Opportunity Structures.' *Organization and Environment* 19 (1): 28–45.

Vanderklippe, Nathan. 2012. 'Federal Documents Spark Outcry by Oil Sands Critics.' *Globe and Mail*, 26 January.

Veldman, Robin Globus, Andrew Szasz, and Randolph Haluza-DeLay, eds. 2013. *How the World's Religions Are Responding to Climate Change: Social Science Investigations*. Abingdon, UK: Routledge.

Warnica, Richard. 2010. 'Bishop Assesses Fallout from Letter.' *Edmonton Journal*, 28 February.

Watts, Michael J. 2005. 'Righteous Oil? Human Rights, the Oil Complex and Corporate Social Responsibility.' *Annual Review of Environment and Resources* 30 (1): 373–407.

Weber, Bob. 2013. 'Environmentalists Fight Exclusion from Hearings.' *Edmonton Journal*, 6 September, A6.

Weber, Bob. 2014. 'Alberta Government Bars Environmentalists from Oilsands Hearings.' *Toronto Star*, 6 May.

Wiebe-Neufeld, Donita. 2013. 'A Different Perspective on Alberta's Oilpatch.' *Canadian Mennonite*, 8 July, 20–1.

Woolley, Alice. 2008. 'Enemies of the State? The Alberta Energy and Utilities Board, Landowners, Spies, a 500kV Transmission Line and Why Procedure Matters.' *Journal of Energy and Resource Law* 26: 234–66.

Yaffe, Barbara. 2008. '"In Fact, Every Canadian Has a Stake in This"; Alberta Oilsands Industry Fights a Public Relations War in Advance of a New Energy Policy in the U.S.' *Vancouver Sun*, 25 November.

Zalik, Anna. 2009. 'Zones of Exclusion: Offshore Extraction, the Contestation of Space and Physical Displacement in the Nigerian Delta and the Mexican Gulf.' *Antipode* 41 (3): 557–82.

15 Alberta's Electricity Future

TIM WEIS, BENJAMIN THIBAULT, AND BYRON MILLER

Solutions exist for many of the problems associated with the use of fossil fuels to generate electricity. While many technical challenges remain, they are not often the limiting factors to the widespread deployment of renewable energy technologies. A clean electricity system needs to be the backbone of a broader transformation in the energy system to power electric vehicles, industrial processes, and building heating. The short-term wealth and technical capacity that non-renewable energy has provided could pave the way to Alberta's becoming a global leader in sustainable energy development. However, the reality of Alberta's present energy system is that investment, infrastructure, policy, and, to a large extent market design are all geared towards high-carbon forms of non-renewable energy development. This is apparent in the rush to develop the oil sands in northeastern Alberta, the environmental impacts of which have attracted significant national and international media and political attention. But it is also apparent in the much less discussed electricity sector's heavy reliance on and systemic orientation towards fossil-fuel-fired thermal generation. Because provincial policies promote centralized fossil-fuel-based energy development, Alberta's wealth of renewable energy resources are under-exploited and represent only a small proportion of energy use in the province.

As the preceding chapters in this book have demonstrated, political and economic barriers present more of an obstacle than does technology. Alberta is heavily entrenched in conventional, fossil fuel development, which to date has been a burden to developing sustainable energy systems. This need not be the case. Jurisdictions that have excelled in renewable energy development have largely done so as a result of concerted government policies (Sovacool 2008). Alberta will not become a clean energy leader, in spite of its endowment of renewable resources, unless governmental institutions demonstrate leadership.

Alberta's Energy Reality

Alberta has relied on coal for the majority of its electricity supply for much of the province's history. In 2010, over 71 per cent of electricity consumed in Alberta outside of the oil sands[1] was generated from coal. By 2013 the percentage had dropped to close to 64, due to the growth of other sources (notably natural gas), but the absolute capacity of coal had actually increased, with a new coal unit commissioned in 2012. Including natural gas (including 'peaking,' 'intermediate,' and industrial 'cogeneration'), Alberta continues to rely on fossil fuels for over 90 per cent of its electricity generation (Alberta Electric System Operator 2014).

Renewable energy in Alberta provides less than 10 per cent of the province's electricity: just over 5 per cent from wind and 3 per cent from hydro, and around 1 per cent from biomass energy (Alberta Electric System Operator 2014). Hydro generation, while fluctuating year over year, has remained relatively constant at just over 3 per cent of supply over the past decade, and wind energy has largely been responsible for the proportion of renewably generated electricity doubling in size from only 5 per cent of supply a decade ago (Alberta Electric System Operator 2014). In spite of these gains, the growth of the overall energy system has meant that greenhouse gas (GHG) emissions from the sector have continued to rise. While Alberta led the country in wind-generated electricity until quite recently, jurisdictions from across the country have surpassed it in both absolute and proportional use of wind. Ontario has very quickly doubled Alberta in installed wind capacity (Canadian Wind Energy Association 2013), and Quebec, a province that already generates almost all of its electricity from non-fossil-fuel-based sources, has contracted for 4,000 MW of wind energy by 2015, almost four times as much as Alberta's current capacity. The development of wind energy in Alberta has been supported by policies from outside the province (such as federal production incentives and even the sale of green credits to California) that no longer exist, and the growth of the past decade is unlikely to continue in the absence of new provincial policy measures such as those we sketch in this chapter.

Impact of Electricity Mix on Alberta's Pollution Profile and Environment

Electricity generation in Alberta emits more air pollution than electricity generation in any other province in Canada. Coal power, while supplying only 16 per cent of the nation's electricity, is responsible for about 77 per cent of Canada's electricity-related GHG emissions, almost half of which come from Alberta (Environment Canada 2011). Although it represents just over 10 per cent of all the electricity generated in Canada, Alberta's coal-dominated system has been responsible for over half of all of the greenhouse gas emissions from the entire Canadian electricity fleet (Environment Canada 2012).

The heavy reliance on coal-fired generation, with high secondary reliance on natural gas, gives Alberta's electricity grid the highest GHG emissions intensity in the country. Alberta's GHG grid intensity in 2009 was 880 tCO_2e/GWh (tonnes of carbon dioxide equivalent per gigawatt hour), nearly five times the national average at 180 tCO_2e/GWh. Alberta's electricity plants emitted 47.8 $MtCO_2e$ in 2009 – more than three times as much as the second-highest-emitting province, Ontario. Ontario has over three-and-a-half times the population of Alberta and successfully phased out its entire coal fleet in 2014 (Environment Canada 2011), despite its coal fleet being the same size as Alberta's only a decade ago.

Contrary to popular perception, oil and gas development is not the only reason for Alberta's high GHG emissions per capita and highly GHG-intensive economy (referring to high GHG emissions per GDP); electricity generation also plays a significant role. Despite pronounced growth in energy-intensive resource extraction projects, particularly the oil sands in northern Alberta, electricity generation remains one of the largest sources of GHG pollution by sector, accounting, in 2009, for just over 20 per cent of the province's annual emissions (Environment Canada 2011),[2] second only to the oil sands. Coal's prevalence in Alberta's electricity generation is also responsible for substantial contributions to key air pollutants, including sulphur oxides (SO_x), nitrogen oxides (NO_x), heavy metals (mercury, cadmium, and lead), and hexachlorobenzene (HCB). Table 15.1 summarizes the proportion of the total 2009 emissions of each pollutant in Alberta that was caused by electricity generation and by coal power specifically.

Table 15.1. Proportions of Alberta's 2009 Emissions Caused by All Electricity Generation and Specifically Coal-Fired Electricity Generation

Pollutant	All Electricity	Coal-Fired Electricity	Problems
SO_x	30.8%	30.8%	Contributes to fine particles and acid rain, as well as respiratory and cardiac problems
NO_x	12.0%	10.3%	Contributes to fine particles, acid rain, ground-level ozone formation (smog), respiratory and cardiac problems, and visible smog
Mercury	56.2%	55.8%	Toxic to humans and wildlife, including nervous system and fetal development impairment; heavy metals also tend to bioaccumulate
Cadmium	36.3%	30.8%	
Lead	11.3%	10.7%	
HCB	76.9%	76.9%	Probable human carcinogen, among other adverse effects

Source: Environment Canada (2013)

Economics of Electricity Generation Mix

Alberta's electricity generation mix affects the size as well as the qualitative characteristics of the province's economy. Costs of electricity influence economic growth by impacting other industrial sectors, but the differences in costs of generation are not as straightforward as many presume. Different-generation technologies can also affect economic disparities and other important social factors within an economy because they determine employment and entrepreneurial opportunities.

Costs of Generation

Accurate comparisons of future costs of generation among electricity-generating technologies are complex. They must account for fuel price projections, direct and indirect subsidies (including permission to treat the environment as a dumping ground for waste products), transmission costs, and how experience with contemporary technologies facilitates future deployment of technologies undergoing rapid cost decreases. Although past investments have resulted in currently low market prices, recent trends indicate that coal power will not continue to be cheap.

Large generating plants are capital-intensive developments whose costs are substantially determined by the prices of materials and equipment such as steel, cement, boilers, rotating equipment, piping, electrical components, and electric wiring. Increases in these prices have greatly increased the cost of building large generating plants; some regions witnessed a 50 per cent increase in costs between 2006 and 2008 alone (World Bank Energy Sector Management Assistance Program 2009). While cost increases have since slowed since the economic downturn that began in 2008, the long-term trend is for conventional large central generating plants to continue to become more expensive. With new federal regulations requiring new coal plants to employ carbon capture and storage (CCS) technology, additional capital costs for these complex projects will add to the expense of coal power.

Without large hydro developments, Albertans pay some of the highest rates for electricity in Canada, although by North American standards rates are relatively low for a predominately coal-based generation mix. Many of Alberta's large existing coal plants were built in the 1970s and 1980s (table 15.2). The relatively low price of coal-fired electricity in Alberta today is a product, to a large degree, of the relatively low capital costs of these older plants, which have largely been amortized and are not required to reduce any of their NOx or SOx emissions until they reach forty years of operating life. Compared to the cost of building new facilities, the combined costs of fuel, operation, and management for these older units are low. But as these plants reach the end of their economic lives in the next twenty years, new federal regulations will force them to decommission or install expensive CCS upgrades. As these plants leave the grid, any new coal-fired plants that replace them will come with the new reality of high capital costs, and the cost of coal-fired electricity will rise substantially.

Indeed, the price of electricity in Alberta is expected to steadily increase over the coming decade (EDC Associates 2011). In addition, expanded transmission infrastructure to support a growing system will be required – with costs that are borne by consumers. Albertans will not be able to take relatively 'cheap coal power' for granted in the future.

The requirements for coal-generated electricity producers to reduce air pollution, including greenhouse gas emissions, will mean that the price of coal power will climb, while the prices for many renewable electricity technologies will continue to decline. Wind energy is already

Table 15.2. Alberta's Coal-Fired Power Plants

Company	Unit Name	Commissioning Year	Capacity (MW)	End of Economic Life under Federal Regulations	Notes
ATCO Power	Battle River 3	1969	150	2019	
Transalta	Sundance 1	1970	280	2019	
Maxim Power	HR Milner 1	1972	150	2019	
Transalta	Sundance 2	1973	280	2019	
ATCO Power	Battle River 4	1975	150	2025	
Transalta	Sundance 3	1976	407	2026	
Transalta	Sundance 4	1977	392	2027	
Transalta	Sundance 5	1978	392	2028	
Transalta	Sundance 6	1980	392	2029	
ATCO Power	Battle River 5	1981	370	2029	
Transalta	Keephills 1	1983	406	2029	
Transalta	Keephills 2	1984	406	2029	
ATCO Power	Sheerness 1	1986	380	2036	Co-owned by TransAlta
Capital Power	Genesee 1	1994	410	2044	
ATCO Power	Sheerness 2	1990	380	2040	Co-owned by TransAlta
Capital Power	Genesee 2	1989	410	2039	
Capital Power	Genesee 3	2005	495	2055	Co-owned by TransAlta
Transalta	Keephills 3	2011	450	2061	

Source: Environment Canada 2011/12; Weis et al. 2012, 9

often competitive with new fossil fuel electricity production. In fact, TransAlta – one of Alberta's largest utilities with a major coal portfolio – forecasts that new coal power with carbon capture and storage (CCS) technology will be at least 30 per cent more expensive than new wind energy. Even without CCS, a modern supercritical coal unit is in fact more expensive than new wind power (EDC Associates 2013). Moreover, driven mostly by the creation of economies of scale for wind turbine manufacturing and distribution, which will offset increases in raw material costs, the capital costs of new wind energy projects are expected to decrease moderately for the next forty years (Global Wind Energy Council 2010). Solar photovoltaic (PV) technology, meanwhile, has shown stunning decreases in price per installed capacity. In Canada, solar PV prices have decreased by an average of 10 per cent annually for the past ten years (Canadian Solar Industry Association 2013) and are likely to continue to do so as PV technology and installers improve efficiency. Of course, the largest advantage that renewable energy sources offer is long-term price certainty. Requiring no fuel, renewables are subject neither to changes in fuel prices, as with natural gas and coal, nor to the risk of price changes resulting from action to curb climate change (pricing of carbon emissions).

Even more complex than projecting capital costs and fuel prices is the effect of transmission infrastructure. Adequate transmission capacity and connectivity are necessary for an efficient, well-functioning electricity system. However, large centralized generation – of any energy type – that is located far from electricity load (demand) necessitates a more expensive transmission network to bring the electricity to market. While this is equally true for very large, remote coal plants and wind farms, heavily polluting coal plants must be located far from major population centres. Many renewable energy technologies, as well as small natural gas generation, can be located much closer to load. Unlike centralized generation, 'distributed energy' lessens the need for large transmission projects, the costs of which are increasing with rising material costs. Large transmission projects also face increasing public scrutiny and opposition.

Liability is another major concern for larger plants that does not, to the same extent, affect renewable plants. Examples include cost overrun liability, such as Ontario tax payers having to pay $15 billion for nuclear plant cost overruns (Winfield et al. 2006), or pollution liability, such as the narrowly avoided breach at the Keephills ash lagoon west of Edmonton in 2008 (Holroyd and Simieritsch 2009).

As stated in Alberta's 2008 Provincial Energy Strategy, 'until recently [renewables] were more expensive, but the rising prices of fossil fuels have leveled the playing field considerably' (Government of Alberta 2008a, 10). In 2007 Alberta Electric System Operator (AESO) calculated that wind and cogeneration were among the cheapest generation options, behind only upgrading existing coal plants without CCS (AESO 2007a) – an option that will no longer be available as a result of Environment Canada regulations. AESO's long-term outlook report in 2014 put wind power very close to combined-cycle natural gas plants for electricity generation cost (AESO 2014, 63). Any cost advantage for combined-cycle natural gas plants presumes a nonexistent or very low carbon tax; the new NDP provincial government and the new Liberal federal government (both elected in 2015) have already signaled their intent to raise carbon taxes. These price comparisons do not account for the environmental and health externalities caused by pollution-intensive coal-fired power; these are real costs to our society, including health care costs. If the full environmental costs of conventional fossil-fuel-fired power are considered, then renewable energy technologies look more and more desirable. As international, national, or provincial policies move towards reducing GHG emissions either by forcing pollution-intensive electricity producers to internalize the cost of climate change to society or by directly mandating reductions in GHG emissions, coal power will further lose cost competitiveness in electricity generation.

In all, the oft-stated cost advantage of coal is either exaggerated or outright untrue. A recent study by Greenstone and Looney (2011) found that the environmental and social costs of coal-fired power are close to 5.3 ¢/kWh, and 1 ¢/kWh for combined cycle gas. Furthermore, a recent study published in the *American Economics Review* suggests that, on average, the cost of the harm produced by burning coal is more than twice as high as the market price of the electricity it produces (Muller et al. 2011). The government of Ontario calculates that the province will save three billion dollars per year as a result of its coal phase-out (Ontario Ministry of Energy 2012).

If these costs were to remain externalized, a transition to renewable energy would be unlikely. While some renewable energy options are currently cost competitive with new fossil fuel generation, no new electricity generation, from whatever source (including new coal), can match the costs of the previous generation of coal plants that have amortized their capital costs, socialized their transmission infrastructure, and externalized the costs of their pollution.

Employment and Economic Quality

It is not only the raw differences in costs of electricity from different generation sources that affect Alberta's economy. The types of costs inherent to each source and their market structures determine who benefits from the revenue from electricity generation, creating significant distributional and employment consequences. Different cost profiles and rates of employment generation are associated with different energy sources. Renewable energy sources, in particular, create more jobs. Large coal facilities are relatively capital intensive, with material and equipment costs being key drivers in the overall cost of coal power. Renewable energy, by contrast, is relatively labour intensive, with a greater proportion of costs going to labour and less to material and equipment costs. For the same amount of electricity generated, renewable energy creates more job-years than coal or natural gas power, ranging from around 20 per cent more from wind power to almost eight times as many job-years from solar PV (Kammen et al. 2004). Likewise, wind and solar PV each create 90 per cent more jobs for each dollar invested than coal, while biomass energy creates over 2.5 times as many jobs (Pollin et al. 2009).

These numbers cannot be mapped directly onto Alberta's economy because some jobs, for both renewable and fossil fuel energy, are in equipment and component manufacturing, which can be done outside the province. Indeed, these form a substantial proportion of wind and solar energy jobs. However, a significant local market for renewable energy, particularly one with some longer-term stability, can encourage more local supply chain, manufacturing, distribution, and research and development as industries seek to locate near market to reduce cost. Moreover, significant jobs exist also for inherently local activities, like retailing, installation, maintenance, and private consulting work. In fact, as renewable energy technology costs decrease, particularly for solar PV, these inherently local economic activities represent a greater proportion of the overall installed system costs (Renewables International 2011).

Despite the difficulty in pegging precise job creation numbers, it is clear that local renewable electricity will employ more people than conventional fossil-fuel-fired electricity – and certainly more people locally. The greater the proportion of renewable energy in our electricity mix, the more revenue from electricity generation will go into employment rather than capital stock, fuel costs, and profits repatriated to other jurisdictions.

Table 15.3 Alberta Energy Royalties, 2007–2009

Royalty revenue collected ($ billions)	2007	2008	2009
Total revenue collected	12.260	11.271	12.176
Oil royalty revenue	1.400	1.655	1.800
Natural gas and byproduct royalty revenue	5.988	5.199	5.834
Oil sands royalty revenue	2.411	2.913	2.973
Coal royalty revenue	0.013	0.014	0.036

Source: Government of Alberta (2010)

Finally, smaller power developments offer more opportunities for more dispersed local ownership of electricity generation by, for example, small businesses, landowners, communities, and cooperatives. With Alberta's power grid heavily reliant on relatively few, very large, very expensive coal plants, electricity generation is concentrated in relatively few hands. Only a few, large, independent power corporations can afford the massive capital investment required for a centralized power generation system. Distributing and democratizing ownership of electricity generation offers social and economic benefits such as less wealth disparity and greater economic resiliency. These benefits improve the quality, strength, and fairness of Alberta's economy.

Intergenerational Inequity

Consumption of non-renewable fossil fuel resources makes them unavailable for future generations and is ultimately unsustainable in the long run. As the representative of the owners of non-renewable energy resources (the citizens of Alberta), the Government of Alberta should be redeeming and saving revenue from their exploitation (table 15.3) for future generations. But to cite just one example, Alberta's $28.9 million royalty revenues from coal pale in comparison to the approximately $2.5 billion in revenue from coal-fired electricity generation in the province (Alberta Energy 2013a).

Determinants of Electricity Generation in Alberta

A number of factors have shaped Alberta's unique energy reality. Alberta is very well positioned for a diverse, resilient mix of energy sources for electricity generation. In addition to its well-known fossil fuel reserves, it has vast and diverse renewable energy resources. Moreover, it has a deregulated electricity system that allows for innovation and deployment of some of the most economically viable renewable electricity generating opportunities in Canada. However, the conventional view in the government is that the 'open market' unilaterally dictates electricity generation choices, giving the sense that the current mix is 'naturally' chosen and economically optimal. In fact, non-market influences and market failings have shaped the province's generation mix, and have tended to benefit large fossil-fuel-fired generation, leading to a less than optimal energy mix in Alberta.

Electricity Resources and Deregulation

In 2001 Alberta deregulated its electricity system. This opened the door for renewable energy development – particularly wind energy in southern Alberta. With the support of federal programs, varying electricity prices, and provincial mechanisms that allow wind energy to act as emission offsets for large greenhouse emitters, many projects were able to compete with traditional electricity sources, giving Alberta an early lead in wind power development in Canada.

Alberta's wind energy resources are among the most accessible land-based wind resources in Canada, with many locations that are already economically viable relative to today's grid price. Alberta has an estimated total of viable wind energy *potential* of 150,000 MW (Solas Energy Consultant) – over ten times its total installed electricity generation capacity (from all sources, renewable and non-renewable), which is approximately 14,500 MW (Alberta Electric System Operator 2013). In comparison, in a more densely populated landscape about half the physical size of Alberta, Germany had already installed over 29,000 MW of wind energy by 2010. According to the German Wind Energy Association (BWE), the country could host 45,000 MW of wind energy capacity by the year 2020, indicating that a major expansion from Alberta's current 1,000 MW of wind energy is certainly technically achievable from a resource capacity point of view.

While Alberta does not have the installed hydro capacity of some Canadian provinces, it was estimated in 2006 that it has the potential to add more than 11,500 MW of economically viable hydro capacity, including both reservoir and run-of-the-river projects (Canadian Hydropower Association 2006). A 2006 estimate indicates that 108,000 GWh of annual energy generation potential is available from Alberta's substantial agricultural biomass and wood residues, amounting to over half of its total current electricity generation from all sources. According to a scan by the Geological Survey of Canada, there is enough geothermal potential for one million times Canada's electricity needs, and Alberta is among the provinces with the best geothermal resource (Munro 2011). Finally, Alberta benefits from some of the best solar resource in the country, sufficient to meet total electricity demand. Again, Germany serves as a working model: although it has significantly weaker annual insolation, Germany was already generating over 11,600 GWh of electricity annually from solar photovoltaic systems in 2010, an amount equivalent to roughly 20 per cent of Alberta's electricity demand.

In spite of the excellent wind energy resource in Alberta, its lead has not lasted, as other provinces have enacted more ambitious goals and supportive policies above and beyond the open-market approach in Alberta. Ontario now has twice the installed wind power capacity; Quebec could soon move Alberta into third position, while Saskatchewan produces a larger proportion of its electricity from wind than does Alberta (Canadian Wind Energy Association 2013).

Barriers to Renewable Energy Development

The laissez-faire approach allows Albertan entrepreneurs to participate in developing renewable energy projects, but also disproportionally favours fossil fuel development. Between 2001 and 2010, 748 MW of new wind and 49 MW of new hydro capacity were added, compared to 1,194 MW of new natural gas, 713 MW of new coal, and 2,340 MW of natural gas combined heat and power systems (Alberta Electric System Operator 2010). Key extra-market forces as well as market failures act as barriers to more rapid renewable energy adoption.

One of the primary barriers to renewable energy development stems from the geography of Alberta's transmission grid and the path dependency it creates. A significant part of the problem is that Alberta's existing electricity transmission system is built for the very large centralized power plants that have dominated the grid since the 1970s. This infrastructure, inherited from the era prior to generation deregulation, favours existing coal mines and coal-fired power plants (not renewable energy resources). This infrastructure bias is amplified with each round of expansion. With costly transmission infrastructure serving large centralized plants, it is easiest to build new power plants and refurbish existing ones at the same locations, thereby minimizing the cost of connecting to the grid. A prime example is the closure in 2010 of the 537 MW Wabamun coal power plant after over fifty years of operation, followed immediately by the commissioning of the 450 MW Keephills 3 power plant on the same lake. As transmission infrastructure ages, still more is spent to replace or upgrade transmission service to these same areas.

Transmission grid constraints significantly hamper the growth of the wind energy sector in particular. Although 7,500 MW of applications for new wind generation were in the Alberta Electric System Operator (AESO) transmission connection queue as of April 2011, AESO estimated that only about 1,100 MW of total capacity would be connected by the end of 2012 (Alberta Electric System Operator 2012).

Under a perfectly functioning free market, unencumbered by the real-world geography of pre-existing transmission networks, all forms of electricity generation would have equal access to the power system regardless of the infrastructure that was built specifically to serve one particular resource. Distributed generation, closer to electricity loads, would realize substantial economic advantages: shorter transmission lines would be less expensive to build and maintain and less electricity would be wasted in long-distance transmission. In reality, however, transmission corridors are always geographically specific, and otherwise viable forms of electricity generation that are inaccessible to transmission corridors cannot reach the market.

Alberta's transmission regulation disaggregates transmission costs from generation. This is intended to open the market for generation, but also means that consumers see the bundled transmission costs, rather than the costs from specific geographic locations. Power plants do not cover most of the costs of transmission, although they are charged a line loss factor (Alberta Electric System Operator 2007b). Transmission costs are rapidly rising. The Alberta Utilities Commission forecasts transmission charges to increase from $14 per MWh in 2011 to $32 per MWh by 2018 (Henton 2013), largely due to the cost of implementing the Alberta Electric System Operator's (AESO's) long-range transmission infrastructure expansion plan. This plan calls for dozens of new transmission infrastructure projects at a total cost projected to range between $13.5 and $16 billion (MacLeod 2011).

Alone among Alberta utilities, Calgary's municipally owned utility, ENMAX, has opposed the major transmission expansion plan. ENMAX has taken some initial steps towards more distributed generation, enabled by its vertically integrated structure: it owns generating plants and is a distributor, which is unique in Alberta's electricity market (MacLeod 2011). ENMAX opened its gas-fired Downtown District Energy Centre in 2010, with heating capacity for up to six million square feet of downtown buildings, and has begun building one of Alberta's largest gas-fired power plants, the Shepard Energy Centre, within the City of Calgary. It has a small program to help finance photovoltaic and micro-wind systems, located within or near cities. If it is able to boost local generation sufficiently, it claims it could even drop its contracts with two coal-fired power plants, according to Rob Harris, vice president of distributed generation at ENMAX (MacLeod 2011) – although there has been no public statement in this regard. With its distributed generation strategy, ENMAX 'aims to bring more variety to the energy supply mix by encouraging generation created either closer to its customers or at its customers' houses, as well as at central power plants' (MacLeod 2011, 2).

Distributed generation does not realize its full advantages in Alberta's market due in part to the pre-existing, large, centralized coal plants that have the advantage of existing transmission capacity, and due to the fact that generation does not pay transmission 'build' costs. Nonetheless, a more distributed generation strategy can offer certain advantages, notably lower losses. It does not eliminate the need for new transmission lines, but does demonstrate the importance of the geographies of different generation strategies, and specifically their cost and efficiency implications.

We have already alluded to another key market failure: environmental and health cost 'externalities' from pollution. As noted above, electricity generated from combustion of fossil fuels, particularly coal combustion, releases substantial amounts of harmful pollution, including GHG emissions that harm the climate and air contaminants that cause health and environmental problems. These power plants can also have other ecological and economic impacts on land use, as well as thermal and chemical water pollution. These are real costs to society, but they are not reflected in the market price of electricity from these types of generation, as the generators are not required to pay for them. Alberta's Specified Gas Emitters Regulation does place some cost on GHG emissions from coal plants – but it only applies to a portion of facility emissions (those that are in excess of a required 12 per cent intensity reduction from their 2003–2005 GHG baseline), and the price applied to these emissions was effectively capped at \$15/tonne CO_2e (with opportunities for emitters to pay less), until 2016, well below the projected cost to society.[3] As such, costs are imposed on society for which the market does not account (see, e.g., Stern 2006). As a result, cleaner energy sources that do not cause these harms are not sufficiently incentivized in the market to achieve a socially, economically, and environmentally optimal generation mix. It is this form of market failure that gave rise to the imposition of an enhanced carbon tax, announced by Alberta's newly elected NDP government in November 2015.

Under Alberta's Progressive Conservative Party governments the province weighed into the electricity market, favouring coal-fired low-carbon generation over other technologies. Rather than promote all low-carbon sources through neutral policies, PC governments offered to subsidize future coal plants by funding carbon capture and storage technology research, initially through $700 million in project funding (as part of an overall $2 billion incentive package to carbon capture and storage technology in the province) (Alberta Energy 2011). Nonetheless, both of the carbon capture and storage projects in the electricity sector were eventually abandoned (Blackwell 2013).[4] Carbon capture and storage policy was an attempt to provide a substantial advantage to a particular method – moreover, a commercially unproven one – of reducing GHGs from coal-fired electricity generation, relative to renewable energy. Renewable energy technologies, with the exception of a small and temporary biomass support mechanism (Alberta Energy 2013b), have never received provincial financial incentives from governments in Alberta, although the current NDP government plans to institute incentives. While it is hard to quantitatively determine an appropriate royalty to reflect the liquidation of publicly owned coal reserves, or even the health care costs resulting from their combustion, the manifestly inadequate coal royalty rates in Alberta provide, effectively, another form of subsidy for coal-fired electricity generation.

All in all, a number of factors and policies have undermined fairness in the Alberta electricity market and the possibility of finding the most beneficial electricity generation mix. These non-market factors and market failures have skewed the proper functioning of the so-called invisible hand.

Impetus for Change

While the current system clearly favours traditional polluting forms of electricity, there is also substantial impetus to change the status quo. In addition to the rising costs described above, a number of emerging social, political, and economic considerations run counter to coal's privileged position in Alberta's electricity grid. These forces push towards renewing Alberta's electricity grid with modern, cleaner technologies.

Provincial Steps toward Action on Climate Change

While Alberta's PC governments actively opposed federal government action on climate change, including, under Premier Ralph Klein, launching an advertising campaign in 2002 against Canada's ratification of the Kyoto Protocol, the global community and, increasingly, Albertans have become more and more aware of the costs of inaction. In response to international campaigns against fossil fuels, the Alberta government announced greenhouse gas reduction targets in 2008 (Government of Alberta 2008b). These were to stabilize GHG emissions by 2020 and reduce emissions by 200 megatonnes (Mt) CO_2e relative to 'business as usual,' or 14 per cent below actual 2005 levels by 2050. To do this, the Progressive Conservatives said they would seek 37 Mt of reduction from 'greening energy production,' along with 139 Mt from the implementation of CCS technology by 2050, although it was never clear how they would achieve these reductions; the government's 'intensity-based' emissions targets were largely ineffective. Expanding the pace of oil sands development or building new coal plants for electricity will cause emissions to rise dramatically in Alberta in the coming years, even with reduced emissions intensity. It is not yet clear, from a technical standpoint, whether CCS projects will succeed in reducing emissions from crude oil production – particularly because enhanced oil recovery is an explicit goal of the CCS projects (as noted in Adkin and Stares, this volume) and because CCS technology reduces power plant efficiency, in turn necessitating the burning of more coal. The July 2014 report of the provincial auditor general on the government's environmental policy performance revealed that, by 2012,

> it was apparent to the department [Environment and Sustainable Resource Development] that the expected reductions from carbon capture and storage will not be achieved. Carbon capture and storage in the 2008 strategy represents the majority of forecasted emission reductions ... However, with only two carbon capture and storage projects planned, the total emissions reductions are expected to be less than 10% of what was originally anticipated. (Auditor General 2014, 40)

Clearly CCS technology alone will not achieve absolute emissions reduction; relying more heavily upon renewable energy development will be an important component of any serious emissions reduction policy. Unless future electricity consumption is met by utilizing the province's tremendous wealth in renewable energy resources, coupled with energy efficiency, the projected additional consumption will contribute significantly to Alberta's GHG emissions and undermine efforts to substantially reduce GHG emissions provincially and nationally. There is, however, opportunity for large GHG reductions from the emissions-intensive electricity grid by phasing out coal-fired generation and substituting renewable sources of electricity in its place. This fact underlies the dramatic shift in Alberta's energy policy announced by the NDP government on 30 November 2015. The provincial government committed to a 100 megaton cap on carbon emissions from the oil sands; a tax of $20 per ton on carbon emissions beginning in 2017, rising to $30 in 2018; a phase-out of coal-fired electricity generation by 2030; a goal to cut methane emissions by almost half by 2025; and incentives to have almost one third of all electricity generated by renewables such as wind and solar (Giovannetti and Jones, 2015). These new policies represent a tectonic shift in Alberta's approach to its GHG emissions problem.

Federal Greenhouse Gas Policy

In 2011 Environment Canada introduced regulations to prevent new conventional coal plants that are not equipped with CCS technology.[5] The regulations do not affect the operations of existing coal plants until they reach between forty-five and fifty years after their commissioning date. Then the regulations require them either to reduce their emissions by two-thirds through carbon-capture and storage or to be retired. All new units that begin generation after 1 July 2015 must match the GHG emission levels of natural gas-fired electricity generation. The regulations' ensure that new unmitigated coal plants will not tie Alberta's grid to high emissions intensity for the long term, although they are largely superfluous given Alberta's new energy policies under the NDP. Implementation of the new NDP energy policies is critically important. If all new coal power plants anticipated under the previous Alberta government's business-as-usual plans were built, with capture and storage of the majority of their GHG emissions, Alberta's electricity emissions would still rise from today's 50 Mt to approximately 55 Mt of CO_2e by 2020.

Civil society is certainly prepared to see that Alberta cleans its electricity generation, as we saw in the case of a recent attempt by a power company to get around the new federal regulation. Maxim Power sought and obtained, in June 2011, expedited approval from the Alberta Utilities Commission (AUC) for the expansion of a conventional coal plant without CCS technology. In a letter to the AUC commissioner the company made clear that it was seeking to beat the coming into effect of the new federal regulations and to obtain a blank cheque to generate highly emitting coal power for forty-five years (*CBC News* 2011). After a public outcry including thousands of letters from Canadians, federal Minister of Environment Peter Kent issued a strong statement that he would not allow the plant ('Ottawa Warns' 2011).

Energy Security

Options exist to move away from the high emissions intensity of coal. Some look to natural gas combustion, which has notably lower emissions, but growth in electricity demand is poised to overtake greenhouse gas emissions reductions due to a lower reliance on coal (IPPSA, 2013). Furthermore, the 2011 report from Premier Ed Stelmach's Council for Economic Strategy surmised that the price volatility and finite nature of natural gas would make greater reliance on natural gas an unsustainable long-term strategy, noting that 'from a long-term public interest perspective ... the competitiveness of Alberta's industry could be at risk if we are entirely reliant on electricity from natural gas. Diversifying the mix of fuels used for electricity generation would make the province less vulnerable' (Premier's Council for Economic Strategy 2011, 80). On industrial competitiveness grounds, the council claimed that reliance on natural gas would be problematic because 'industrial users account for 80% of electricity use in Alberta' and that 'electricity is an essential input to oil sands extraction.' While the council incorrectly estimated industrial electricity use – Alberta Energy (2012) puts industrial electricity consumption at 51.8 per cent of the provincial total – it was correct about the importance of electricity to the provincial economy. Indeed, non-residential consumption – including farm, commercial, and industrial usage – accounts for 82.3 per cent of the province's electricity consumption (Alberta Energy 2012).

Moreover, Alberta's economic growth puts tremendous upward pressure on energy demand. Current demand projections indicate that there will be an electricity shortfall in Alberta in the future. While there remains an abundance of fossil fuels in Alberta in the near term, the province is not insulated from price volatility on the global market, and has very little infrastructure in place for meeting future energy demands when fossil fuel reserves begin to decline or become expensive again. Renewable energy investment opportunities offer enormous dividends in energy security and supply reliability. In particular, by supplying electricity without fuel costs to the grid, renewable electricity generation can buffer projected electricity price volatility.

Reputation

Some large traditional oil companies, particularly major oil sands players, have recognized the double benefit of economic opportunities and reputational gains from investing in renewable energy development. In certain situations, these companies would like to do more, but have found it difficult in Alberta's electricity generation context to justify such investment. Unable to make a wind farm project work in 2012, Dick Williams, president of Shell Wind Energy, was quoted as saying: 'You can have the transmission interconnection and you can have the turbines, but you still need to sell the power. We need a power purchase agreement in place and it needs to be a long-term agreement. The Alberta market is a merchant market where you don't have these long-term agreements' (Rieger 2012). For these companies, policies to support cleaner electricity generation can help them develop viable renewable energy projects that can diversify their energy portfolio, reduce their carbon footprint, and assist their marketing efforts.

Alberta's Energy Futures

Alberta is at the crossroads of key generational decisions in electricity supply. Projections suggest the demand for electricity could double within the next twenty years. By then, almost half of the province's coal capacity will have reached the end of its forty-five-to-fifty-year operating window. In the short term, the rapid pace of oil sands development is likely to dictate much of the energy supply needs in the province, given that development's high energy consumption.

To a large extent, the Alberta government's environmental, natural resource, and even economic and financial approaches have amounted to hitching its wagon to the largely unfettered exploitation of non-renewable energy resources. While this approach has resulted in relatively short-term booms, it has also meant that there has been no coordinated planning around future energy systems. Given the magnitude of the

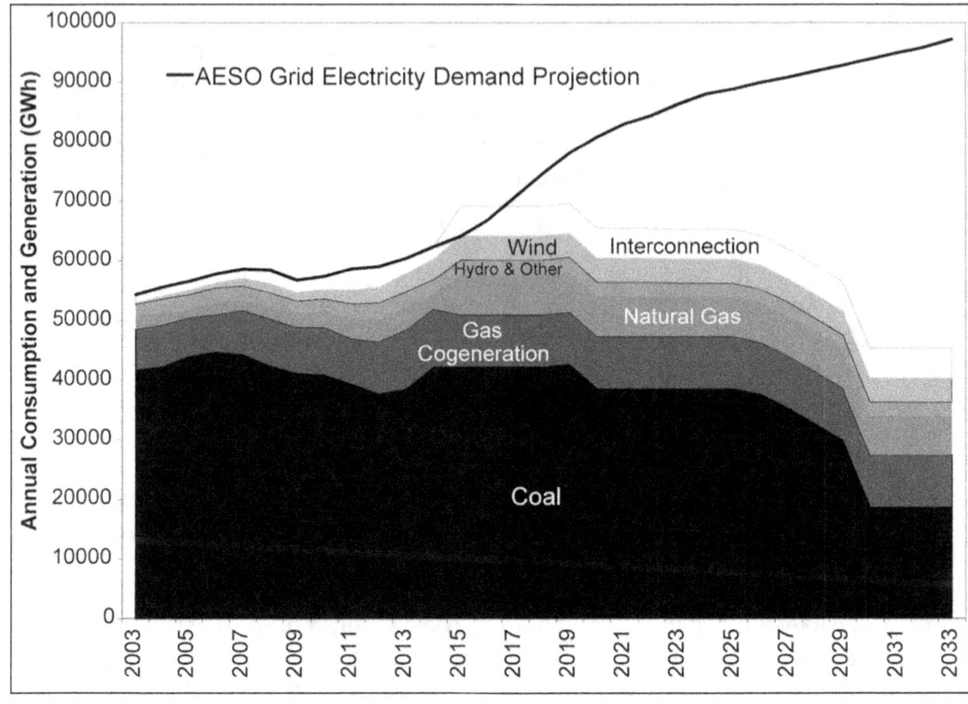

Figure 15.1. Alberta projected electricity consumption by fuel source. (*Source*: Weis and Bell 2009.)

changes that the province's electricity system is facing, the province must now prepare for a transition to renewables, including where transmission lines are built and how the market should be adapted to make such changes. Indeed, even the (Progressive Conservative) Premier's Council for Economic Strategy argued in 2011 that 'the province cannot rely on market forces alone to produce enough low-emission, reliable, competitive electricity to support its long-term economic strategy' (2011, 81).

Among the reasons for Alberta's traditionally slow pace of change may be the disproportionate level of non-residential electricity consumption in the province. In most provinces, electricity is roughly evenly consumed by the residential, commercial, and industrial sectors. In Alberta, however, residential electricity consumption accounts for only 18 per cent of total electricity consumption, with the industrial sector consuming over 50 per cent of the electricity generated – not counting the considerable electricity generated and consumed on industrial sites. The dominance of the industrial sector, and in particular a relatively small number of large industrial consumers, has resulted in its disproportionate influence on electricity policy and planning, as is generally the case with petroleum producers in petro-states (Karl 1997; Carter and Zalik, this volume).

Alberta's Clean Energy Opportunity

In spite of Alberta's entrenched, fossil-fuel-laden electricity system, and the enormous pressures to expand electricity generation over the coming decades, Alberta is endowed with enough clean energy resources to drastically reduce GHG and other emissions from its electricity system, phase out conventional coal completely, and transition towards a significantly more sustainable electricity system. Such a transition will require big changes, not only in terms of technology and infrastructure investment priorities, but also in terms of the vision and calculus of its political leaders, who traditionally have been fixated on fossil fuel exploitation even when desirable and viable alternatives to serve the common good abound. As the Premier's Council on Economic Strategy argued in 2011:

> While short-term economics and environmental goals might suggest converting from coal to natural gas for electricity generation in Alberta, over the long term this has the double effect of 'stranding' Alberta's huge coal supplies and reducing the amount of gas that could be put to higher-value use. Investments in clean coal technology will have substantial payoffs if we succeed in realizing value from our 37 billion tonnes of coal. (2011, 28).

Clearly, concern over 'stranding' Alberta's coal resources was top of mind in 2011. But since then the energy policy landscape has shifted. Today, CCS for coal-fired electricity generation is a silver bullet no more as technical limitations, carbon pricing, and significantly lower natural gas price forecasts have all combined to dramatically erode support for the technology in Alberta. TransAlta Corporation cancelled the CCS component of the Keephills 3 coal-fired power plant in 2012, even though provincial funding was available. In 2013 the province of Alberta itself cancelled funding for the Swan Hills Synfuels synthetic gas plant. Premier Alison Redford stated at the time, "We're not going to continue to push things [CCS] if the private sector's telling us they don't make sense" (Blackwell 2013, B5).

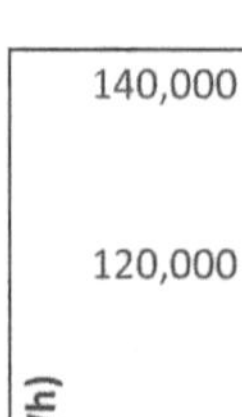

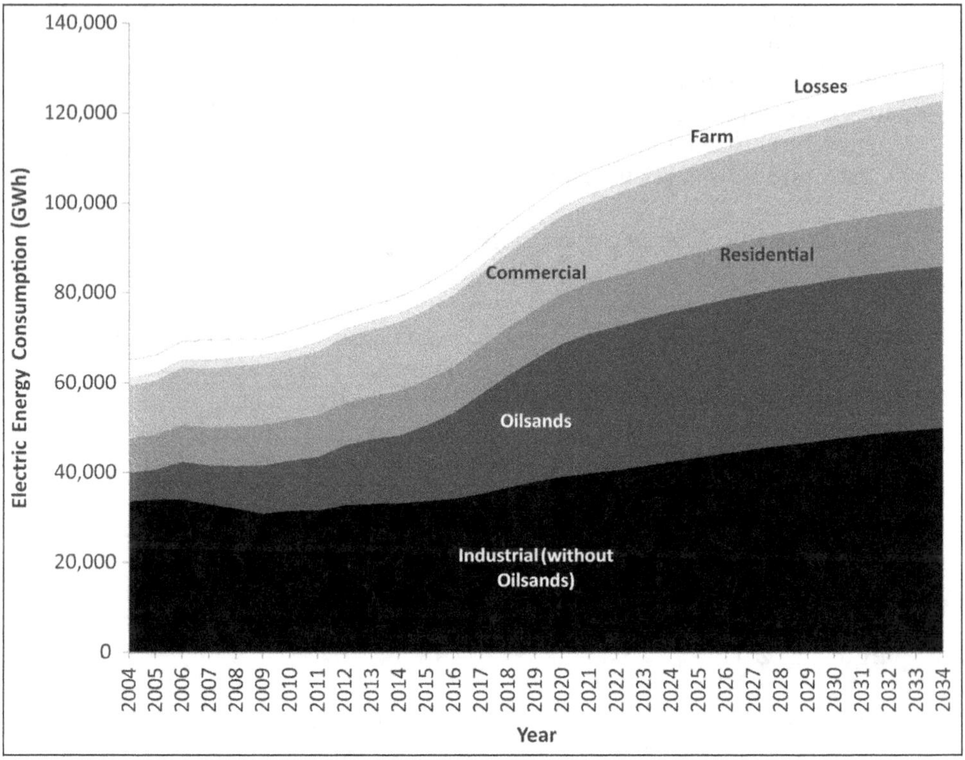

Figure 15.2. Historic and projected total Alberta electricity usage by sector. (*Source:* Pembina Institute, based on data from EDC Associates Ltd., quarterly forecast update, third quarter 2008, 19 August 2008.)

Note: The figure does not include electricity generated onsite by large industry (such as oil sands companies) and used within their operations, but it does include any surplus electricity generated by those industries and sold into the grid. In 2007 about 10,000 GWh of electricity were generated by industry and, because the electricity was used onsite, it is not included in the usage depicted.

Active planning is necessary now to secure stable generation and to buffer price shocks from old coal going offline. This is a challenge, but also an opportunity. Transitioning to renewable energy offers a variety of benefits, including relative insulation against rising and volatile fuel prices, improved cost trends, and far fewer health and environmental costs from economic externalities. Beginning the transition now will hasten future cost improvements of renewable energy and the efficiency with which these technologies work.

The need for a rapid transition away from coal in particular is clearer than ever in Alberta. A major study by the Pembina Institute that compared expected electricity consumption in the province over the next twenty years with its renewable resources, and examined the renewable energy technologies that are already commercially available, showed that alternatives to coal are practical and achievable (Bell and Weis 2009). This research detailed two scenarios: 'pale green' and 'green.' Relying on relatively modest renewables-and-efficiency-oriented strategies, the pale green scenario illustrates that that no new coal-generation plants are required. Neither are other slow-to-deploy and highly capital-intensive systems required, such as nuclear. Should Albertans set their sights higher, much more is possible. The green scenario shows that even if projected electricity consumption is exceeded by 25 per cent, demand could still be supplied by renewable and transitional technology energy options. Not only are no new coal or nuclear plants needed, but existing coal plants could be phased out, as is being done in Ontario, and/or clean energy could be exported to neighbouring jurisdictions. This ambitious approach would create opportunities for Alberta to clean up its electricity supply and add additional jobs. Both scenarios are achievable using existing technologies and conservative estimates of adoption rates. For example, in the case of biogas, the green scenario assumes that over the next twenty years Alberta installs only 14 per cent of what Germany installed between 1996 and 2007. For solar photovoltaics (PV), the amount over twenty years is one quarter of what Germany installed in 2007 alone.

Numerous jurisdictions are in the midst of such transformative changes, proving they are technically and politically possible. Denmark has produced over 20 per cent of its national electricity supply from wind since the early 2000s. Germany is phasing out its entire nuclear fleet by 2018, and aiming to achieve 100 per cent renewable electricity sources by 2050. While Nova Scotia is a much smaller province, it had a grid proportionally similar to Alberta's at the turn of the century and has implemented laws that will reduce its coal-fired electricity generation by almost 50 per cent by the year 2020. Meanwhile, Ontario is on track to phase out its entire coal fleet by 2014, and has achieved reduced emissions while diversifying supply (Weis 2013).

Although there are many differences between Ontario and Alberta, there are some lessons to be learned in the Alberta context. One of these is the importance of preparing consumers and the electorate for the reality that electricity prices will rise in the future regardless of energy choice. As aging generating stations are forced to comply with emissions regulations and begin to retire, new facilities, regardless of the technology they use, will be more expensive.

Ontario's renewable energy policies have succeeded in fostering a new industry, including the development, manufacturing, service, and integration of renewables. In the Alberta context, an outcomes-based approach – notably the long-term reduction of greenhouse gas emissions – may be the most effective approach to restructuring energy supply and demand. This concept has been developed by the Pembina Institute as a Clean Electricity Standard (Thibault and Weis, 2013). Moving Alberta's electricity sector in an environmentally sustainable direction will require clear and dedicated political leadership at all levels of government and civil society, including provincial ministries, municipal and county governments, and neighbourhood planning institutions. For too long Alberta has had the dubious distinction of having the dirtiest electricity production portfolio in Canada. This need not be our future and, indeed, we have taken the first steps toward a low-emission electricity system.

NOTES

1 The oil sands operations frequently generate their own electricity through natural gas co-generation, as they have a huge demand for heat.

2 This excludes emissions from fuel extraction sectors that are indirectly but clearly attributable to electricity generation, particularly coal mining.

3 In April 2013 the Redford government tested the idea of increasing the emissions intensity target from 12 per cent to 40, and the price for exceeding the target to $40 per tonne of $C0_2e$, but energy economists consider this price to be much lower than the price needed to bring about a the GHG reductions required to meet Alberta's own goals. See Dyer (2013).

4 The political objectives of this investment choice are discussed in Adkin and Stares, this volume.

5 The regulation was published in the *Canada Gazette* on 27 August 2011. See http://canadagazette.gc.ca/rp-pr/p1/2011/2011-08-27/html/reg1-eng.html.

REFERENCES

Alberta Electric System Operator. 2007a. *Long-Term Transmission System Planning, AESO Stakeholder Consultation*. 16 November. http://www.aeso.ca/downloads/Nov_16_Long_Term_Transmission_Stakeholder_Presentation-_for_posting.pdf.

Alberta Electric System Operator. 2007b. *Comparison of Transmission Regulation, 2004–2007*. http://www.aeso.ca/downloads/Comparison_of_Transmission_Regulation_2004-2007.pdf.

Alberta Electric System Operator. 2010. *Annual Market Statistics*. http://www.aeso.ca/downloads/AESO_2010_Market_Stats.pdf.

Alberta Electric System Operator. 2012. *Wind Power in Alberta*. http://poweringalberta.com/wp-content/uploads/2010/09/Wind-Power-in-Alberta_2012-06_FINAL.pdf.

Alberta Electric System Operator. 2013. Annual Market Statistics. http://www.aeso.ca/downloads/AESO_2013_Market_Stats.pdf.

Alberta Electric System Operator. 2014. AESO 2014 Long-Term Outlook. AESO: Calgary. http://www.aeso.ca/downloads/AESO_2014_Long-term_Outlook.pdf.

Alberta Energy. 2011. *CCS Major Initiatives*. http://www.energy.alberta.ca/Initiatives/1897.asp.

Alberta Energy. 2012. *Electricity Statistics*. http://www.energy.alberta.ca/Electricity/682.asp.

Alberta Energy. 2013a. *Coal Statistics*. http://www.energy.alberta.ca/coal/643.asp.

Alberta Energy. 2013b. *Bioenergy Producer Credit Program Frequently Asked Questions*. http://www.energy.alberta.ca/News/1110.asp.

Auditor General of Alberta. 2014. *Report of the Auditor General of Alberta*. July. Edmonton: Government of Alberta.

Bell, Jeff, and Tim Weis. 2009. *Greening the Grid: Powering Alberta's Future with Renewable Energy*. Pembina Institute. http://www.pembina.org/pub/1763.

Blackwell, Richard. 2013. 'Alberta Cancels Funding for CCS Project. *Globe and Mail*, 26 February, B5.

Bramley, Matthew. 2011. *To Hit Climate Target, Ottawa Would Have to Work 10 Times Harder*. http://www.pembina.org/blog/561.

Canadian Hydropower Association. 2006. *Study of the Hydropower Potential in Canada*. Report prepared by ÉEM Sustainable Management for the Canadian Hydropower Association, Ottawa.

Canadian Solar Industry Association. 2013. *Canadian Solar PV Market*. http://

www.cansia.ca/market-intelligence/market-research/canadian-solar-pv-market.

Canadian Wind Energy Association. 2013. *Canadian Wind Farms*. http://www.canwea.ca/farms/index_e.php.

CBC News. 2011. 'Alberta OK's Coal Power Plant, Angering Eco Groups.' 11 August. http://www.cbc.ca/news/canada/edmonton/story/2011/08/11/alberta-maxim-power-coal-plant-approved.html.

Dyer, Simon. 2013. 'What You Need to Know about Alberta's 40/40 Carbon Pricing Proposal.' Pembina Institute, 5 April. http://www.pembina.org/blog/707?goback=.gde_4208965_member_229763143.

EDC Associates. 2011. *Quarterly Forecast Update: First Quarter 2011*. 14 March, 117.

EDC Associates. 2013. *Trends in GHG Emissions in the Alberta Electricity Market – Impact of Fuel Switching to Natural Gas*. 2 May, 7.

Environment Canada. 2011. *National Inventory Report 1990–2009, Part 3*. http://unfccc.int/files/national_reports/annex_i_ghg_inventories/national_inventories_submissions/application/zip/can-2011-nir-16may.zip.

Environment Canada. 2011/12. *Reduction of Carbon Dioxide Emissions from Coal-Fired Generation of Electricity Regulations* (SOR/2012-167). http://www.ec.gc.ca/lcpe-cepa/eng/regulations/detailReg.cfm?intReg=209.

Environment Canada. 2013. *National Pollutant Release Inventory*. http://ec.gc.ca/pdb/websol/emissions/ap/ap_query_e.cfm.

Giovannetti, Justin and Jeffrey Jones. 2015. 'Alberta Carbon Plan a Major Pivot in Environmental Policy,' *The Globe and Mail*, November 22. http://www.theglobeandmail.com/news/alberta/alberta-to-release-climate-change-policy-at-edmonton-science-centre/article27433002/

Global Wind Energy Council. 2010. *Global Wind Energy Outlook 2010*. http://www.gwec.net/fileadmin/documents/Publications/GWEOpercent2020 10percent20final.pdf.

Government of Alberta. 2008a. *Launching Alberta's Energy Future: Provincial Energy Strategy*. http://www.energy.alberta.ca/Org/pdfs/AB_Provincial EnergyStrategy.pdf.

Government of Alberta. 2008b. *Alberta's 2008 Climate Change Strategy*. http://environment.gov.ab.ca/info/library/7894.pdf.

Government of Alberta. 2010. *Energy Economics: Understanding Royalties*. http://www.energy.alberta.ca/Org/pdfs/Energy_Economic.pdf.

Greenstone, Michael, and Adam Looney. 2011. A *Strategy for America's Energy Future: Illuminating Energy's Full Costs*. http://www.brookings.edu/~/media/Files/rc/papers/2011/05_energy_greenstone_looney/05_energy_greenstone_looney.pdf.

Henton, Darcy. 2013. 'Transmission Charges Set to Double, Report Warns.'

Calgary Herald, 14 February. http://www.calgaryherald.com/technology/Power+transmission+charges+double+Albertans+report+warns/7962009/story.html.

Holroyd, Peggy, and Terra Simieritsch. 2009. *The Waters That Bind Us: Transboundary Implications of Oil Sands Development.*

https://www.pembina.org/reports/watersthatbindus-report.pdf.

Independent Power Producers Society of Alberta. 2013. *Trends in GHG Emissions in the Alberta Electricity Market.* http://www.ippsa.com/IP_pdfs/Analysis%20of%20GHG%20Emissions%20in%20the%20Alberta%20Electricity%20Market%20-%20May%202,%202013.pdf.

Kammen, Daniel, Kamal Kapadia, and Matthias Fripp. 2004. *Putting Renewables to Work: How Many Jobs Can the Clean Energy Industry Generate?* http://rael.berkeley.edu/sites/default/files/very-old-site/renewables.jobs.2006.pdf.

Karl, Terry Lynn. 1997. *The Paradox of Plenty: Oil Booms and Petro-States*. Berkeley: University of California Press.

MacLeod, Steve. 2011. 'Power from the People: Will Enmax's Plan for Distributed Generation Take Off?' *Alberta Venture*, 1 August. http://albertaventure.com/2011/08/power-from-the-people/.

Muller, Nicholas Z., Robert Mendelsohn, and William Nordhaus. 2011. 'Environmental Accounting for Pollution in the United States Economy.' *American Economic Review* 1015: 1649–75.

Munro, Margaret. 2011. 'Hot Rocks Could Fulfill Nation's Power Needs.' *Edmonton Journal*, 14 September. http://www.edmontonjournal.com/technology/rocks+could+fulfil+nation+power+needs+report/5399108/story.html.

Ontario Ministry of Energy. 2012. *Results Based Plan Briefing Book 2011–2012.* http://www.energy.gov.on.ca/docs/en/ENERGY%202011-12%20RBP%20EN.pdf.

'Ottawa Warns Pending Emissions Rules Will Be Enforced.' 2011. Canadian Press, 9 September. http://www.ctv.ca/CTVNews/TopStories/20110909/peter-kent-emissions-110909/.

Pollin, Robert, James Heintz, and Heidi Garrett-Peltier. 2009. *The Economic Benefits of Investing in Clean Energy: How the Economic Stimulus Program and New Legislation Can Boost U.S. Economic Growth and Employment*. Amherst: Political Economy Research Institute, University of Massachusetts. http://www.peri.umass.edu/fileadmin/pdf/other_publication_types/green_economics/economic_benefits/economic_benefits.PDF.

Premier's Council for Economic Strategy. 2011. *Shaping Alberta's Future.* http://premier.alberta.ca/plansinitiatives/economic/RPCES_ShapingABFuture_Report_web2.pdf.

Rieger, Jamie. 2012. 'Shell Delivers Bad News – Wild Steer Butte Project Put on Hold.' *The 40-Mile County Commentator.* Bow Island Commentator/ Alta Newspaper Group Limited Partnership. 14 March. http://www.bowislandcommentator.com/component/content/article/1185-shell-delivers-bad-news-wild-steer-butte-project-put-on-hold.html (accessed 1 May 2013).

Renewables International. 2011. *Cost of Turnkey PV in Germany Drops*. January 14. http://www.renewablesinternational.net/cost-of-turnkey-in-germany-drops/150/510/29911/.

Sovacool, Benjamin. 2008. 'The Importance of Comprehensiveness in Renewable Electricity and Energy-Efficiency Policy.' *Energy Policy* 37: 1529–41.

Stern, Nicholas. 2006. *Stern Review on the Economics of Climate Change*. HM Treasury. http://webarchive.nationalarchives.gov.uk/+/http:/www.hm-treasury.gov.uk/independent_reviews/stern_review_economics_climate_change/stern_review_report.cfm.

Thibault, Benjamin, and Tim Weis. 2013. *Clean Electricity Thought Leader Forum: A Made-in-Alberta Proposal to Green the Grid.* http://pubs.pembina.org/reports/tlf-clean-electricity-standard-white-paper.pdf.

Weis, Tim. 2013. 'Recognizing Progress in Ontario's New Long-Term Energy Plan.' Pembina Institute, 19 December. http://www.pembina.org/blog/767.

Weis, Tim, Ben Thibault, P.J. Partington, Sachi Gibson, and Kristi Anderson. 2012. 'The High Costs of Cheap Power: Pollution from Coal-Fired Electricity in Canada.' June. Drayton Valley, AB: Pembina Institute.

Winfield, Mark, Alison Jamison, Rich Wong, and Paulina Czajkowski. 2006. *Nuclear Power in Canada: An Examination of Risks, Impacts and Sustainability.* Drayton Valley, AB: Pembina Institute. http://pubs.pembina.org/reports/Nuclear_web.pdf.

World Bank. Energy Sector Management Assistance Program. 2009. *Study of Equipment Prices in the Power Sector*. http://www.esmap.org/esmap/sites/esmap.org/files/TR122-09_GBL_Study_of_Equipment_Prices_in_the_Power_Sector.pdf.

16 Alberta, Fossil Capitalism, and the Political Ecology of Change

LAURIE E. ADKIN AND BYRON MILLER

The issues we face are profound ones, going beyond the ideological conflicts that have occupied the world for so long, over who should run the industrial machine, and who should reap the benefits. Now we are being asked: How much energy does it take to run the industrial machine? Where must the energy come from? Where is the machine going? And what happens to the people who live in the path of the machine?

Judge Thomas Berger,
Report on the MacKenzie Valley Pipeline Inquiry, vol. 1, 1977, 1

The democratization of the Alberta petro-state will require, in the first instance, a shift away from reliance on fossil fuel revenues and a turn towards other, less volatile revenue streams, for as long as the provincial state is dependent upon fossil fuel revenues the fossil fuel industry will have inordinate leverage over it. Governments of Alberta have, since the discovery of massive oil and gas reserves, relied predominantly upon foreign capital for these resources' extraction, and have institutionalized a rentier state that is highly vulnerable to global market forces beyond the control of the provincial state. Conservative governments have, moreover, supported the trade and investment agreements that have served to reduce the options available to the provincial state to forge an alternative path of development. While the ideological orientations of particular parties, or leaders, play a very key role in explaining these political decisions, the economy that has been structured over many decades, along with the province's political institutions, now constitute the terrain of action for both the state and civil society. In this chapter we trace the deepening social and ecological contradictions of the province's carbon-extraction-driven economic model, as seen in its worsening fiscal vulnerabilities, growing income inequality, insecurity of livelihoods, conflict with Aboriginal rights, and rising greenhouse gas emissions. We review the evidence that significant segments of civil society have become more critical of the petro-state and more open to reforms in such areas as taxation, resource rents, pace of oil sands development, and environmental regulation. These cracks in the foundations of the petro-state model have indeed widened over the course of the Stelmach, Redford, and Prentice governments, i.e., since 2006. At the same time, we recognize the heavy weight of economic structuring and the interests vested in the status quo—factors that must be considered in any political-ecological program of change. We stop short, however, of offering an assessment of the actions taken to date by the newly-elected NDP government, as – at the time of writing—much is still in play.

Dependence on Oil Revenue

In 2012, once again, the government of Alberta proclaimed a budget deficit attributed to depressed oil and gas prices, and targeted public services for funding cuts and 'austerity.' In a televised address to Albertans 24 January 2013, Premier Redford stated: 'Today, 30 per cent of our budget is funded by revenues from oil and gas. This means we are vulnerable to swings in resource prices – as we have seen with natural gas prices in the past, and now [in] the price we receive for Alberta oil.' In the same broadcast she said that the six billion dollars in royalty revenues from bitumen that the government had foregone (due to the unexpectedly low price obtained by Alberta bitumen in the US market in 2012) was equivalent to the province's entire education budget. In this way, she directly linked oil revenues to the financing of public goods.

Jim Prentice, Alison Redford's successor, took office in September 2014, just after the US price for West Texas Intermediate (WTI) crude oil had begun its precipitous fall from about $100/barrel, reaching $47/barrel in January 2015.[1] His priority concern rapidly shifted from securing pipeline access for Alberta bitumen to 'tidewater' and foreign markets, to the management of a domestic fiscal crisis. In early 2015 the Alberta government was forecasting a $7 billion budget deficit due to low oil prices, and calling upon Albertans, once again, to tighten their belts. The new premier claimed that Albertans had been 'living beyond their means'[2] and that an effective 9 per cent reduction of program

Figure 16.1. Protestor greeting Alison Redford during an election campaign stop in Edmonton in April 2012. (*Photo*: Mike Hudema.)

spending would be required to manage the deficit caused by the loss of oil revenue. Most of this reduction would be found in the three largest areas of program expenditure: health, education, and social services.

Although the Progressive Conservatives (PCs) claimed that such a fall in royalty revenue could not have been foreseen, fluctuations of this sort have long characterized Alberta's political economy. Stuart Landon and Constance Smith (2010) have shown that Alberta has since 1981 had significantly greater revenue volatility than any other Canadian province. Social democratic critics of the government insist that these repeated boom-and-bust cycles are the product of excessive reliance for operating revenue on oil and gas royalties. In other words, Alberta under the PCs was seen by its critics on the left to be behaving as a typical petro-state, so wholly identified with the interests of an oil sector dominated by foreign multinationals that it would consider no alternatives to austerity measures inflicted upon its own citizens.

Even before the 2012 'bust,' advisors appointed by the government itself had called for a shift away from dependence on oil revenues. The Premier's Council for Economic Strategy (appointed by Premier Stelmach) recommended in May 2011 that the province stop using rent from non-renewable resources to pay for ongoing government programs, and instead divert that revenue into a 'shaping the future fund' that would invest in 'companies, projects and infrastructure needed to grow the economy, ... research, better electricity transmission, public-private partnerships, or direct investments in fledgling companies.'[3] The chair of the council observed that Alberta's taxes were excessively low compared to other provinces and that higher taxes could be phased in gradually to replace revenue from resource rents. Such recommendations have been forthcoming for some years from other economists of varying ideological stripes (Adachi et al. 2002; Flanagan 2011; Gibbins and Roach 2006; Hodgson 2013; Landon and Smith 2010; Mintz 2013; Shiell and Busby 2008). In addition to proposals for the investment of resource royalties, or for raising revenue from other sources, calls have been made for many years for the government to increase its 'take' from resource exploitation.

Income Tax

Proposals to create a more balanced and predictable fiscal regime have included the replacement of the 10 per cent flat income tax – implemented in 2000 – with a progressive income tax structure. Bruce Campbell of the Canadian Centre for Policy Alternatives observed in 2013 that Alberta was the only Canadian province to have a flat tax. As a result, low-and middle-income Albertans pay a higher income tax rate than they would in most other provinces while Albertans in the highest income bracket pay by far the lowest provincial income tax rate in the country' (2013, 4). Parkland Institute researchers have also shown that the flat tax has contributed to Alberta's dramatically growing income inequality, shifting the tax burden to middle- and lower-income-earning brackets. Using 2010 income data, they calculate that the province could have brought in an additional $1.8 billion that year had it used the progressive income tax system that was in place prior to 2000 (Stunden Bower et al. 2013, 14; see also Flanagan and Stunden Bower 2014).

Interestingly, despite the widely held stereotype outside the province of Albertans as extreme libertarians when it comes to personal taxation, there is extensive survey evidence showing that a majority of Albertans would be willing to pay higher taxes to support public services. Of 1,207 Albertans surveyed by the Population Research Laboratory at the University of Alberta in spring 2012, 40 per cent said that they agreed or strongly agreed that they would be willing to pay higher taxes to improve public services. Twenty-five per cent had no opinion, and 35 per cent disagreed or strongly disagreed. Looking at how household income affected the answers, the researchers found that people in households earning between $60,000 and $150,000 were the most willing to pay higher taxes (about 45 per cent), and even 41 per cent of those earning more than $150,000 a year were willing to pay higher taxes. Lower-income households (under $60,000) were most likely to agree that people with higher incomes should pay a greater share of taxes (i.e., supported a progressive form of income tax) (74 per cent), but a substantial majority of respondents in higher-income households ($60,000 to $150,000) also agreed with this principle. Interestingly, only the highest-income households (over $150,000) were less favourable to progressive income taxation, but even within this group 44 per cent agreed.

A February 2013 public opinion survey of 1,014 Albertans commissioned by the Alberta Federation of Labour and conducted by Environics found that 72 per cent of Albertans favoured returning to a progressive income tax; 77 per cent supported increasing taxes paid by corporations and those who earn more than $200,000 per year. Only 17 per cent of those polled were in support of a provincial sales tax.[4] Analysing the poll, Conference Board of Canada economists Alicia Macdonald and Todd Crawford claimed that 'a five per cent sales tax would more than eliminate the current deficit, leave Alberta in a competitive position relative to the other provinces and provide some much needed stability to government revenues.'[5] The regressive nature of sales taxes, however, makes them one of the less attractive options for overcoming reliance on fossil fuel revenues.

Polls have also shown that Albertans care strongly about the quality of their public services. A survey of 507 Albertans commissioned by the Canada West Foundation and carried out by the Population Research Laboratory (PRL) between 22 and 27 September 2005 asked respondents how they felt the province's surplus oil and gas revenue should be spent. Responses showed very strong support for program spending, particularly to improve health care. Ranked less highly, but also viewed as priorities, were K–12 education, the elimination of health premiums, investment in research into alternative energy, and post-secondary education (in that order, but not far apart). There was comparatively very little support for *lowering* corporate taxes (Berdahl 2006; Krahn 2005). Another survey, conducted by the provincial government to learn Albertans' views about how the surplus of 2004–05 should be spent, found very similar results. Sociologist Harvey Krahn summarized the result of the government's survey in these terms: 'The quarter million citizens who registered their opinion gave advice very similar to that received in the 2005 Canada West Foundation survey. Specifically, much higher proportions of Albertans stated that spending on education, health care, and other government programs and services was a high priority, compared to providing rebates to individual Albertans or investing in the Heritage Trust Fund' (2005, 5–6; see also *Edmonton Journal*, 20 October 2004, A3). Krahn commented that, in Alberta, 'citizens have quite a different perspective from that of their government about how the windfall royalties should be used' (1).

In the 2015 election campaign, both the Progressive Conservative Party and the New Democratic Party proposed to make adjustments to the flat income tax rate. Following its victory in the May election, the NDP government increased tax rates for incomes above $125,000 on an incremental basis, beginning at 12 per cent and capping at a rate of 15 per cent for incomes over $300,000. According to the government, only 7 per cent of tax filers will be affected by these rate increases, which are expected to generate an additional $450 million in 2015–16 and an additional $906 million in 2016–17.[6]

Corporate Taxes

In his 2012 study for the Parkland Institute, David Campanella observed:

> The past decade has seen corporate income tax rates in Alberta reduced dramatically, both federally and provincially. In 1997, these rates were 29.12% and 15.5% respectively, and by 2011 had dropped to 18.5% and 10% with further reductions planned for 2012. These cuts to the tax rate were mirrored by an enormous rise in corporate profits. As a share of the provincial economy, from 1989 to 2008 corporate profits more than doubled their take from 9.6% to 22.6%, and before the Great Recession had grown in real dollars more than fivefold. The lower tax rates, however, meant that government revenue from the corporate sector stagnated. (12)

Using the government's own figures, the Parkland Institute calculated that by increasing the corporate tax rate from 10 to 12 per cent, Alberta could have collected an additional $840 million in 2012 (Stunden Bower et al. 2013, 13). The provincial New Democratic Party (NDP) adopted this measure in its 2015 election platform, promising to increase the corporate tax rate to 12 per cent if elected, in order to help finance spending on health, education, and elder care.[7] The NDP was able to draw attention to the gap between popular opinion and Progressive Conservative Party policy as revealed by the results of a survey of public opinion that the PC government itself conducted in February 2015 (Government of Alberta 2015). The survey purported to consult Albertans as to how they wanted their government to address the projected $5 billion budget deficit. According to the government's report on the results, 40,513 online responses were received. Of these, 69 per cent supported raising corporate tax and 58 per cent supported the reinstatement of a progressive income tax. The *least* popular measures, once again, were the implementation of a provincial sales tax and the reintroduction of health care premiums. The option of increasing royalty rates was not included in the survey. In its March 2015 budget,[8] the Prentice government added two tax brackets (for incomes over $100,000) and did not propose a sales tax. However, it ignored Albertans' support for a higher corporate tax rate, instead choosing to keep the flat tax rate of 10 per cent. In addition, the government proposed a 'health care contribution levy' at graduated rates for incomes over $50,000, up to a maximum of $1,000 annually. It added to this a swath of other taxes and user fee increases expected to generate, in total, $14.8 billion in revenue. The NDP interpreted this as a budget that would force average Albertans to pay for the deficit rather than the large corporations and the wealthiest income earners.

Premier Jim Prentice, like his predecessors, rejected calls for significant fiscal restructuring of corporate taxes or royalty rates, sticking to the belief that a low tax environment would attract capital to (or keep it in) the province.[9] Revenue has also been lost through significant federal and provincial subsidies to the oil industry since the 1980s. According to Bruce Campbell (2013, 44): 'Subsidies to the petroleum industry by both levels of government totalled over $2.8 billion in 2008, with [corporations based in] Alberta receiving $2.1 billion, or 73% of all subsidies. Roughly half came from the federal government and half from the provincial government. Most seek to increase exploration and development activities through a mix of tax breaks and royalty reductions.' The Prentice government also pushed back its deadlines for revising the 2008 greenhouse gas emissions reduction strategy, with the premier saying that the low oil price environment made this the wrong time to 'damage our industrial competitiveness' by 'laying on costs, including regulatory costs' on the energy sector.[10]

Increasing Albertans' Share of Oil Rents[11]

Possibly unique to the Alberta petro-state are the exceptionally poor terms of its contract with the multinational oil industry. As discussed in Adkin and Stares (this volume), at the height of the 2006 oil price boom and under pressure to respond to evidence of the growing gap between corporate and government shares of oil and gas rents, Premier Stelmach appointed a Royalty Review Panel, which concluded in 2007 that

> Albertans do not receive their fair share from energy development and they have not, in fact, been receiving their fair share for quite some time. Royalty rates and formulas have not kept pace with changes in the resource base, world energy markets and conditions in other energy-rich jurisdictions. Albertans own the resources. The onus is on government to re-balance the royalty and tax system to ensure a fair share is collected both currently and as circumstances change. (Alberta Royalty Review Panel 2007, 4)

And as if this indictment of the government's competence as a manager of the resources were not enough, the panel further stated that 'there is an absence of accountability from the government to the owners of the resource' (5). The panel strongly implied that it was dealing with a captured state, lecturing the government on its proper obligations to govern in the public interest: 'The resources do not belong to the developers; they belong to the people. This is the Panel's viewpoint and the viewpoint of this report. We believe this viewpoint should also become part of the government and departmental culture with respect to this issue. That this is not the case was demonstrated several times at various points throughout the Panel's analysis' (5).

The panel recommended that the government increase Albertans' share of the 'total take' of revenue from the oil sands from 47 per cent to 64 per cent, with more modest increases for conventional oil and natural gas revenue shares (7). It acknowledged that the new revenue share targets would remain lower than 'the total take of several competing jurisdictions,' but deemed them to be 'fair in the context of the Alberta resource base and market opportunities' (8). The panel predicted that implementation of its recommendations would increase the government's revenues by 20 per cent (8). Despite the evidence produced, its recommendations were eventually (with some exceptions) rejected, following a decline in oil exploration leases in 2008 that was blamed on the proposed royalty structure reform, despite its coincidence with the onset of the global financial crisis and recession. The strong backlash from energy sector companies and the decision of some of them to back the upstart, further-right Wildrose Party rather than the PCs led the government to capitulate (Busby et al. 2011; Plourde 2010; Weir 2010).

Successive Parkland Institute studies published since the Royalty Review Panel reported have made the case that Albertans are not receiving their 'fair share' of the wealth produced from the extraction of non-renewable resources. David Campanella (2012, 7) calculated that 'since 1986, more than $285 billion worth of bitumen and synthetic crude oil have been produced from the tar sands. From those resources the oil companies have netted approximately $260 billion dollars in pre-tax profits, while the public has received less than $25 billion in return ... That means roughly 6% of the total value extracted from the tar sands has gone to the public through royalties and land sales.' Stunden Bower et al. (2013, 8) found that if the Government of Alberta had collected 35 per cent of oil revenue in 2012–13 instead of 10 per cent, it would have had $22.3 billion more in its coffers. Jim Roy (2015), an expert on royalty policy, authored a report for Parkland that concluded that Alberta lost

$13.5 billion in oil and gas revenues between 2009 and 2014 because of changes to its royalty structure and its 'failure to manage the pace of oil sands development.'[12] Roy also noted that

> Alberta has a low royalty rate compared to rates in countries that have similar quality of resource. Alberta takes from 25% to 40% of profit, equivalent to 10% of gross revenue. Venezuela, which has the most comparable resource, takes 40% of gross revenue – four times as much as Alberta. Saudi Arabia takes 85% of profit and Norway takes 80% of profit – both three times as much as Alberta. Newfoundland/Hibernia takes 30% to 50% of profit plus 7.5% of gross revenue – twice as much as Alberta.

Despite the weight of such comparative research, a succession of PC governments refused to increase royalty rates for oil producers to a level on par with rates in other petro-states. While the explanation offered for the generous royalty regime some twenty years ago was the need to attract foreign investment – particularly in the high-cost extraction of bitumen – this argument has long ceased to hold.[13] The government's critics ask what Albertans have to lose by raising royalty rates: are the oil companies, so heavily invested in the Athabasca tar sands mines, really going to pull up stakes and relocate to Venezuela? In 2006, in response to the royalty review, companies did threaten to move investment to Saskatchewan and BC, and from time to time there have been freezes on spending (for oil sands project expansions) and reductions in spending on exploration due to slumps in global oil prices.[14] Yet at the height of the oil price boom in 2006, many Albertans thought that a slow-down in oil sands investment would be a good thing; the provincial economy faced inflation (particularly in construction and housing costs), labour shortages, demands for new infrastructure that municipalities could not meet, social problems caused by an increasingly skewed distribution of income, as well as all of the environmental damage and threats to Aboriginal communities that have been documented in the preceding chapters. Many analysts share the view that the pace of oil and bitumen extraction and exports should be slowed down to reduce costs and maximize the province's long-term revenues (Campanella and Bower 2013; Plourde 2009, 2010; Roy 2015).

Notwithstanding the refusal of PC governments to increase royalty rates, there is substantial evidence that – at least until the most recent collapse in global oil prices and the hard-times claims of the oil sands producers – many Albertans supported the reform of the royalty regime. The 2012 Population Research Laboratory survey cited above asked respondents if they agreed or disagreed that 'Alberta's oil and gas royalty rates should be raised.' Fifty per cent agreed or strongly agreed with this statement, while 25 per cent had no opinion either way. These results suggest that only a minority – perhaps a quarter – of the population opposed increases to royalty rates – even when oil prices had fallen from their 2008 peak. The 2013 Environics poll found even stronger evidence for the view that most Albertans did not agree with the direction their governments had taken on oil royalty rates. In this poll, 71 per cent agreed with the statement that Albertans are not getting their fair share of royalty revenue (this number fell to 64 per cent in Calgary). Clearly, it was not public opinion that was holding governments back from implementing reforms to the royalty regime over this period.

Saving and Investing Alberta's Oil Revenue

There is consensus across the political spectrum that Alberta's governments have done a poor job of saving and investing resource revenues since at least the mid-1980s. Associates of the centrist Canada West Foundation, a policy think tank based in Calgary,[15] have repeatedly called upon the government of Alberta to put aside more revenue in its Heritage Savings Trust Fund (HSTF). Economists with the conservative C.D. Howe Institute have made a similar case, comparing Alberta unfavourably to Norway (Shiell and Busby 2008, 11). In a study for the C.D. Howe Institute published in 2010, economists Stuart Landon and Constance Smith observed that

> the Alberta Heritage Savings Trust Fund's ability to stabilize revenues is quite limited: when the fund was established, it received a fixed percentage of resource revenues each year, but this practice was ended in 1987, although the fund has received ad hoc contributions from general revenues in several recent years … In a report commissioned by the Alberta Minister of Finance, Tuer (2002) proposes that the AHSTF be redesigned to stabilize the impact of volatile resource revenues on the province's budget, but this has not been done. There is also the Alberta Sustainability Fund, created in 2003 and designed to stabilize revenues but it, along with the AHSTF, has been subject to considerable discretion in terms of contributions and withdrawals. (2010, 15–16).

To deal with the problem of resource revenue volatility, these economists advocate the creation of a resource revenue stabilization fund 'into which a large percentage of the most volatile component of revenues – resource revenues – [would be] deposited. If withdrawals from the fund were a fixed proportion of the fund's total assets, there would be a considerable reduction in the volatility of government budgetary revenues' (21).

The Alberta Federation of Labour reminds us that the Alberta government ceased making payments into the HSTF in the early 1990s in lockstep with the rise of neoliberal ideology within the PC Party. In keeping with the Klein government's mantra that 'government has no business being in business,' Heritage Fund investments in Alberta Government Telephones were sold to Telus, the mortgages held by the fund were sold to private institutions, and the HSTF was no longer used to invest in economic diversification projects (AFL 2011).

> Those sales deprived the fund of continuing revenues and provided only a one-time injection of $1.6 billion, and show the extent to which privatization robs governments of ongoing cash: investments in AGT and Alberta Mortgage Corporation were worth a total of $4.37 billion in 1988. Alberta did not make a single deposit to the Heritage Fund between 1987 and 2005. Through the 1990s, all yearly income from our initial investments was spent. Over that time, Alberta lost $7 billion to inflation alone. The Calgary Chamber of Commerce calculates that had Alberta continued to save 30 per cent of resource revenues in the Heritage Fund, it would now be worth $128 billion. (AFL 2011; see also Mumey and Osterman 1990)

Bruce Campbell contrasts the approximately $16 billion in the Alberta HSTF in 2013 to the $664 billion in assets held by Norway's Government Pension Fund Global, attributing the difference to the Norwegian governments' more egalitarian, ecological, and future-oriented commitments (2013, 7–9). (The Norwegian GPFG reports its value as US $893 billion as of February 2015. The September 2014 value of Alberta's Heritage Savings Trust Fund was reported to be CAD $17.4 billion, or approximately US $14 billion.) Significantly, Alberta's resource sectors are heavily dominated by large multinational corporations, whereas Norway has retained majority state control over its petroleum production and has retained autonomy with regard to the rate of exploitation and markets (19). Alberta and Canada, on the other hand, have been constrained in these regards by the terms of the Free Trade Agreement with the United States and its successor, the North American Free Trade Agreement.

The AFL and the Parkland Institute support reinvestment in a savings fund, but argue that the conditions for this must include a larger government take from resource revenues and a restructured tax system (rather than cuts to program spending). Moreover, the AFL and environmental organizations have advocated that the savings fund be used much in the way proposed by ecological economist Herman Daly (1996), that is, as a 'capital liquidation fund' used to finance transition to a post-carbon economy:

> The Heritage Fund should be the primary vehicle for building a sustainable, green economy. Because its revenues are directly derived from our oil and gas wealth, we should be using the fund for the eventual and inevitable time when fossil fuels are no longer the energy source of choice for the majority of the world's population. The resource bounty we all enjoy now is a privilege, but also implies great responsibilities to the environment and future generations in the context of climate change.
>
> The Heritage Fund could be used to finance the growth of a green energy sector in Alberta and across the country. The Fund could be used as a revolving, no-interest source of cash for municipalities and other nonprofit entities to invest in green infrastructure in ways they cannot afford to do right now.
>
> The Heritage Fund could be used to start an Albertan-owned public Crown Corporation focused on green energy, or could also be used to invest in projects that are only marginally economical now but show great promise in the future, such as rail links between major cities or between cities and suburbs.
>
> It could be used on a low-interest basis by the private sector and even by other provinces to build green energy infrastructure, including research and development, grid upgrades and even venture capital. None of this is wild, radical or impossible – it could begin to happen tomorrow, as long as we are also prepared to have a grown-up conversation about taxes, royalties, and what kind of legacy we want to leave to future generations of Albertans. (AFL 2011)

This is, indeed, the dream of many citizens of the province, and of people beyond our province: the potential to slow the extraction of bitumen, to conserve natural gas, to ration (globally) the use of oil for essential purposes (where there are no current substitutions), and to use the revenues from fossil fuel exports to radically accelerate a green transition. In April 2009 the Alberta Federation of Labour, along with Greenpeace and the Sierra Club Prairie Chapter, released a report entitled *Green Jobs: It's Time to Build Alberta's Future*, which argued that, with a reorientation of investment away from fossil fuel subsidies, the government could lay the groundwork for the creation of tens of thousands of jobs within two to seven years (Thompson 2009).

Understanding the Inertia of Alberta's Petro-Economy

All these arguments notwithstanding, PC governments reiterated at every opportunity their determination to continue the province's dependence on oil revenue. Energy Minister Ken Hughes told media in February 2013 that his department was 'working on an aggressive new plan to upgrade Alberta's resources in Canada and get them to coastal ports for export "however we can."'[16] The government called this its Oil Market Diversification Strategy. Following in the steps of Ed Stelmach, Premier Redford visited her provincial counterparts, federal politicians, and United States policy makers to secure support and approval (from the Obama administration) for pipelines that would carry Alberta's diluted bitumen or synthetic crude to multiple shipping terminals or refineries. Her government opened new offices in Ottawa (estimated to cost the government $850,000 annually) (Wingrove 2012), India, and Singapore to promote Alberta's exports, and the ministry responsible for intergovernmental and international relations (IIR) received a 61 per cent increase in its budget from 2009 to 2011 (Kleiss 2013). According to then minister for IIR Cal Dallas, 'A high priority of the government is obviously to realize full value for bitumen and related energy products ... Clearly we need to message [*sic*] about the things that we're doing in Alberta to create and maintain that social licence to be able to market around the world' (quoted in Kleiss 2013). The government placed a $30,000 ad in the *New York Times* 17 March 2013, assuring the American public that the Keystone XL (extension) pipeline from Alberta to the Gulf of Mexico was 'the choice of reason.' In essence, the Redford and Prentice governments used the same methods as their predecessors to combat international calls for the boycott of tar sands oil.

Why have Alberta governments continued to strongly support, subsidize, and advance the interests of the oil and gas industry when alternative revenue generation and viable green job strategies exist? The explanation encompasses the ideological factors explained in other chapters in this volume, vested elite interests, and the structured dependency of employment and revenue generation on the status quo economic strategy. Here we focus on the third explanation, as the others have been discussed elsewhere.

Far more than the number of new jobs that would need to be created is at stake. Indeed, analysts of the industry generally recognize that oil and gas employment generation is not as large, proportionally, as its contribution to GDP would suggest. In what is still widely regarded to be the most comprehensive analysis of the economic impacts of Alberta's oil and gas industry, Mansell and Schlenker (2006) acknowledged that, as of December 2004, 'direct employment [in the oil and gas industry] represented only about 6 percent of total Alberta employment (24).' This is not to say that the employment effects of oil and gas are insignificant. On the contrary, 'when the employment in other sectors that is attributable to oil and gas activity is taken into account, it is clear that the total employment impact is many times larger' (24). Mansell and Schlenker pointed to professional, scientific, and technical services, manufacturing, and construction as sectors that are closely linked to expenditures in the oil and gas sector. Overall, they estimated that *45 per cent* of Alberta's jobs are either directly or indirectly related to oil and gas (29). Moreover, and perhaps most significantly, average earnings in the oil and gas sector are the highest in the Alberta economy.

More recent data indicate the continued heavy weight of employment in the 'natural resource sector' in the province's economy since 2004. A 2013 report by the Institute for Public Economics at the University of Alberta states that 'the portion of Alberta's labour force employed by the resource sector has grown from five per cent in 2000 to 8.2 per cent in 2012. The 117.6 per cent increase in resource sector employment significantly exceeds the 32.9 per cent increase in the total labour force of the Province during that period' (5–6). The most recent bust in oil prices reveals, once again, the skewed occupational, earnings, and gender characteristics of Alberta's 'oil economy.' Average crude oil prices fell from $108.00 in June 2014 to $47.00 in January 2015 (rounded averages). After a modest recovery to $63.00 in May 2015, the average price fell further, to $30.00 in January 2016. Given the current oversupply of crude oil in global markets, along with longer-term cost factors and the growth in renewable energy supply, high profit margins for oil sands producers may not return for a long time, if ever. These conditions have produced a freeze on investment in projects-in-development, resulting in large-scale layoffs in the resource industry and construction sectors.[17] Hardest hit have been young men in blue-collar jobs; approximately 14,500 of the lowest-income earners in the natural resources sector were laid off between September 2014 and December 2015 (Younglai 2016). Altogether, more than 30,000 jobs in this sector were lost in 2015 (Younglai 2016), and layoffs have continued into 2016. Concurrent with the loss of 73,000 full-time jobs in the province in 2015 was a rise in part-time employment (up by 38,000), suggesting that some of the workers laid off in the resource and construction sectors were turning to part-time work (Statistics Canada 2016). This is a sector in which female labour predominates, and wages are much lower (see chapter 1, this volume). Alberta's unemployment rate rose from 4.7 per cent in 2014 to 7.9 per cent in February 2016 (Government of Alberta 2016).

As the current employment crisis suggests, the difficulty with alternative job creation strategies in such an economy is twofold. First, what is at stake is not the creation of jobs in the abstract, but particular, specific, jobs. There is no guarantee that a person with a good and relatively secure job in the current economy will find equivalent employment in a green economy. Given the proportion of the Alberta economy dependent on oil and gas, there is tremendous aversion, on the part of many individuals as well as the government, to the risks and uncertainty transforming the economy would entail. Second, there is no guarantee – indeed, it is unlikely – that jobs in an alternative green economy will pay as well as those in the resource extraction sector. In September 2014, before the industry began to pull back on investment, earnings in the natural resources sector in Alberta reached an average of $2,144 per week – higher than the wages in any other sector (Younglai 2016). Substantial year-end bonuses are common in the oil and gas industry, as evidenced by the 2014 Hays Oil and Gas Global Survey that found that 42.8 per cent of oil and gas employees received a bonus (Lauletta 2014). For the richest and most powerful members of society – the 1 per cent – the oil and gas economy has been especially lucrative. While the top 1 per cent of tax filers across the twenty-five largest Canadian metropolitan areas took in 9 per cent of all income in 2010, in Calgary – the command and control centre of the Canadian oil and gas industry – that figure was 28 per cent, by far the highest share taken by the 1 per cent in any Canadian metropolitan area. High salaries in the energy and construction sectors have pushed mean income figures well above the Canadian average for many years (Statistics Canada 2015). Given both the perceived security of Alberta's petro-economy – volatility of the price of oil notwithstanding – and the high-paying jobs associated with it, a concerted policy shift away from a petro-economy is a very hard sell, especially to the most economically and politically powerful.

None of the above is meant to imply that the benefits of a restructured fiscal regime and a greener economy might not in the medium term, and certainly for future generations, provide equally desirable forms of well-being and security as those currently provided by the hydrocarbon-driven economy. One form this is expected to take is a diminution in income inequality. Indeed, a 2013 analysis by the Parkland Institute and the Canadian Centre for Policy Alternatives shows, in the words of its title, that 'Alberta is Canada's most unequal province, and Calgary the most unequal city.' As mentioned previously, the oil economy has been profoundly gendered; the booms have not produced nearly as many high-paying jobs for women as they have for men. Alberta has the country's highest gender wage gap (Lahey 2015). The PCs' refusal, over a long period, to invest in child care provision is a factor in the decision of many women to seek part-time work. As Sara Dorow (2015) and others (including O'Shaughnessy and Döğu, this volume) have documented, the lowest-paid service sector work with the most insecure conditions of employment is also significantly racialized. A more diversified economy and affordable early childhood education and care services are conditions for a more egalitarian distribution of income and better conditions of employment for a greater number. The new government will need to make the case for the benefits of a rapid green transition, and to invest every available resource in consultation and planning processes to make this happen.

Beyond the question of job creation is the question of the province's economic growth. Mansell and Schlenker stressed that the oil and gas sector has had a larger impact on provincial GDP than it has had on employment. They calculated that the oil and gas industry, including both conventional oil and gas and oil sands, was responsible for $1.5 trillion in provincial GDP from 1971 to 2004 and that 'without the oil and gas industry, provincial GDP would have been only 47 percent of what it actually was in 2004' (2006, 29). GDP is important, above and beyond employment, because it represents economic activity that can be taxed. From 2000 to 2004, the province's non-renewable resource revenue (which is derived almost entirely from oil and gas) 'totaled over $44 billion (or an average of $8.8 billion annually), representing an average of just over 32 percent of total Alberta revenues. To put these government revenues in perspective it might be noted that in order to obtain the same revenues as achieved in 2004, a provincial sales tax of 16 percent would be required' (Mansell and Schlenker 2006, 20). With oil and gas revenue representing almost a third of the provincial revenue stream (at the time of Mansell and Schlenker's analysis), not to mention significant tax revenues that accrue to the federal government, politicians – both provincial and federal – are loath to pursue policies that could reduce tax revenues. These revenues, while far less than what they would be under a royalty regime such as Norway's, nonetheless have allowed Alberta to operate without a sales tax (the only Canadian province not to have one), without health care premiums (since 2009), without liquor control board revenues, without operating debt (from 2004 to 2012) and associated interest payments, without capital taxes, without payroll taxes, and with the lowest general corporate tax rate in Canada (10 per cent), the lowest gasoline taxes in Canada, and a flat income tax (10 per cent) that foregoes additional revenue that could be collected from higher-income households. Moreover, without oil and gas revenues the province could not have dispensed a one-time 'prosperity bonus' (aka 'Ralph Bucks') of $400 dollars to each and every resident of Alberta, except prisoners, in 2006.

The capacity to offer relatively good public services in combination with very low taxes is what the provincial government has referred to as the 'Alberta Advantage.' Perhaps more accurately, the 'Alberta Advantage' could be referred to as the 'Alberta Free Lunch,' with the cannibalization of non-renewable resource revenues allowing successive Progressive Conservative governments to offer a combination of good-quality public services and low taxes virtually impossible to match in any other North American jurisdiction. This combination was undoubtedly key to the Alberta Progressive Conservatives' ability to hold power continuously from 1971 to 2015, and the prospect of a long period of low oil prices and reduced state revenues from resource rents is destabilizing for the neoliberal regime.

The fact that the energy sector accounts for more than two-thirds of exports in Alberta's heavily export-oriented economy (Government of Alberta 2012a) and a quarter of Canada's exports (Eyford 2013) is a third factor reinforcing Alberta's petro-economy. The rapid growth of Alberta's economy (a 26.5 per cent increase in GDP from 2000 to 2010, according to the provincial government [Government of Alberta 2011, 8]) cannot be understood apart from the province's deliberate strategy to attract foreign capital investment to the oil patch – and the investment has been massive. According to the Government of Alberta

(2012a), over $116 billion was invested in the Alberta oil sands from 2000 to 2010; an even greater amount was invested in conventional oil production. As a result, Alberta's exports soared by 64 per cent from 2001 to 2011, 71 per cent of which ($66 billion) were energy exports. Of Alberta's 2011 crude oil production, only 22 per cent stayed in Alberta, 16 per cent went to the rest of Canada, 2 per cent was sold offshore, and the remaining 60 per cent went to the United States (Government of Alberta 2012b). Without a strong export base, the overall Alberta economy would not have grown to more than a fraction of its present size (Mansell and Schlenker 2006, iii).

Resisting Fossil Capitalism: The Political Ecology of Change in a Globally Connected World

Despite Alberta's dependence on the resource rents and well-paying jobs provided by the oil and gas industry, and despite the techniques employed by the Progressive Conservative government to maintain hegemony (see chapters 6 and 17, this volume), popular support for Alberta's model of development is not as solid as the PCs' forty-four-year rule suggests. First, the first-past-the-post electoral system plays an important role in creating the false perception that Albertans are homogenously conservative. In fact, as the polling data we have reviewed indicate, many of the policies of Alberta's government do not align with public opinion. Combined with vote splitting among multiple left-of-centre parties, the FPTP system has permitted the PCs to win a majority of seats in every legislative assembly since 1971. In the twelve provincial elections held between 1971 and 2012, the PCs won more than 57 per cent of the vote only four times – most recently in 2001. In three elections they won narrow majorities – between 51 and 53 per cent of the vote – and in five elections they garnered a minority of the vote – between 44 and 47 per cent. Growing public awareness of the cronyism and other abuses of power described in chapter 17, as well as questioning of the PCs' competence to manage budgets and to adequately fund public services, have corrosive effects on the party's hegemony. In the 2012 election the PCs lost substantial support to the upstart libertarian-right Wildrose Party in their traditional rural strongholds, but captured 'strategic' votes from more centrist voters who feared a Wildrose Party victory.

Once Jim Prentice had won the leadership of the PCP and his party had captured all four seats in by-elections held in October 2014, the Wildrose Party leader, then leader of the Opposition, and eight members of her caucus agreed to 'reunite the right.' In the middle of the legislative session, she and her colleagues (re)joined the PCs, leaving the Wildrose caucus with five members. The legislative assembly was left even more heavily unbalanced, with seventy-two of eighty-seven seats held by PCs and the remaining fifteen divided among three other parties and one independent MLA. Rather than shoring up the governing PCs, however, this move provoked a backlash among Wildrose supporters against their former leader, and created an unexpected opening for the centre-left NDP. For the first time, the province's electorate appears to be dividing along the left-right lines that, historically, have characterized politics in other jurisdictions.

Beyond Alberta's borders, its political ecology is contested within Canada. Alberta's export-based petro-economy has major ramifications for the rest of the country. Central Canadian politicians decry the negative impacts that a high Canadian petro-dollar has on Canada's manufacturing, based largely in Ontario and Quebec. Federal New Democratic Party leader Thomas Mulcair argued that Canada has suffered from 'Dutch Disease': a dollar overinflated by western Canada's oil boom that in turn reduces the competitiveness of central Canada's manufacturing exports (Beltrame 2012). As mentioned in chapter 1, a number of studies – including two from the OECD and one from Industry Canada – lend support to Mulcair's claims. Mulcair's position has been criticized as 'divisive,' pitting different regions of Canada against each other; yet these conflicts among regional economic interests have much to do with Canadian federalism and the Conservative government's support for both devolution of powers to the provinces and the interests of the resource extraction industries. As a result of the Natural Resources Transfer Acts, passed in 1930, Alberta, Saskatchewan, and Manitoba have exclusive jurisdiction over the development of their natural resources. Moreover, with the shift of economic power and population westward, and the federal Conservative Party's electoral support heavily based in the west, less consideration has been given to the concerns of central Canadian manufacturing than at any other time in recent history. The net result has been unchecked development of the oil industry.

Both Premier Redford and the federal minister of natural resources, Joe Oliver, condemned Thomas Mulcair for a speech he made in Washington on 13 March 2013 that declared his party's strong support for 'sustainable development' and criticized the Harper government's handling of the environmental problems associated with the oil sands. Redford, as well as the conservative premier of Saskatchewan, Brad Wall, accused Mr Mulcair of 'betraying' Canada's economic interests by failing to unreservedly support the Keystone XL pipeline project, which would substantially expand pipeline capacity to the United States (*CBC News*, 18 March 2013).

Meanwhile, the federal Conservative government not only established secretive mechanisms for co-ordinating provincial, federal, and corporate lobbying efforts to win approval for the Keystone XL pipeline in the USA,[18] but – in the wake of the Northern Gateway Pipeline Review Panel hearings – gave its own chief negotiator on comprehensive land claims, Doug Eyford, the additional assignment of finding solutions to First Nations' opposition to the pipeline projects (Eyford 2013; Scoffield 2013). The Harper government gambled that Canadians would subordinate concern over the 'abstract,' 'distant' threat of climate change to the 'imperative' of economic growth and employment, and that environmental harms associated with the oil sands could be localized and minimized by the promise of better ('world class') environmental monitoring and enforcement of regulations. Likewise, Prime Minister Harper tried to dampen the uproar in British Columbia about oil supertankers navigating coastal waters by reassuring citizens that the supertankers will be 'world class.' As mentioned above, he also sought ways to accommodate First Nations sufficiently to permit resource exploitation across the country to proceed unhindered by Aboriginal treaty rights and land claims. And still, for every fire that has been dampened by an industrial relations corporation or a memorandum of agreement (see Zalik, this volume), another has been lit. The global People's Climate March, held in 162 countries on 22 September 2014, represented the largest climate change demonstrations to that date. The Idle No More movement, proliferating across Canada and internationally in 2012, shows no signs of abating, and may well represent the best hope for halting the growth of CO_2 emissions in Canada (see Palmater 2013).

On 20 March 2013 an alliance of First Nations from territories in Canada and the United States held a press conference to announce their united determination to stop both the Northern Gateway and the Keystone XL pipelines from being built (*CBC News*, 20 March 2013). Chief Martin Louie of the Nadleh Whut'en First Nation in northern BC declared: 'If we have to keep going to court, we'll keep doing that ... We're the ones that's [*sic*] going to save whatever we have left of this Earth.'

The assembled leaders added their signatures to Save the Fraser Declaration and the International Treaty to Protect the Sacred from Tar Sands Projects, which they intended to register with the United Nations. In addition to court actions, Aboriginal leaders and environmental activists have declared their willingness to practise civil disobedience.

Demonstrating the truly dialectical nature of this conflict over the future path of development, the Conservative government responded to the escalation of environmental and indigenous activism by framing the latter as a threat to Canada's economic and financial stability, or security. A 'Critical Infrastructure Intelligence Assessment' prepared for the government by the RCMP in January 2014 (but not leaked to the press until January 2015) characterized organizations like Greenpeace, Tides Canada, and Sierra Club as an 'anti-petroleum movement' and claimed that 'violent anti-petroleum extremists' are 'a realistic criminal threat to Canada's petroleum industry.'[19] Disturbingly, the RCMP report identified civil disobedience as non-peaceful activity. The Conservative government's subsequent Bill C-51 (Security of Canada Information Sharing Act), introduced 30 January 2015, identified 'interference with the economic or financial stability of Canada,' or 'interference with critical infrastructure,' as 'activity that undermines the security of Canada' (sec. 2).

We find ourselves locked in struggle with the same 'industrial machine' that Thomas Berger warned of almost four decades ago, only now the ecological stakes are far higher. It should come as no surprise that, in Canada, effective resistance to the machine – in the form of protest, civil disobedience, and legal challenges – is being led by those whose culture is most closely rooted in the land and whose survival is most imminently threatened: indigenous peoples.

Globally, resistance comes from many locations. European publics have not been impressed by the road shows of Conservative politicians, intended to persuade them of Canada's excellent environmental record. Canada has repeatedly been awarded 'worst in class' standing in environmental forums. A very telling blow to Alberta's promotional strategy occurred in March 2013, when Germany's largest scientific research organization cancelled its participation in a research consortium in Alberta that was developing bitumen processing technology (and to which the government of Alberta had granted $25 million over five years). The reason was German public opinion. Nor have governments and publics of the global south been impressed by Canada's role in international climate change negotiations. Civil society actors around the globe who will not stand down in their opposition to the further expansion of bitumen production from the oil sands and its implications for climate change have proven to be effective adversaries of Canada's petro-elite. Resistance springs up at multiple points along the very arteries of oil distribution, seeking to slow its progress or stop it altogether.

Ecology and the Alberta Petro-State

One of the driving concerns of this book is global warming, caused by the greenhouse gas emissions of human economies. Although the governments of Alberta and Canada frequently repeat that Alberta's oil sands production (using 2010 emission figures) accounts for only 7 per cent of Canadian GHG emissions and less than 0.15 per cent of global emissions,[20] environmentally concerned actors frame the problem differently. The CCPA, the Pembina Institute, and many ENGOs, Canadian and international, emphasize instead that the oil sands are the fastest-growing source of industrial carbon emissions in Canada; that the government and industry plan to double (or more) oil sands production from 2012 to 2022; and that 'the resulting emissions will cancel out every other effort in Canada to reduce climate pollution' (including the phasing out of coal-fired plants in Ontario).[21] Because Alberta derives most of its electricity from coal (see Weis et al., this volume), and because of its oil sands mining and refining, it now emits more than 37 per cent of Canada's carbon emissions (Alberta Climate Leadership Panel 2015, 23). Alberta's GHG emissions grew by 37 per cent from 1990 to 2007 as oil sands production expanded. The report of the Climate Panel appointed by the Notley government in 2015 notes that if oil extraction, coal consumption, and other components of economic growth proceed without any new checks on greenhouse gas emissions, Alberta will account for 60 per cent of the total growth in Canada's emissions between 2013 and 2030 (Alberta Climate Leadership Panel 2015, 23).

While Alberta's governments have, since the passage of its Climate Change Emissions Management Act in 2007, proclaimed that Alberta is one of the few North American jurisdictions to have 'imposed a price on carbon,' this claim referred to its very weak cap-and-trade system for large industrial emitters.[22] This system included the option of buying offset credits from other sectors or paying a fee to the Climate Change and Emissions Management Fund of $15 CAD per tonne of CO_2 that exceeded an emissions-intensity target. This approach yielded very poor results. As noted above, in absolute terms, the province's emissions simply continued to rise. To achieve meaningful reduction targets (for example, of 25 per cent below 1990 level by 2020, and 80–95 per cent below 1990 level by 2050), organizations like the Pembina Institute have called on Alberta governments to implement much higher prices for carbon (Bramley et al. 2011). Pembina has advocated moving to a comprehensive carbon tax while removing the offsets option, and slowing the rate of approvals of new oil sands and other industrial projects so that environmental limits are not surpassed.

In relation to the last point, it is worth remembering scientist Jim Hansen's warning, published in the *New York Times* on 9 May 2012, that 'If we burn the oil contained in the tar sands, we will release 240 gigatons of carbon into the atmosphere, or 120 ppm.'[23] Hansen claims that the tar sands 'contain twice the amount of carbon dioxide emitted by global oil use in our entire history.' If we burn this, plus the other sources of non-conventional oil and gas, and coal, there will be no hope of keeping CO_2 in the atmosphere below 500 ppm. The concentration was estimated to be 393 ppm in 2014, and 350 ppm was considered to be the safety threshold ten years ago. Even more starkly, McGlade and Ekins' 2015 analysis, published in the highly respected journal *Nature*, shows that 99 per cent of Canada's 'unconventional' oil must be regarded as unburnable if global warming is to be limited to 2 C. These messages have not been lost on climate change activists around the globe, which helps to explain why the Athabasca tar sands and their associated pipelines have faced such intense opposition.

Chapters 12 and 13 describe the environmental effects and risks created by Alberta's fossil fuel extraction for other North American jurisdictions as well – particularly Montana, Idaho, and British Columbia. The web of nodes and arteries of fossil fuel extraction, refining, and transportation spans all scales. Our analyses could be extended to the sites of crude oil refining and petrochemicals production in the Great Lakes Basin, including their environmental conditions and the forms of local resistance to environmental harms and risks. Zalik, in chapter 11, argues that it is also important to study corporate strategies across multiple sites, even those as different as the Niger Delta and northern Alberta.

Within Canada, other provinces are critical of the federal government's failures to regulate GHG emissions from the energy sector and have not passively accepted the Alberta government's effective refusal to do its share. Foreign and environmental policies driven by the federal Conservatives' 'energy superpower' ambitions (withdrawal of

Canada from the Kyoto Accord, obstructionism at conferences of the parties to the United Nations Convention on Climate Change, arguments with the European Union over fuel directives, weakening of domestic environmental legislation, subsidization of oil producers, etc.) are associated by many Canadians with the privileging of 'Alberta's' interests over those of other regions (although – as Adkin and Stares argue in this volume – it is really the interests of transnational energy companies that are privileged above all else by the policies of the Alberta and Canadian petro-states).

Finally, it is important to remember that the privileging of corporate interests in resource extraction over all other social or environmental considerations has had enormous environmental consequences for Alberta on a local scale. Some of these consequences are described in chapters 4, 5, 9, 10, and 15. Determined opposition to oil sands expansion on the part of First Nations, using both legal and media strategies, has put up substantial roadblocks to new project approvals for mines and pipelines. The new NDP government will have a difficult time reconciling its commitment to a respectful relationship with First Nations with continued reliance on fossil fuel production as the centrepiece of the provincial economy.

Conclusion: Moving Beyond Fossil Capitalism and the Alberta Petro-State

Why have Alberta's governments permitted the exploitation of fossil fuels *as they have*, with the attendant environmental, economic, and political consequences? To understand the basis of the Alberta petro-state, as well as the potential for social and political change, we need to consider the structure of the Alberta economy, Alberta's political institutions, the ideological orientations of the governing parties, the changing demographics of Alberta's population, the reality of global circuits of trade, production, and consumption in energy resources, the role of transnational oil companies, federal politics, and developments beyond Canada's borders. These factors intersect in different ways at particular junctures to delimit policy options and to alter the strategic terrain upon which actors seek different outcomes. Greenhouse gas reduction strategies in the electricity generation sector, for instance, appear to be politically viable in Alberta given the small role that electricity generation plays in the provincial economy and its insignificance as an export. The oil sands, however, are an entirely different matter. Thus the oil and gas corporations and NDP government have reached agreement rather easily on a 'climate change' action plan that will involve the rapid phase-out of coal-fired electricity generation while permitting emissions from the oil sands to rise until 2030.[24] It appears that pushing the NDP government to move more rapidly to phase out reliance on bitumen exports and direct resources to the building of a low-carbon economy will require a greater mobilization of Alberta's civil society, in conjunction with developments at other scales.

Transition from an economy based on oil and gas extraction and export is partly a matter of choice for Albertans, and partly a matter of necessity. The industry with which Albertans have for so long been encouraged to identify is in fact not 'Albertan,' but transnational. What drives its investment decisions has little to do with the provision of long-term, sustainable livelihoods for any particular community in Alberta, or ecologically sustainable production and consumption. Actors in innumerable locations, as we have suggested, are both reinforcing and contesting the social and ecological relationships of fossil capitalism. Producers, refiners, distributors, consumers, and citizens are all implicated in the geographically extensive flows of capital and carbon that are fossil capitalism's lifeblood, and all points along the commodity chain represent opportunities to intervene and alter the governance of the system. With the world now confronting this unprecedented ecological crisis, signalling the arrival of a post-carbon order whose social relations have yet to be determined, only the very short-sighted could imagine that Albertans will find economic stability and well-being in a rear-guard defence of a fossil-fuel dominated economy. Albertans can have a bright and prosperous future in a renewable energy economy. The challenge lies in making the transition.

NOTES

1 US Energy Information Administration, 'Short-Term Energy Outlook,' 10 February 2015, http://www.eia.gov/forecasts/steo/report/prices.cfm.

2 For one occurrence of this phrase, see Premier Prentice's teleconference interview with journalists from Washington, DC, 3 February 2015, https://soundcloud.com/your-alberta/premier-prentice-teleconference-from-washington-feb-3-2015/ (accessed 16 February 2015).

3 David Emerson, as reported by Keith Gerein, 'Alberta Must Invest Its Energy Wealth to Prepare For Future: Economic Council,' *Edmonton Journal,*

5 May 2011. The council's report may be found at http://alberta.ca/premierscounciileconomicstrategy.cfm.

4 Alberta Federation of Labour, 'Albertans Reject Austerity,' press release, 4 March 2013, http://www.afl.org/index.php/Press-Release/albertans-reject-austerity.html; Sheila Pratt, 'Alberta Unions Denounce 'Klein-Style Cuts' Coming in Budget,' *Edmonton Journal*, 5 March 2013, http://www.edmontonjournal.com/news/Alberta+unions+denounce+Klein+style+cuts+coming+budget/8046535/story.html (accessed 22 March 2013).

5 The authors are quoted in a Canadian Press story published on the website of *MacLean's*, 26 February 2013, http://www2.macleans.ca/2013/02/26/conference-board-of-canada-calls-on-alberta-to-introduce-provincial-sales-tax/.

6 For more information about the income tax changes, see http://www.alberta.ca/budget-revenue.cfm.

7 The NDP government followed through on this promise, increasing the general corporate tax rate from 10 per cent to 12 per cent in Bill 2, *An Act to Restore Fairness to Public Revenue*, introduced in June 2015. The tax rate for small business remains 3 per cent. For more details of the tax rate changes, see http://www.albertacanada.com/business/overview/competitive-corporate-taxes.aspx. The government estimates that the new tax rate will bring in an additional $250 million in revenue in 2015–16 and an additional $450 million in 2016–17 (http://www.alberta.ca/budget-revenue.cfm).

8 Government of Alberta Budget, March 2015, http://www.alberta.ca/Budget.cfm (accessed 4 May 2015).

9 Government of Alberta, 'Competitive Corporate Tax Rates,' http://www.albertacanada.com/business/overview/competitive-corporate-taxes.aspx (accessed 20 March 2013). See Premier Prentice's 4 February 2015 speech to the US Chamber of Commerce (http://www.alberta.ca/us-chamber-of-commerce.cfm); *CBC News*, 'Jim Prentice Will Not Raise Corporate Tax to Compensate for Low Oil Prices,' 3 February 2015, http://www.cbc.ca/1.2944201.

10 The premier is quoted in Josh Wingrove, 'Alberta Premier Jim Prentice Says Oil Industry Needs to Remain Competitive,' *Globe and Mail*, 16 October 2014.

11 Economic rent is what is left after the costs of production and a 'normal' rate of profit are subtracted from the total revenue earned from the sale of the resource (AGA 2007, 98). Campanella says that a normal rate of profit in the oil and gas industry is assumed to be 10 per cent of the investment. 'Profits earned above 10% are termed "excess," as they represent the portion of economic rent accruing to the industry rather than the owners. As

the chair of the Royalty Review Panel, Bill Hunter, noted, "As Albertans we own 100 per cent of the resource, and we should expect nothing less than 100 per cent of the rent." Indeed, capturing less than the entirety of the economic rent is equivalent to the provision of a corporate subsidy' (2012, 11). Bill Hunter is quoted in *CBC News*, 'Premier Won't Be Bullied into Royalty Decision,' 20 September 2007, http://www.cbc.ca/canada/edmonton/story/2007/09/20/stelmachroyalty.html.

12 It is not even certain that the Ministry of Energy has had sufficient numbers of qualified staff to fully collect the royalty payments owing the government. The Auditor General of Alberta's *Annual Report* for 2001–02 stated: 'The Ministry needs assurance on well and production data to ensure the completeness and accuracy of Crown royalty revenues, and to develop energy policies' (79; http://www.oag.ab.ca/files/oag/ar2001-02.pdf) . Again in its 2006–07 *Annual Report*, the Office of the Auditor General warned: 'Without assurance over volumetric data the Department cannot support a conclusion that all royalties due under the existing regime are being collected' (67; http://www.oag.ab.ca/files/oag/2006-2007_Annual_Report_Vol_2.pdf). This report also noted that the Ministry had insufficient staff to authenticate the production data being submitted by producers (63–73).

13 For an explanation of why this is the case, see Max Fawcett, 'Is Investment in the Oil Sands Slowing the Industry Down?' *Alberta Venture*, 16 December 2013, http://albertaventure.com/2013/12/oil-sands-investment-transform-alberta/.

14 Comparatively low oil prices since 2008 (especially since mid-2014, brought on, in part, by the enormous expansion of production of shale liquids in the USA since 2000) and large cost over-runs in the construction of new mines and upgraders have slowed investment in new capacity in the Alberta oil sands.

15 The CWF's occasional papers, authored by different academics, reflect a variety of political viewpoints, although the economic approach of its leading figures tends to be neoliberal.

16 Quoted in Karen Kleiss, 'Energy Minister Lays Out Alberta's New Oil Strategy,' *Edmonton Journal*, 10 February 2013, http://www.edmontonjournal.com/news/edmonton/Energy+minister+lays+Alberta+strategy/7945829/story.html (accessed 20 March 2013).

17 According to the Petroleum Labour Market Information Division of Enform (an upstream oil and gas industry agency with funding from industry associations and the Canadian government): 'Following the record spending of $35.7 billion in 2014, oil sands capital investment dropped by 30% in 2015 and is not expected to recover before 2020. Capital expendi-

tures (CAPEX) are focused on bringing projects that are currently under construction to completion' (PetroLMI 2016, 5).

18 Early in 2010 the federal and Alberta governments 'struck up a secret, high-level committee ... to coordinate the promotion of the oilsands with Canada's most powerful industry lobby group [the CAPP]' (Martin Lukacs, 'Alberta, Ottawa, Oil Lobby Formed Secret Committee,' *Toronto Star*, 12 March 2012, http://www.thestar.com/news/canada/2012/03/12/alberta_ottawa_oil_lobby_formed_secret_committee.html).

19 This document was available at http://bit.ly/1La1sGx as of 18 February 2015.

20 See, for example, the 'Oil Sands' website of the Government of Alberta, http://www.oilsands.alberta.ca/ghg.html (accessed 24 March 2013). Federal Environment Minister Peter Kent characterized the oil sands' GHGs emissions as being insignificant in Steven Chase, 'Peter Kent's Green Agenda: Clean Up Oil Sands' Dirty Reputation,' *Globe and Mail*, 6 January 2011, http://www.theglobeandmail.com/news/politics/peter-kents-green-agenda-clean-up-oil-sands-dirty-reputation/article560974/.

21 Tzeporah Berman, an environmental author and co-founder of ForestEthics, Sarah Winterton, acting executive director of Evironmental Defence, Steven Guilbeault, deputy director of Equiterre, and Ben West, oil-sands campaign director of ForestEthics Advocacy, co-signatories of an article published in the *Globe and Mail* on 19 February 2013, http://www.theglobeandmail.com/commentary/washington-is-right-canada-must-confront-its-climate-neglect/article8798658/ (accessed 24 March 2013).

22 The Prentice government relabelled Alberta's system a 'carbon levy system,' which is probably a more accurate description of the program than 'cap and trade.'

23 http://www.nytimes.com/2012/05/10/opinion/game-over-for-the-climate.html.

24 For details of the NDP government's proposals to date, see its 'Climate Leadership' webpage, http://www.alberta.ca/climate.cfm.

REFERENCES

Adachi, Rod, Wendy Armstrong, Arlene Chapman, Greg Flanagan, Jason Foster, Tom Fuller, Trevor Harrison, Tim Johnston, Mel McMillan, and Dean Neu. 2002. *Advantaged No More: How Low Taxes Flattened Alberta's Future*. 18 March. Edmonton: Parkland Institute.

Adkin, Laurie. 2014. 'The End of Alison: The Multiple Meanings of the Redford Resignation.' *Alberta Views*, 17, no. 4. https://albertaviews.ab.ca/2014/05/27/the-end-of-alison-2/.

Alberta Climate Leadership Panel. 2015. *Climate Leadership: Report to Minister*. 20 November. http://www.alberta.ca/documents/climate/climate-leadership-report-to-minister.pdf.

Alberta Federation of Labour (AFL). 2007. 'AFL Sets Up Temporary Foreign Worker Advocate Office.' Media release, 3 May. http://www.afl.org/index.php/Press-Release/aflsetsuptemporaryforeign.html.

Alberta Federation of Labour (AFL). 2011. 'Failing to Save for Our Future: Heritage Fund Hampered by Backwards Financial Planning.' 1 January. http://www.afl.org/index.php/January-2011/failing-to-save-for-our-future-heritage-fund-hampered-by-backwards-financial-planning.html.

Alberta Royalties Review Panel. 2007. *Our Fair Share*. Final Report. 18 September. Edmonton: Government of Alberta.

Anielski, Mark. 2007. *The Economics of Happiness: Building Genuine Wealth*. Gabriola Island, BC: New Society.

Beltrame, Julian. 2012. 'Thomas Mulcair's Dutch Disease Warning Supported by OECD Report.' *National Post*, 13 June. http://news.nationalpost.com/2012/06/13/thomas-mulcairs-dutch-disease-warning-supported-by-oecd-report/.

Berdahl, Loleen. 2006. 'What Do Albertans Think? Saving and Alberta Public Opinion.' In *Seizing Today and Tomorrow: An Investment Strategy for Alberta's Future*, ed. Roger Gibbins and Robert Roach, 25–36. Calgary: Canada West Foundation.

Boychuk, Regan. 2010. 'Misplaced Generosity: Extraordinary Profits in Alberta's Oil and Gas Industry.' 25 November. Edmonton: Parkland Institute.

Bramley, Matthew, Marc Huot, Simon Dyer, and Matt Horne. 2011. *Responsible Action? An Assessment of Alberta's Greenhouse Gas Policies*. December. Drayton Valley, AB: Pembina Institute.

Busby, Colin, Benjamin Dachis, and Bev Dahlby. 2011. 'Rethinking Royalty Rates: Why There Is a Better Way to Tax Oil and Gas Development.' C.D. Howe Institute commentary no. 333, September. http://www.cdhowe.org/pdf/commentary_333.pdf.

Campanella, David. 2012. *Misplaced Generosity: Update 2012: Extraordinary Profits in Alberta's Oil and Gas Industry*. 15 March. Edmonton: Parkland Institute.

Campanella, David, and Shannon Bower. 2013. *Taking the Reins: The Case for Slowing Alberta's Bitumen Production*. April. Edmonton: Parkland Institute.

Campbell, Bruce. 2013. *The Petro-Path Not Taken: Comparing Norway with Canada and Alberta's Management of Petroleum Wealth*. January. Ottawa: Canadian Centre for Policy Alternatives.

Cosh, Colby. 2012. 'Cue the Outrage. Not.' *Maclean's*, 125, no. 7 (27 February), 23–4.

Dabbs, Frank. 2006. 'Ralph Klein's Real Legacy.' *Alberta Views*, September, 25–9.

Daly, Herman. 1996. *Beyond Growth: The Economics of Sustainable Development*. Boston: Beacon Press.

Dorow, Sara. 2015. "Gendering Energy Extraction in Fort McMurray.' In *Alberta Oil and the Future of Democracy in Canada*, ed. Meenal Shrivastava and Lorna Stefanick, 275–94. Edmonton: University of Athabasca Press.

Flanagan, Greg. 2011. *Fixing What's Broken: Fair and Sustainable Solutions to Alberta's Revenue Problems*. Edmonton: Parkland Institute.

Flanagan, Greg, and Shannon Stunden Bower. 2014. *The Way Forward: Progressive Income Tax in Alberta*. March. Edmonton: Parkland Institute.

Geels, Frank. 2014. 'Regime Resistance against Low-Carbon Transitions: Introducing Politics and Power in the Multi-Level Perspective.' *Theory, Culture & Society* 31 (5): 21–40.

Gibbins, Roger, and Robert Roach, eds. 2006. *Seizing Today And Tomorrow: An Investment Strategy for Alberta's Future*. Calgary: Canada West Foundation.

Goldberg, Ellis, Erik Wibbels, and Eric Mvukiyehe. 2008. 'Lessons from Strange Cases: Democracy, Development, and the Resource Curse in the U.S. States.' *Comparative Political Studies* 41 (4/5): 477–514.

Government of Alberta. 2011. *Highlights of the Alberta Economy 2011*. December. Edmonton: Government of Alberta.

Government of Alberta. 2012a. *Oil Sands Economic Benefits*. http://www.oilsands.alberta.ca/FactSheets/Economic_FSht_May_2012_Online.pdf.

Government of Alberta. 2012b. *Alberta's Energy Industry: an Overview*. http://www.energy.alberta.ca/Org/pdfs/Alberta_Energy_Overview.pdf.

Government of Alberta. 2015. *Alberta's Fiscal Situation: Budget 2015 and Beyond. What We Heard (Public Consultations December 2014–March 2015)*. c. March. http://finance.alberta.ca/publications/budget/budget2015/Budget-2015-Consultation-Report.pdf.

Government of Alberta. Treasury Board and Finance. 2016. 'Labour Market Notes.' 11 March. http://www.finance.alberta.ca/aboutalberta/labour-market-notes/current-labour-market-notes.pdf.

Hodgson, Glen. 2013. 'Alberta's Fiscal Model Needs a Fresh Re-think.' 23 January. Conference Board of Canada. http://www.conferenceboard.ca/economics/hot_eco_topics/default/13S01S22/alberta_s_fiscal_model_needs_a_fresh_reSthink.aspx.

Homer-Dixon, Thomas. 2013. 'The Tar Sands Disaster.' *New York Times*, op-ed, 31 March. http://www.nytimes.com/2013/04/01/opinion/the-tar-sands-disaster.html?_r=0.

Institute for Public Economics. 2013. *An Examination of Alberta Labour Markets.* July. Edmonton: Department of Economics, University of Alberta.

Kleiss, Karen. 2013. 'Alberta to Open Strategic Offices in India, Singapore.' *Calgary Herald,* 14 March. html#ixzz2OVOmU9DJhttp://www.calgaryherald.com/news/alberta/Alberta+open+strategic+offices+India+Singapore/8094384/story.html#ixzz2OVOHIdxJ.

Krahn, Harvey. 2005. 'Save or Spend? Albertans' Preferences Regarding the Year-End Surplus.' Investing Wisely Project discussion paper series no. 2. October. Calgary: Canada West Foundation. http://cwf.ca/pdf-docs/publications/October2005-Save-or-Spend-Albertans%E2%80%99-Preferences-Regarding-the-Year-End-Surplus.pdf.

Lahey, Kathleen. 2015. *The Alberta Disadvantage: Gender, Taxation, and Income Inequality*. March. Edmonton: Parkland Institute. http://s3-us-west-2.amazonaws.com/parkland-research-pdfs/thealbertadisadvantage.pdf.

Landon, Stuart, and Constance Smith. 2010. 'Energy Prices and Alberta Government Revenue Volatility.' C.D. Howe Institute commentary no. 313. November. Toronto: C.D. Howe Institute.

Lauletta, Kelli. 2014. 'Oil and Gas Companies Increasing Benefits, Bonuses.' *Oil Online,* 14 July. http://www.oilonline.com/industry-news/workforce-trends/oil-gas-companies-increasing-benefits-bonuses.

Lemphers, Nathan, and Dan Woynillowicz. 2012. *In the Shadow of the Boom: How Oilsands Development is Reshaping Canada's Economy*. Drayton Valley: Pembina Institute.

Macdonald, Alicia, and Todd Crawford. 2013. *Opportunity Lost? Alberta Is Facing Short- and Long-Term Financial Challenges Despite Its Oil Wealth.* February. Ottawa: Conference Board of Canada.

Mansell, Robert, and Ron Schlenker. 2006. *Energy and the Alberta Economy: Past and Future Impacts and Implications*. Paper no. 1 of the Alberta Energy Futures Project, Institute for Sustainable Energy, Environment and Economy, 15 December. Calgary: University of Calgary.

McGlade, Christophe, and Paul Ekins. 2015. 'The Geographical Distribution of Fossil Fuels Unused when Limiting Global Warming to 2C.' *Nature* 517 (8): 187–202.

Mcmillan, Mel. 2006. 'Investing for Alberta's Future: Improving the Use of Non-Renewable Resource Revenue in a Resource Rich Province.' In *Seizing Today and Tomorrow: An Investment Strategy for Alberta's Future*, ed. Roger Gibbins and Robert Roach, 73–82. Calgary: Canada West Foundation.

Miller, Byron. 2013. 'Spatialities of Mobilization: Building and Breaking Relationships.' In *Spaces of Contention: Spatialities and Social Movements*, ed. Walter Nicholls, Byron Miller, and Justin Beaumont, 285–98. Farnham, UK/Burlington, USA: Ashgate.

Mintz, Jack. 2013. 'Don't Count on Oil.' *Financial Post*, 21 January. http://opinion.financialpost.com/2013/01/21/jackSmintzSdontScountSonSoil/.

Mitchell, Timothy. 2011. *Carbon Democracy: Political Power in the Age of Oil*. London: Verso.

Mumey, Glen, and Joseph Osterman. 1990. 'Alberta Heritage Fund: Measuring Value and Achievement.' *Canadian Public Policy* 16 (1): 29–50.

Palmater, Pamela. 2013. 'Idle No More: What Do We Want and Where Are We Headed?' *rabble.ca*, 4 January. http://rabble.ca/blogs/bloggers/pamela-palmater/2013/01/what-idle-no-more-movement-really#.UOdLK-5zFMY.facebook (accessed 8 January 2013).

Parkland Institute and Canadian Centre for Policy Alternatives. 2013. 'Alberta Is Canada's Most Unequal Province, and Calgary the Most Unequal City.' Media release, 28 January. http://parklandinstitute.ca/media/comments/alberta_is_canadas_most_unequal_province/.

PetroLMI. 2016. *Labour Demand Outlook to 2020: Oil Sands Construction, Maintenance and Operations*. February. Calgary: PetroLMI/ENFORM. http://www.careersinoilandgas.com/media/243763/oil_sands_outlook_online.pdf

Plourde, André. 2009. 'Oil Sands Royalties and Taxes in Alberta: An Assessment of Key Developments since the Mid-1990s.' *Energy Journal* 30 (1): 111–39.

Plourde, André. 2010. 'On Properties of Royalty and Tax Regimes in Alberta's Oil Sands.' Energy Policy 38: 4652–62.

Roy, Jim. 2015. 'Billions Forgone: The Decline in Alberta Oil and Gas Royalties.' Fact sheet, 23 April. Edmonton: Parkland Institute. http://parklandinstitute.ca/research/summary/billions_forgone (accessed 4 May 2015).

Sayers, Anthony M., and David K. Stewart. 2011. 'Is This the End of the Tory Dynasty? The Wildrose Alliance in Alberta Politics.' School of Public Policy research papers, vol. 4, no. 6, May. University of Calgary.

Schiell, Leslie, and Colin Busby. 2008. 'Greater Saving Required: How Alberta Can Achieve Fiscal Sustainability from its Resource Revenues.' C.D. Howe Institute commentary no. 263, May. Toronto: C.D. Howe Institute.

Scoffield, Heather. 2013. 'Harper Names Envoy for First Nations Concerns on Pipelines and Energy.' *Globe and Mail*, 19 March. http://www.theglobeandmail.com/news/national/harper-names-envoy-for-first-nations-concerns-on-pipelines-and-energy/article9945566/.

Statistics Canada. 2015. 'Median Total Income, by Family Type, by Province and Territory (All Census Families).' CANSIM, table 111-0009 (last modified 06-26 June 2015). http://www.statcan.gc.ca/tables-tableaux/sum-som/l01/cst01/famil108a-eng.htm.

Statistics Canada. 2016. 'Employment by Age, Sex, Type of Work, Class of Worker and Province (Monthly) (Alberta).' Data from tables 282-0087 and 282-0089 (last modified 11 March 2016). http://www.statcan.gc.ca/tables-tableaux/sum-som/l01/cst01/labr66j-eng.htm.

Stunden Bower, Shannon, Trevor Harrison, and Greg Flanagan. 2013. *Stabilizing Alberta's Revenues: A Common Sense Approach*. February. Edmonton: Parkland Institute.

Taylor, Amy. 2006. 'Klein Shortchanging Albertans and Putting Environment at Risk.' *Edmonton Journal*, 25 March. http://www.pembina.org/op-ed/1217.

Taylor, Amy, Chris Severson-Baker, Dan Woynillowicz, Mark S. Winfield, and Mary Griffiths. 2004. *When the Government Is the Landlord*. July 1. Drayton Valley, AB: Pembina Institute. http://www.pembina.org/pub/171.

Thompson, David. 2008. 'Saving for the Future: Fiscal Responsibility and Budget Discipline in Alberta.' April. Edmonton: Parkland Institute.

Thompson, David. 2009. *Green Jobs: It's Time to Build Alberta's Future*. Edmonton: Sierra Club Prairie Chapter/Greenpeace/Alberta Federation of Labour. http://www.sierraclub.ca/national/media/green-jobs.pdf.

Weir, Erin. 2010. 'Alberta Royalty Cuts Set a Dangerous Precedent.' *Globe and Mail*, 7 April. http://www.theglobeandmail.com/report-on-business/industry-news/energy-and-resources/alberta-royalty-cuts-set-a-dangerous-precedent/article1210370/.

Woynillowicz, Dan, and Nathan Lemphers. 2012. *In the Shadow of the Boom: How Oilsands Development is Reshaping Canada's Economy*. 30 May. Drayton Valley, AB: Pembina Institute.

Younglai, Rachelle. 2016. 'Alberta's Blue-Collar Workers Worst Off in Oil Slump.' *Globe and Mail*, January 17.

17 Democracy and the Albertan Petro-State

LAURIE E. ADKIN

A number of commentators have in recent years labelled Alberta a petro-state with the aim of drawing attention to the ruling party's disregard for political accountability and transparency, or for serving the public interest as defined by citizens themselves. The well-known Canadian journalist Andrew Nikiforuk, in particular, has referred to both the Albertan and recent Canadian governments as petro-states or as 'petro-tyrannies.' In his 2008 book *Tar Sands: Dirty Oil and the Future of a Continent*, Nikiforuk argues that trends in Alberta and Canada exemplify the 'First Law of Petropolitics,' that is, that 'the price of oil and the quality of freedom invariably travel in opposite directions' in an oil-dominated country (153). 'Neither Texans nor Canadians are exempt,' says Nikiforuk, from the inverse relationship between oil wealth and democracy (153).

'The First Law of Petropolitics' is a phrase associated with the American author Thomas L. Friedman, who wrote in 2006 that

> the higher the average global crude oil price rises, the more free speech, free press, free and fair elections, an independent judiciary, the rule of law, and independent political parties are eroded. And these negative trends are reinforced by the fact that the higher the price goes, the less petrolist leaders are sensitive to what the world thinks or says about them. Conversely – according to the First Law of Petropolitics – the lower the price of oil, the more petrolist countries are forced to move towards a political system and a society that is [*sic*] more transparent, more sensitive to opposition voices, and more focused on building the legal and educational structures that will maximize their people's ability – both men's and women's – to compete, start new companies, and attract investments from

> abroad. The lower the price of crude oil falls, the more petrolist leaders are sensitive to what outside forces think of them. (31)

Friedman, in turn, was drawing upon the work of political scientist Michael L. Ross, who used data from 1971 to 1997 on 113 states to track correlations among economic and democratic variables (Ross 2001). Ross' statistical experiments showed that dependence on rent-generating oil exports does have a negative impact on democratic indicators (this influence was even stronger for non-fuel mineral exports); however, he also found that 'oil and mineral wealth cause greater damage to democracy in poor countries than in rich ones' (343), and historical and cultural factors remain highly significant in explaining differences in regime types (345–6). Ross then asked what the causal links between oil wealth and quality of democracy might be. He examined a number of possible causes of authoritarian tendencies in oil-exporting states. These included

> a 'rentier effect,' which suggests that resource-rich governments use low tax rates and patronage to relieve pressures for greater accountability; a 'repression effect,' which argues that resource wealth retards democratization by enabling governments to boost their funding for internal security; and a 'modernization effect,' which holds that growth based on the export of oil and minerals fails to bring about the social and cultural changes that tend to produce democratic government. (328)

Ross breaks the 'rentier effect' down into three further subtheses: the 'taxation effect,' the 'spending effect,' and the 'group formation effect.' The taxation effect suggests that 'when governments derive sufficient revenues from the sale of oil, they are likely to tax their populations less heavily or not at all, and the public in turn will be less likely to demand accountability from – and representation in – their government' (332). The 'spending effect' refers to the ability of the ruling party to dispense patronage, thereby manipulating or conditioning the behaviour of actors in civil society. 'Oil wealth,' says Ross, 'may lead to greater spending on patronage, which in turn dampens latent pressures for democratization' (333). Closely related to the 'spending effect' is the 'group formation effect.' According to Ross, this 'implies that when oil revenues provide a government with enough money, the government will use its largesse to prevent the formation of social groups that are independent from the state and hence that may be inclined to demand political rights' (334).

Ross updated his research on the politics of oil-exporting states in a 2012 book, *The Oil Curse: How Petroleum Wealth Shapes the Development of Nations*. He reports that by 2006 there were fifty-seven oil-producing countries (defined as states that generate at least a hundred dollars per capita from oil and gas in a given year), as compared to only a dozen in 1960 (10).[1] Looking for correlations among variables for data from 170 countries (over a period of fifty years), Ross finds statistical evidence for a number of relationships. 'Today,' he says, compared to the years before 1980, 'the oil states are 50 percent more likely to be ruled by autocrats and more than twice as likely to have civil wars as the non-oil states. They are also more secretive, more financially volatile, and provide women with fewer economic and political opportunities' (1–2). The countries where these characteristics are most pronounced are in the Middle East, and the turning point, he proposes, was the sudden rise and increased volatility of oil prices in the 1970s, along with the nationalization of oil companies in many third world countries at that time. Once ruling elites had direct access to, and control over, enormous oil revenues, they were empowered to govern in ways that were less accountable to citizens, more repressive of opposition, and made use of patronage to secure consent or quiescence.

According to Ross, petroleum revenue bestows such powers upon rulers because of its particular qualities: its scale, source, (in)stability, and secrecy (5). First, 'the scale of oil revenues can be massive,' enabling governments to 'silence dissent.' The availability of such wealth may also provoke uprisings by sections of the population that feel they are not getting their fair share. Second, oil revenue permits governments to rely far less on taxation to finance their projects and priorities. Here, Ross repeats an argument found in the earlier petro-state literature, to the effect that oil revenue insulates ruling elites from 'public pressure,' or accountability to citizen-taxpayers. The actual connection here between the spending of revenue from the people's resource rents (in lieu of revenue derived from income and other taxes) and government's release from accountability to citizens is not specified, and seems to assume a lot about the institutional context (e.g., that there are no elections through which citizens can hold elites accountable for their spending of oil revenues, no independent media to inform citizens, etc.). One would also expect citizens to be concerned about accountability if it is apparent that the elite is 'squandering' oil wealth. Third, changes in world oil prices or in a country's oil reserves (rate of production) can result in sudden and major fluctuations in government revenue, which Ross and others suggest may explain why the governments of oil states 'frequently squander their resource wealth.' Again, the steps that lead to this outcome are not spelled out and seem to be assumed. There are measures governments can take to stabilize their revenues and plan their spending; the cyclical availability of oil wealth does not seem to be a sufficient explanation for why governments may 'squander' it. The fourth quality of oil revenue that Ross considers to be significant and distinct from other kinds of revenue is its 'secrecy.' By this Ross means that 'governments often collude with international oil companies to conceal their transactions, and use their own national oil companies to hide both revenues and expenditures' (6).

Other characteristics of oil production and revenue that have been observed by economists are also mentioned by Ross. These include the capital-intensive nature of oil extraction, which means that it generates relatively few employment benefits for the communities in which it is located, but on the other hand may burden these communities with large social and environmental externalities. While this is often the case with regard to conventional oil extraction, as well as off-shore operations, some of these assumptions do not hold for the kind of mega-scale bitumen mining that takes place in Alberta. In particular, due to the massive scale of operations in the Alberta oil sands, labour shortages have been generated throughout the province for certain categories of workers. It is true, on the other hand, that local communities like Fort McMurray and Aboriginal communities in the oil sands region or downstream of it have suffered many negative social and environmental effects from the extraction booms. Ross also mentions the effects of oil export revenues on a country's exchange rates (referred to in earlier chapters in this volume as 'Dutch Disease'), noting that these may 'reduce the size of manufacturing and agricultural sectors.' In some contexts (depending upon the sexual division of labour), this can bring about loss of employment opportunities for women. However, he credits the four properties of oil revenues described above (large scale; source being rent, rather than taxation; instability; secrecy or non-transparency of where the money goes) as being the most significant causes of the 'conflict-ridden, autocratic' nature of governance in oil states.

One sees substantial agreement between the causal explanations Ross offers for his statistical findings in 2012 and in 2001. He admits that the correlations he finds are most evident in cases in 'the developing world' (although there are significant variations among these countries), and that 'not all states with oil are susceptible to the curse' (2012, 2).

> Countries like Norway, Canada, and Great Britain, which have high incomes, diversified economies, and strong democratic institutions, have extracted lots of oil and had few ill effects. The United States – which for much of its history has been both the world's leading oil producer and the world's leading oil consumer – has also been an exception in most ways. Petroleum wealth is overwhelmingly a problem for low-and middle-income countries, not rich, industrialized ones. (2)

Yet (putting aside questions about what constitutes 'strong democratic institutions') this statement suggests alternative explanations for the differences in how oil states have managed revenue and made decisions about developmental paths. That is, it seems to be the nature of the *political-economic structures* (industrialized versus agrarian; diversified versus staples dependent) or the *political institutions* (strongly democratic or weakly democratic) that existed *prior to* the discovery of oil resources that explain the governance and developmental outcomes. The questions that then arise are: What factors determined the nature of these structures or institutions? And, despite the democratically corrosive effects of large-scale resource rents, what factors maintain the relative stability of 'strong democratic institutions' in certain countries?

One further observation regarding the above quotation from Ross is that it exemplifies the narrow focus of the petro-state literature on selected indicators of democracy, in relation to oil revenue or prices, while ignoring the ecological implications of petro-state governance. Reliance upon the extraction of fossil fuels and the revenues earned from their export and consumption is indeed a 'problem' for oil-producing and -exporting states, because it locks them into a model of development that is ecologically destructive at multiple scales. The macro scale of analysis has the further limitation of obscuring or minimizing the socially inequitable effects of oil production *within* 'rich industrialized' countries. The contributions to this book have identified such inequities in the Alberta context.

Earlier work by political scientist Terry Lynn Karl (1997) compared a number of cases (in particular, Venezuela, Iran, Nigeria, Algeria, and Indonesia) to see whether oil dependence could explain authoritarianism in petro-states.[2] She concluded that 'dependence on a leading export commodity has a profound impact on contractual relations, property rights, the relative importance of markets versus states, the degree of internationalization of the economy, the opportunities for technological innovation, the relative power of organized interests, the structure of taxation, the prerogatives accorded different state agencies, and the symbolic content of the state' (239).

On the other hand, Karl recognized that there was a significant exception to most generalizations about the characteristics of petro-states: Norway. The Norwegian exception, along with other differences among her cases, led her to argue also that the governance directions that would be taken by petro-states following the discovery of their oil wealth would depend *upon the character of the political institutions that were already in place* (236–42). She attributes the comparative wisdom with which Norwegian governments have managed oil revenues to that country's highly professional, autonomous, and efficient civil service and strong mechanisms of government accountability dating back to the eighteenth century (213–21). Norway's policy environment was exceptional, Karl argues, in that it 'emphasized caution in the face of change, respect for standard operating procedures, segmentation according to issue area, consensus building, and egalitarianism' (217). Thus, when oil and gas fields were discovered, 'in contrast to all other oil-exporting countries, multinational companies were forced to bargain with the representatives of a highly developed state bureaucracy who felt no strong need for a qualitatively new revenue base. Organizing a framework for controlling the oil industry required a high degree of sophistication in planning and administration, which Norway, unlike other oil exporters, possessed in abundance' (217).

Although it may be tempting to latch on to oil revenue as an all-encompassing explanation for a range of authoritarian practices, political-institutional differences among oil-exporting jurisdictions tell us that far more complex explanations are required to account for the nature of state-society relations in any particular case. Friedman's widely influential 2006 article in *Foreign Policy* elevated the general observation that many oil-rich states have undemocratic regimes to a 'law' that reduces complex political phenomena to the price of oil. In his view, 'one can actually correlate rises and falls in the price of oil with rises and falls in the pace of freedom in petrolist countries' (2006, 32). Friedman further claims that 'geology trumps ideology' as an explanation for regime types (35). Yet in the same article he acknowledges that only certain oil-producing countries really fit into his category of 'petrolist states.' In parentheses, he says: 'Countries that have a lot of crude oil but were well-established states, with solid democratic institutions and diversified economies before their oil was discovered – Britain, Norway, the United States, for example – would not be subject to the First Law of Petropolitics' (31).

Democracy and Oil in Alberta

Friedman's claims – like Ross's – are questionable on a number of grounds, starting with the criteria used to classify states by their 'democratic' or 'authoritarian' institutions. As mentioned in the first chapter of this book, a small number of researchers have begun to investigate the 'petrolist' properties of jurisdictions within North America that Friedman considered to be 'democratic' and 'diversified.' When it comes to Alberta, Andrew Nikiforuk (2008), among others, identifies many twists and turns in the politics of Alberta as well as the federal Conservative governments (since 2006) that may be interpreted as trending towards authoritarianism (lack of transparency and accountability to the public, privileging of corporate interests over those of democracy and environmental sustainability). Meenal Shrivastana and Lorna Stefanick conclude from their 2012 study of politics in Alberta, using comparisons to Canada as a whole, that 'the negative effects of oil on democratic development is [*sic*] not confined to the Global South' (20). These authors see some grounds for viewing Alberta – at least within the Canadian context – as having an 'exceptional' democratic deficit. They suggest that even entrenched liberal democratic institutions may be eroded by a state's heavy reliance on oil rents.

Political scientist Ian Urquhart, while vehemently rejecting the characterization of Alberta as a petro-state in any way comparable to others in that category, such as Iran, Nigeria, or Venezuela, nevertheless agrees that Alberta's governments have actively privileged the interests of the oil companies (2010).

> In spite of our high ranking as a free country, paranoia seems to animate how Alberta's governing party runs our affairs. Dissent is a four-letter word; the Progressive Conservatives frown upon actions that might inform or facilitate dissent, such as more transparency in governing, easier access to public information and more opportunities for citizen participation in administrative affairs. Government secrecy is impossible to understand without seeing it as an effort to protect petroleum in Alberta. (2010, 28)

Other chapters in this volume provide many examples of the antidemocratic nature of the Albertan and Canadian petro-states. In the Alberta case, we see how successive governments have systematically disregarded public opinion, notwithstanding their populist rhetoric and claim to be defending the interests of all Albertans. This includes public opinion about the province's share of resource rents and the need for fiscal restructuring. The evidence strongly supports the conclusion that Alberta's governments have failed to distinguish the interests of large energy corporations in the accumulation of profits from the conditions for citizens' well-being. As Larry Pratt recounted in *The Tar Sands: Syncrude and the Politics of Oil* (1976), this trend began with the Social Credit governments in the 1950s, soon after the discovery of Alberta's large oil reserves; it continued under the Conservatives, notwithstanding their initial 'province-building' impetus (see also Brownsey 2005).

While Pratt, as a political economist, explained the numerous ways in which the rapid development of the tar sands – more or less handed over to multinational oil corporations by the Conservative government in the 1970s – conflicted with Albertans' interests in sovereignty over the pace and terms of resource development, an environmental lawyer provided examples of the government's overriding of public opinion concerning development of the resource during the same period. Philip Elder explained in a 1974 article that the Environmental Conservation Authority (ECA) created by the Social Credit government in 1970 was initially given a wide mandate to review policies and hold public inquiries. However, these powers were reined in by the Conservative government in its amendments to the Environment Conservation Act in 1972 (Elder 1974, 410). Thereafter, institutions would be reformed as necessary to protect the imperatives of fossil-fuel extraction against too much democracy.

A public advisory committee appointed by the ECA, comprised of representatives from sixty-five 'organizations, institutions, and groups,' held a meeting in 1973 at which it took positions on the tar sands that might sound familiar to today's opponents of the expansion of bitumen mining:

> For a start, it passed resolutions which called for, among other things, more research into the environmental problems of developing the Athabasca tar sands; the release of the civil service report which called for a slowdown in tar sands development until technological and environmental problems have been solved; funding for urban transit and waste recycling programs; and legislation to discourage wasteful consumption of energy.
>
> During the same meeting, members of a panel roundly attacked the proposed development of the Athabasca tar sands from political, technological, and environmental points of view. On the latter two points, the

primary misgiving was the overwhelming lack of knowledge. (Elder 1974, 413, as reported in the *Calgary Herald* on 7, 8 December 1973)

Citizens were particularly concerned about lack of environmental impact data for the Syncrude project, which had already been approved by the Department of Environment. Elder commented:

> The DOE has certified to the ERCB that the project will not breach any Alberta pollution laws. One wonders to whom the government listens, if a consultant's report and a confidential report of an interdepartmental committee of civil servants to the same effect are both ignored. If it can state with confidence that provincial pollution standards will not be breached, while lacking an in-depth study of possible environmental impact of a massive project, surely the government's fundamental philosophy is wrong (427).

Elder also provided an analysis of public participation under the terms of the Energy Resources Conservation Board (ERCB), concluding, for a number of reasons, that 'the ERCB hearing does not offer much scope for the airing of environmental concerns' (425).[3] Subsequent studies of environmental regulation in Alberta have documented a similar pattern, in which the 'public interest' is routinely equated with the interests of the resource-extractive industries, even when clear conflicts are evident or have been expressed through consultation processes.[4] In other policy domains, too, critics have identified many contradictions in the hegemonic or 'official' discourse of governments and business associations, which represents the privatization of public goods or the low royalty and corporate tax rates as 'advantages' or guarantees of security and prosperity for the people of Alberta. (Many examples were provided in Adkin and Miller, this volume; see also Davidson and Gismondi 2011). Despite the never-ending stream of 'public consultations' initiated by the Government of Alberta, these have been largely exercises in legitimation (and information gathering for the government) rather than vehicles of meaningful citizen participation in policy making.

The model of economic development in Alberta has never been egalitarian, of course. Throughout this book (and particularly in chapter 16) references have been made to the regressive nature of the flat-rate income tax, low corporate tax rates, and the loss of enormous (foregone) resource revenue that might have been invested in transition to a low-carbon economy that could provide good, sustainable livelihoods for the province's residents, along with more universal public goods (such as early childhood education and care, home care, and other services). According to Statistics Canada economists Philip Cross and Geoff Bowlby, profits in Alberta more than doubled from $23.5 billion in 2002 to $53.1 billion in 2005, directly accounting for over half of all income growth (GDP growth) in the province. 'Most of this increase,' the authors stated, 'reflects the soaring price of oil and gas exports' (2006, 3.2). Researchers at the Parkland Institute calculate that 'the inflation-adjusted average income for the bottom 99% of income earners in the province increased a modest 13% between 1982 and 2011, from $41,749 to $48,800. Meanwhile, the average incomes of the top 1% and top 0.1% grew 93% and 149%, respectively' (Campanella et al. 2014, 32).

Moreover, the privileging of investment in non-renewable resource extraction has deepened gender inequality (Lahey 2015). Oil sands expansion has generated demand for skilled trades, construction, and equipment operator jobs that fall largely to men – drawing young men out of high school and bringing in a predominantly male mobile labour force from other provinces (Paskey 2011; Toneguzzi 2012). Workers in these sectors earn comparatively high wages. In June 2014 Statistics Canada reported that Albertan non-farm workers had the highest average weekly wages in Canada, at $1,141.70 (Toneguzzi 2014). But such averages obscure the real polarization happening in the labour market. When one takes into account that Alberta also has the lowest minimum wage in the country, and that wages are often at this level in the service sector, where women predominate, it is not surprising that in the Wood Buffalo municipality (where Fort McMurray is located) the median earning for males in 2006 was $78,761, while for females it was $26,111 (Davidson and Gismondi 2014, 30).

In regard to relations with Aboriginal peoples, Alberta governments have a long history of bulldozing over land claims and through consultation processes in order to promote resource 'development.' The Lougheed governments' promotion of oil and gas development as well as other resource-based industries throughout the 1970s and 1980s accelerated the process of dispossession of First Nations (Martin-Hill 2004, 2008; Price 1977).[5] This is not to say that certain bands have not made contractual agreements of various kinds with corporations to wrest benefits from resource 'development,' as discussed in the chapters in this book by Parlee and Zalik (see also Altamirano-Jiménez 2004; Slowey 2008; Slowey and Stefanick 2015; Taylor and Friedel 2011). Yet, given the enormous pressures upon Aboriginal peoples, the absence of a right to 'say no' to development (see Stendie and Adkin, this volume), and the costs involved in engagement in consultations and judicial challenges, questions remain about the meaning of free consent.

A comprehensive discussion of Alberta's political system, democratic rights and freedoms, and First Nations sovereignty rights is beyond the scope of this chapter; its more limited objective is to consider the usefulness of petro-state theses for understanding the political ecology and governance of Alberta. While it is evident that Alberta is not (for reasons having to do with history, geography, and its position within the global economy) Saudi Arabia or Nigeria, it is perfectly legitimate and important to ask how oil interests and political hegemony have become entwined in Alberta. Understanding this reality is essential to mapping paths to a more ecologically sustainable and democratic future.

The election of a social democratic party to government in May 2015 has been read by many as promising democratic reforms, and indeed, a number of such reforms have been implemented at the time of writing. One important step was Bill 1, An Act to Renew Democracy in Alberta, introduced in the first sitting of the legislature and passed in June 2015. The act bans corporate and union donations to political parties. The Minister of Justice, in speaking to the act, said, "It puts the power back in the hands of Alberta citizens rather than those with the deepest pockets in terms of determining the political future of this province" (Kathleen Ganley, quoted in Bennett 2015). The government has also appointed a panel to review the practices and criteria used in the appointment of individuals to public boards and agencies – historically a key lever for political patronage benefiting members and supporters of the Progressive Conservative Party. Notably, the NDP premier appointed a Minister Responsible for Democratic Renewal and established an all-party Select Special Ethics and Accountability Committee to review the Election Act and other legislation. Supporters of these steps have nevertheless pressed the new government to go much further than Bill 1 to democratize political processes and institutions (Cournoyer 2016; Public Interest Alberta 2016; Urquhart 2016). The reversion by the NDP government to status quo approaches to the organization of public consultations in several policy areas has disappointed many who hoped for new, more meaningful forms of citizen involvement in policy making. (See, for example, comments by political scientist David Kahane in Pratt 2015.)

None of these developments may be examined in detail in this chapter, due to their ongoing nature and to space limitations. However, through its examination of the anti-democratic practices and institutional norms that have characterized Alberta's petro-politics in recent decades, this chapter provides a roadmap to the kinds of reforms that civil society should demand of its new government. Indeed, the actions taken or not taken by the NDP government to democratize Alberta's political system and (as suggested in chapter 16) to transform its economic model of development constitute an 'experiment' of great interest to all first world jurisdictions confronted with the multiple crises of fossil capitalism. Given the powerful entrenched interests and the economic structures in place when the social democrats were elected in Alberta – not to mention the current recession brought about by low oil prices – how, and how far, will the new government proceed to mobilize public support for democratic and ecological reforms?

Disassembling Hegemonic Governance in the Alberta Context

Insofar as Alberta is characterized by 'democratic deficits,' how are these related to the large place of fossil-fuel extraction and export in its political economy? Given that Alberta is a 'first world' context with formally liberal democratic political institutions and a unique political culture, we can expect that its 'petro-politics' will differ from the kinds of cases that receive primary attention in the petro-state literature. Armed conflict provoked by oil extraction and revenue distribution like that evident in Nigeria and other contexts is not part of our picture. It is arguable, however, that a polarization has happened at the national level in Canada between, on the one hand, civil society actors prepared to practise civil disobedience (or even sabotage) to stop the expansion of fossil fuel production/consumption and, on the other hand, governments prepared to use legislative and police powers to suppress this opposition to their 'energy superpower' project (Cheung 2015).

In this section I consider some of the techniques of governance routinely used by Alberta's Progressive Conservative (PC) Party, highlighting the ways in which they are distinct from, or resemble, Ross's rentier effects. For example, the 'spending effect' identified by Ross is similar to the idea of patron-client practices, in which the ruling party dispenses (or withholds) patronage (appointments, contracts, grants, and capital spending), thereby manipulating or conditioning the behaviour of actors in civil society. It is important to remember, however, that the use of these 'techniques' is not solely explained by the province's reliance upon oil revenue, although each has been practised by the ruling elite in relation to the oil interests that predominate in Alberta's political economy. At the same time, governance techniques may assume particular forms because of the imperative to protect oil interests.

Techniques of Governance

- *disciplining* powers exercised through control over revenue and spending decisions and over the bureaucracy, as well as through legislation (e.g., the province's labour laws, federal immigration policy with regard to temporary foreign workers) and the judiciary (e.g., rulings on Aboriginal land claims and traditional or treaty rights);
- *design of institutions* (such as the electoral system, party financing rules) and decision-making processes (degree of government secrecy, creation and constitution of quasi-judicial regulatory bodies such as the Energy Resources Conservation Board or its successors) in such a way as to systematically protect and privilege the interests and priorities of the ruling party and its core constituency;
- *patron-client practices* (employing political control over budgetary resources, as well as regulatory decisions);
- *reproduction of a hegemonic 'common sense'* through discursive practices (public identification with particular values, symbols, and identities, such as what it means to be an Albertan; what constitutes the 'public interest'; framing critics as illegitimate outsiders; the withholding of information);
- *democratic legitimation* (e.g., occasional prosecutions of corporations for environmental or labour code violations, periodic elections, democratic rhetoric; public consultations);
- *normalization* of hegemonic knowledge and ideas through use of mainstream media (diffusing government-or-corporate-provided information and issue frames; spending on 'public information' campaigns, advertising, lobbying other governments);
- *(dis)organization of civil society* through practices such as the creation or de-funding of research bodies; the designation of 'legitimate' stakeholders/interests for membership of quasi non-governmental organizations (QUANGOs) that administer various functions for the government, or in the design of government-initiated public participation processes. These practices contribute (along with the

discursive strategies referred to above) to the construction of certain collective identities and prevent or disrupt the formation of others.

It should be noted that the same practices (e.g., the creation of a QUANGO) have multiple hegemonic functions, and so, in the following discussion of techniques, a particular action may be listed under more than one category. Also, the examples given are illustrative, rather than constituting a comprehensive account of hegemonic governance.

DISCIPLINING

The term 'disciplining' is more appropriate to the Alberta context than the term 'repression' as used by Ross, because for him the term refers to such practices as state-sponsored assassinations, firing on protestors, detention without trial, or the building up of armed forces. (Armed forces are a federal jurisdiction in Canada. We have seen their use in the suppression of First Nations' occupations and blockades.) The key point here is that overt force is not required to effectively 'discipline' or habituate a population to a highly passive form of citizenship. While policing is used in Canada to prevent and to contain street protests and to monitor and arrest activists, less visible forms of disciplining have predominated in 'normal' times.

The Klein years (1992–2006) were characterized by intolerance of dissent, with the government's actions against its critics (civil servants, academics, leaders of civil society organizations) including dismissals, sanctions imposed on civil servants, threatened de-funding, public denunciations, and curtailed legislative sessions (Adkin 1995; Dabbs 2006; Harrison and Laxer 1995; Lisac 1995; Taft 2007). Such tactics have continued under subsequent governments, particularly with regard to critics of the expansion of the oil sands (environmentalists, health professionals). In addition to making accusations of disloyalty, government discourse has occasionally associated social movement activists with threats to public order and safety, or even terrorism (Le Billon and Carter 2012).[6]

PC governments have become particularly vulnerable, since the mid-2000s, to criticisms of economic mismanagement, wasteful spending on perks for politicians, and corruption with regard to donations to the party. Following the 2008 provincial election, the chief electoral officer, Lorne Gibson, reported evidence of 'illegal campaign contributions and proposed more than 100 ways to run elections better' (Pratt 2010, 38). In an example of 'disciplining' practices, Progressive Conservative (government) members of the Standing Committee of Legislative Offices later unanimously opposed the chief electoral officer's reappointment. Two years later, Mr Gibson filed a law suit against the government of Alberta, claiming unfair dismissal and asking for $450,000, or roughly two years' salary and benefits.

Of course, beyond the role of governments in shaping the institutional, juridical, and fiscal conditions of political struggle, the broader political economy exercises its own powerful disciplining effects. The dependence (directly or indirectly) of many workers on revenue produced in the resource sectors, as Connie Davidsen and Adkin and Miller point out in their contributions to this volume, promotes forms of self-censorship, competition, misrepresentation, and instrumentalism. The potential is created for pitting workers in a resource-extraction-dominated private sector against workers in the public sector who can more easily envisage a transition to a model of economic development that relies on human services and information technologies, and is fuelled by non-fossil sources of energy. Even within institutions we can see such divides; the post-secondary education sector – which has been subjected to repeated, chaotic assaults by PC governments – tends to split between faculties that are deeply invested (and have been encouraged, by the structuring of research funding, to specialize) in technology development for fossil fuels exploitation, mining, forestry, and downstream industries and services, on the one hand, and faculties that prioritize social, cultural, and ecological knowledge and skills, on the other hand. Driven by competition for scarce resources, their division is destructive of relationships and forms of governance that value non-instrumental communication (dialogue), cooperation, or 'speaking truth to power.' The implantation of the energy sector's imperatives within Alberta's institutions of higher learning has made it all the more difficult for them to speak with one voice in defence of post-secondary education's mandate to serve the public good rather than narrow, market-driven interests.

INSTITUTIONAL DESIGN

While Alberta's Conservative governments have conducted a non-stop spectacle of 'consultations' with the public and selected representatives of stakeholder groups, their policy-making processes have in reality been secretive and heavily biased towards the interests of key stakeholders. Consistent with Ross's observation of the secrecy that characterizes policy making in third world petro-states, very little regarding the decision-making process in Alberta has been transparent. Commissioned research and the results of public consultations are often not made public, or only in mediated forms such as consultants' summaries of surveys or public input (at hearings or via websites) that are vetted by the relevant ministry before being published.

Various regulatory agencies, as well as ministries, of the provincial government have some form of mandate to consult or engage the public. The 1974 article by Phil Elder referred to above reviews a period when government was beginning to introduce public hearings, polling, or advisory committees as means of learning about public opinion. The scope for participation varied, with the Environmental Conservation Authority hearings permitting citizens to be 'heard fairly and fully' although they 'did not have a major innovative impact on public policy') (423), at one end of the scale, and the ERCB hearing – with its narrow criteria for participation ('any person who may be directly and adversely affected' by a proposed project) and heavy weighting towards decisions already taken by the Ministry of Environment – at the other end of the scale. Elder also characterized public participation in the Ministry of Environment's Citizen Advisory Committee on Resource Development as 'token' (423). The ECA, as Elder noted, was brought under closer supervision by the minister of the environment under the Lougheed governments, and had lost most of its autonomy by the time the Klein government announced its elimination in 1994. Notably, two keen observers of environmental policy in the province remarked at about this time that 'one of the striking features of the Klein revolution in Alberta is the increasing extent to which public participation in the province has become symbolic and manipulative' (McInnes and Urquhart 1995, 247). More recent research on environmental consultation processes in Alberta has drawn similar conclusions about the opportunities for citizens to meaningfully influence policy outcomes (Adkin 2014; Bowness and Hudson 2014; Fluet and Krogman 2009; Garvin, this volume; Hoberg and Phillips 2011; Urquhart 2005).

One documented example of government secrecy benefiting the oil and gas sector was provided in the 2006–07 Annual Report of Alberta's auditor general, Fred Dunn. In this report, the AG revealed that the Energy Ministry had kept from the public for at least three years research showing that the collection of royalties had fallen below the government's targeted share of rent, during which time it had failed to collect one billion dollars or more in revenue (Office of the Auditor General 2007). According to the AG, 'sound analysis of Albertans' most valuable physical asset does not appear to have led to timely action' (92).

The government might have had a reason for collecting less than its targeted rent, he remarked, but the public deserved to know what was going on. Journalist Sheila Pratt quoted Dunn as saying that the energy minister could 'raise or lower the 25 per cent revenue target. But he should make those decisions public. That's called being accountable' (quoted in Pratt 2010, 37). (Such problems had already been identified by the auditor general's office in 2001–02.)

Many questions were raised by this report, from the capacity of the Ministry to effectively monitor and collect the revenue owing to the public to the performance of the government as the representative of the owners of the resource. The lack of rigorous oversight of royalty collection and accounting certainly exemplifies the lack of transparency that Ross sees as a primary characteristic of petro-states, and that Pratt (1976), Urquhart (2010), and Adkin (2014), among others, have observed in relation to the protection of the interests of Alberta's fossil fuel industries. Mr Dunn's recommendations included that the Ministry of Energy 'clearly describe and publicly state the objectives and targets of Alberta's royalty regimes'; 'improve the planning, coverage, and internal reporting of its royalty review work'; 'improve its annual performance measures that indicate royalty regime results'; 'periodically report on the province's royalty regimes,' providing 'information to owners, MLAs, and stakeholders about the performance and issues for Alberta's royalty regimes'; and 'demonstrate the Department's capacity and methodology to analyze its royalty regimes' (Office of the Auditor General 2007, 16). Similar concerns were raised shortly thereafter by the government-appointed Royalty Review Panel. The government's response to these criticisms demonstrates the use of disciplining techniques, as it attempted to deprive the Auditor-General's Office of resources, restrict his independence, and discredit the report by suggesting that he had overstepped his mandate (Pratt 2010).

Another example of lack of transparency and accountability, discussed in chapter 6, was the government's failure to track the greenhouse gas emissions of large emitters – a failure that was eventually revealed to the public by a different auditor general in 2014.

PATRON-CLIENT RELATIONS (CRONYISM)

Consulting, legal, marketing/branding, and other firms whose owners are close to the PC Party have benefited from government contracts.[7] Such clientelistic patterns have been reported over time by journalists and other observers of Alberta politics, but they have not been comprehensively documented. Corporate executives have privileged access to politicians, and ethics rules permit individuals to rotate between corporate and government positions (Lisac 2013). Over the years there has been a notable circulation of the same individuals among positions in the energy companies, the ministries of Environment and Energy, and other government-appointed positions.[8] Illustrative examples of party/government/corporate circulation of personnel include the appointments of

- the former president of the Canadian Association of Petroleum Producers (CAPP), Gerald Protti, as the first head of the Alberta Energy Regulator;
- former PC environment minister Lorne Taylor as director of the board of the Alberta Environmental Monitoring, Evaluation and Reporting Agency (AEMERA), established in 2014;
- former CAPP president David Manning as Alberta's representative in Washington, DC;
- Richard Masson, former Nexen and Shell executive and former director of oil sands policy and development with Alberta Energy, as CEO of the Alberta Petroleum Marketing Commission;
- PC Party insider Doug Mitchell as CEO of the Alberta Economic Development Authority from 1994 to 2002 (an organization created by Premier Klein in 1994 to enable 'a closer relationship between government and business')[9] as well as a member of Premier Stelmach's 2007 task force on value-added and technology commercialization;
- the former chief operating officer of Syncrude, Jim Carter, as a member of the board of directors of the Alberta Research Council, the Alberta Carbon Capture and Storage Development Council, and the Climate Change and Emissions Management Corporation (CCEMC);
- the former CEO of Syncrude, Eric Newell, as a member of the board of governors of the University of Alberta and later its chancellor, and as the chair of the board of directors of the CCEMC;
- former Suncor Energy VP Heather Kennedy as assistant deputy minister in charge of the Oilsands Secretariat, reporting to the Treasury Board;
- PC Party fundraiser, campaign manager, and vice president Douglas Goss as chair of the board of governors of the University of Alberta.

Political party and campaign financing rules have enabled a 'you scratch my back, I'll scratch yours' relationship between the ruling party and individual clients (e.g., appointees) or business interests. The chief electoral officer who was 'disciplined' for his report on the 2008 election had found nine cases of illegal campaign donations to the PCs in 2007 (Wingrove 2011). As of 2011, Alberta Justice had not prosecuted a single one. Prosecution was made difficult due the fact that 'Alberta law had no requirement for a political party to record when donations are received (making enforcement of annual donation limits impossible)' (Wingrove 2011). Nor, the chief electoral officer complained, did Alberta require parties to record donations to leadership contests, unlike other provinces. (This was changed with the passage of Bill 7, the Election Accountability Amendment Act, in December 2012 ['Alberta Legislature' 2012; Legislative Assembly 2012].)

In 2013 a judicial inquiry was underway into contributions made to the PC's 2012 electoral campaign by an Alberta billionaire, executives in his company, and members of his family.[10] As Mark Lisac (2013) points out, wealthy individuals and large corporations have been well-positioned to make such donations (union donations to the NDP pale by comparison), filling the campaign coffers of the PCs (and, in 2012, the Wildrose Party). Bill 7 failed to limit party spending, did not end corporate donations, and did not lower the thirty-thousand-dollar donation limit (to bring limits on private influence over politicians in line with the rules in other provinces and at the federal level).

It was further revealed in 2012, thanks to a CBC investigation, that publicly funded organizations (including some of the province's universities, municipalities, and health authorities) had been making contributions to the PC Party, although this was not permitted under existing rules. This often took the form of purchasing tickets to PC Party fundraisers – a practice viewed as obligatory if an organization or firm wanted to remain in the government's good books (Cosh 2012). In January 2013 Alberta's chief electoral officer announced fines in forty-five cases of illegal donations in 2010 and 2011.[11] Another case of a PC government appointee making donations to the PC Party from public funds came to light in July 2014 (Rusnell and Russell 2014).

We should not be too surprised to find that practices like this are ubiquitous in a jurisdiction where the same party formed the government from 1971 to 2015. Alberta may not be a one-party state, strictly speaking, but – as political scientist Steven Patten (2015) points out – it has functioned, historically, as a 'one-party dominant' political system.

By this he means that there has been 'a melding of party and state, a situation in which the quality of democratic life is compromised by the fact that the elected legislature is marginalized and the governing party has the capacity to shape political discourse and popular understandings of the public interest' (257). The predominance of the energy industry in the provincial economy – which the Social Credit and Progressive Conservative parties promoted – dealt the conservatives a strong hand of political cards, whether in the form of patronage resources or political support of the entrenched interests attached to the 'single resource' extractive model of development.

NORMALIZATION OF GOVERNMENT DISCOURSE

In addition to the government's control of information generated by the bureaucracy, or through its polling of public opinion, the Conservatives created their own public relations department, known as the Public Affairs Bureau (PAB). First established by Premier Lougheed in 1973, the PAB came to play a strongly partisan role within the Executive Council under the direction of the premier (see the comments of political scientists Jonathon Rose, John Saroski, and Nelson Wiseman in Rusnell and Russell 2013; Sampert 2005; Adkin and Stares, this volume). The PAB's key roles are described in the 2011–12 annual report of the Executive Council as 'coordinating with ministries and organizations to promote Alberta's energy, immigration, employment, investment and tourism potential to the world,' including 'developing a newsletter for corporate consumers and investors, conducting Clean Energy Speakers Bureau training sessions, leading the cross-ministry approach to Canadian consulate oil sands advocacy meetings and organizing media tours' (11). Its spending for the year ending 31 March 2012 was approximately $14.2 million, while another category in the Executive Council's budget, called 'Promoting Alberta,' accounted for about $3.5 million (15).

The government has also launched campaigns to sway public opinion, either within the province or beyond its borders. Some of this spending is documented in Adkin and Stares, this volume. In addition, the government has access, of course, to all of the communications resources of its ministries (each of which has communications officers) as well as media coverage of all of its announcements, speeches by ministers and MLAs, and so on. Conservative government discourse on energy-related issues has closely parallelled that of the business associations in the sector.

(DIS)ORGANIZATION OF CIVIL SOCIETY AND CONSTRUCTION OF COLLECTIVE IDENTITIES

Through the structuring of citizen participation in governance processes, budgetary control, withholding of information from the public, and frequent reorganization of the regulatory system, the government continually *(dis)organizes civil society.* There are similarities here to what Ross (2001) calls the 'group formation' effect; however, the term '(dis) organization' encompasses not only the resources and political opportunity structures for citizens to associate and organize, but also the processes that construct identities and understandings of citizenship. For example, the state actively organizes civil society by creating new institutions of a research, consultative, or advisory nature to deal with the problems generated by dependence on oil and gas extraction (e.g., research institutes, advisory panels, task forces, councils, roundtables,[12] QUANGOs). 'Stakeholder bodies' are created to manage citizen involvement in a policy area over shorter or longer periods of time. Bodies like the Regional Steering Committee for the Lower Athabasca Land-Use Plan, the Oil Sands Multi-Stakeholder Committee, or the Multi-Stakeholder Roundtable for the 2007 public consultation on climate change policy are established as short-term advisory bodies and then dissolved. Energy-sector-heavy 'partnerships' like the Alberta Energy Research Institute, Alberta Innovates-Technology Futures, the Clean Air Strategic Alliance, the Cumulative Environmental Management Association, or Climate Change Central are created to support government priorities and/or manage conflict over longer periods.

The government determines the criteria of eligibility to participate, and defines stakeholder roles in both ad hoc and more institutionalized forms of public involvement, thereby constructing collective identities. The construction and authorization of differentiated 'stakeholders' (business sectors or 'industry,' 'agriculture,' 'ENGO,' 'Aboriginal,' 'municipal,' etc.) in any particular policy domain serves to frame what is at stake in the conflict (the interests and trade-offs). In public consultation processes, hegemonic understandings of possible outcomes are reproduced by the government's construction of interests and identification of policy options. Space for critical scrutiny of the ends of the model of development itself is not 'designed in,' although critical voices sometimes infiltrate the process. Business priorities are typically presented as being identical to the public interest in 'development,' 'economic growth,' 'job creation,' 'Albertans' prosperity,' or, possibly, Canadians' 'energy security,' while the social and environmental consequences of these priorities, as well as alternatives to them, are generally excluded from consideration. 'ENGOs' may be included in stakeholder consultations for their environmental knowledge, but not for their economic or social knowledge; here, corporate representatives and bureaucrats are assumed to be the experts. Aboriginal groups have either been included only in token ways, or redirected to separate government-to-government consultation tracks (in which government has, effectively, delegated corporations to conduct the consultation).[13]

Participation that is conditional upon evidence that individuals will be personally affected – either in monetary or health terms – prevents intervenors from speaking to the decision-making process, the ecological risks or harms a project poses, or other broader societal concerns. The rules used by bodies like the ERCB, Alberta Utilities Commission, or Alberta Energy Regulator (AER) require individuals to speak not as citizens concerned about the overall model of development, including the necessity of ever-expanding fossil fuels extraction, but as property owners or individuals with personal health concerns. The AER has, moreover, tried to prevent citizens from acting collectively through environmental associations (like the Oil Sands Environmental Coalition) by applying a strict interpretation of 'persons.' Under Section 73 of the Environmental Protection and Enhancement Act (EPEA) and Section 109 of the Water Act, a 'person' who may be 'directly affected' by a pipeline, a gas well, or other energy project is entitled to submit a 'Statement of Concern' to the director of the ERCB/AER (Fluker 2011; *Pembina Institute v. Alberta* 2013, 2).[14] The AER oversees 'hearings' at which 'stakeholders' and project proponents present concerns and responses. It decides whether or not applicants requesting a hearing will be given standing, and whether a hearing will be held. The process is legalistic and weighted towards the expertise of the proponent (whose technical evidence citizens must examine and contest if they are not convinced of the safety of the project) (AER 2013). The AER appears to be implementing its directives on public hearings in an even more restrictive fashion than the ERCB, having denied standing to environmental organizations or First Nations at least ten times since 2013 (Weber 2014a). When the Pembina Institute and the Fort McMurray Environmental Association challenged their exclusion from a hearing concerning oil sands expansion, a Queen's Bench judge ruled that the AER's decision violated the norms of 'fair and open procedure,' the spirit of the legislation, and the principles of natural justice (*Pembina Institute v. Alberta* 2013).

Lastly, the government of Alberta has, over many decades, discursively constructed 'Albertan' identity. All governments, of course, do the same thing with regard to the construction of national identities, and other subnational jurisdictions (states, provinces, or even municipalities) often seek to create and reproduce distinctive identities (to attract investment or tourism, or to create citizen identification with particular parties or economic interests). Governments since the Social Credit years have constructed an 'Albertan' identity not merely in terms of 'distinctiveness' or 'difference,' in relation to other provincial identities, but also in terms of a fighting, feisty defence of provincial interests from the predations and insults of hostile outsiders. Larry Pratt, an astute observer of Alberta politics, observed in the mid-1970s that 'western resentment has many legitimate roots. But it is fast becoming the illegitimate blunt sword of those who would play off one part of Canada against the others, keeping old wounds open and old animosities alive, diverting attention from real injustices to imaginary enemies. It is an old and vicious game and far too many Canadians seem prepared to play it indefinitely' (1976, 111–12). The interests best served by the Albertans-under-siege-by-the-feds discourse, he argued, were those of the multinational oil companies that successfully played off one level of government against another during critical negotiations about the terms of oil sands exploitation in 1973–75, as well as those of the privileged white settler population benefiting from the dispossession of northern Aboriginal peoples. "In their perennial obsession with east-west confrontations, most Canadians have conveniently overlooked the fact that the real lines of social, economic, and political exploitation within the country increasingly follow a north-south pattern" (112). Although, as argued earlier in this chapter (and elsewhere in this book), the 'real lines' of exploitation also follow class and gender patterns, Pratt draws our attention here to the racist and colonial dimensions of Canada's political ecology.

As Adkin and Stares argue, the practice of opposing 'Albertans' or 'the interests of Albertans' to 'non-Albertans' (outsiders) and their (hostile-or-harmful-to-Alberta) interests continued under PC governments. The reproduction of the 'us/them' binary in Albertan political discourse is very important to understanding how support for successive conservative governments has been sustained over time. There have been some alterations in government discourse since the 1970s, however. First, the enemies of 'Albertans'' interests have come to include environmental opponents of oil sands expansion, fracking, and new pipeline construction, and these encompass not only provincial environmental associations but also NGOs connected to the global climate justice movement. Second, the election of a Conservative government at the federal level in 2006 – a government that shares the world view of the Alberta PCs – deflated much of the 'anti-federal government' rhetoric of the Alberta PCs. The federal Conservatives, under Stephen Harper's leadership, employed the same tactic of labelling environmental organizations as 'foreign funded' or serving 'foreign interests,' as described in several chapters in this book. Moreover, the sizeable environmental movement in British Columbia and its opposition to pipelines connecting the tar sands to the west coast have made it more difficult for Alberta governments to single out 'central Canadian' interests as the enemy of Alberta's future prosperity. In response to the complaints of governments in Quebec and Ontario about the effects of Canada's petrodollar on their manufacturing industries, or on the achievement of national greenhouse gas reduction goals, the Redford and Prentice governments claimed that Alberta's energy resources benefit the whole country and sought agreement on a new 'Canada energy strategy'.[15]

Although the political construction of 'Albertan identity' may be cracking at the seams – and the May 2015 election which returned a majority NDP government may indicate the potential for a new articulation of Albertan identity – discourse homogenizing 'Albertan interests' has been produced, unrelentingly, by conservative governments and corporate entities. The elements of this cannot be fully described here, but the gist – as Davidson and Gismondi (2011, 2014) also conclude – is that energy development is the only path available to Albertans. The metaphors of oil as the 'lifeblood' of the province's economy and pipelines as its critical circulatory system are frequently employed. In the 1970s, as Larry Pratt described in *The Tar Sands: Syncrude and the Politics of Oil,* the justifications given for the rapid exploitation of the tar sands included ensuring *Canada's* energy supply, notwithstanding the contradiction that most of the crude produced was slated for export to the United States. The massive provincial investment in the industry undertaken by the Lougheed governments was pursued in the name of kick-starting rapid industrial development and making Alberta a power player within Confederation (redressing past grievances/western alienation from 'central Canada').

Environmental, anti-poverty, feminist, Aboriginal, and other voices within civil society must fight continually against the tide of official, corporate, and mainstream media representations of Albertans' interests as they try to establish the feasibility and desirability of other possible developmental paths. Solidarities must be constructed among workers, farmers, ranchers, small business owners, Aboriginal communities, and others around different conceptions of citizenship, common interest, or public good. As Larry Pratt recognized, questions must be asked about who benefits from the wealth produced by this model of growth, and who is disproportionately harmed. Is there no more ecologically sustainable and egalitarian model available to us?

From Laws of Petro-Politics to the Comparative Study of State-Society Relations

The techniques of governance described in this chapter are certainly not unique to Alberta; they belong to the repertoire of all capitalist states. Oil and gas revenues have made possible state reliance upon rent- rather than taxation-based income, and one could make the case that the weight of fossil fuel extraction in the provincial economy has had specific effects on class structure and identities, the nature of civil society, and so on. Oil has unquestionably created an enormous resource for the ruling party. The revenue from oil that flowed to the state when the Social Credit Party was in office, and to its successor, the Progressive Conservatives, provided both parties with exceptional means to entrench their bases of political support. However, we are still left with a number of questions. Why – despite the evidence presented even by mainstream economists and some Conservative allies – have successive PC governments declined to extract maximal revenue from the oil sands? Why have they chosen to reduce corporate taxes and, consequently, state revenue, despite repeated cycles of fiscal boom and bust? Why have they poured revenue into 'clean energy' branding campaigns and CCS, rather than into replacing coal-fired electricity with renewable energies and reducing dependence on bitumen production? These decisions and others have generated the opposition we now see at a global level against Alberta's fossil-fuel-driven model of development, and have necessitated the use of the disciplining and other techniques described above to repair the ruptures in hegemonic discourse, weaken and divide critics, and keep the (increasingly unreliable) 'machine' running.

To answer these questions we must look at multiple factors shaping the province's current political institutions and state-society relations.

As Karl suggested, it is not merely the existence of oil reserves or the properties of oil revenue but the nature of the society so endowed that explains the political ecology and governance of Alberta. In addition to the study of state-society relations, we need to keep in play the nature and implications of Alberta's positioning within the global economy, as argued by Carter and Zalik and Adkin and Miller in this volume. In particular, how have Alberta's governments mediated between the demands and expectations of the international oil industry and the needs and interests of the citizens they represent, or (in the case of the Canadian government), the First Nations for whom they have constitutional or fiduciary duties?

Interestingly, Larry Pratt's analyses of Alberta's political economy were written before 'resource curse' theory became influential. Rather, in the 1970s the leading approach to the study of oil-producing states was neo-Marxism, with its interest in the contracts that producing states (often former colonies) were able to negotiate with the multinational corporations that dominated commodities markets (see, e.g., Barnet and Müller 1974; Emmanuel 1976; Evans 1971; Laux and Molot 1978; Magdoff 1969; Moran 1974; Onimode 1978; Tanzer 1969). These theorists looked to the rise of multinational corporations, backed by geopolitically powerful states, and to the role of post-colonial elites (comprador ruling classes), or to other problems facing post-colonial governments (bargaining capacity, inherited economic structure – including single-commodity-based economies and lack of investment capital for diversification) to explain the highly unequal relationships that developed between local governments and multinational corporations. Resource curse theory may have been a displacement of the problem from the power relationships and strategic decisions of political actors to the properties of the resource itself.

Three critical junctions in the province's development path were the rise of the Social Credit movement in the 1930s, the provincial-federal-multinational corporation negotiations over the terms of oil sands development in 1973–75, and the later conversion of the Alberta PC Party leadership to neoliberal ideology in the early 1990s. The Social Credit era – in turn rooted in the composition of the province's immigrants, discussed in chapter 1 – entrenched the libertarian, individualist political culture that differentiated Alberta's electorate so significantly from its counterpart in Saskatchewan, where the cooperative movement and the CCF took root. Alberta's government did not consider public ownership as a framework for developing natural resources; instead, it invited international mining and petroleum companies to 'take over' (Pratt 1976, 30). The Lougheed government later relied – also for ideological reasons, in part – upon the same multinational corporations to develop the tar sands, and essentially ceded provincial sovereignty over the terms of their exploitation. The enormous scale of the investment made by the Lougheed (and Canadian) governments in the exploitation of bitumen made it extremely difficult for the government to later change course (Pratt 1976, 30). The Lougheed government initially attempted to play the role of a 'developmental' state, but these goals fell away in the specific conditions in which that government found itself, as Pratt argued in the mid-1970s (referring to the obstacles to an industrialization-led economic diversification strategy). That project was, in any case, profoundly colonial and anti-ecological, embodying assumptions about the value and aims of 'economic growth' that we should question today. With its neoliberal turn under the leadership of Ralph Klein, the PC Party had no inclination whatsoever to contemplate even regulation of the pace of exploitation, let alone public ownership or government steering of investment towards renewable energy alternatives. The incapacity of civil society and political actors to resist this neoliberal turn or, subsequently, to establish a new societal consensus around an ecological and egalitarian program of reforms goes a long way towards explaining the terms upon which oil and other resources have been exploited in Alberta since the early 1990s.

None of this is meant to imply that a social democratically oriented Alberta state would have turned away from the opportunity to exploit the oil sands; Norway's governments have, after all, exploited their offshore resources. Norway's fiscal regime may not be dominated by transnational oil firms, but social democratic regimes have not rejected extractive models of economic growth in favour of ecologically sustainable, steady-state economies; Norway has not escaped the conflict between fossil-fuel production and global ecological imperatives (Cumbers 2012). Comparative political ecology analysis pushes us to go further than the petro-state literature has, to date, in incorporating the ecological as well as the social and political dimensions of development.

We cannot trace a simple, direct line from oil wealth to the erosion of democracy; various measures of democracy are, indeed, weak and eroded in very many liberal democracies where neoliberal regimes have dominated for decades. The 2012 Democratic Audit of the United Kingdom, for example, used more than forty comparative datasets to compare the UK to other advanced industrialized countries in the OECD on a range of democratic indicators (Wilks-Heeg et al. 2012). The authors found that 'in virtually every case, the UK ranks below the EU-15 average' (11). With regard to the blurring of lines between corporate interests and political parties' representation of public interests, it was reported that 'the proportion of major UK corporations which have direct connections to MPs, either in the forms of directorships, consultancies or shareholdings,' was 'many times greater than that found in other established democracies' (16) (including Canada).[16]

When one looks at key indicators of democracy, the Nordic countries (including Norway) continue to far outperform other OECD countries, in large part because of their more active commitments to social equality. Social democratic regimes are clearly more democratic than neoliberal regimes, even following a degree of recommodification of labour in the 'Nordic model' since the 1980s (Hermann and Hofbauer 2007; Lyngstad 2008). The UK and Norway are both petro-states, but they have had significantly different state-society relations, underpinning a neoliberal regime in the first case and a social democratic regime in the second. Such comparisons return us to the observation made in the first section of this chapter that the historical development of social classes and political institutions is foundational to understanding why states that are the recipients of substantial oil revenues might manage these revenues – and the social and environmental conditions of their production – very differently. Karl referred to the particular culture of the Norwegian civil service to explain that country's 'exceptionalism' among petro-states.[17] As any comparative politics scholar who has studied welfare state regimes would recognize, the historical formation of the Scandinavian labour movements is another key explanation for the social democratic orientation of these societies (O'Connor and Olsen 1998). Indeed, in regard to the Alberta-Norway comparison it is interesting that Alberta's civil service took a position on the development of the tar sands that was not so different from the perspective Karl described for the Norwegian civil service (see Pratt 1976). In the Alberta case, however, the civil servants were overruled by their political masters, leaving one to ask if the 'civil society' explanation for the different paths (the strength of the union movement in the Norwegian case and its weakness in the Alberta case) is not one of the keys to explaining the divergence.

In a very fruitful comparison of the United Kingdom's and Norway's management of North Sea oil revenue, Andrew Cumbers observes:

> A critical factor in Norway has been the strength of the labour movement and the broad social democratic consensus that has underpinned the country's development since 1945. Over half of the workforce in Norway is unionised, with nearly 75 per cent of workers covered by collective bargaining agreements, compared to only around 25 percent in the United Kingdom. Although the United Kingdom had a strong labour movement in the 1970s, there was never quite the social consensus achieved [as] in Norway; at the same time business and financial interests have always been stronger in the United Kingdom. UK employment relations remained largely voluntaristic, while in Norway, unions became much more enshrined as social partners in the formal state apparatus. (Cumbers 2012, 228)

In the UK, the interests of finance capital have dominated historically within the Conservative Party and the state (Leys 1989), and with the defeat of the workers' movement under Thatcherism in the 1980s the way was cleared for a 'regime of rapid depletion of North Sea oil resources' by multinational corporations (led by BP and Shell). British neoliberal governments did not seek to use oil revenue for financing long-term, employment-creating, environmentally sustainable development (i.e., as for Ralph Klein, there was no 'plan'); instead, they used it to cover budgetary deficits and to reduce taxes for business and the middle class. Along with the drive to ensure a 'a friendly and accommodating' investment environment for the multinational oil companies (Woolfson et al. 1997, quoted in Cumbers 2012, 229), 'trade unions [were] left out of the policy loop between 1979 and 1997; local communities affected by oil development and local business interests were also given little voice' (Cumbers 229). The similarities to Alberta – particularly since the beginning of the Klein governments – are evident.

Indeed, bringing Alberta into this comparison of first world oil-producing states, we find many similarities with the British case (as well as some obvious, significant differences). In 2012, the percentage of workers in Canada who were members of unions was about 30 per cent; Alberta had the lowest unionization rate in Canada, at 23.5 per cent.[18] The historical weakness of organized labour in Alberta (and Canada), and other characteristics of its civil society – indeed, the nature of state-society relations overall – are critically important to understanding the ways in which this particular petro-state has managed the province's enormous wealth of natural resources. The settler colonial context adds another dimension to state-society relations in Alberta and to the conditions and consequences of fossil-fuel extraction. Randolph Haluza-DeLay and Angela Carter (this volume) cover much ground in mapping the characteristics of Alberta's civil society, but more work remains to be done. Overall, the chapters in this book have documented a great deal of the political economy of Alberta over the last twenty-five years, and this work will be useful for future comparisons of Alberta with other first world petro-states.

A political ecology framework of analysis necessarily broadens the scope of concerns and relationships that has characterized the petro-state literature to date. A historically contextualized, comparative approach permits a richer understanding of relationships between natural resource endowments and models of development, or forms of political governance, than the petro-state theses have typically offered. Friedman's assertion that 'geology trumps ideology' is simply not borne out by comparative research. On the contrary, the path chosen by successive governments of Alberta was not preordained by an accident of geology, but built from multiple decisions taken over time. These decisions were shaped, significantly, by ideology, calculations by political decision makers of what kinds of economic development might be possible (given the province's geographic location within the Canadian federation and the global economy), and the political identities and capacities of civil society.

The unique political ecology of Alberta has deep roots, going back to the colonial settlement of the province and the displacement of Aboriginal peoples from their ancestral territories, and to the formation of the post-colonial elite and its relationship to 'Central Canada.' The hybridization that occurred between a political culture permeated with 'nativist,' promethean, colonial, and modernizing worldviews, on the one hand, and the Klein era's neoliberal ideology, on the other hand, resulted in the neoliberal petro-state that has driven investment in fossil fuel extraction for the last twenty-five years. But this Alberta is changing; more than half its population of four million now lives in two cities and is increasingly multi-ethnic and cosmopolitan. A large number have arrived fairly recently from other parts of Canada. Moreover, the global context in which Albertans evaluate their options is changing. Peak oil, climate change, environmental and indigenous peoples' campaigns, and what some view as a deepening and widening legitimation crisis of neoliberal capitalism and US hegemony, shape our perceptions of what is possible, necessary, or desirable. The old consensus is rupturing in all kinds of ways. One thing that is clear is that the interests driving the fossil fuels industries are in conflict with democratic goals, which means that democratization is both a condition and an outcome of a societal transition to a low-carbon model of development. The chapters in this book have suggested multiple ways in which our political system needs to be democratized in order for meaningful ecological citizenship to be realized.

NOTES

1 According to Ross's calculations, Canada produced oil/gas income per capita of $2,530 in 2009 (2012, 21). The oil and gas sector in Alberta reported $27 billion in revenue in 2011–12 (Government of Alberta, http://www.albertacanada.com/business/industries/og-about-the-industry.aspx (accessed 6 August 2014).) With a population of 4.083 million in 2014, that gives Alberta a per capita oil income of $6,612.00.
2 It is significant to Karl's analysis that her focus is 'the so-called capital-deficient oil exporters,' a subset that includes Mexico, Algeria, Indonesia, Nigeria, Venezuela, Iran, Trinidad-Tobago, Ecuador, Gabon, Oman, Egypt, Syria, and Cameroon, but not Saudi Arabia, Kuwait, Libya, Qatar, and the United Arab Emirates (1997, 17).
3 Notably, at this time, the entire proceedings of the hearings supervised by the ECA were made available to the public. This was not the case in subsequent decades. In 2015, the NDP government released an archive of the submissions to its Climate Change Panel.
4 See, for example, Fluet and Krogman (2009), Pratt and Urquhart (1994), Richardson et al. (1993), and Davidson and Gismondi (2011).
5 See also publications of the Canadian Institute of Resources Law.
6 Eighty-six public intellectuals signed a letter sent to Alberta newspapers in October 2009 defending Greenpeace activists from the government's association of their civil disobedience actions with terrorism. See 'Civil Disobedience a Far Cry from Terrorism,' *Edmonton Journal*, 26 October 2009, http://www2.canada.com/edmontonjournal/news/letters/story.html?id=494dbd32-7576-49a7-a305-cb4aee22b831.
7 For example, one of two PR firms hired by the Stelmach government to develop the 'Alberta brand' was Edmonton-based Calder-Bateman Communications (Howell 2008). According to the Calgary publication *Fast Forward Weekly*, Margaret Bateman was a former managing director of the government's Public Affairs Bureau, the department that awarded the contract. She was also listed as a member of Premier Stelmach's election cam-

paign team (http://www.ffwdweekly.com/article/news-views/news/rebranding-alberta-contract-awarded-tory-friendly-/).

8 Moreover, the regulatory functions of the Environment and Energy ministries have been effectively merged where these concern 'energy' industry approvals. Examples of such merger include the formation of a Climate Change Bureau within the Ministry of Environment, staffed by bureaucrats transferred over from the Ministry of Energy in December 1999 (senior bureaucrats in the Ministry of Environment, interview by L.E. Adkin, 2013), and the appointment of a 'director of policy and regulatory alignment for both Alberta Environment and Alberta Energy' (Weber 2014a).

9 Alberta Economic Development Authority, http://aeda.alberta.ca/who-we-are/about-aeda.aspx (accessed 31 July 2014). See also the mention of Doug Mitchell in David Ebner and Dawn Walton, 'Billionaire Oilers Owner Katz Gave $430,000 to Alberta PCs," *Globe and Mail*, 24 October 2012, http://www.theglobeandmail.com/news/politics/billionaire-oilers-owner-katz-gave-430000-to-alberta-pcs/article4647260/.

10 Dawn Walton, 'Outside Help Called In to Probe Alberta PC Election Funding,' *Globe and Mail*, 8 January 2013, http://www.theglobeandmail.com/news/politics/outside-help-called-in-to-probe-alberta-pc-election-funding/article7042881/.

11 *CBC News*, 'Elections Alberta Orders Illegal Donations to PCs Returned,' 31 January 2013, http://www.cbc.ca/news/canada/edmonton/story/2013/01/30/edmonton-elections-alberta-illegal-political-donations.html.

12 Marc Lisac's (1995) account of Ralph Klein's use of roundtables, pp. 144–5, 153–9, is an interesting read.

13 The Government of Alberta has struggled for some time to develop protocols to fulfil its constitutional duty to consult with First Nations concerning development projects that may have negative effects on traditional or treaty rights. Such consultations are a condition for government approval of these projects to be validated by the courts. Aboriginal groups in Alberta have increasingly resisted participation in such consultations on the grounds that, in the absence of an effective Aboriginal veto power, they ultimately serve only to advance development projects. See, for example, Klinkenberg (2014) and Weber (2014b).

14 The Alberta Energy Regulator took over, in 2013, the responsibilities of the ERCB (now dissolved), as well as the energy-related regulatory duties of the Ministry of Environment and Sustainable Resource Development.

15 This approach has been continued, as of March 2016, by Alberta's NDP government. See Canada's Premiers 2015.

16 There are no comparable data, so far as I am aware, for Alberta.

17 I am not aware of any historical study of the culture of Alberta's civil service; this would be a useful undertaking. Evidently, the composition and characteristics of the civil service will have changed, along with its expansions and contractions, since 1905. The dominance of conservative governments since the 1930s may make the Alberta case quite different from jurisdictions where changes of ruling party are normal events.

18 Employment and Social Development Canada, 'Work – Unionization Rates,' http://www4.hrsdc.gc.ca/.3ndic.1t.4r@-eng.jsp?iid=17 (accessed 1 August 2014).

REFERENCES

Adkin, Laurie. 1995. 'Life in Kleinland: Democratic Resistance to "Folksy Fascism."' *Canadian Dimension* 29 (2): 31–42.

Adkin, Laurie. 2014. 'Making Climate Change Policy in Alberta.' Paper presented to the Canadian Political Science Association, 29 May, Brock University, St. Catharines.

Alberta Energy Regulator (AER). 2013. 'The Hearing Process for the Alberta Energy Regulator.' Manual 003, June. http://www.aer.ca/documents/manuals/Manual003.pdf (accessed 7 August 2014).

'Alberta Legislature: Election Accountability Amendment Act Would Expose Illegal Donations,' 2012. HuffPost Alberta, 20 November. http://www.huffingtonpost.ca/2012/11/20/alberta-legislature-election-donations_n_2168002.html.

Altamirano-Jiménez, Isabel. 2004. 'North American First Peoples: Slipping Up into Market Citizenship?' *Citizenship Studies* 8 (4): 349–65.

Barnet, Richard J., and Ronald E. Müller. 1974. *Global Reach: The Power of the Multinational Corporations.* New York: Simon and Shuster.

Bennett, Dean. 2015. 'Alberta Passes Bill Banning Political Donations from Corporations and Unions.' *Globe and Mail,* 23 June. http://www.theglobeandmail.com/news/alberta/alberta-passes-bill-banning-political-donations-from-corporations-and-unions/article25074664/.

Bowness, Evan, and Mark Hudson. 2014. 'Sand in the Cogs? Power and Public Participation in the Alberta Tar Sands.' *Environmental Politics* 23 (1): 59–76.

Brownsey, Keith. 2005. 'Alberta's Oil and Gas Industry in the Era of the Kyoto Protocol.' In *Canadian Energy Policy and the Struggle for Sustainable Development,* ed. G. Bruce Doern, 200–22. Toronto: University of Toronto Press.

Campanella, David, Bob Barnetson, and Angella MacEwen. 2014. *On the Job: Why Unions Matter in Alberta*. May. Edmonton: Parkland Institute. http://

s3-us-west-2.amazonaws.com/parkland-research-pdfs/Unions_Matter_Report_(low_res).pdf.

Cheung, Carmen K.M. 2015. 'Submission to the Standing Committee on Public Safety and National Security concerning Bill C-51, the Anti-Terrorism Act, 2015.' March. Vancouver: British Columbia Civil Liberties Association. https://bccla.org/wp-content/uploads/2015/03/BCCLA-Submissions-on-C-51-For-website.pdf.

Cosh, Colby. 2012. 'Cue the Outrage. Not.' *Maclean's*, 125, no. 7 (27 February), 23–4.

Cournoyer, Dave. 2016. '10 Ways to Renew Democracy in Alberta.' Blog post, February. http://daveberta.ca/2016/02/10-ways-to-renew-democracy-in-alberta/.

Cross, Philip, and Geoff Bowlby. 2006. 'The Alberta Economic Juggernaut: The Boom on the Rose.' *Canadian Economic Observer*, 19, no. 9 (September), 3.1–3.12.

Cumbers, Andrew. 2012. 'North Sea Oil, the State and Divergent Development in the United Kingdom and Norway,' In *Flammable Societies: Studies on the Socio-economics of Oil and Gas*, ed. John-Andrew McNeish and Owen Logan, 221–42. London: Pluto.

Dabbs, Frank. 2006. 'Ralph Klein's Real Legacy.' *Alberta Views*, September, 25–9.

Davidson, Debra J., and Mike Gismondi. 2011. *Challenging Legitimacy at the Precipice of Energy Calamity*. New York: Springer.

Davidson, Debra, and Mike Gismondi. 2014. 'Footprints in the Oilsands: A Double Diversion with Global Repercussions.' *Sociological Imagination* 50 (1): 13–34.

Elder, P.S. 1974. 'The Participatory Environment in Alberta.' *Alberta Law Review* 12: 403–30.

Emmanuel, Arghiri. 1976. 'The Multinational Corporations and Inequality of Development.' *International Social Science Journal* 28 (4): 754–72.

Evans, Peter B. 1971. 'National Autonomy and Economic Development: Critical Perspectives on Multinational Corporations in Poor Countries.' *International Organization* 25 (3): 675–88.

Fluet, Colette, and Naomi Krogman. 2009. 'The Limits of Integrated Resource Management in Alberta for Aboriginal and Environmental Groups: The Northern East Slopes Sustainable Resource and Environmental Management Strategy.' In *Environmental Conflict and Democracy in Canada*, ed. Laurie E. Adkin, 123–39. Vancouver: UBC Press.

Fluker, Shaun. 2011. 'Public Participation at the Alberta Energy Resources Conservation Board.' *Resources Newsletter*, no. 111, Canadian Institute of Resources Law, University of Calgary.

Friedman, Thomas L. 2006. 'The First Law of Petropolitics,' *Foreign Policy* 154 (May/June): 28–36.

Goldberg, Ellis, Erik Wibbels, and Eric Mvukiyehe. 2008. 'Lessons from Strange Cases: Democracy, Development, and the Resource Curse in the U.S. States.' *Comparative Political Studies* 41 (4/5): 477–514.

Harrison, Trevor, and Gordon Laxer, eds. 1995. *The Trojan Horse: Alberta and the Future of Canada*. Montreal: Black Rose.

Hermann, Christoph, and Ines Hofbauer. 2007. 'The European Social Model: Between Competitive Modernisation and Neoliberal Resistance.' *Capital & Class* 93: 125–39.

Hoberg, George, and Jeffrey Phillips. 2011. 'Playing Defence: Early Responses to Conflict Expansion in the Oil Sands Policy Subsystem.' *Canadian Journal of Political Science* 44 (3): 507–27.

Homer-Dixon, Thomas. 2013. 'The Tar Sands Disaster.' *New York Times*, op-ed, 31 March. http://www.nytimes.com/2013/04/01/opinion/the-tar-sands-disaster.html?_r=0.

Karl, Terry Lynn. 1997. *The Paradox of Plenty: Oil Booms and Petro-States*. Berkeley: University of California Press.

Klinkenberg, Marty. 2014. 'First Nations Chiefs Boycott Alberta Government over Consultation Plan.' *Edmonton Journal*, 28 August.

Laux, Jeanne, and Maureen Appel Molot. 1978. 'Multinational Corporations and Economic Nationalism: Conflict over Resource Development in Canada.' *World Development* 6 (6): 837–49.

Le Billon, Philippe, and Angela Carter. 2012. 'Securing Alberta's Tar Sands: Resistance and Criminalization on a New Energy Frontier.' In *Natural Resources and Social Conflict: Towards Critical Environmental Security*, ed. Matthew A. Schnurr and Larry A. Swatuk, 170–92. London: Palgrave Macmillan.

Legislative Assembly, Province of Alberta. 2012. 'The Election Accountability Amendment Act.' http://www.assembly.ab.ca/net/index.aspx?p=bills_status&selectbill=007&legl=28&session=1 (accessed 7 August 2014).

Leys, Colin. 1989. *Politics in Britain*. Toronto: University of Toronto Press.

Lisac, Mark. 1995. *The Klein Revolution*. Edmonton: NeWest.

Lisac, Marc. 2007. *Insight into Government*, 22, no. 1 (week ending August 24).

Lisac, Mark. 2013. 'Money Talks: The Problems with Alberta's Lax Party Funding Rules.' *Alberta Views*, 16, no. 1. https://albertaviews.ab.ca/2013/01/15/money-talks/.

Lyngstad, Roiv. 2008. 'The Welfare State in the Wake of Globalization: The Case of Norway.' *International Social Work* 51 (1): 69–81.

Magdoff, Harry. 1969. *The Age of Imperialism: The Economics of U.S. Foreign Policy*. New York: Monthly Review.

Martin-Hill, Dawn. 2004. 'Resistance, Determination and Perseverance of the Lubicon Cree Women.' In *In the Way of Development: Indigenous Peoples, Life Projects & Globalization*, ed. Mario Blaser, Harvey A. Feit, and Glenn McRae, 313–30. New York: Zed Books in association with International Development Research Centre, Ottawa.

Martin-Hill, Dawn. 2008. *The Lubicon Lake Nation: Indigenous Knowledge and Power*. Toronto: University of Toronto Press.

McInnes, John, and Ian Urquhart. 1995. 'Protecting Mother Earth or Business?: Environmental Politics in Alberta.' In *The Trojan Horse: Alberta and the Future of Canada*, ed. Trevor Harrison and Gordon Laxer, 239–53. Montreal: Black Rose.

Moran, Theodore H. 1974. *Multinational Corporations and the Politics of Dependence: Copper in Chile*. Princeton: Princeton University Press.

Nikiforuk, Andrew. 2007. 'Is Canada the Latest Emerging Petro-Tyranny?' *Globe and Mail*, 11 June.

Nikiforuk, Andrew. 2008. *Tar Sands: Dirty Oil and the Future of a Continent*. Vancouver: David Suzuki Foundation/Greystone.

Nikiforuk, Andrew. 2011. 'Biggest Silent Election Issue: Oil's Erosion of Canada.' *The Tyee*, 21 April. http://thetyee.ca/Opinion/2011/04/21/SilentElectionIssue/ (accessed 11 July 2012).

Nikiforuk, Andrew. 2012. 'Petro State per Usual: Reading Alberta's Election.' *The Tyee*, 25 April. http://thetyee.ca/Opinion/2012/04/25/Reading-Albertas-Election/ (accessed 11 July 2012).

O'Connor, Julia S., and Gregg M. Olsen, eds. 1998. *Power Resource Theory and the Welfare State: A Critical Approach*. Toronto: University of Toronto Press.

Office of the Auditor-General, Government of Alberta. 2007. *Annual Report of the Auditor General of Alberta 2006–2007*. Vol. 1 of 2. 19 September. http://www.oag.ab.ca/files/oag/2006-2007_Annual_Report_Vol_1.pdf.

Onimode, Bade. 1978. 'Imperialism and Multinational Corporations: A Case Study of Nigeria.' *Journal of Black Studies* 9 (2): 207–32.

Oremus, Will. 2012. 'Saudi Arabia. Nigeria. Venezuela. Canada? Is Our Neighbor to the North Becoming a Jingoistic Petro-State?' *Slate Magazine*, 20 January. http://www.slate.com/articles/news_and_politics/politics/2012/01/canadian_tar_sands_is_our_neighbor_to_the_north_becoming_a_jingoistic_petro_state_.2.html (accessed 11 July 2012).

Paskey, Janice. 2011. 'Where Are the Boys?' *Alberta Views*, 14, no. 7, 34–8. http://www.albertaviews.ab.ca/2011/08/25/where-are-the-boys/.

Public Interest Alberta. 2016. 'Strengthening Our Democracy.' Public Interest Alberta Democracy Task Force Submission to Alberta's Select Special

Ethics and Accountability Committee. February. http://pialberta.org/sites/default/files/Democracy%20Task%20Force%20Submission%20FINAL.pdf.

Patten, Steve. 2015. 'The Politics of Alberta's One-Party State.' In *Transforming Provincial Politics: The Political Economy of Canada's Provinces and Territories in a Neoliberal Era*, ed. Bryan Evans and Charles Smith, 255–83. Toronto: University of Toronto Press.

Pembina Institute v. Alberta (Environment and Sustainable Resource Development), [2013]. ABQB 567. Court of Queen's Bench of Alberta. Reasons for Judgment of the Honourable Mr. Justice R. P. Marceau. Filed 1 October. https://www.pembina.org/reports/abqb567-reasons-for-judgment.pdf (accessed 6 August 2014).

Pratt, Larry. 1976. *The Tar Sands: Syncrude and the Politics of Oil*. Edmonton: Hurtig.

Pratt, Larry. 1977. 'The State and Province-Building: Alberta's Development Strategy.' In *The Canadian State: Political Economy and Political Power*, ed. Leo Panitch, 133–62. Toronto: Toronto University Press.

Pratt, Larry, and Ian Urquhart. 1994. *The Last Great Forest: Japanese Multinationals and Alberta's Northern Forests*. Edmonton: NeWest.

Pratt, Sheila. 2010. 'An Audit Too Far: The Stelmach Government Pledged Accountability – Until Fred Dunn Discovered Their Billion-Dollar Giveaway.' *Alberta Views*, 13, no. 8, 34–9.

Pratt, Sheila. 2015. 'Climate Change Panel Ready to Hear from Albertans.' *Edmonton Journal*, 31 August. http://edmontonjournal.com/news/politics/climate-change-panel-ready-to-hear-from-albertans.

Price, Richard T. 1977. 'Indian Land Claims in Alberta: Politics and Policy-Making (1968–77).' Master's thesis, University of Alberta.

Richardson, Mary, Joan Sherman, and Michael Gismondi. 1993. *Winning Back the Words: Confronting Experts in an Environmental Public Hearing*. Toronto: Garamond.

Ross, Michael L. 2001. 'Does Oil Hinder Democracy?' *World Politics* 53 (April): 325–61.

Ross, Michael L. 2012. *The Oil Curse: How Petroleum Wealth Shapes the Development of Nations*. Princeton: Princeton University Press.

Rusnell, Charles, and Jennie Russell. 2013. 'Redford Ramps Up Politicization of Government Communications.' *CBC News*, 21 March. http://www.cbc.ca/news/canada/edmonton/story/2013/03/20/edmonton-redford-politicizes-communications.html.

Rusnell, Charles, and Jennie Russell. 2014. 'Health Executive Involved in Illegal Political Donations Resigns.' *CBC News*, 24 July 24. http://www.cbc

.ca/news/canada/edmonton/health-executive-involved-in-illegal-political-donations-resigns-1.2716156.

Sampert, Shannon. 2005. 'King Ralph, the Ministry of Truth, and the Media in Alberta,' In *The Return of the Trojan Horse: Alberta and the New World (Dis) Order*, ed. Trevor Harrison 37–51. Montreal: Black Rose.

Shrivastava, Meenal, and Lorna Stefanick. 2012. 'Do Oil and Democracy Only Clash in the Global South?: Petro Politics in Alberta, Canada.' *New Global Studies* 6 (1), article 5: 1–27.

Slowey, Gabrielle, 2008. *Navigating Neoliberalism: Self-Determination and the Mikisew Cree First Nation.* Vancouver: UBC Press.

Slowey, Gabrielle, and Lorna Stefanick. 2015. 'Development at What Cost? First Nations, Ecological Integrity, and Democracy." In *Alberta Oil*, ed. Meenal Shrivastava and Lorna Stefanick, 195–24. Edmonton: University of Athabasca Press.

Taft, Kevin. 2007. *Democracy Derailed: A Breakdown of Government Accountability in Alberta – And How to Get It Back on Track.* Red Deer, AB: Red Deer Press.

Tanzer, Michael. 1969. *The Political Economy of International Oil and the Underdeveloped Countries.* Boston: Beacon.

Taylor, Alison, and Tracy Friedel. 2011. 'Enduring Neoliberalism in Alberta's Oil Sands: The Troubling Effects of Private-Public Partnerships for First Nation and Métis Communities.' *Citizenship Studies* 15 (6–7): 815–35.

Toneguzzi, Mario. 2012. 'Alberta Has Low Post-Secondary Enrolment.' *Calgary Herald*, 19 March. http://blogs.calgaryherald.com/2012/03/19/alberta-has-low-post-secondary-enrolment/.

Toneguzzi, Mario. 2014. 'Alberta Wages Highest in Canada.' *Calgary Herald*, 26 June. http://www.calgaryherald.com/business/alberta+wages+highest+canada/9976608/story.html.

Urquhart, Ian. 2005. 'Alberta's Land, Water, and Air: Any Reasons Not to Despair?' In *The Return of the Trojan Horse: Alberta and the New World (Dis)Order*, ed. Trevor Harrison, 136–55. Montreal: Black Rose.

Urquhart, Ian. 2010. 'Petrostate?' *Alberta Views*, 13, no. 8, 26–32. http://www.albertaviews.ab.ca/2011/10/04/petrostate/.

Urquhart, Ian. 2016. 'Let's Finish the Job of Renewing Democracy in Alberta.' Op-ed in *Calgary Herald*, 20 February. http://calgaryherald.com/opinion/columnists/urquhart-lets-finish-the-job-of-renewing-democracy-in-alberta.

Weber, Bob. 2014a. 'Alberta Silent on Decisions about Who Speaks at Energy Hearings.' Canadian Press, 4 August. http://www.cbc.ca/news/canada/calgary/alberta-silent-on-decisions-about-who-speaks-at-energy-hearings-1.2727091.

Weber, Bob. 2014b. 'Alberta First Nations Shut Out of Oilsands Planning Review: Lawyers.' Canadian Press, 28 August. http://www.huffingtonpost.ca/2014/08/26/alberta-first-nations-oilsands-review_n_5713647.html.

Wilks-Heeg, Stuart, Andrew Blick, and Stephen Crone. 2012. *How Democratic Is the UK? The 2012 Audit*. Liverpool: Democratic Audit. http://www.democraticaudit.com/wp-content/uploads/2013/06/auditing-the-uk-democracy-the-framework-2.pdf.

Wingrove, Josh. 2011. 'Fired Chief Electoral Officer Sues Province of Alberta.' *Globe and Mail*, 8 March. http://www.peterspolitics.com/news/read_news.html?news=76352&target=%2Fnews%2Findex.html%3Fstart%3D478 (accessed 7 August 2014).

Woolfson, C., J. Foster, and M. Beck. 1997. *Paying for the Piper: Capital and Labour in Britain's Offshore Oil Industry*. London: Mansell.

Contributors

Laurie E. Adkin is associate professor in the Department of Political Science and also teaches in the Environmental Studies BA program at the University of Alberta. She is the author of *Politics of Sustainable Development: Citizens, Unions, and the Corporations* (Black Rose Books, 1998) and editor and co-author of *Environmental Conflict and Democracy in Canada* (UBC Press, 2009).

Shelley M. Alexander is associate professor, Department of Geography, University of Calgary. She has conducted field-based and GIS analysis of large carnivore ecology and studied human-wildlife conflict in the Canadian Rockies since 1990.

Angela V. Carter is assistant professor in the Department of Political Science at the University of Waterloo. She researches comparative environmental policy regimes surrounding oil developments in key Canadian cases.

Benjamin Courteau is director of the Associated Students of the University of Montana, Office of Transportation, Missoula, Montana. (This is a student-operated transit system and bicycle program.) He holds an M.A. in international studies from the University of Denver (Josef Korbel School of International Studies) with a concentration in sustainable development and human rights.

Conny Davidsen is associate professor of environmental policy and governance in the Department of Geography, University of Calgary. Her research interest focuses on how political dynamics and media influences shape environmental discourses and their policy processes from local to global. Apart from her work in the Canadian oil/tar sands, past and present projects have led her to Ecuador, Peru, Mexico, British Columbia, and Nunavut.

Göze Doğu is a doctoral candidate in the Department of Sociology at the University of Alberta. Her research and area of work expand on social and environmental impact assessments of large-scale infrastructure projects and due diligence analysis of public and private companies.

Danah Duke is executive director of the Miistakis Institute, a non-profit research institute focused on land-use and ecosystem management, affiliated with Mount Royal University in Calgary, Alberta.

Theresa Garvin is professor of human geography in the Department of Earth and Atmospheric Sciences at the University of Alberta. She conducts community-based research on perceptions of environmental issues, built environments, and geographies of health.

Randolph Haluza-DeLay is associate professor in the Department of Sociology, The King's University, Edmonton. He received his doctorate in education from the University of Western Ontario in 2007. He has co-edited two books: *Speaking for Ourselves: Environmental Justice in Canada* (UBC Press, 2009) and *How the World's Religions are Responding to Climate Change* (Routledge, 2013).

Steven A. Kennett is an independent policy consultant based in Calgary. From 1992 to 2007 he was a research associate at the Canadian Institute of Resources Law, University of Calgary, and from 2007 to 2009 he was a senior policy analyst with the Pembina Institute.

Byron Miller is associate professor of geography and adjunct associate professor of planning at the University of Calgary.

Sara O'Shaughnessy earned her doctorate in rural sociology from the University of Alberta. She has extensive experience in community-based research (CBR), having worked on projects in communities from Fort McMurray in northern Alberta to Fort Portal in western Uganda.

Brenda Parlee is associate professor and Canada Research Chair in the

Faculty of Native Studies and in the Department of Resource Economics and Environmental Sociology, Faculty of Agriculture, Life, and Environmental Sciences, University of Alberta.

Monique Passelac-Ross is a recently retired senior fellow of the Canadian Institute of Resources Law, University of Calgary. Her research focuses on the rights of Aboriginal people in the face of resource development.

Noah Purves-Smith created Ursudio Multimedia after completing a course in computer engineering at the University of Victoria and studying art fundamentals at the University of Calgary. A gifted designer and programmer, Noah uses his talents to create effective web applications, designs, multimedia content, and print advertising.

Michael Quinn is associate vice president of research, scholarship and community engagement as well as the Talisman Energy Chair in environmental sustainability at Mount Royal University in Calgary. Mike's work focuses on interdisciplinary approaches to complex problems with the ultimate goal of sustaining social-ecological systems. He applies these approaches to road ecology, regional ecohydrology, human-wildlife conflict, and transboundary ecosystem management.

Brittany J. Stares holds a BA (Honours) in political science. She was the managing editor of *Curb Magazine* at the City-Region Studies Centre, University of Alberta, before undertaking a master's degree in the Philanthropy and Nonprofit Leadership Program at Carleton University.

Brad Stelfox, PhD, is a landscape ecologist with the ALCES Group, Forem Consulting (Calgary) and adjunct professor in the Department of Biological Sciences, University of Alberta, and the Department of Environmental Design, University of Calgary.

Larissa Stendie is the Energy and Climate Campaigner for Sierra Club BC. She has an MPhil in Culture, Environment and Sustainability from the Centre for Development and Environment of the University of Oslo, Norway, where she worked in communications and research for The Lancet – University of Oslo Commission on Global Governance for Health, the Arne Næss Chair projects, and Oslo Sustainability Summits.

Benjamin Thibault holds a JD from Harvard Law School. He works primarily on renewable energy and electricity policy design and analysis in Alberta, as well as legal and regulatory analysis related to environment and energy issues across Canada. Ben has also worked on international human rights law and environmental law in Canada and abroad.

Mary-Ellen Tyler is associate professor of environmental science in the Faculty of Environmental Design, University of Calgary. Her research is related to regional landscape planning and social ecological systems.

Nickie Vlavianos is associate professor with the Faculty of Law, University of Calgary, and an affiliated faculty member with the Institute for Sustainable Energy, Environment and Economy and the Canadian Institute of Resources Law. She teaches and researches in the areas of energy, environmental, natural resources, and property law.

Tim Weis is a professional engineer and holds a PhD in environmental sciences from l'Université du Québec à Rimouski. He has over thirteen years of experience working on renewable energy issues, ranging from project assessments to community consultations to policy research and advocacy. He is the past director for renewable energy and efficiency policy at the Pembina Institute, former Alberta regional director for the Canadian Wind Energy Association, and an adjunct professor in the Department of Mechanical Engineering at the University of Alberta.

Anna Zalik is an associate professor in the Faculty of Environmental Studies, York University, Toronto, where she teaches in the areas of global environmental politics and critical development studies. Her research focuses on the political economy and political ecology of oil, gas, and other extractives, centring on their spatial and social relations with historical and contemporary colonialism.

Index

Page numbers with (f) refer to figures and illustrations; page numbers with (t) refer to tables.

www.ingramcontent.com/pod-product-compliance
Lightning Source LLC
LaVergne TN
LVHW020434080826
844660LV00033B/1284
9781442612587